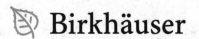

Frontiers in Mathematics

This series is designed to be a repository for up-to-date research results which have been prepared for a wider audience. Graduates and postgraduates as well as scientists will benefit from the latest developments at the research frontiers in mathematics and at the "frontiers" between mathematics and other fields like computer science, physics, biology, economics, finance, etc. All volumes are online available at SpringerLink.

More information about this series at http://www.springer.com/series/5388

Nicuşor Costea • Alexandru Kristály •
Csaba Varga

Variational and Monotonicity Methods in Nonsmooth Analysis

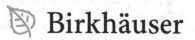

Nicuşor Costea
Department of Mathematics and Computer Science
University Politehnica of Bucharest
Bucharest, Romania

Alexandru Kristály
Department of Economics
Babeş-Bolyai University
Cluj-Napoca, Romania

Institute of Applied Mathematics
Óbuda University
Budapest, Hungary

Csaba Varga
Department of Mathematics
Babeş-Bolyai University
Cluj-Napoca, Romania

ISSN 1660-8046   ISSN 1660-8054   (electronic)
Frontiers in Mathematics
ISBN 978-3-030-81670-4   ISBN 978-3-030-81671-1   (eBook)
https://doi.org/10.1007/978-3-030-81671-1

Mathematics Subject Classification: 35A15, 35A16, 47J30, 49J52, 35B38, 58E05, 35Q91

This book is published under the imprint Birkhäuser, www.birkhauser-science.com, by the registered company Springer Nature Switzerland AG.
The registered company address is: Gewerbestrasse 11, 6330 Cham, Switzerland

To my wife Diana and my son Nicholas.

N. Costea

To my wife Tünde and my children Marót, Bora, Zonga, and Bendegúz.

A. Kristály

To my wife Ibolya, my son Csaba, and my sister Irma.

Cs. Varga

Preface

The present book provides a comprehensive presentation of a wide variety of nonsmooth problems arising in nonlinear analysis, game theory, engineering, mathematical physics, and contact mechanics. The subject matter of the monograph had its genesis in the early works of F. Clarke, who paved the way for the modern development of nonsmooth analysis.

Our initial aim is to cover various topics in nonsmooth analysis, based mainly on variational methods and topological arguments. The present work includes recent achievements, mostly obtained by the authors during the last 15 years (four main parts, divided into 13 chapters), putting them into the context of the existing literature.

Part I contains fundamental mathematical results concerning convex and locally Lipschitz functions. Together with the appendices, this background material gives the book a self-contained character.

Part II is devoted to variational techniques in nonsmooth analysis and their applications, providing various existence and multiplicity results for differential inclusions, hemivariational inequalities both on bounded and unbounded domains. The set of results for unbounded domains is the first systematic material in the literature, which requires deep arguments from variational methods and group-theoretical arguments in order to regain certain compactness properties.

Part III deals with variational and hemivariational inequalities treated via topological methods. By using fixed point theorems and KKM-type approaches, various existence and localization results are established including Nash-type equilibria on curved spaces and inequality problems governed by set-valued maps of monotone type.

Part IV contains several applications to nonsmooth mechanics. Using the theoretical results from the previous parts we are able to provide weak solvability for various mathematical models which describe the contact between a body and a foundation. We consider the antiplane shear deformation of elastic cylinders in contact with an insulated foundation, the frictional contact between a piezoelectric body and an electrically conductive foundation, and models with nonmonotone boundary conditions for which we derive a variational formulation in terms of bipotentials.

At the end of each chapter we listed those references that are quoted in that part.

We really hope the monograph will be useful, providing further ideas for the reader.

Bucharest, Romania   Nicuşor Costea
Cluj-Napoca, Romania and Budapest, Hungary   Alexandru Kristály
Cluj-Napoca, Romania   Csaba Varga

Acknowledgments

We would like to acknowledge the constant support of Dr. Thomas Hempfling (Executive Editor, Mathematics, Birkhäuser) and Daniel Ignatius Jagadisan and Sarah Annette Goob (Springer) for their initial advises and editorial guidance.

We are extremely grateful to our colleagues and students who helped over the years to put us on the right path on several topics. We thank B. Breckner, M. Csirik, F. Faraci, C. Farkas, H. Lisei, A. Matei, I. Mezei, M. Mihăilescu, G. Moroşanu, D. Motreanu, R. Precup, P. Pucci, B. Ricceri, Á. Róth, I. Rudas, D. Stancu-Dumitru, K. Szilák, and O. Vas for their comments and critics.

Research of N. Costea was partially supported by CNCS-UEFISCDI Grant No. PN-III-P1-1.1-TE-2019-0456.

Research of A. Kristály was supported by the National Research, Development and Innovation Fund of Hungary, financed under the K_18 funding scheme, Project No. 127926.

Bucharest, Romania   Nicuşor Costea
Cluj-Napoca, Romania and Budapest, Hungary   Alexandru Kristály
Cluj-Napoca, Romania   Csaba Varga

Contents

Acronyms

X^*	Dual space of X
$\langle \cdot, \cdot \rangle$	Scalar product in the duality X^*, X
$s - X$	X endowed with the strong topology
$w - X^*$	X^* endowed with the weak topology
$w^* - X^*$	X^* endowed with the weak-star topology
φ^*	The conjugate function of φ
$D(\varphi)$	The effective domain of the φ
$\partial \varphi(u)$	The convex subdifferential of φ at u
$\phi^0(u; v)$	The generalized directional derivative of ϕ at u in the direction v
$\partial_C \phi(u)$	The Clarke subdifferential of ϕ at u
$\rightarrow$	Strong convergence
$\rightharpoonup$	Weak convergence
$\overrightarrow{}$	Weak-star convergence
∇u	The gradient of u
Δu	The Laplacian of u
$\Delta_p u$	The p-Laplacian of u
$\Omega \subset \mathbb{R}^N$	Open bounded connected subset of $\mathbb{R}^N$
$\partial \Omega$, Γ	The boundary of Ω
$L^p(\Omega)$	The Lebesgue space of p-integrable functions
$L^{p(\cdot)}(\Omega)$	Variable exponent Lebesgue space
$L^\infty(\Omega)$	The space of essentially bounded functions
$L^\phi(\Omega)$	Orlicz space
$C^k(\Omega)$	The space of k times continuously differentiable functions
$C^\infty(\Omega)$	The space of indefinite differentiable functions
$C_0(\Omega)$	The space of continuous functions with compact support in Ω
$W^{1,p}(\Omega)$, $W_0^{1,p}(\Omega)$	Sobolev spaces
$W^{1,p(\cdot)}(\Omega)$, $W_0^{1,p(\cdot)}(\Omega)$	Variable exponent Sobolev spaces

$W^{1,\phi}(\Omega)$, $W_0^{1,\phi}(\Omega)$	Orlicz-Sobolev spaces
$a := b$	a takes by definition the value b
a.e	Almost everywhere
s.t.	Such that

Part I

Mathematical Background

Convex and Lower Semicontinuous Functionals

1.1 Basic Properties

We briefly present basic properties of convex functions, the direct method in the calculus of variations as well as the variational principle of Ekeland. For further results and a comprehensive treatment of the subject we refer the reader to Borwein & Zhu [1], Brezis [2], Niculescu & Persson [4] and Rockafellar [5]. Unless otherwise stated, throughout this chapter, X denotes a real Banach space. For a functional $\varphi : X \to (-\infty, +\infty]$, we denote by $D(\varphi)$ the *effective domain* of φ, that is,

$$D(\varphi) := \{u \in X : \varphi(u) < +\infty\}.$$

The *epigraph* of φ is the set

$$\mathrm{epi}(\varphi) := \{(u, \lambda) \in X \times \mathbb{R} : \varphi(u) \leq \lambda\}.$$

Definition 1.1 A functional $\varphi : X \to (-\infty, +\infty]$ is said to be *lower semicontinuous (l.s.c.)* if for every $\lambda \in \mathbb{R}$ the set

$$[\varphi \leq \lambda] := \{u \in X : \varphi(u) \leq \lambda\}$$

is closed.

We recall next some well-known elementary facts about l.s.c. functionals.

© The Author(s), under exclusive license to Springer Nature Switzerland AG 2021
N. Costea et al., *Variational and Monotonicity Methods in Nonsmooth Analysis*,
Frontiers in Mathematics, https://doi.org/10.1007/978-3-030-81671-1_1

(i) The functional φ is l.s.c. if and only if epi(φ) is closed in $X \times \mathbb{R}$;

(ii) φ is l.s.c. if and only if for every sequence $\{u_n\}$ in X such that $u_n \to u$ we have

$$\liminf_{n\to\infty} \varphi(u_n) \geq \varphi(u);$$

(iii) If φ_1 and φ_2 are l.s.c., then $\varphi_1 + \varphi_2$ is l.s.c.;

(iv) If $(\varphi_i)_{i \in I}$ is a family of l.s.c. functionals then their *superior envelope* is also l.s.c., that is, the functional φ defined by

$$\varphi(u) := \sup_{i \in I} \varphi_i(u)$$

is l.s.c.;

(v) If X is compact and φ is l.s.c., then $\inf_{u \in X} \varphi(u)$ is achieved.

Definition 1.2 A function $\varphi : X \to (-\infty, +\infty]$ is said to be *convex* if

$$\varphi(\lambda u + (1 - \lambda)v) \leq \lambda\varphi(u) + (1 - \lambda)\varphi(v), \quad \forall u, v \in X, \ \forall \lambda \in [0, 1].$$

We have the following elementary properties of convex functionals:

(i) The functional φ is convex if and only if epi(φ) is a convex set in $X \times \mathbb{R}$;

(ii) If φ is a convex functional, then for every $\lambda \in \mathbb{R}$ the set $[\varphi \leq \lambda]$ is convex. The converse is not true in general;

(iii) If φ_1 and φ_2 are convex, then $\varphi_1 + \varphi_2$ is convex;

(iv) If $(\varphi_i)_{i \in I}$ is a family of convex functionals then their superior envelope is also convex, that is, the function φ defined by

$$\varphi(u) := \sup_{i \in I} \varphi_i(u).$$

is convex.

The following theorem provides useful information regarding the continuity of convex functionals.

Theorem 1.1 *Let $\varphi : X \to (-\infty, +\infty]$ be a convex functional such that $\varphi \not\equiv +\infty$. Then the following statement are equivalent:*

(i) *φ is bounded above in a neighborhood of u_0;*

(ii) *φ is continuous at u_0;*

(iii) *int(epi(φ)) $\neq \emptyset$;*

(iv) *int($D(\varphi)$) $\neq \emptyset$ and $\varphi|_{\text{int}(D(\varphi))}$ is continuous.*

Proof $(i) \Rightarrow (ii)$ Taking a translation if necessary, we may assume that $u_0 = 0$ and $\varphi(0) = 0$. Let U be a neighborhood of 0 such that $\varphi(u) \leq M$ for all $u \in U$. Fix $\varepsilon \in (0, M]$ and let $V := (\varepsilon/M)U \cap (-\varepsilon/M)U$ be a symmetric neighborhood of 0. Let $u \in V$ be fixed. Then $(M/\varepsilon)u \in U$ and

$$\varphi(u) \leq \frac{\varepsilon}{M}\varphi\left(\frac{M}{\varepsilon}\right) + \left(1 - \frac{\varepsilon}{M}\right)\varphi(0) \leq \frac{\varepsilon}{M}M = \varepsilon. \tag{1.1}$$

On the other hand, $(-M/\varepsilon)u \in U$ and

$$0 = \varphi(0) \leq \frac{1}{1 + (\varepsilon/M)}\varphi(u) + \frac{\varepsilon/M}{1 + (\varepsilon/M)}\varphi\left(-\frac{\varepsilon}{M}u\right)$$

$$\leq \frac{1}{1 + (\varepsilon/M)}\varphi(u) + \frac{\varepsilon/M}{1 + (\varepsilon/M)}M,$$

thus showing that

$$\varphi(u) \geq -\varepsilon. \tag{1.2}$$

From (1.1) and (1.2) we deduce that

$$|\varphi(u)| \leq \varepsilon, \quad \forall u \in V,$$

therefore φ is continuous at $u = 0$.

$(ii) \Rightarrow (i)$ Follows directly from the continuity of φ at u_0.

$(i) \Rightarrow (iii)$ Let U be a neighborhood of u_0 such that $\varphi(u) \leq M$ for all $u \in U$. Then $U \subset \text{int}(D(\varphi))$ and

$$\{(u, \lambda) \in X \times \mathbb{R} : u \in U, M < \lambda\} \subset \text{epi}(\varphi),$$

which shows that $\text{int}(\text{epi}(\varphi)) \neq \varnothing$.

$(iii) \Rightarrow (i)$ Fix $(u, \lambda) \in \text{int}(\text{epi}(\varphi))$. Then there exist a neighborhood U of u and $\varepsilon > 0$ such that

$$U \times [\lambda - \varepsilon, \lambda + \varepsilon] \subset \text{epi}(\varphi).$$

Then $U \times \{M\} \subset \text{epi}(\varphi)$ for $M \in [\lambda - \varepsilon, \lambda + \varepsilon]$, therefore

$$\varphi(u) \leq M, \quad \forall u \in U.$$

$(i) \Rightarrow (iv)$ Again, without loss of generality, we may assume $u_0 = 0$. Let U be a neighborhood of u_0 such that $\varphi(u) \leq M$, for all $u \in U$. Then $U \subset D(\varphi)$, therefore $\text{int}(D(\varphi)) \neq \varnothing$.

For the second statement, fix $u \in \text{int}(D(\varphi))$. Due to the convexity of $D(\varphi)$ there exists $\lambda > 1$ such that $v_0 := \lambda u \in D(\varphi)$. Set

$$V := u + \frac{\lambda - 1}{\lambda} U.$$

Then V is a neighborhood of u and any $w \in V$ satisfies $w := u + \frac{\lambda-1}{\lambda} v$ for some $v \in U$. Thus

$$\varphi(w) = \varphi\left(\frac{1}{\lambda} v_0 + \frac{\lambda - 1}{\lambda} v\right) \leq \frac{1}{\lambda} \varphi(v_0) + \frac{\lambda - 1}{\lambda} \varphi(v)$$

$$\leq \frac{1}{\lambda} \varphi(v_0) + \frac{\lambda - 1}{\lambda} M =: M_0.$$

This shows that φ is bounded above on a neighborhood of u and, since $(i) \Leftrightarrow (ii)$, it follows that φ is continuous at u.

$(iv) \Rightarrow (i)$ Pick any $u \in \text{int}(D(\varphi))$. Then φ is continuous at u, therefore it is also bounded above on a neighborhood of u. □

The following theorem identifies the kind of continuity of a convex functional on the interior of the effective domain.

Theorem 1.2 (Lipschitz Property of Convex Functionals) *Let* $\varphi : X \to (-\infty, +\infty]$ *be a proper convex l.s.c. functional. Then* φ *is locally Lipschitz on* $\text{int}(D(\varphi))$.

Proof The proof will be carried out in 3 steps as follows.

Step 1. *If* φ *is locally bounded above at* $u_0 \in \text{int}(D(\varphi))$, *then* φ *is locally bounded at* u_0.
    Assume $\varphi(u) \leq M$ in $B(u_0, r) \subset \text{int}(D(\varphi))$. Then for each $u \in B(u_0, r)$ the element $v := 2u_0 - u \in B(u_0, r)$ and

$$\varphi(u_0) \leq \frac{\varphi(u) + \varphi(v)}{2} \leq \frac{\varphi(u) + M}{2},$$

thus proving that $\varphi(u) \geq 2\varphi(u_0) - M$, i.e., φ is also locally bounded below at u_0.
Step 2. *If* φ *is locally bounded at* $u_0 \in \text{int}(D(\varphi))$, *then* φ *is locally Lipschitz at* u_0.
    Assume $|\varphi(u)| \leq M$ for all $u \in B(u_0, 2r)$, fix $u, v \in B(u_0, r)$, $u \neq v$ and define

$$d := \|v - u\| \text{ and } w := v + \frac{r}{d}(v - u).$$

Then $w \in B(u_0, 2r)$ and, since $v = \frac{d}{d+r}w + \frac{r}{d+r}u$, we have

$$\varphi(v) \leq \frac{d}{d+r}\varphi(w) + \frac{r}{r+d}\varphi(u).$$

Thus

$$\varphi(v) - \varphi(u) \leq \frac{d}{d+r}(\varphi(w) - \varphi(u)) \leq \frac{2Md}{r} = \frac{2M}{r}\|v - u\|.$$

Step 3. φ is locally Lipschitz on $\mathrm{int}(D(\varphi))$.

In view of the previous two steps, we only need to show that φ is locally bounded above. For each $n \geq 1$ define

$$E_n := [\varphi \leq n].$$

Then E_n is closed for each $n \geq 1$ due to the lower semicontinuity of φ and

$$\mathrm{int}(D(\varphi)) \subset \bigcup_{n=1}^{\infty} E_n.$$

It follows, by the Baire Category Theorem, that $\mathrm{int}(E_{n_0}) \neq \varnothing$ for some $n_0 \geq 1$. Suppose $B(u_0, r) \subset \mathrm{int}(E_{n_0})$. Then φ in bounded above by n_0 on $B(u_0, r)$. Since $\mathrm{int}(D(\varphi))$ is open, if $u \neq v \in \mathrm{int}(D(\varphi))$, then there exists $\mu > 1$ such that $w := u + \mu(v - u) \in \mathrm{int}(D(\varphi))$. Then the set

$$U := \left\{ \frac{1}{\mu}w + \frac{\mu - 1}{\mu}b : b \in B(u_0, r) \right\}$$

is a neighborhood of $v \in \mathrm{int}(D(\varphi))$. Thus, for any $z \in U$ one has

$$\varphi(z) \leq \frac{1}{\mu}\varphi(w) + \frac{\mu - 1}{\mu}n_0,$$

so φ is locally bounded above. $\square$

Proposition 1.1 *Assume that $\varphi : X \rightarrow (-\infty, +\infty]$ is convex and l.s.c. in the strong topology. Then φ is weakly l.s.c., i.e., it is lower semicontinuous in the weak topology τ_w of X.*

Proof For every $\lambda \in \mathbb{R}$ the set

$$[\varphi \le \lambda] := \{u \in X : \varphi(x) \le \lambda\}$$

is convex and (strongly) closed. Then, by Theorem A.5 it is weakly closed and thus φ is weakly l.s.c. $\square$

1.2   Conjugate Convex Functions and Subdifferentials

Definition 1.3 Let $\varphi : X \to (-\infty, +\infty]$ be a proper functional. We define the *conjugate function* $\varphi^* : X^* \to (-\infty, +\infty]$ to be

$$\varphi^*(\zeta) := \sup_{u \in X} \{\langle \zeta, u \rangle - \varphi(u)\}.$$

Note that φ^* is convex and l.s.c. on X^*. In order to check this we point out that for each $u \in X$, the functional $\zeta \mapsto \langle \zeta, u \rangle - \varphi(u)$ is convex and continuous, therefore l.s.c. on X^*. In conclusion $\varphi^*(\zeta)$ is convex and l.s.c., being the superior envelope of these functionals.

Remark 1.1 We have the inequality

$$\langle \zeta, u \rangle \le \varphi(u) + \varphi^*(\zeta), \quad \forall u \in X, \forall \zeta \in X^*, \tag{1.3}$$

which is called *Young's inequality*.

Theorem 1.3 *Assume that $\varphi : X \to (-\infty, +\infty]$ is convex l.s.c and proper. Then φ^* is proper, and in particular, φ is bounded below by an affine continuous function.*

Proof Fix $u_0 \in D(\varphi)$ and $\lambda_0 < \varphi(u_0)$. Applying the Strong Separation Theorem in the space $X \times \mathbb{R}$ with $A := \text{epi}(\varphi)$ and $B := \{(u_0, \lambda_0)\}$ we obtain the existence of a closed hyperplane $H : [\Lambda = \alpha]$ that is strictly separating A and B. Since $X \ni u \mapsto \Lambda(u, 0)$ is a linear and continuous functional on X, it follows that there exists $\zeta \in X^*$ such that

$$\Lambda(u, 0) := \langle \zeta, u \rangle,$$

and

$$\Lambda(u, \lambda) = \langle \zeta, u \rangle + \lambda \Lambda(0, 1), \quad \forall (u, \lambda) \in X \times \mathbb{R}.$$

There exists $\varepsilon > 0$ such that

$$\Lambda(u_0, \lambda_0) + \varepsilon \leq \alpha \leq \Lambda(u, \lambda) - \varepsilon, \quad \forall(u, \lambda) \in \text{epi}(\varphi),$$

which leads to

$$\langle \zeta, u_0 \rangle + \lambda_0 \Lambda(0, 1) < \alpha < \langle \zeta, u \rangle + \varphi(u)\Lambda(0, 1), \quad \forall u \in D(\varphi).$$

It follows that $\Lambda(0, 1) > 0$ (just set $u = u_0$). Moreover,

$$\left\langle -\frac{1}{\Lambda(0, 1)} \zeta, u \right\rangle - \varphi(u) < -\frac{\alpha}{\Lambda(0, 1)}, \quad \forall u \in D(\varphi).$$

Setting $\xi := -\zeta/\Lambda(0, 1)$ and $\beta := \alpha/\Lambda(0, 1)$ we conclude that $\varphi^*(\xi) < +\infty$ and

$$\varphi(u) \geq \langle \xi, u \rangle + \beta. \qquad \square$$

If we iterate the operation $*$, we obtain a function φ^{**} defined on X^{**}. Instead, we choose to restrict φ^{**} to X, that is we define

$$\varphi^{**}(u) := \sup_{\zeta \in X^*} \{\langle \zeta, u \rangle - \varphi^*(\zeta)\} \quad (u \in X).$$

Definition 1.4 For a given functional $\varphi : X \to \mathbb{R}$ the limit (if it exists)

$$\lim_{t \searrow 0} \frac{\varphi(u + tv) - \varphi(u)}{t}, \tag{1.4}$$

is called the *directional derivative* of φ at u in the direction v and it is denoted by $\varphi'(u; v)$. The function φ is called *Gateaux differentiable* at $u \in X$ if there exists $\zeta \in X^*$ such that

$$\varphi'(u; v) = \langle \zeta, v \rangle, \quad \forall v \in X. \tag{1.5}$$

In this case ζ is called the *Gateaux derivative* (or *gradient*) of φ at u and it is denoted by $\nabla\varphi(u)$.

We point out the fact that, if the convergence in (1.4) is uniform w.r.t. v on bounded subsets, then φ is said *Fréchet differentiable* at u and ζ in (1.5) is denoted by $\varphi'(u)$ (*the Fréchet derivative*). Needless to say that if φ is Fréchet differentiable at u, then it is also Gateaux differentiable at u and the two derivatives coincide, whereas the converse is not true in general.

Definition 1.5 Let $\varphi : X \to (-\infty, +\infty]$ be a proper convex l.s.c. functional. Then the *subdifferential* of φ at $u \in D(\varphi)$ is the (possibly empty) set

$$\partial\varphi(u) := \left\{ \zeta \in X^* : \langle \zeta, v - u \rangle \leq \varphi(v) - \varphi(u), \forall v \in X \right\},$$

and $\partial\varphi(u) := \varnothing$ if $u \notin D(\varphi)$.

In general, $\partial\varphi$ is a set-valued map from X into X^*. An element of $\partial\varphi(u)$, if any, is called *subgradient* of φ at u. As usual, the *domain* of $\partial\varphi$, denoted $D(\partial\varphi)$, is the set of all $u \in X$ for which $\partial\varphi(u) \neq \varnothing$.

Let us provide the following simple (but important) examples.

Example 1.1 Consider $\varphi(u) := \|u\|$. It is easy to check

$$\varphi^*(\zeta) = \begin{cases} 0, & \text{if } \|\zeta\|_* \leq 1, \\ +\infty, & \text{otherwise.} \end{cases}$$

It follows that

$$\varphi^{**}(u) = \sup_{\substack{\|\zeta\|_* \leq 1 \\ \zeta \in X^*}} \langle \zeta, u \rangle,$$

therefore $\varphi^{**} = \varphi$. Moreover,

$$\partial\varphi(u) = \begin{cases} B_{X^*}, & \text{if } u = 0, \\ \frac{J(u)}{\|u\|}, & \text{otherwise,} \end{cases}$$

where B_{X^*} is the closed unit ball of X^* and J is the normalized duality mapping i.e.,

$$J(u) := \left\{ \zeta \in X^* : \|\zeta\| = \|u\| \text{ and } \langle \zeta, u \rangle = \|u\|^2 \right\}.$$

For more details regarding the duality mapping check out Chap. 5.

Example 1.2 Given a nonempty set $K \subset X$, we set

$$I_K(u) := \begin{cases} 0, & \text{if } u \in K, \\ +\infty, & \text{otherwise.} \end{cases}$$

The function I_K is called *indicator function of* K. Note that I_K is proper if and only if $K \neq \varnothing$, I_K is convex if and only if K is a convex set and I_K is l.s.c. if and only if K is closed.

The conjugate function

$$I_K^*(\zeta) := \sup_{u \in K} \langle \zeta, u \rangle,$$

is called the *supporting function of K*.

It is readily seen that $D(\partial I_K) = K$, $\partial I_K(u) = 0$ for each $u \in \text{int}(K)$ and

$$\partial I_K(u) = N_K(u) = \{\zeta \in X^* : \langle \zeta, v - u \rangle \le 0, \ \forall v \in K \}.$$

Recall that for any boundary point $u \in K$ the set $N_K(u)$ is the normal cone of K at u.

Example 1.3 Let $\varphi : X \to (-\infty, \infty]$ be convex and Gateaux differentiable at u. Then $\partial \varphi(u) = \{\nabla \varphi(u)\}$.

Indeed, due to the convexity of φ we have

$$\varphi(u + t(v - u)) \le t\varphi(v) + (1 - t)\varphi(u), \quad \forall v \in X, \forall t \in [0, 1].$$

Thus

$$\frac{\varphi(u + t(v - u)) - \varphi(u)}{t} \le \varphi(v) - \varphi(u),$$

and letting $t \searrow 0$ we get that $\nabla \varphi(u) \in \partial \varphi(u)$.

For the converse inclusion, let $\zeta \in \partial \varphi(u)$ be fixed. Then

$$\langle \zeta, w - u \rangle \le \varphi(w) - \varphi(u), \quad \forall w \in X,$$

Taking $w := u + tv$ we get

$$\langle \zeta, v \rangle \le \frac{\varphi(u + tv) - \varphi(u)}{t}, \quad \forall v \in X, \forall t > 0.$$

Letting $t \searrow 0$ we obtain

$$\langle \zeta, v \rangle \le \langle \nabla \varphi(u), v \rangle, \quad \forall v \in X.$$

Replacing v with $-v$ in the above relation we get that $\zeta = \nabla \varphi(u)$.

Proposition 1.2 *Let $\varphi : X \to (-\infty, +\infty]$ be a proper convex l.s.c. functional. Then $u \in D(\varphi)$ is a global minimizer of φ if and only if $0 \in \partial \varphi(u)$.*

Proof The point $u \in D(\varphi)$ is a global minimizer of φ if and only if

$$0 \leq \varphi(v) - \varphi(u), \quad \forall v \in X.$$

But,

$$\langle 0, v - u \rangle = 0, \quad \forall v \in X,$$

thus showing that $0 \in \partial \varphi(u)$.   □

We point out the fact that there is a close relation between $\partial \varphi$ and $\partial \varphi^*$ as it can be seen from the following result.

Theorem 1.4 *Let X be a reflexive space and $\varphi : X \to (-\infty, +\infty]$ be a proper convex functional. Then the following assertions are equivalent:*

(i) $\zeta \in \partial \varphi(u)$;
(ii) $\varphi(u) + \varphi^*(\zeta) = \langle \zeta, u \rangle$;
(iii) $u \in \partial \varphi^*(\zeta)$.

In particular, $\partial \varphi^* = (\partial \varphi)^{-1}$ and $\varphi^{**} = \varphi$.

Proof According to Young's inequality we have

$$\varphi^*(\zeta) + \varphi(u) \geq \langle \zeta, u \rangle, \quad \forall u \in X, \forall \zeta \in X^*,$$

and equality takes place if and only if $0 \in \partial \phi(u)$, with $\phi(u) = \varphi(u) - \langle \zeta, u \rangle$. Hence (i) and (ii) are equivalent. On the other hand, if (ii) holds, then ζ is a global minimizer of $\xi \mapsto \varphi^*(\xi) - \langle \xi, u \rangle$, therefore $u \in \partial \varphi^*$. Hence $(ii) \Rightarrow (iii)$. Since (i) and (ii) are equivalent for φ^* we may write (iii) as

$$\varphi^*(\zeta) + \varphi^{**}(u) = \langle \zeta, u \rangle.$$

Thus, in order to complete the proof it suffices to prove that $\varphi^{**} = \varphi$. We show this in two steps as follows:

Step 1. *If $\varphi \geq 0$, then $\varphi^{**} = \varphi$.*
         One can easily check that $\varphi^{**}(u) \leq \varphi(u)$ for all $u \in X$. Assume by contradiction there exists $u_0 \in X$ such that $\varphi^{**}(u_0) < \varphi(u_0)$. We apply the Strong Separation Theorem in $X \times \mathbb{R}$ with $A := \text{epi}(\varphi)$ and $B := (u_0, \varphi^{**}(u_0))$. As in

the proof of Theorem 1.3 there exist a closed hyperplane $H : [\Lambda = \alpha]$ strictly separating A and B and $\zeta \in X^*$ such that

$$\langle \zeta, u \rangle + \lambda\Lambda(0, 1) > \alpha > \langle \zeta, u_0 \rangle + \varphi^{**}(u_0)\Lambda(0, 1), \quad \forall(u, \lambda) \in \mathrm{epi}(\varphi). \quad (1.6)$$

It follows that $\Lambda(0, 1) \geq 0$ (fix $u \in D(\varphi)$ and let $\lambda \to +\infty$). We cannot deduce that $\Lambda(0, 1) > 0$ as we may have $\varphi(u_0) = +\infty$. For a fixed $\varepsilon > 0$, since $\varphi \geq 0$ we get using (1.6)

$$\langle \zeta, u \rangle + (\Lambda(0, 1) + \varepsilon)\varphi(u) \geq \alpha, \quad \forall u \in D(\varphi).$$

Thus

$$\varphi^*(\xi) \leq \beta,$$

for $\xi := -\frac{\zeta}{\Lambda(0,1)+\varepsilon}$ and $\beta := -\frac{\alpha}{\Lambda(0,1)+\varepsilon}$. The definition of $\varphi^{**}(u_0)$ then implies that

$$\varphi^{**}(u_0) \geq \langle \xi, u_0 \rangle - \varphi^*(\xi) \geq \langle \xi, u_0 \rangle - \beta,$$

and this shows that

$$\langle \zeta, u_0 \rangle + (\Lambda(0, 1) + \varepsilon)\varphi^{**}(u_0) \geq \alpha,$$

which obviously contradicts the second inequality of (1.6).

Step 2. $\varphi^{**} = \varphi$.

According to Theorem 1.3, $D(\varphi^*) \neq \varnothing$, therefore we can fix $\zeta_0 \in D(\varphi^*)$ and define

$$\phi(u) := \varphi(u) - \langle \zeta_0, u \rangle + \varphi^*(\zeta_0).$$

Then ϕ is convex, proper, l.s.c. and satisfies $\phi \geq 0$ and, due to Step 1, $\phi^{**} = \phi$. On the other hand

$$\phi^*(\zeta) = \varphi^*(\zeta + \zeta_0) - \varphi^*(\zeta_0),$$

and

$$\phi^{**}(u) = \varphi^{**}(u) - \langle \zeta_0, u \rangle + \varphi^*(\zeta_0).$$

Using the fact that $\phi^{**} = \phi$ it follows at once that $\varphi^{**} = \varphi$. $\square$

Proposition 1.3 *If $\varphi : X \to (-\infty, +\infty]$ is proper, convex and l.s.c., then $D(\partial\varphi)$ is a dense subset of $D(\varphi)$.*

Proposition 1.4 *If $\varphi : X \to (-\infty, +\infty]$ is proper, convex and l.s.c., then $\mathrm{int} D(\varphi) \subset D(\partial\varphi)$.*

Theorem 1.5 *Let $\varphi : X \to (-\infty, +\infty]$ is proper, convex and l.s.c. functional. Then the following conditions are equivalent:*

(i) $\frac{\varphi(u)}{\|u\|} \to +\infty$ *as* $\|u\| \to +\infty$.
(ii) $R(\partial\varphi) = X^*$ *and* $(\partial\varphi)^{-1} = \partial\varphi^*$ *maps bounded sets into bounded sets;*

Proof $(i) \Rightarrow (ii)$ If (i) holds, then for each $\zeta \in X^*$ the functional $\phi : X \to (-\infty, +\infty]$ defined by

$$\phi(u) := \varphi(u) - \langle \zeta, u \rangle$$

is convex, l.s.c. and coercive, therefore it attains its infimum on X (see Corollary 1.1 in the next section). Thus $0 \in \partial\phi(u) = \partial(\varphi(u) - \langle \zeta, u \rangle)$ or, equivalently, $\zeta \in \partial\varphi(u)$. Moreover, if $\{\zeta\}$ remains in a bounded subset of X^*, then so does $\{(\partial\varphi)^{-1}(\zeta)\}$.
$(ii) \Rightarrow (i)$ By Young's inequality we have

$$\varphi(u) \geq \langle \zeta, u \rangle - \varphi^*(\zeta), \quad \forall u \in X, \forall \zeta \in X^*. \tag{1.7}$$

Fix $u \in X$ and let $\zeta_0 \in X$ be such that $\|\zeta_0\| = \|u\|$ and $\langle \zeta_0, u \rangle = \|u\|^2$. Then taking $\zeta_1 := \frac{\lambda}{\|u\|}\zeta_0$ in (1.7) we get

$$\varphi(u) \geq \lambda\|u\| - \varphi^*\left(\frac{\lambda}{\|u\|}\zeta_0\right), \quad \forall u \in X, \forall \lambda > 0,$$

which combined with the fact that φ^* and $\partial\varphi^*$ map bounded sets into bounded sets yields the desired conclusion. $\qquad\square$

Definition 1.6 A *bipotential* is a functional $B : X \times X^* \to (-\infty, +\infty]$ satisfying the following conditions:

(i) for any $u \in X$, if $D(B(u, \cdot)) \neq \emptyset$, then $B(u, \cdot)$ is proper convex l.s.c.; for any $\zeta \in X$, if $D(B(\cdot, \zeta)) \neq \emptyset$, then $B(\cdot, \zeta)$ is proper convex l.s.c.;
(ii) $B(u, \zeta) \geq \langle \zeta, u \rangle$ for all $u \in X$ and all $\zeta \in X^*$;
(iii) $\zeta \in \partial B(\cdot, \zeta)(u) \Leftrightarrow u \in \partial B(u, \cdot)(\zeta) \Leftrightarrow B(u, \zeta) = \langle \zeta, u \rangle$.

1.3   The Direct Method in the Calculus of Variations

Theorem 1.6 *Let M be a topological Hausdorff space, and suppose that $\phi : M \to (-\infty, +\infty]$ satisfies the Borel-Heine compactness condition, that is, for any $\alpha \in \mathbb{R}$ the set*

$$[\phi \leq \alpha] := \{u \in M : \phi(u) \leq \alpha\}, \tag{1.8}$$

is compact.

  Then ϕ is uniformly bounded from below on M and attains its infimum. The conclusion remains valid if, instead of (1.8), we assume that any sub-level-set $[\phi \leq \alpha]$ is sequentially compact.

Proof Suppose (1.8) holds. We may assume that $\phi \not\equiv +\infty$. Let

$$\alpha_0 := \inf_M \phi \geq -\infty,$$

consider a sequence $\{\alpha_n\}$ such that $\alpha_n \searrow \alpha_0$, as $n \to \infty$ and let $K_n := [\phi \leq \alpha_n]$. By assumption, each K_n is compact and nonempty. Moreover, $K_{n+1} \subset K_n$ for all $n \in \mathbb{N}^*$. By the compactness of K_n there exists a point $u \in \bigcap_{n \in \mathbb{N}} K_n$, satisfying

$$\phi(u) \leq \alpha_n, \quad \forall n \geq n_0.$$

Taking the limit as $n \to \infty$ we obtain that

$$\phi(u) \leq \alpha_0 = \inf_M \phi,$$

and the claim follows.

  If instead of (1.8) each $[\phi \leq \alpha]$ is sequentially compact, we choose a *minimizing sequence* $\{u_n\}$ in M such that $\phi(u_n) \to \alpha_0$. Then for any $\alpha > \alpha_0$ the sequence $\{u_n\}$ will eventually lie entirely within $[\phi \leq \alpha]$. The sequential compactness of $[\phi \leq \alpha]$ ensures that $\{u_n\}$ will accumulate at a point $u \in \bigcap_{\alpha > \alpha_0} [\phi \leq \alpha]$ which is the desired minimizer. $\square$

Remark 1.2 If $\phi : M \to \mathbb{R}$ satisfies (1.8), then for any $\alpha \in \mathbb{R}$ the set

$$\{u \in M : \phi(u) > \alpha\} = M \setminus [\phi \leq \alpha]$$

is open, that is, ϕ is *lower semicontinuous*. Respectively, if each $[\phi \leq \alpha]$ is sequentially compact, then ϕ will be sequentially lower semicontinuous.

  Conversely, if ϕ is sequentially lower semicontinuous and for some $\bar{\alpha} \in \mathbb{R}$, the set $[\phi \leq \bar{\alpha}]$ is (sequentially) compact, then $[\phi \leq \alpha]$ will be (sequentially) compact for all $\alpha \leq \bar{\alpha}$ and again the conclusion of Theorem 1.6 will be valid.

Theorem 1.7 *Suppose that X is a reflexive Banach space with norm $\|\cdot\|$, and let $M \subset X$ be a weakly closed subset of X. Suppose $\phi : M \to (-\infty, +\infty]$ is coercive on M with respect to X, that is,*

$$\phi(u) \to +\infty \text{ as } \|u\| \to \infty, \quad (u \in M),$$

and sequentially weakly lower semicontinuous on M with respect to X, that is, for any $u \in M$, any sequence $\{u_n\}$ in M such that $u_n \rightharpoonup u$ we have

$$\phi(u) \leq \liminf_{n \to \infty} \phi(u_n).$$

Then ϕ is bounded from below on M and attains its infimum in M.

Proof Let $\alpha_0 := \inf_M \phi$ and assume $\{u_n\}$ is a minimizing sequence in M, that is, $\phi(u_n) \to \alpha_0$, as $n \to \infty$. By coerciveness, $\{u_n\}$ is bounded in X and, since X is reflexive, the Eberlein-Šmulian theorem ensures the existence of $u \in X$ such that $u_n \rightharpoonup u$. But M is weakly closed, therefore $u \in M$, and the weak lower semicontinuity of ϕ shows that

$$\phi(u) \leq \liminf_{n \to \infty} \phi(u_n) = \alpha_0,$$

i.e., u is a global minimizer of ϕ. $\square$

A direct consequence of Proposition A.8 and Theorem 1.7 is the following.

Corollary 1.1 *Let X be a reflexive Banach space and let $K \subset X$ be a nonempty, closed and convex subset of X. Let $\phi : K \to (-\infty, +\infty]$ be a proper convex l.s.c. function such that*

$$\lim_{\substack{u \in K \\ \|u\| \to +\infty}} \phi(u) = +\infty. \tag{1.9}$$

Then ϕ achieves its minimum on K, i.e., there exists some $u_0 \in K$ such that

$$\phi(u_0) = \inf_K \phi.$$

Proof Fix any $u \in K$ such that $\phi(u) < +\infty$ and consider the set

$$\tilde{K} := \{v \in K : \phi(v) \leq \phi(u)\}.$$

Then $\tilde{K}$ is closed, convex and bounded and thus it is weakly compact. On the other hand, ϕ is also l.s.c. in the weak topology τ_w. It follows that ϕ achieves its minimum on $\tilde{K}$, i.e., there exists $u_0 \in \tilde{K}$ such that

$$\phi(u_0) \leq \phi(v), \quad \forall v \in \tilde{K}.$$

If $v \in K \setminus \tilde{K}$, we have $\phi(u_0) \leq \phi(u) < \phi(v)$. Thus $\phi(u_0) \leq \phi(v)$, $\forall v \in K$. $\quad\square$

1.4 Ekeland's Variational Principle

Theorem 1.8 (Ekeland's Variational Principle [3]) *Let (X, d) be a complete metric space and let $\phi : X \rightarrow (-\infty, +\infty]$ be a proper, lower semicontinuous and bounded from below functional. Then for every $\varepsilon > 0$, $\lambda > 0$, and $u \in X$ such that*

$$\phi(u) \leq \inf_X \phi + \varepsilon,$$

there exists an element $v \in X$ such that

(*i*) $\phi(v) \leq \phi(u)$;
(*ii*) $d(v, u) \leq \frac{1}{\lambda}$;
(*iii*) $\phi(w) \geq \phi(v) - \varepsilon\lambda d(w, v)$, $\forall w \in X$.

Proof It suffices to prove our assertion for $\lambda = 1$. The general case is obtained by replacing d by an equivalent metric λd. We now construct inductively a sequence $\{u_n\}$ as follows: $u_0 = u$, and assuming that u_n has been defined, we set

$$S_n := \{w \in X : \phi(w) + \varepsilon d(w, u_n) \leq \phi(u_n)\},$$

and consider two possible cases:

(a) $\inf_{S_n} \phi = \phi(u_n)$. Then define $u_{n+1} := u_n$;
(b) $\inf_{S_n} \phi < \phi(u_n)$. Then choose $u_{n+1} \in S_n$ such that

$$\phi(u_{n+1}) < \inf_{S_n} \phi + \frac{1}{2}\left(\phi(u_n) - \inf_{S_n} \phi\right) = \frac{1}{2}\left(\phi(u_n) + \inf_{S_n} \phi\right) < \phi(u_n). \qquad (1.10)$$

We prove next that $\{u_n\}$ is a Cauchy sequence. In fact, if (a) ever occurs, then $\{u_n\}$ is stationary for sufficiently large n and the claim follows. Otherwise,

$$\varepsilon d(u_n, u_{n+1}) \leq \phi(u_n) - \phi(u_{n+1}). \qquad (1.11)$$

Adding (1.11) from n to $m - 1 > n$ we get

$$\varepsilon d(u_n, u_m) \leq \phi(u_n) - \phi(u_m).$$ (1.12)

Note that $\{\phi(u_n)\}$ is a decreasing and bounded from below sequence of real numbers, hence it is convergent, which combined with (1.12) shows that $\{u_n\}$ is indeed Cauchy. Since X is complete, there exists $v \in X$ such that $v := \lim_{n \to \infty} u_n$. In order to complete the proof we show that v satisfies $(i) - (iii)$. Setting $n = 0$ in (1.12) we have

$$\varepsilon d(u, u_m) + \phi(u_m) \leq \phi(u),$$ (1.13)

and letting $m \to \infty$ we get

$$\varepsilon d(u, v) + \phi(v) \leq \phi(u).$$ (1.14)

In particular, this shows that (i) holds. On the other hand,

$$\phi(u) - \phi(v) \leq \phi(u) - \inf_X \phi < \varepsilon,$$

which together with (1.14) shows that (ii) holds.

Now, let us prove (iii). Fixing n in (1.12) and letting $m \to \infty$ yields $v \in S_n$, therefore

$$v \in \bigcap_{n \geq 0} S_n.$$

But, for any $w \in \bigcap_{n \geq 0} S_n$ we have

$$\varepsilon d(w, u_{n+1}) \leq \phi(u_{n+1}) - \phi(w) \leq \phi(u_{n+1}) - \inf_{S_n} \phi.$$ (1.15)

It follows from (1.10) that

$$\phi(u_{n+1}) - \inf_{S_n} \phi \leq \phi(u_n) - \phi(u_{n+1}),$$

therefore

$$\lim_{n \to \infty} \left(\phi(u_{n+1}) - \inf_{S_n} \phi \right) = 0.$$

Taking the limit as $n \to \infty$ in (1.13) we get $\varepsilon d(w, v) = 0$, hence

$$\bigcap_{n \geq 0} S_n = \{v\}.$$ (1.16)

One can easily check that the family $\{S_n\}$ is nested, i.e., $S_{n+1} \subset S_n$, thus for any $w \neq v$ it follows from (1.16) that $w \notin S_n$, for sufficiently large n. Thus,

$$\phi(w) + \varepsilon d(w, u_n) > \phi(u_n).$$

Letting $n \to \infty$ we arrive at (iii).   □

Corollary 1.2 *Let (X, d) be a complete metric space with metric d and let $\phi : X \to (-\infty, +\infty]$ be a proper, lower semicontinuous and bounded from below functional. Then for every $\varepsilon > 0$ and every $u \in X$ such that*

$$\phi(u) \leq \inf_X \phi + \varepsilon,$$

there exists an element $u_\varepsilon \in X$ such that

(i) $\phi(u_\varepsilon) \leq \phi(u)$;
(ii) $d(u_\varepsilon, u) \leq \sqrt{\varepsilon}$;
(iii) $\phi(w) \geq \phi(u_\varepsilon) - \sqrt{\varepsilon}d(w, u_\varepsilon), \forall w \in X$.

References

1. J.M. Borwein, Q.J. Zhu, *Techniques of Variational Analysis* (Springer, Berlin, 2005)
2. H. Brezis, *Opérateurs Maximaux Monotones et Semigroupes de Contractions dans un Espace de Hilbert* (North-Holland, Amsterdam, 1973)
3. I. Ekeland, On the variational principle. J. Math. Anal. Appl. **47**, 324–353 (1974)
4. C.P. Niculescu, L.-E. Persson, *Convex Functions and Their Applications* (Springer, Berlin, 2006)
5. R.T. Rockafellar, *Convex Analysis* (Princeton University Press, Princeton, 1969)

Locally Lipschitz Functionals

2

2.1 The Generalized Derivative and the Clarke Subdifferential

In this part we focus our attention to the theory of locally Lipschitz functionals developed by Clarke [6]. For further details we refer the reader to Carl, Le & Motreanu [4], Chang [5], Kristály [15], Ledyav and Zhu [17], Motreanu & Panagiotopoulos [19] and Naniewicz & Panagiotopoulos [20]. Unless otherwise stated, throughout this section X denotes a real Banach space.

Definition 2.1 A function $f : X \to \mathbb{R}$ is called *locally Lipschitz* if every point $u \in X$ possesses a neighborhood $N_u \subset X$ such that

$$|f(u_1) - f(u_2)| \leq K \|u_1 - u_2\|, \quad \forall u_1, u_2 \in N_u,$$

for a constant $K > 0$ depending on N_u.

Definition 2.2 The *generalized directional derivative* of the locally Lipschitz function $f : X \to \mathbb{R}$ at the point $u \in X$ in the direction $v \in X$ is defined by

$$f^0(u; v) := \limsup_{\substack{w \to u \\ t \searrow 0}} \frac{f(w + tv) - f(w)}{t}.$$

A natural question arises:

(Q_1): *What is the relationship between the generalized directional derivative $f^0(u; v)$ and the derivative notions from classical analysis?*

© The Author(s), under exclusive license to Springer Nature Switzerland AG 2021
N. Costea et al., *Variational and Monotonicity Methods in Nonsmooth Analysis*,
Frontiers in Mathematics, https://doi.org/10.1007/978-3-030-81671-1_2

The next two results make connections of this kind. Suppose first that the classical (one-sided) directional derivative of a function $f : X \rightarrow \mathbb{R}$ exists, i.e.,

$$f'(u; v) := \lim_{t \searrow 0} \frac{f(u + tv) - f(u)}{t}.$$

Proposition 2.1 *If $f : X \rightarrow \mathbb{R}$ is a continuously differentiable function, then*

$$f^0(u; v) = f'(u; v), \quad \forall u, v \in X. \tag{2.1}$$

Proof Fix $u, v, w \in X$. For $t > 0$ sufficiently small, the function $g(t) := f(w + tv)$ is continuously differentiable with derivative $g'(t) = f'(w + tv; v)$. By the classical mean value theorem, there exists $s \in (0, t)$ such that

$$\frac{f(w + tv) - f(w)}{t} = \frac{g(t) - g(0)}{t} = g'(s) = f'(w + sv; v).$$

Now, if $w \rightarrow u$ in X and $t \rightarrow 0$ in $\mathbb{R}$, due to the continuity of the differential of f, the desired relation yields.   □

Proposition 2.2 *If $f : X \rightarrow \mathbb{R}$ is convex and l.s.c., then f is locally Lipschitz and the following equality holds*

$$f^0(u; v) = f'(u; v), \quad \forall u, v \in X.$$

Proof According to Theorem 1.2 any convex l.s.c. functional is locally Lipschitz on the interior of its domain and since $X = \text{int}(X)$ it follows that f is locally Lipschitz on X.

The convexity of $f : X \rightarrow \mathbb{R}$ guarantees the existence of the one-sided directional derivative $f'(u; v)$. Fix $u, v \in X$ and an arbitrary small number $\delta > 0$ such that the Lipschitz condition

$$|f(w) - f(u)| \leq K \|w - u\|, \quad \forall w \in B(u, \delta).$$

Due to Definition 2.2 and to the convexity of f, one has

$$f^0(u; v) = \lim_{\varepsilon \searrow 0} \sup_{\|w-u\|<\varepsilon\delta} \sup_{0<t<\varepsilon} \frac{f(w + tv) - f(w)}{t}$$

$$= \lim_{\varepsilon \searrow 0} \sup_{\|w-u\|<\varepsilon\delta} \frac{f(w + \varepsilon v) - f(w)}{\varepsilon}.$$

Since f is locally Lipschitz, then

$$\left| \frac{f(w + \varepsilon v) - f(w)}{\varepsilon} - \frac{f(u + \varepsilon v) - f(u)}{\varepsilon} \right| \leq 2\delta K,$$

for $\|w - u\| < \varepsilon \delta$ and $\varepsilon \in (0, 1)$. One has

$$f^0(u; v) \leq \lim_{\varepsilon \searrow 0} \frac{f(u + \varepsilon v) - f(u)}{\varepsilon} + 2\delta K = f'(u; v) + 2\delta K.$$

If δ tends to 0 we have $f^0(u; v) = f'(u; v)$. □

Useful properties of the generalized directional derivative are given below.

Proposition 2.3 *Let $f : X \to \mathbb{R}$ be a locally Lipschitz function. Then*

(i) *For every $u \in X$ the function $f^0(u; \cdot) : X \to \mathbb{R}$ is positively homogeneous and subadditive (therefore convex) and satisfies*

$$|f^0(u; v)| \leq K \|v\|, \quad \forall v \in X. \tag{2.2}$$

Moreover, it is Lipschitz continuous on X with the Lipschitz constant K, where $K > 0$ is a Lipschitz constant of f near u.

(ii) *$f^0(\cdot; \cdot) : X \times X \to \mathbb{R}$ is upper semicontinuous.*

(iii) *$f^0(u; -v) = (-f)^0(u; v), \quad \forall u, v \in X.$*

Proof (i) Let $\lambda > 0$. The positive homogeneity of $f^0(u; \cdot)$ follows from

$$f^0(u; \lambda v) = \limsup_{\substack{w \to u \\ t \searrow 0}} \frac{f(w + t\lambda v) - f(w)}{t} = \limsup_{\substack{w \to u \\ t \searrow 0}} \frac{\lambda(f(w + t\lambda v) - f(w))}{t\lambda}$$

$$= \lambda f^0(u; v).$$

Relation (2.2) follows easily from Definition 2.2. To verify the subadditivity of $f^0(u; \cdot)$ let $v_1, v_2 \in X$ be fixed. One has

$$f^0(u; v_1 + v_2) = \limsup_{\substack{w \to u \\ t \searrow 0}} \frac{f(w + t(v_1 + v_2)) - f(w)}{t}$$

$$\leq \limsup_{\substack{w \to u \\ t \searrow 0}} \frac{f(w + t(v_1 + v_2)) - f(w + tv_2)}{t} + \limsup_{\substack{w \to u \\ t \searrow 0}} \frac{f(w + tv_2) - f(w)}{t}$$

$$\leq f^0(u; v_1) + f^0(u; v_2).$$

For arbitrary $v_1, v_2 \in X$, using the Lipschitz constant K on a neighborhood of u, we obtain

$$f(w + tv_1) - f(w) = f(w + tv_1) - f(w + tv_2) + f(w + tv_2) - f(w)$$
$$\leq K\|v_1 - v_2\|t + f(w + tv_2) - f(w),$$

if w is close to u and $t > 0$ is sufficiently small. Then

$$f^0(u; v_1) \leq K\|v_1 - v_2\| + f^0(u; v_2).$$

Interchanging v_1 and v_2, assertion (i) is now completely verified.

(ii) To prove the upper semicontinuity of $f^0(\cdot, \cdot)$, let $\{u_n\}$ and $\{v_n\}$ be sequences in X such that $u_n \to u \in X$ and $v_n \to v \in X$, as $n \to \infty$. Let us fix sequences $\{w_n\} \subset X$ and $\{t_n\} \subset \mathbb{R}_+^*$, with $\|w_n - u_n\| + t_n < \frac{1}{n}$ and

$$f^0(u_n; v_n) \leq \frac{f(w_n + t_n v_n) - f(w_n)}{t_n} + \frac{1}{n}.$$

Then

$$f^0(u_n; v_n) - \frac{1}{n} \leq \frac{f(w_n + t_n v) - f(w_n) + f(w_n + t_n v_n) - f(w_n + t_n v)}{t_n}$$
$$\leq \frac{f(w_n + t_n v) - f(w_n)}{t_n} + K\|v_n - v\|,$$

where $K > 0$ is the Lipschitz constant of f around u. Letting $n \to \infty$, one has

$$\limsup_{n \to \infty} f^0(u_n; v_n) \leq \limsup_{n \to \infty} \frac{f(w_n + t_n v) - f(w_n)}{t_n} \leq f^0(u; v).$$

(iii) Fix $u, v \in X$. Then

$$f^0(u; -v) = \limsup_{\substack{w \to u \\ t \searrow 0}} \frac{f(w - tv) - f(w)}{t} = \limsup_{\substack{w \to u \\ t \searrow 0}} \frac{-f(w - tv + tv) + f(w - tv)}{t}$$

$$= (-f)^0(u; v). \qquad \square$$

Definition 2.3 Let $f : X \to \mathbb{R}$ be a locally Lipschitz function. The *Clarke subdifferential* $\partial_C f(u)$ of f at a point $u \in X$ is the subset of the dual space X^* defined as follows

$$\partial_C f(u) := \left\{ \zeta \in X^* : \langle \zeta, v \rangle \leq f^0(u; v), \ \forall v \in X \right\}.$$

In the following result we prove the most important properties of the Clarke subdifferential.

Proposition 2.4 *Let* $f : X \rightarrow \mathbb{R}$ *a locally Lipschitz function. Then the following assertions are true:*

(i) *For every* $u \in X$, $\partial_C f(u)$ *is a nonempty, convex and weak*-compact subset of* X^*. *Moreover,*

$$\|\zeta\|_* \leq K, \quad \forall \zeta \in \partial_C f(u),$$

with $K > 0$ *the Lipschitz constant of* f *near* u.

(ii) *For every* $u \in X$, $f^0(u; \cdot)$ *is the support function of* $\partial_C f(u)$, *i.e.,*

$$f^0(u; v) = \max\{\langle \zeta, v \rangle : \zeta \in \partial_C f(u), \forall v \in X\}.$$

(iii) *The set-valued map* $\partial_C f : X \rightsquigarrow X^*$ *is closed from* $s - X$ *into* $w^* - X^*$;
In particular, if X *is finite dimensional, then* $\partial_C f$ *is an u.s.c. set valued map;*

(iv) *The set-valued map* $\partial_C f : X \rightsquigarrow X^*$ *is u.s.c. from* $s - X$ *into* $w^* - X^*$.

Proof   (i) Proposition 2.3 and the Hahn-Banach Theorem ensure that there exists $\zeta \in X^*$ satisfying

$$\langle \zeta, v \rangle \leq f^0(u; v), \ \forall v \in X.$$

Hence $\partial_C f(u) \neq \emptyset$. The convexity of $\partial_C f(u)$ follows easily from Definition 2.3. Let $\zeta \in \partial_C f(u)$ and $v \in X$ be fixed. Using Definition 2.3, relation (2.2) and Proposition 2.3 we obtain

$$-K\|v\| \leq -(-f)^0(u; \ v) \leq \langle \zeta, v \rangle \leq f^0(u; v) \leq K\|v\|.$$

Therefore $|\langle \zeta, v \rangle| \leq K\|v\|$, and one has

$$\|\zeta\|_* \leq K, \ \forall \zeta \in \partial_C f(u). \tag{2.3}$$

Since $\partial_C f(u)$ is weak* closed, the boundedness in (2.3) and the Banach-Alaoglu Theorem guarantee that $\partial_C f(u)$ is weak*compact in X^*.

(ii) Suppose by contradiction that there exists $v \in X$ with

$$f^0(u; v) > \max\{\langle \zeta, v \rangle : \zeta \in \partial_C f(u)\}. \tag{2.4}$$

Again, the Hahn-Banach theorem ensures the existence of $\xi \in X^*$ such that

$$\langle \xi, v \rangle = f^0(u; v) \text{ and } \langle \xi, w \rangle \leq f^0(u; w), \ \forall w \in X.$$

Therefore $\xi \in \partial_C f(u)$, which contradicts (2.4).

(iii) Fix some $v \in X$ and assume that the sequences $\{u_n\} \subset X$ and $\{\zeta_n\} \subset X^*$ are such that $u_n \to u$ in X and $\zeta_n \in \partial_C f(u_n)$, with $\zeta_n \to \zeta$ in X^*. Taking into account Proposition 2.3 (ii) and passing to the limit in the inequality $\langle \zeta_n, v \rangle \leq f^0(u_n; v)$ we obtain

$$\langle \zeta, v \rangle \leq \limsup_{n \to \infty} f^0(u_n; v) \leq f^0(u; v),$$

which shows that $\zeta \in \partial_C f(u)$.

(iv) We need to prove that for all $u, v \in X$ and $\varepsilon > 0$ there exists $\delta > 0$ such that whenever $\xi \in \partial_C f(v)$ and $\|v - u\| < \delta$, one can find $\zeta \in \partial_C f(u)$ satisfying

$$|\langle \zeta - \xi, v \rangle| < \varepsilon.$$

Arguing by contradiction, assume this is not the case, i.e., there exist $u, v \in X$, $\varepsilon_0 > 0$ and sequences $\{u_n\} \subset X$ and $\{\zeta_n\} \subset X^*$, with $\zeta_n \in \partial_C f(u_n)$ such that

$$\|u_n - u\| \leq \frac{1}{n} \text{ and } |\langle \zeta_n - \xi, v \rangle| \geq \varepsilon_0, \ \forall \xi \in \partial_C f(u). \tag{2.5}$$

From (i) we deduce $\|\zeta_n\|_* \leq K$, for sufficiently large n. Thus, up to a subsequence, $\zeta_n \to \zeta$ for some $\zeta \in X^*$. Then the assertion (iii) implies $\zeta \in \partial_C f(u)$, which clearly contradicts (2.5). $\qquad\square$

Proposition 2.5 Let $f, g : X \to \mathbb{R}$ be a locally Lipschitz functions. The following assertions hold:

(i) For every $\lambda \in \mathbb{R}$ one has

$$\partial_C (\lambda f)(u) = \lambda \partial_C f(u), \ \forall u \in X;$$

(ii) For all $u \in X$

$$\partial_C (f + g)(u) \subset \partial_C f(u) + \partial_C g(u).$$

Proof (i) Clearly the relation holds $\lambda \geq 0$. Thus it suffices to justify it for $\lambda = -1$, which follows actually from Proposition 2.3-(iii).

(ii) Since the support functions of the left- and right-hand side (evaluated in a fixed point $v \in X$) are $(f + g)^0(u; v)$ and $f^0(u; v) + g^0(u; v)$, respectively, it suffices to prove that

$$(f + g)^0(u; v) \leq f^0(u; v) + g^0(u; v).$$

This is in fact a straightforward consequence of Definition 2.2.   □

Proposition 2.6 *Let $f : X \to \mathbb{R}$ be a locally Lipschitz function. Then*

(i) *If f is Gateaux differentiable at $u \in X$, then its Gateaux derivative $\nabla f(u)$ belongs to $\partial_C f(u)$.*

(ii) *If, in addition, X is convex and $f : X \to \mathbb{R}$ is a convex function, then the generalized gradient $\partial_C f(u)$ coincides with the subdifferential of f at u in the sense of convex analysis, for every $u \in X$. Moreover, $f^0(u; v)$ coincides with the usual directional derivative $f'(u; v)$ for every $v \in X$.*

Proof (i) The Gateaux derivative $f'(u)$ satisfies

$$\langle \nabla f(u), v \rangle := f'(u; v) = \lim_{t \searrow 0} \frac{f(u + tv) - f(u)}{t} \leq f^0(u; v), \; \forall v \in X.$$

Therefore, by Definition 2.3 this means that $\nabla f(u) \in \partial_C f(u)$.

(ii) This is in fact a consequence of Proposition 2.4-(ii), Proposition 2.2 and of the convexity of f.   □

Remark 2.1 If $f : X \to \mathbb{R}$ is continuously differentiable at $u \in X$, then $\partial_C f(u) = \{f'(u)\}$. More generally, f is strictly differentiable at $u \in X$ if and only if f is locally Lipschitz near u and $\partial_C f(u)$ reduces to a singleton which is necessarily the strict derivative of f at u.

Theorem 2.1 (Lebourg's Mean Value Theorem [16]) *Let X be an open subset of a Banach space X and u, v be two points of X such that the line segment $[u, v] = \{(1 - t)u + tv : 0 \leq t \leq 1\} \subset X$. If $f : X \to \mathbb{R}$ is a locally Lipschitz function, then there exist $w \in (u, v)$ and $\zeta \in \partial_C f(w)$ such that*

$$f(v) - f(u) = \langle \zeta, v - u \rangle.$$

Proof The function $g : [0, 1] \to \mathbb{R}$ given by $g(t) := f(u + t(v - u))$ is locally Lipschitz. First, we shall prove that

$$\partial_C g(t) \subset \langle \partial_C f(u + t(v - u)), v - u \rangle.$$

Since the above closed convex sets are actually intervals in $\mathbb{R}$, it suffices to prove that

$$\max\{\partial_C g(t)s\} \leq \max\{s\langle \partial_C f((u + t(v - u), v - u)\},$$

for $s = \pm 1$. To this end, we point out that

$$\max\{\partial_C g(t)s\} = g^0(t; s) = \limsup_{\substack{\tau \to t \\ \lambda \searrow 0}} \frac{g(\tau + \lambda s) - g(\tau)}{\lambda}$$

$$= \limsup_{\substack{\tau \to t \\ \lambda \searrow 0}} \frac{f(u + (\tau + \lambda s)(v - u)) - f(u + \tau(v - u))}{\lambda}$$

$$\leq \limsup_{\substack{w \to u + t(v-u) \\ \lambda \searrow 0}} \frac{f(w + \lambda s(v - u)) - f(w)}{\lambda} = f^0(u + t(v - u); s(v - u))$$

$$= \max\{\langle \partial_C f(u + t(v - u)v, s(v - u)\rangle\}.$$

Now, if we introduce the function $h : [0, 1] \to \mathbb{R}$ given by $h(t) := g(t) + t(f(u) - f(v))$ then $h(0) = h(1) = f(u)$. But this implies that h has a local minimum or maximum at some $t_0 \in (0, 1)$. By Propositions 2.4-(ii) and 3.2 we have

$$0 \in \partial_C h(t_0) \subset \partial_C g(t_0) + f(u) - f(v).$$

Therefore, for $w := u + t_0(v - u)$, the following inclusion holds

$$f(v) - f(u) \in \partial_C g(t_0) \subset \langle \partial_C f(w), v - u \rangle.$$

$\square$

Definition 2.4 A locally Lipschitz function $f : X \to \mathbb{R}$ is said to be *regular* at $u \in X$, if the one-sided directional derivative $f'(u; v)$ exists for all $v \in X$ and $f^0(u; v) = f'(u; v)$. The function f is regular on X, if f is regular in every point $u \in X$.

Theorem 2.2

(i) *Let X, Y be two real Banach space and $F : X \to Y$ a continuously differentiable mapping and let $g : Y \to \mathbb{R}$ be a locally Lipschitz function. Then one has*

$$\partial_C(g \circ F)(u) \subset \partial_C g(F(u)) \circ F'(u), \quad \forall u \in X. \tag{2.6}$$

Equality holds in (2.6) if, for instance, g is regular at $F(u)$.

(ii) *Let $f : X \to \mathbb{R}$ and $h : \mathbb{R} \to \mathbb{R}$ be two locally Lipschitz functions. Then*

$$\partial_C(h \circ f)(u) \subset \overline{co}^{w^*}(\partial_C h(f(u)) \cdot \partial_C f(u)), \quad \forall u \in X, \tag{2.7}$$

where the notation $\overline{co}^{w^*}$ stands for the weak*-closed convex hull. Furthermore, if h continuously differentiable at $f(u)$ or if h is regular at $f(u)$ and f is continuously differentiable at u, in (2.7) the equality holds and the symbol $\overline{co}$ becomes superfluous.

Proof

(i) By Proposition 2.4-(ii) it suffices to show that

$$(g \circ F)^0(u; v) \leq \max\left\{\langle \zeta, F'(u)v \rangle : \zeta \in \partial_C g(F(u))\right\}, \ \forall v \in X. \tag{2.8}$$

Due to Lebourg's mean value theorem (see Theorem 2.1), one has

$$(g \circ F)(w + tv) - (g \circ F)(w) = \langle \xi, F(w + tv) - F(w) \rangle,$$

for some $\xi \in \partial_C g(y)$ and $y \in (F(w), F(w + tv))$. Then, the classical mean value theorem guarantees that $F(w + tv) - F(w) = tF'(x)v$, for a point $x \in (w, \ w + tv)$. Thus we obtain

$$(g \circ F)^0(u; v) = \limsup_{\substack{w \to u \\ t \searrow 0}} \frac{(g \circ F)(w + tv) - (g \circ F)(w)}{t} = \limsup_{\substack{w \to u \\ t \searrow 0}} \langle \xi, F'(u)v \rangle$$

$$\leq \max\{\langle \zeta, F'(u)v \rangle : \zeta \in \partial_C g(F(u))\}.$$

Suppose now that g is regular at $F(u)$. By Proposition 2.4-(ii) and the regularity assumption for every $v \in X$ we get

$$\max_{\zeta \in \partial_C g(F(u))} \langle \zeta, F'(u)v \rangle = g^0(F(u); F'(u)v) = g'(F(u); F'(u)v)$$

$$= \lim_{t \searrow 0} \frac{g(F(u) + tF'(u)v) - g(F(u))}{t} = \lim_{t \searrow 0} \frac{(g \circ F)(u + tv) - (g \circ F)(u)}{t}$$

$$\leq (g \circ F)^0(u; v),$$

which yields the equality in (2.6).

(ii) Fix $v \in X$. Applying twice Theorem 2.1, for every w close to $u \in X$ and sufficiently small $t > 0$ one gets the existence of $\xi \in \partial_C h(s) \subset \mathbb{R}$, $s \in (f(w), \ f(w + tw))$ and $\zeta \in \partial_C f(x)$ with $x \in (w, w + tv) \subset X$ such that

$$(h \circ f)(w + tv) - (h \circ f)(w) = \xi(f(w + tv) - f(w)) = \xi \langle \zeta, tv \rangle. \tag{2.9}$$

Then, according to Proposition 2.4-(iii),

$$(h \circ f)^0(u; v) \leq \max\{\xi \langle \zeta, v \rangle : \zeta \in \partial_C f(u), \xi \in \partial_C h(f(u))\}.$$

Then the inclusion (2.7) holds true. The proof of the assertion regarding the equality in (2.7) follows in a similar manner as in the statement (i). In the mentioned cases the symbol $\overline{co}$ is not necessary in (2.7) due to Remark 2.1.   □

Proposition 2.7 *Let* $\phi : [0, 1] \to X$ *be a function of class* C^1 *and let* $f : X \to \mathbb{R}$ *be a locally Lipschitz function. Then the function* $h : [0, 1] \to \mathbb{R}$ *given by* $h = f \circ \phi$ *is differentiable a.e.* $t \in [0, 1]$ *and*

$$h'(t) \leq \max\left\{\langle \zeta, \phi'(t) \rangle : \zeta \in \partial_C f(\phi(t))\right\}.$$

Proof The function h is clearly locally Lipschitz, thus differentiable for a.e. $t \in [0, 1]$. Suppose that h is differentiable at $t = t_0$. Then

$$h'(t_0) = \lim_{\lambda \to 0} \frac{f(\phi(t_0 + \lambda)) - f(\phi(t_0))}{\lambda} = \lim_{\lambda \to 0} \frac{f(\phi(t_0) + \phi'(t_0)\lambda + o(\lambda)) - f(\phi(t_0))}{\lambda}$$

$$= \lim_{\lambda \to 0} \frac{f(\phi(t_0) + \phi'(t_0)\lambda) - f(\phi(t_0))}{\lambda}$$

$$\leq \limsup_{\substack{s \to 0 \\ \lambda \searrow 0}} \frac{f(\phi(t_0) + s + \phi'(t_0)\lambda) - f(\phi(t_0) + s)}{\lambda}$$

$$= f^0(\phi(t_0); \phi'(t_0)) = \max\{\langle \zeta, \phi'(t_0) \rangle : \zeta \in \partial_C f(\phi(t_0))\}.$$

□

Theorem 2.3 *Suppose that* X, Y *are two Banach spaces,* X *is reflexive and* $X \hookrightarrow Y$, *i.e.* $X \subset Y$ *and the embedding mapping is continuous, and assume that* X *is dense in* Y. *Let* $f : Y \to \mathbb{R}$ *be a locally Lipschitz continuous function and let* $\widehat{f} = f|_X$. *Then for every* $u \in X$, *one has* $\partial_C \widehat{f}(u) \subset \partial_C f(u)$.

In order to prove this result, we need the following lemma.

Lemma 2.1 *Suppose that in Theorem 2.3* f *is convex. Then for every* $u \in X$ *we have* $\partial_C \widehat{f}(u) = \partial_C f(u)$.

Proof In this case, the generalized gradient $\partial_C f(u)$ is the same as the subdifferential in the convex analysis. By definition, it is easy to see that $\partial f(u) \subset \partial \widehat{f}(u)$. But we know that $\partial \widehat{f}(u) \cap Y^* \subset \partial f(u)$. In fact, if $\zeta \in \partial \widehat{f}(u) \cap Y^*$, then

$$\langle \zeta, v - u \rangle + \widehat{f}(u) \leq \widehat{f}(v), \tag{2.10}$$

for each $v \in X$. The fact that X is dense in Y, $\zeta \in Y^*$ and f is continuous in Y guarantee the extension of the inequality (2.10) to all $v \in Y$, i.e.

$$\langle \zeta, v - u \rangle + f(u) \leq f(v), \quad \forall v \in Y.$$

This means $\zeta \in \partial f(u)$. Since X is reflexive, Y^* is dense in X^*, so that $\partial f(u)$ is dense in $\partial \widehat{f}(u)$ in the *weak**-topology of X^*. For every $\zeta \in \partial \widehat{f}(u)$ there exists $\zeta_n \in \partial f(u)$ such that $\langle \zeta_n, v \rangle \to \langle \zeta, v \rangle$ for every $v \in X$. But,

$$|\langle \zeta_n, v \rangle| \leq \|\zeta_n\|_{Y^*} \|v\|_Y \leq K \|v\|_Y,$$

provided by Proposition 2.3, then

$$|\langle \zeta, v \rangle| \leq K \|v\|_Y.$$

This implies that ζ may be continuously extended onto Y. Thus $\zeta \in \partial \widehat{f}(u) \cap Y^* \subset \partial f(u)$.

$\square$

Proof of Theorem 2.3 It is clear that the function $v \mapsto f^0(u; v)$ is convex and continuous on Y and $f^0(u, \cdot)|_X \geq \widehat{f}^0(u, \cdot)$. Since the generalized gradients coincide with the convex subdifferentials, the conclusion of this theorem follows directly from Lemma 2.1. $\square$

We close this section with some properties of partial Clarke subdifferentials.

Proposition 2.8 *Let $h : X_1 \times X_2 \to \mathbb{R}$ be a locally Lipschitz function which is regular at $(u, v) \in X_1 \times X_2$. Then the following hold:*

(i) *$\partial_C h(u, v) \subseteq \partial_C^1 h(u, v) \times \partial_C^2 h(u, v)$, where $\partial_C^1 h(u, v)$ denotes the (partial) generalized gradient of $h(\cdot, v)$ at the point u, and $\partial_C^2 h(u, v)$ that of $h(u, \cdot)$ at v.*

(ii) *$h^0(u, v; w, z) \leq h_1^0(u, v; w) + h_2^0(u, v; z)$ for all $w, z \in X$, where $h_1^0(u, v; w)$ (resp. $h_2^0(u, v; z)$) is the (partial) generalized directional derivative of $h(\cdot, v)$ (resp. $h(u, \cdot)$) at the point $u \in \mathbb{R}$ (resp. $v \in \mathbb{R}$) in the direction $w \in \mathbb{R}$ (resp. $z \in \mathbb{R}$).*

Proof

(i) Fix $\zeta := (\zeta_1, \zeta_2) \in \partial_C h(u, v)$. It suffices prove that ζ_1 belongs to $\partial_C^1 h(u, v)$, which is equivalent to show that for every $w \in X$ one has $\langle \zeta_1, w \rangle \leq h_1^0(u, v; w)$. The latter coincides with

$$h_1'(u, v; w) = h'(u, v; w, 0) = h^0(u, v; w, 0),$$

which clearly majorizes $\langle \zeta_1, w \rangle$.

(*ii*) Let us fix $w, z \in X$. Proposition 2.4-(ii) ensures that there exists $\zeta \in \partial_C h(u, v)$ such that

$$h^0(u, v; w, z) = \langle \zeta, (w, z) \rangle.$$

By (i) we have $\zeta = (\zeta_1, \zeta_2)$, where $\zeta_i \in \partial_C^i h(u, v)$ ($i \in \{1, 2\}$), and using the definition of the generalized gradient, we obtain

$$h^0(u, v; w, z) = \langle \zeta_1, w \rangle + \langle \zeta_2, z \rangle \leq h_1^0(u, v; w) + h_2^0(u, v; z).$$

$\square$

Remark 2.2 It is worth to note that in general we have no equality in Proposition 2.8 (b). Indeed, let us consider for instance $h : \mathbb{R}^2 \to \mathbb{R}$, defined by $h(u, v) := \max\{|u|^{5/2}, |v|^{5/2}\}$. It is clear that h is regular on $\mathbb{R}^2$, but for every $\alpha, \beta > 0$, $h^0(\alpha, \alpha; \beta, \beta) = h_1^0(\alpha, \alpha; \beta) = h_2^0(\alpha, \alpha; \beta) = 5\alpha^{3/2}\beta/2$.

2.2 Nonsmooth Calculus on Manifolds

In this section we present some basic notions and results from the subdifferential calculus on Riemannian manifolds, developed by Azagra et al. [2] and Ledyaev and Zhu [17]. Moreover, following Kristály [15], two subdifferential notions are introduced based on the cut locus, and we establish an analytical characterization of the limiting/Fréchet normal cone on Riemannian manifolds (see Corollary 2.1) which plays a crucial role in the study of Nash-Stampacchia equilibrium points, see Sect. 9.1. Before doing this, we first recall those elements from Riemannian geometry which will be used in the sequel; we mainly follow do Carmo [11].

Let (M, g) be a connected m-dimensional Riemannian manifold, $m \geq 2$ and let $TM = \cup_{p \in M}(p, T_p M)$ and $T^*M = \cup_{p \in M}(p, T_p^* M)$ be the tangent and cotangent bundles to M. For every $p \in M$, the Riemannian metric induces a natural Riesz-type isomorphism between the tangent space $T_p M$ and its dual $T_p^* M$; in particular, if $\xi \in T_p^* M$ then there exists a unique $W_\xi \in T_p M$ such that

$$\langle \xi, V \rangle_{g,p} = g_p(W_\xi, V) \text{ for all } V \in T_p M. \tag{2.11}$$

Instead of $g_p(W_\xi, V)$ and $\langle \xi, V \rangle_{g,p}$ we shall write simply $g(W_\xi, V)$ and $\langle \xi, V \rangle_g$ when no confusion arises. Due to (2.11), the elements ξ and W_ξ are identified. With the above notations, the norms on $T_p M$ and $T_p^* M$ are defined by

$$\|\xi\|_g = \|W_\xi\|_g = \sqrt{g(W_\xi, W_\xi)}.$$

The generalized Cauchy-Schwartz inequality is also valid, i.e., for every $V \in T_p M$ and $\xi \in T_p^* M$,

$$|\langle \xi, V \rangle_g| \leq \|\xi\|_g \|V\|_g. \tag{2.12}$$

Let $\xi_k \in T_{p_k}^* M$, $k \in \mathbb{N}$, and $\xi \in T_p^* M$. The sequence $\{\xi_k\}$ converges to ξ, denoted by $\lim_k \xi_k = \xi$, when $p_k \to p$ and $\langle \xi_k, W(p_k) \rangle_g \to \langle \xi, W(p) \rangle_g$ as $k \to \infty$, for every C^∞ vector field W on M.

Let $h : M \to \mathbb{R}$ be a C^1 functional at $p \in M$; the differential of h at p, denoted by $dh(p)$, belongs to $T_p^* M$ and is defined by

$$\langle dh(p), V \rangle_g = g(\operatorname{grad}h(p), V) \text{ for all } V \in T_p M.$$

If $(x^1, \ldots, x^m)$ is the local coordinate system on a coordinate neighborhood (U_p, ψ) of $p \in M$, and the local components of dh are denoted $h_i = \frac{\partial h}{\partial x_i}$, then the local components of $\operatorname{grad}h$ are $h^i = g^{ij} h_j$. Here, g^{ij} are the local components of g^{-1}.

Let $\gamma : [0, r] \to M$ be a C^1 path, $r > 0$. The length of γ is defined by

$$L_g(\gamma) = \int_0^r \|\dot{\gamma}(t)\|_g dt.$$

For any two points $p, q \in M$, let

$$d_g(p, q) = \inf\{L_g(\gamma) : \gamma \text{ is a } C^1 \text{ path joining } p \text{ and } q \text{ in } M\}.$$

The function $d_g : M \times M \to \mathbb{R}$ is a metric which generates the same topology on M as the underlying manifold topology. For every $p \in M$ and $r > 0$, we define the open ball of center $p \in M$ and radius $r > 0$ by

$$B_g(p, r) = \{q \in M : d_g(p, q) < r\}.$$

Let us denote by ∇ the unique natural covariant derivative on (M, g), also called the Levi-Civita connection. A vector field W along a C^1 path γ is called parallel when $\nabla_{\dot{\gamma}} W = 0$. A C^∞ parameterized path γ is a geodesic in (M, g) if its tangent $\dot{\gamma}$ is parallel along itself, i.e., $\nabla_{\dot{\gamma}} \dot{\gamma} = 0$. The geodesic segment $\gamma : [a, b] \to M$ is called minimizing if $L_g(\gamma) = d_g(\gamma(a), \gamma(b))$.

Standard ODE theory implies that for every $V \in T_p M$, $p \in M$, there exists an open interval $I_V \ni 0$ and a unique geodesic $\gamma_V : I_V \to M$ with $\gamma_V(0) = p$ and $\dot{\gamma}_V(0) = V$. Due to the 'homogeneity' property of the geodesics (see do Carmo [11, p. 64]), we may define the exponential map $\exp_p : T_p M \to M$ as $\exp_p(V) = \gamma_V(1)$. Moreover,

$$d \exp_p(0) = \operatorname{id}_{T_p M}. \tag{2.13}$$

Note that there exists an open (starlike) neighborhood $\mathcal{U}$ of the zero vectors in TM and an open neighborhood $\mathcal{V}$ of the diagonal $M \times M$ such that the exponential map $V \mapsto \exp_{\pi(V)}(V)$ is smooth and the map $\pi \times \exp : \mathcal{U} \to \mathcal{V}$ is a diffeomorphism, where π is the canonical projection of TM onto M. Moreover, for every $p \in M$ there exists a number $r_p > 0$ and a neighborhood $\tilde{U}_p$ such that for every $q \in \tilde{U}_p$, the map $\exp_q$ is a C^∞ diffeomorphism on $B(0, r_p) \subset T_q M$ and $\tilde{U}_p \subset \exp_q(B(0, r_p))$; the set $\tilde{U}_p$ is called a *totally normal neighborhood* of $p \in M$. In particular, it follows that every two points $q_1, q_2 \in \tilde{U}_p$ can be joined by a minimizing geodesic of length less than r_p. Moreover, for every $q_1, q_2 \in \tilde{U}_p$ we have

$$\| \exp_{q_1}^{-1}(q_2) \|_g = d_g(q_1, q_2). \tag{2.14}$$

The *tangent cut locus* of $p \in M$ in $T_p M$ is the set of all vectors $v \in T_p M$ such that $\gamma(t) = \exp_p(tv)$ is a minimizing geodesic for $t \in [0, 1]$ but fails to be minimizing for $t \in [0, 1 + \varepsilon)$ for each $\varepsilon > 0$. The *cut locus of* $p \in M$, denoted by C_p, is the image of the tangent cut locus of p via $\exp_p$. Note that any totally normal neighborhood of $p \in M$ is contained into $M \setminus C_p$.

Let (M, g) be an m-dimensional Riemannian manifold and let $f : M \to \mathbb{R} \cup \{+\infty\}$ be a lower semicontinuous function with $\mathrm{dom}(f) \neq \emptyset$. The *Fréchet-subdifferential* of f at $p \in \mathrm{dom}(f)$ is the set

$$\partial_F f(p) = \{dh(p) : h \in C^1(M) \text{ and } f - h \text{ attains a local minimum at } p\}.$$

The following properties are adaptations of earlier Euclidean results to Riemannian manifolds.

Proposition 2.9 ([2, Theorem 4.3]) *Let (M, g) be an m-dimensional Riemannian manifold and let $f : M \to \mathbb{R} \cup \{+\infty\}$ be a lower semicontinuous function, $p \in \mathrm{dom}(f) \neq \emptyset$ and $\xi \in T_p^* M$. The following statements are equivalent:*

(i) $\xi \in \partial_F f(p)$;

(ii) *For every chart* $\psi : U_p \subset M \to \mathbb{R}^m$ *with* $p \in U_p$, *if* $\zeta = \xi \circ d\psi^{-1}(\psi(p))$, *we have that*

$$\liminf_{v \to 0} \frac{(f \circ \psi^{-1})(\psi(p) + v) - f(p) - \langle \zeta, v \rangle_g}{\|v\|} \geq 0;$$

(iii) *There exists a chart* $\psi : U_p \subset M \to \mathbb{R}^m$ *with* $p \in U_p$, *if* $\zeta = \xi \circ d\psi^{-1}(\psi(p))$, *then*

$$\liminf_{v \to 0} \frac{(f \circ \psi^{-1})(\psi(p) + v) - f(p) - \langle \zeta, v \rangle_g}{\|v\|} \geq 0.$$

In addition, if f is locally bounded from below, i.e., for every $q \in M$ there exists a neighborhood U_q of q such that f is bounded from below on U_q, the above conditions are also equivalent to

(*iv*) *There exists a function $h \in C^1(M)$ such that $f - h$ attains a global minimum at p and $\xi = dh(p)$.*

The *limiting subdifferential* and *singular subdifferential* of f at $p \in M$ are the sets

$$\partial_L f(p) = \{\lim_k \xi_k : \xi_k \in \partial_F f(p_k), \ (p_k, f(p_k)) \to (p, f(p))\}$$

and

$$\partial_\infty f(p) = \{\lim_k t_k \xi_k : \xi_k \in \partial_F f(p_k), \ (p_k, f(p_k)) \to (p, f(p)), t_k \to 0^+\}.$$

Proposition 2.10 ([17]) *Let (M, g) be a finite-dimensional Riemannian manifold and let $f : M \to \mathbb{R} \cup \{+\infty\}$ be a lower semicontinuous function. Then, we have*

(*i*) $\partial_F f(p) \subset \partial_L f(p)$, $p \in \text{dom}(f)$;
(*ii*) $0 \in \partial_\infty f(p)$, $p \in M$;
(*iii*) *If $p \in \text{dom}(f)$ is a local minimum of f, then $0 \in \partial_F f(p) \subset \partial_L f(p)$.*

Proposition 2.11 ([17, Theorem 4.8 (Mean Value Inequality)]) *Let $f : M \to \mathbb{R}$ be a continuous function bounded from below, let V be a C^∞ vector field on M and let $c : [0, 1] \to M$ be a curve such that $\dot{c}(t) = V(c(t))$, $t \in [0, 1]$. Then for any $r < f(c(1)) - f(c(0))$, any $\varepsilon > 0$ and any open neighborhood U of $c([0, 1])$, there exists $m \in U$, $\xi \in \partial_F f(m)$ such that $r < \langle \xi, V(m) \rangle_g$.*

Proposition 2.12 ([17, Theorem 4.13 (Sum Rule)]) *Let (M, g) be an m-dimensional Riemannian manifold and let $f_1, \ldots, f_H : M \to \mathbb{R} \cup \{+\infty\}$ be lower semicontinuous functions. Then, for every $p \in M$ we have either $\partial_L(\sum_{l=1}^{H} f_l)(p) \subset \sum_{l=1}^{H} \partial_L f_l(p)$, or there exist $\xi_l^\infty \in \partial_\infty f_l(p)$, $l = 1, \ldots, H$, not all zero such that $\sum_{l=1}^{H} \xi_l^\infty = 0$.*

The *cut-locus subdifferential* of f at $p \in \text{dom}(f)$ is defined as

$$\partial_{cl} f(p) = \{\xi \in T_p^* M : f(q) - f(p) \geq \langle \xi, \exp_p^{-1}(q) \rangle_g \text{ for all } q \in M \setminus C_p\},$$

where C_p is the cut locus of the point $p \in M$. Note that $M \setminus C_p$ is the maximal open set in M such that every element from it can be joined to p by exactly one minimizing geodesic, see Klingenberg [13, Theorem 2.1.14]. Therefore, the cut-locus subdifferential is well-defined, i.e., $\exp_p^{-1}(q)$ makes sense and is unique for every $q \in M \setminus C_p$. We first prove

Theorem 2.4 ([15]) *Let (M, g) be a Riemannian manifold and $f : M \to \mathbb{R} \cup \{+\infty\}$ be a proper, lower semicontinuous function. Then, for every $p \in \mathrm{dom}(f)$ we have*

$$\partial_{cl} f(p) \subset \partial_F f(p) \subset \partial_L f(p).$$

Moreover, if f is convex, the above inclusions become equalities.

Proof The last inclusion is standard, see Proposition 2.10-(i). Now, let $\xi \in \partial_{cl} f(p)$, i.e., $f(q) - f(p) \geq \langle \xi, \exp_p^{-1}(q) \rangle_g$ for all $q \in M \setminus C_p$. In particular, the latter inequality is valid for every $q \in B_g(p, r)$ for $r > 0$ small enough, since $B_g(p, r) \subset M \setminus C_p$ (for instance, when $B_g(p, r) \subset M$ is a totally normal ball around p). Now, by choosing $\psi = \exp_p^{-1} : B_g(p, r) \to T_p M$ in Proposition 2.9-(ii), one has that $f(\exp_p v) - f(p) \geq \langle \xi, v \rangle_g$ for all $v \in T_p M$, $\|v\| < r$, which implies $\xi \in \partial_F f(p)$.

Now, we assume in addition that f is convex, and let $\xi \in \partial_L f(p)$. We are going to prove that $\xi \in \partial_{cl} f(p)$. Since $\xi \in \partial_L f(p)$, we have that $\xi = \lim_k \xi_k$ where $\xi_k \in \partial_F f(p_k)$, $(p_k, f(p_k)) \to (p, f(p))$. By Proposition 2.9-$(ii)$, for $\psi_k = \exp_{p_k}^{-1} : \tilde{U}_{p_k} \to T_{p_k} M$ where $\tilde{U}_{p_k} \subset M$ is a totally normal ball centered at p, one has that

$$\liminf_{v \to 0} \frac{f(\exp_{p_k} v) - f(p_k) - \langle \xi_k, v \rangle_g}{\|v\|} \geq 0. \tag{2.15}$$

Now, fix $q \in M \setminus C_p$. The latter fact is equivalent to $p \in M \setminus C_q$, see Klingenberg [13, Lemma 2.1.11]. Since $M \setminus C_q$ is open and $p_k \to p$, we may assume that $p_k \in M \setminus C_q$, i.e., q and every point p_k is joined by a unique minimizing geodesic. Therefore, $V_k = \exp_{p_k}^{-1}(q)$ is well-defined. Now, let $\gamma_k(t) = \exp_{p_k}(t V_k)$ be the geodesic which joins p_k and q. Then (2.15) implies that

$$\liminf_{t \to 0^+} \frac{f(\gamma_k(t)) - f(p_k) - \langle \xi_k, t V_k \rangle_g}{\|t V_k\|} \geq 0. \tag{2.16}$$

Since f is convex, one has that $f(\gamma_k(t)) \leq t f(\gamma_k(1)) + (1 - t) f(\gamma_k(0))$, $t \in [0, 1]$, thus, the latter relations imply that

$$\frac{f(q) - f(p_k) - \langle \xi_k, \exp_{p_k}^{-1}(q) \rangle_g}{d_g(p_k, q)} \geq 0.$$

Since $f(p_k) \to f(p)$ and $\xi = \lim_k \xi_k$, it yields precisely that

$$f(q) - f(p) - \langle \xi, \exp_p^{-1}(q) \rangle_g \geq 0,$$

i.e., $\xi \in \partial_{cl} f(p)$, which concludes the proof. $\square$

Remark 2.3 If (M, g) is a Hadamard manifold (i.e., simply connected, complete Riemannian manifold with nonpositive sectional curvature), then $C_p = \emptyset$ for every $p \in M$; in this case, the cut-locus subdifferential agrees formally with the convex subdifferential in the Euclidean setting.

Let $K \subset M$ be a closed set. Following Ledyaev and Zhu [17], the *Fréchet-normal cone* and *limiting normal cone* of K at $p \in K$ are the sets

$$N_F(p; K) = \partial_F \delta_K(p) \quad \text{and} \quad N_L(p; K) = \partial_L \delta_K(p),$$

where δ_K is the indicator function of the set K, i.e., $\delta_K(q) = 0$ if $q \in K$ and $\delta_K(q) = +\infty$ if $q \notin K$.

The following result—which is one of our key tools to study Nash-Stampacchia equilibrium points on Riemannian manifolds—it is know for Hadamard manifolds only, see Li, López and Martín-Márquez [18] and it is a simple consequence of the above theorem. To state this result, we recall that a set $K \subset M$ is *geodesic convex* if every two points $p, q \in K$ can be joined by a unique geodesic segment whose image belongs entirely to K.

Corollary 2.1 *Let (M, g) be a Riemannian manifold, $K \subset M$ be a closed, geodesic convex set, and $p \in K$. Then, we have*

$$N_F(p; K) = N_L(p; K) = \partial_{cl} \delta_K(p) = \{\xi \in T_p^* M : \langle \xi, \exp_p^{-1}(q) \rangle_g \leq 0 \text{ for all } q \in K\}.$$

Proof Applying Theorem 2.4 to the indicator function $f = \delta_K$, we have that $N_F(p; K) = N_L(p; K) = \partial_{cl} \delta_K(p)$. It remains to compute the latter set explicitly. Since $K \subset M \setminus C_p$ (note that the geodesic convexity of K assumes itself that every two points of K can be joined by a unique geodesic, thus $K \cap C_p = \emptyset$) and $\delta_K(p) = 0$, $\delta_K(q) = +\infty$ for $q \notin K$, one has that

$$\xi \in \partial_{cl} \delta_K(p) \Leftrightarrow \delta_K(q) - \delta_K(p) \geq \langle \xi, \exp_p^{-1}(q) \rangle_g \text{ for all } q \in M \setminus C_p$$

$$\Leftrightarrow 0 \geq \langle \xi, \exp_p^{-1}(q) \rangle_g \text{ for all } q \in K,$$

which ends the proof. □

Let $U \subset M$ be an open subset of the Riemannian manifold (M, g). We say that a function $f : U \to \mathbb{R}$ is *locally Lipschitz at* $p \in U$ if there exist an open neighborhood $U_p \subset U$ of p and a number $C_p > 0$ such that for every $q_1, q_2 \in U_p$,

$$|f(q_1) - f(q_2)| \leq C_p d_g(q_1, q_2).$$

The function $f : U \to \mathbb{R}$ is *locally Lipschitz* on (U, g) if it is locally Lipschitz at every $p \in U$.

Fix $p \in U$, $v \in T_pM$, and let $\tilde{U}_p \subset U$ be a totally normal neighborhood of p. If $q \in \tilde{U}_p$, following [2, Section 5], for small values of $|t|$, we may introduce

$$\sigma_{q,v}(t) = \exp_q(tw), \quad w = d(\exp_q^{-1} \circ \exp_p)_{\exp_p^{-1}(q)} v.$$

If the function $f : U \to \mathbb{R}$ is locally Lipschitz on (U, g), then

$$f^0(p; v) = \limsup_{q \to p,\, t \to 0^+} \frac{f(\sigma_{q,v}(t)) - f(q)}{t}$$

is called the *Clarke generalized derivative of f at $p \in U$ in direction $v \in T_pM$*, and

$$\partial_C f(p) = \mathrm{co}(\partial_L f(p))$$

is the *Clarke subdifferential of f at $p \in U$*, where 'co' stands for the convex hull. When $f : U \to \mathbb{R}$ is a C^1 functional at $p \in U$ then

$$\partial_C f(p) = \partial_L f(p) = \partial_F f(p) = \{df(p)\}, \tag{2.17}$$

see [2, Proposition 4.6]. Moreover, when (M, g) is the standard Euclidean space, the Clarke subdifferential and the Clarke generalized gradient agree, see Sect. 2.1.

One can easily prove that the function $f^0(\cdot; \cdot)$ is upper-semicontinuous on $TU = \cup_{p \in U} T_pM$ and $f^0(p; \cdot)$ is positive homogeneous and subadditive on T_pM, thus convex. In addition, if $U \subset M$ is geodesic convex and $f : U \to \mathbb{R}$ is convex, then

$$f^0(p; v) = \lim_{t \to 0^+} \frac{f(\exp_p(tv)) - f(p)}{t}, \tag{2.18}$$

see Claim 5.4 and the first relation on p. 341 of [2], similarly to Proposition 2.2 on normed spaces.

Proposition 2.13 ([17, Corollary 5.3]) *Let (M, g) be a Riemannian manifold and let $f : M \to \mathbb{R} \cup \{+\infty\}$ be a lower semicontinuous function. Then the following statements are equivalent:*

  (*i*) *f is locally Lipschitz at $p \in M$;*
 (*ii*) *$\partial_C f$ is bounded in a neighborhood of $p \in M$;*
(*iii*) *$\partial_\infty f(p) = \{0\}$.*

Proposition 2.14 *Let* $f, g : M \to \mathbb{R} \cup \{+\infty\}$ *be two proper, lower semicontinuous functions. Then, for every* $p \in \text{dom}(f) \cap \text{dom}(g)$ *with* $\partial_{cl} f(p) \neq \emptyset \neq \partial_{cl} g(p)$ *we have* $\partial_{cl} f(p) + \partial_{cl} g(p) \subset \partial_{cl}(f+g)(p)$. *Moreover, if both functions are convex and* f *is locally bounded, the inclusion is equality.*

Let $f : U \to \mathbb{R}$ be a locally Lipschitz function and $p \in U$. We consider the *Clarke 0-subdifferential* of f at p as

$$\partial_0 f(p) = \{\xi \in T_p^* M : f^0(p; \exp_p^{-1}(q)) \geq \langle \xi, \exp_p^{-1}(q) \rangle_g \text{ for all } q \in U \setminus C_p\}$$

$$= \{\xi \in T_p^* M : f^0(p; v) \geq \langle \xi, v \rangle_g \text{ for all } v \in T_p M\}.$$

Theorem 2.5 ([15]) *Let* (M, g) *be a Riemannian manifold,* $U \subset M$ *be open,* $f : U \to \mathbb{R}$ *be a locally Lipschitz function, and* $p \in U$. *Then,*

$$\partial_0 f(p) = \partial_{cl}(f^0(p; \exp_p^{-1}(\cdot)))(p) = \partial_L(f^0(p; \exp_p^{-1}(\cdot)))(p) = \partial_C f(p).$$

Proof The proof will be carried out in several steps as follows.

Step 1. $\partial_0 f(p) = \partial_{cl}(f^0(p; \exp_p^{-1}(\cdot)))(p)$.
It follows from the definitions.

Step 2. $\partial_0 f(p) = \partial_L(f^0(p; \exp_p^{-1}(\cdot)))(p)$.
The inclusion " $\subset$ " follows from Step 1 and Theorem 2.4. For the converse, we notice that $f^0(p; \exp_p^{-1}(\cdot))$ is locally Lipschitz in a neighborhood of p; indeed, $f^0(p; \cdot)$ is convex on $T_p M$ and $\exp_p$ is a local diffeomorphism on a neighborhood of the origin of $T_p M$. Now, let $\xi \in \partial_L(f^0(p; \exp_p^{-1}(\cdot)))(p)$. Then, $\xi = \lim_k \xi_k$ where $\xi_k \in \partial_F(f^0(p; \exp_p^{-1}(\cdot)))(p_k)$, $p_k \to p$. By Proposition 2.9-(ii), for $\psi = \exp_p^{-1} : \tilde{U}_p \to T_p M$ where $\tilde{U}_p \subset M$ is a totally normal ball centered at p, one has that

$$\liminf_{v \to 0} \frac{f^0(p; \exp_p^{-1}(p_k) + v) - f^0(p; \exp_p^{-1}(p_k)) - \langle \xi_k((d \exp_p)(\exp_p^{-1}(p_k))), v \rangle_g}{\|v\|} \geq 0.$$

$$(2.19)$$

In particular, if $q \in M \setminus C_p$ is fixed arbitrarily and $v = t \exp_p^{-1}(q)$ for $t > 0$ small, the convexity of $f^0(p; \cdot)$ and relation (2.19) yield that

$$f^0(p; \exp_p^{-1}(q)) \geq \langle \xi_k((d \exp_p)(\exp_p^{-1}(p_k))), \exp_p^{-1}(q) \rangle_g.$$

Since $\xi = \lim_k \xi_k$, $p_k \to p$ and $d(\exp_p)(0) = \text{id}_{T_p M}$ (see (2.13)), we obtain that

$$f^0(p; \exp_p^{-1}(q)) \geq \langle \xi, \exp_p^{-1}(q) \rangle_g,$$

i.e., $\xi \in \partial_0 f(p)$. This concludes Step 2.

Step 3. $\partial_0 f(p) = \partial_C f(p)$.

First, we prove the inclusion $\partial_0 f(p) \subset \partial_C f(p)$. Here, we follow Borwein and Zhu [3, Theorem 5.2.16], see also Clarke, Ledyaev, Stern and Wolenski [7, Theorem 6.1]. Let $v \in T_p M$ be fixed arbitrarily. The definition of $f^0(p; v)$ shows that one can choose $t_k \to 0^+$ and $q_k \to p$ such that

$$f^0(p; v) = \lim_{k \to \infty} \frac{f(\sigma_{q_k, v}(t_k)) - f(q_k)}{t_k}.$$

Fix $\varepsilon > 0$. For large $k \in \mathbb{N}$, let $c_k : [0, 1] \to M$ be the unique geodesic joining the points q_k and $\sigma_{q_k, v}(t_k)$, i.e, $c_k(t) = \exp_{q_k}(t \exp_{q_k}^{-1}(\sigma_{q_k, v}(t_k)))$ and let $U_k = \bigcup_{t \in [0,1]} B_g(c_k(t), \varepsilon t_k)$ its (εt_k)–neighborhood. Consider also a C^∞ vector field V on U_k such that $\dot{c}_k(t) = V(c_k(t))$, $t \in [0, 1]$. Now, applying Proposition 2.11 with $r_k = f(c_k(1)) - f(c_k(0)) - \varepsilon t_k$, one can find $m_k = m_k(t_k, q_k, v) \in U_k$ and $\xi_k \in \partial_F f(m_k)$ such that $r_k < \langle \xi_k, V(m_k) \rangle_g$. The latter inequality is equivalent to

$$\frac{f(\sigma_{q_k, v}(t_k)) - f(q_k)}{t_k} < \varepsilon + \langle \xi_k, V(m_k)/t_k \rangle_g.$$

Since f is locally Lipschitz, $\partial_F f$ is bounded in a neighborhood of p, see Proposition 2.13, thus the sequence $\{\xi_k\}$ is bounded on TM. We can choose a convergent subsequence (still denoting by $\{\xi_k\}$), and let $\xi_L = \lim_k \xi_k$. From construction, $\xi_L \in \partial_L f(p) \subset \partial_C f(p)$. Since $m_k \to p$, according to (2.13), we have that $\lim_{k \to \infty} V(m_k)/t_k = v$. Thus, letting $k \to \infty$ in the latter inequality, the arbitrariness of $\varepsilon > 0$ yields that

$$f^0(p; v) \le \langle \xi_L, v \rangle_g.$$

Now, taking into account that $f^0(p; v) = \max\{\langle \xi, v \rangle_g : \xi \in \partial_0 f(p)\}$, we obtain that

$$\max\{\langle \xi, v \rangle_g : \xi \in \partial_0 f(p)\} = f^0(p; v) \le \langle \xi_L, v \rangle_g \le \sup\{\langle \xi, v \rangle_g : \xi \in \partial_C f(p)\}.$$

Hörmander's result (see [7]) shows that this inequality in terms of support functions of convex sets is equivalent to the inclusion $\partial_0 f(p) \subset \partial_C f(p)$.

For the converse, it is enough to prove that $\partial_L f(p) \subset \partial_0 f(p)$ since the latter set is convex. Let $\xi \in \partial_L f(p)$. Then, we have $\xi = \lim_k \xi_k$ where $\xi_k \in \partial_F f(p_k)$ and $p_k \to p$. A similar argument as in the proof of Theorem 2.4 (see relation (2.16)) gives that for every $q \in M \setminus C_p$ and $k \in \mathbb{N}$, we have

$$\liminf_{t \to 0^+} \frac{f(\exp_{p_k}(t \exp_{p_k}^{-1}(q))) - f(p_k) - \langle \xi_k, t \exp_{p_k}^{-1}(q) \rangle_g}{\|t \exp_{p_k}^{-1} q\|} \ge 0.$$

Since $\| \exp_{p_k}^{-1} q \| = d_g(p_k, q) \geq c_0 > 0$, by the definition of the Clarke generalized derivative f^0 and the above inequality, one has that

$$f^0(p_k; \exp_{p_k}^{-1}(q)) \geq \langle \xi_k, \exp_{p_k}^{-1}(q) \rangle_g.$$

The upper semicontinuity of $f^0(\cdot; \cdot)$ and the fact that $\xi = \lim_k \xi_k$ imply that

$$f^0(p; \exp_p^{-1}(q)) \geq \limsup_k f^0(p_k; \exp_{p_k}^{-1}(q)) \geq \limsup_k \langle \xi_k, \exp_{p_k}^{-1}(q) \rangle_g = \langle \xi, \exp_p^{-1}(q) \rangle_g,$$

i.e., $\xi \in \partial_0 f(p)$, which concludes the proof of Step 3.   □

2.3   Subdifferentiability of Integral Functionals

This section is concerned with the study of the Clarke subdifferential of integral functionals. First we consider the case of functionals defined on function spaces (Lebesque or Orlicz) on a bounded domain, see Clarke [6] and Costea et al. [9]. In the case of Lebesgue spaces on unbounded domains we present a result due to Kristály [14].

For this let T a positive complete measure space with $\mu(T) < \infty$. We denote by $L^p(T, \mathbb{R}^m)$ the space of p-integrable functions, where $p \geq 1$, $m \geq 1$.

Let $j : T \times \mathbb{R}^m \to \mathbb{R}$ be a function such that $j(\cdot, y) : T \to \mathbb{R}$ is measurable for every $y \in \mathbb{R}$ and satisfies either

$$|j(x, y_1) - j(x, y_2)| \leq k(x)|y_1 - y_2|, \quad \forall y_1, y_2 \in \mathbb{R}^m \text{ and a.e. } x \in T, \tag{2.20}$$

for a function $k \in L^q(T)$ with $\frac{1}{p} + \frac{1}{q} = 1$, or $j(x, \cdot) : \mathbb{R}^m \to \mathbb{R}$ is locally Lipschitz for a.e. $x \in T$ and there is a constant $c > 0$ such that

$$|\zeta| \leq c(1 + |y|^{p-1}), \text{ for a.e. } x \in T, \forall y \in \mathbb{R}^m, \forall \zeta \in \partial_C^2 j(x, y). \tag{2.21}$$

The notation $|\cdot|$ used in (2.20), (2.21) stands for the Euclidian norm in $\mathbb{R}^N$, while $\partial_C^2 j(x, y)$ in (2.21) denotes the generalized gradient of j with respect to the second variable.

We are now in position to handle the functional $J : L^p(T, \mathbb{R}^m) \to \mathbb{R}$ defined by

$$J(u) := \int_T j(x, u(x)) \, d\mu, \quad \forall u \in L^p(T, \mathbb{R}^m), \tag{2.22}$$

The following two cases will be of particular interest in applications:

(i) $T := \Omega$ and $\mu := dx$ for some bounded domain $\Omega \subset \mathbb{R}^N$;
(ii) $T := \partial\Omega$ and $\mu := d\sigma$.

Theorem 2.6 ([1]) *Under either (2.20) or (2.21) the function* $J : L^p(T, \mathbb{R}^m) \to \mathbb{R}$ *defined in (2.22) is Lipschitz continuous on bounded subsets of* $L^p(T, \mathbb{R}^m)$ *and satisfies*

$$\partial_C J(u) \subset \int_T \partial_C^2 j(x, u(x)) \, d\mu, \quad \forall u \in L^p(T, \mathbb{R}^m), \tag{2.23}$$

in the sense that for every $\zeta \in \partial_C J(u)$ *there exists* $\xi \in L^q(T, \mathbb{R}^m)$, *such that*

$$\langle \zeta, v \rangle = \int_T \xi(x) v(x) \, d\mu, \quad \forall v \in L^p(T, \mathbb{R}^m),$$

and

$$\xi(x) \in \partial_C^2 j(x, u(x)), \text{ for a.e. } x \in T.$$

Moreover, if $j(x, \cdot)$ *is regular at* $u(x)$ *for a.e.* $x \in T$, *then* J *is regular at* u *and (2.23) holds with equality.*

Proof The first step of the proof is to check that J is Lipschitz continuous. Suppose that (2.20) is verified. Then using the Hölder inequality it is straightforward to establish that J is Lipschitz continuous on $L^p(T, \mathbb{R}^m)$.

Assume now that (2.21) holds. For a fixed number $r > 0$ and arbitrary elements $u, v \in L^p(T, \mathbb{R}^m)$ with $\|u\|_{L^p} < r$, $\|v\|_{L^p} < r$ we have

$$|J(u) - J(v)| \leq \int_T |j(x, u(x)) - j(x, v(x))| \, d\mu$$

$$\leq c_1 \int_T (1 + |u(x)|^{p-1} + |v(x)|^{p-1}) |u(x) - v(x)| \, d\mu$$

$$\leq c_1 \left(\int_T \left(1 + |u(x)|^{p-1} + |v(x)|^{p-1} \right)^{\frac{p}{p-1}} d\mu \right)^{\frac{p-1}{p}} \|u - v\|_{L^p}$$

$$\leq c_2 \left(\int_T (1 + |u(x)|^p + |v(x)|^p) \, d\mu \right)^{\frac{p-1}{p}} \|u - v\|_{L^p}$$

$$\leq c_3 \|u - v\|_{L^p},$$

with the constants $c_1, c_2, c_3 > 0$ where c_3 depends on r. The inequalities above have been derived by using the Lebourg's mean value theorem, i.e. Theorem 2.1, assumption (2.21) and Hölder inequality. The Lipschitz property on bounded sets for J is thus verified.

The map $x \mapsto j_2^0(x, u(x); v(x))$ is measurable on T. Since $j(x, \cdot)$ is continuous, we may express $j_2^0(x, u(x); v(x))$ as the upper limit of

$$\frac{j(x, y + \lambda v(x)) - j(x, y)}{\lambda},$$

where $\lambda \searrow 0$ taking rational value and $y \to u(x)$ taking values in a countable dense subset of $\mathbb{R}^m$. Thus $j_2^0(x, u(x); v(x))$ is measurable as the "countable limsup" of measurable functions of x.

The next task is to prove (2.23). To this end we are firstly concerned with the proof of the inequality

$$J^0(u; v) \leq \int_T j_2^0(x, u(x); v(x)) \, d\mu, \ \forall u, v \in L^p(T, \mathbb{R}^m). \tag{2.24}$$

Assuming (2.20), it is permitted to apply Fatou's lemma that leads directly to (2.24). Suppose now that the assumption (2.21) is satisfied. Thus using again Theorem 2.1 we obtain

$$\frac{j(x, u(x) + \lambda v(x)) - j(x, u(x))}{\lambda} = \langle \xi_x, v(x) \rangle,$$

for some $\xi_x \in \partial_C j(x, u^*(x))$ and $u^* \in [u(x), u(x) + \lambda v(x)]$. We can now also use Fatou's lemma to obtain (2.24).

The final step, that we only sketch is to pass from (2.24) to (2.23).

Here the essential thing is to observe that, in view of (2.24), any $\zeta \in \partial_C J(u)$ belongs to the subdifferential at $0 \in L^p(T, \mathbb{R}^m)$ (in the sense of convex analysis) of the convex function on $L^p(T, \mathbb{R}^m)$ mapping $v \in L^p(T, \mathbb{R}^m)$ to

$$\int_T j_2^0(x, u(x); v(x)) \, d\mu \in \mathbb{R}. \tag{2.25}$$

The properties and the subdifferentiation in Ioffe and Levin [12] applied to (2.25) yield (2.23). Finally, we are dealing with the regularity assertion in the statement. Under either of hypotheses (2.20) or (2.21) we may apply Fatou's lemma to get, if the regularity of $j(x, \cdot)$ at $u(x)$ is imposed,

$$\liminf_{\lambda \searrow 0} \frac{J(u + \lambda v) - J(u)}{\lambda} \geq \int_T j_2'(x, u(x); v(x)) \, d\mu = \int_T j_2^0(x, u(x); v(x)) \, d\mu.$$

Combining with (2.24) it is readily seen that $J'(u; v)$ exists and $J'(u; v) = J^0(u; v)$, whenever $v \in L^p(T, \mathbb{R}^m)$, which means the regularity of J at u. Moreover we induced the equality:

$$J^0(u; v) = \int_T j_2'(x, u(x); v(x)) \, d\mu, \ v \in L^p(T, \mathbb{R}^m).$$

If we combine the right-hand side (2.23), the regularity assumption for $j(x, \cdot)$ with the above formula we get

$$\langle \zeta, v \rangle = \int_T \langle \xi(x), v(x) \rangle \, d\mu \leq J^0(u; v), \ \forall v \in L^p(T, \mathbb{R}^m),$$

therefore $\zeta \in \partial_C J(u)$. This completes the proof.   □

The next result appears in the paper of Costea et al. [9] and it is a generalization of the above result of Aubin-Clarke in the sense that Orlicz spaces are taken instead of Lebesgue space.

For this let $\varphi : \mathbb{R} \to \mathbb{R}$ be an admissible function which satisfies

$$1 < \varphi^- \leq \varphi^+ < \infty, \tag{2.26}$$

let Φ the N-function generated by φ and assume $h : \Omega \times \mathbb{R} \to \mathbb{R}$ is a function which is measurable with respect to the first variable and satisfies one of the following conditions:

(h_1) there exists $\alpha \in L^{\Phi^*}(\Omega)$ s.t. for a.e. $x \in \Omega$ and all $t_1, t_2 \in \mathbb{R}$

$$|h(x, t_1) - h(x, t_2)| \leq \alpha(x)|t_1 - t_2|;$$

(h_2) there exist $c > 0$ and $\beta \in L^{\Phi^*}(\Omega)$ s.t. for a.e. $x \in \Omega$ and all $t \in \mathbb{R}$ a

$$|\xi| \leq \beta(x) + c\varphi(|t|), \ \ \forall \xi \in \partial_2 h(x, t).$$

Define next $H : L^{\Phi}(\Omega) \to \mathbb{R}$ via the instruction

$$H(u) := \int_\Omega h(x, u(x)) \, dx. \tag{2.27}$$

Theorem 2.7 ([9]) *Assume either (h_1) or (h_2) holds. Then, the functional H defined in (2.27) is Lipschitz continuous on bounded domains of $L^{\Phi}(\Omega)$ and*

$$\partial_C H(u) \subseteq \int_\Omega \partial_C^2 h(x, u(x)) \, dx, \ \ \forall u \in L^{\Phi}(\Omega), \tag{2.28}$$

in the sense that for every $\xi \in \partial_C H(u)$ there exists $\zeta \in L^{\Phi^*}(\Omega)$ such that $\zeta(x) \in \partial_C^2 h(x, u(x))$ for a.e. $x \in \Omega$ and

$$\langle \xi, v \rangle = \int_\Omega \zeta(x) v(x) \, dx, \ \ \forall v \in L^{\Phi}(\Omega).$$

*Moreover, if $h(x, \cdot)$ is regular at $u(x)$ for a.e. $x \in \Omega$, then H is regular at u and (2.28)
holds with equality.*

Proof Let M be a bounded domain of $L^{\Phi}(\Omega)$ and let $u_1, u_2 \in M$. If (h_1) holds, then the
Hölder-type inequality for Orlicz spaces shows that

$$|H(u_1) - H(u_2)| \le 2|\alpha|_{\Phi^*}|u_1 - u_2|_{\Phi},$$

hence H is Lipschitz continuous on M.

If (h_2) is assumed, then Lebourg's Mean Value Theorem ensures that there exist $w \in L^{\Phi}(\Omega)$ and $\tilde{\zeta} \in L^{\Phi^*}(\Omega)$ such that $w(x)$ lies on the open segment of endpoints $u_1(x)$ and
$u_2(x)$, $\tilde{\zeta}(x) \in \partial_C h(x, w(x))$ for a.e. $x \in \Omega$ and

$$h(x, u_1(x)) - h(x, u_2(x)) = \tilde{\zeta}(x)\,(u_1(x) - u_2(x))\,, \text{ for a.e. } x \in \Omega.$$

According to Clément et al. [8, Lemma A.5],

$$w \in L^{\Phi}(\Omega) \to \varphi(|w|) \in L^{\Phi^*}(\Omega),$$

which combined with the Hölder-type inequality for Orlicz spaces leads to

$$|H(u_1) - H(u_2)| \le \int_{\Omega} |h(x, u_1(x)) - h(x, u_2(x))|\,\mathrm{d}x = \int_{\Omega} |\tilde{\zeta}(x)||u_1(x) - u_2(x)|\,\mathrm{d}x$$

$$\le \int_{\Omega} (\beta(x) + c\varphi(|w(x)|))\,|u_1(x) - u_2(x)|\,\mathrm{d}x$$

$$\le 2\,(|\beta|_{\Phi^*} + c\,|\varphi(|w|)|_{\Phi^*})\,|u_1 - u_2|_{\Phi}.$$

In order to prove that H is Lipschitz continuous on M we only need to show that
$|\varphi(|w|)|_{\Phi^*}$ is bounded above by a constant independent of u_1 and u_2. Clearly we may
assume $|\varphi(|w|)|_{\Phi^*} > 1$. The fact that (see Clément et al. [8, Corollary C.7])

$$\frac{1}{\varphi^+} + \frac{1}{(\varphi^{-1})^-} = 1,$$

ensures that

$$1 < |\varphi(|w|)|_{\Phi^*} \le |\varphi(|w|)|_{\Phi^*}^{\frac{\varphi^+}{\varphi^+-1}} = |\varphi(|w|)|_{\Phi^*}^{(\varphi^{-1})^-} \le \int_{\Omega} \Phi^*(\varphi(|w|))\mathrm{d}x.$$

Using Young's inequality, we have

$$\Phi^*(\varphi(t)) \leq \Phi(t) + \Phi^*(\varphi(t)) = t\varphi(t) \leq \int_t^{2t} \varphi(s)\,ds \leq \Phi(2t),$$

and from the Δ_2-condition we get

$$\int_\Omega \Phi^*(\varphi(|w|))\,dx \leq c_1 + c_2 \int_\Omega \Phi(|w|)\,dx.$$

Fix $M > 1$ such that $|v|_\Phi \leq M$, for all $v \in \mathcal{M}$. Obviously $|w|_\Phi \leq M$ and the above estimates show that

$$|\varphi(w)|_{\Phi^*} \leq c_1 + c_2 M^{\varphi^+}.$$

The definition of the generalized directional derivative shows that the map $x \mapsto h^0(x, u(x); v(x))$ is measurable on Ω. Moreover, each of the conditions (h_1), (h_2) implies the integrability of $h^0(x, u(x); v(x))$. Let us check now that

$$H^0(u; v) \leq \int_\Omega h^0(x, u(x); v(x))\,dx, \ \forall u, v \in L^\Phi(\Omega). \tag{2.29}$$

If (h_1) is assumed, then (2.29) follows directly from Fatou's lemma. On the other hand, if we assume (h_2) to hold, then by Lebourg's mean value theorem we can write

$$\frac{h(x, u(x) + tv(x)) - h(x, u(x))}{t} = \zeta(x)v(x),$$

for some $\zeta \in L^{\Phi^*}(\Omega)$ satisfying $\zeta(x) \in \partial_2 h(x, \tilde{u}(x))$ for a.e $x \in \Omega$, with $\tilde{u}(x)$ lying in the open segment of endpoints $u(x)$ and $u(x) + tv(x)$, respectively. Again, Fatou's lemma implies (2.29).

In order to prove (2.28) let us fix $\xi \in \partial_2 H(u)$. Then (see, e.g., Carl et al. [4, Remark 2.170]) $\xi \in \partial H^0(u; \cdot)(0)$, where ∂ stands for the subdifferential in the sense of convex analysis. The latter and relation (2.29) show that ξ also belongs to the subdifferential at 0 of the convex map

$$L^\Phi(\Omega) \ni v \mapsto \int_\Omega h^0(x, u(x); v(x))\,dx,$$

and (2.28) follows from the subdifferentiation under the integral for convex integrands (see, e.g., Denkowski et al. [10]).

For the final part of the theorem, let us assume that $h(x, \cdot)$ is regular at $u(x)$ for a.e. $x \in \Omega$. Then, we can apply Fatou's lemma to get

$$H^0(u; v) = \limsup_{\substack{w \to u \\ t \searrow 0}} \frac{H(w + tv) - H(w)}{t} \geq \liminf_{t \searrow 0} \frac{H(u + tv) - H(u)}{t}$$

$$\geq \int_\Omega \liminf_{t \searrow 0} \frac{h(x, u(x) + tv(x)) - h(x, u(x))}{t} \, dx$$

$$= \int_\Omega h'(x, u(x); v(x)) \, dx = \int_\Omega h^0(x, u(x); v(x)) \, dx \geq H^0(u; v),$$

which shows that the directional derivative $H'(u; v)$ exists and

$$H'(u; v) = H^0(u; v) = \int_\Omega h^0(x, u(x); v(x)) dx, \quad \forall v \in L^\Phi(\Omega).$$

$\square$

In the last part of this section we prove an inequality for integral functionals defined on unbounded domain.

Let $f : \mathbb{R}^N \times \mathbb{R} \to \mathbb{R}$ be a measurable function which satisfies the following growth conditions:

There exist $c > 0$ and $r \in (p, p^*)$ such that

(f_1) $|f(x, s)| \leq c(|s|^{p-1} + |s|^{r-1})$, for a.e. $x \in \mathbb{R}^N$, $\forall s \in \mathbb{R}$;
(f_1') the embedding $X \hookrightarrow L^r(\mathbb{R}^N)$ is compact.

Let $F : \mathbb{R}^N \times \mathbb{R} \to \mathbb{R}$ be the function defined by

$$F(x, t) := \int_0^t f(x, s) ds, \quad \text{for a.e. } x \in \mathbb{R}^N, \forall s \in \mathbb{R}. \tag{2.30}$$

For a.e. $x \in \mathbb{R}^N$ and for every $t, s \in \mathbb{R}$, we have:

$$|F(x, t) - F(x, s)| \leq c_1 |t - s| \left(|t|^{p-1} + |s|^{p-1} + |t|^{r-1} + |s|^{r-1} \right), \tag{2.31}$$

where c_1 is a constant which depends only on c, p and r. Therefore, the function $F(x, \cdot)$ is locally Lipschitz and we can define the generalized directional derivative, i.e.,

$$F_2^0(x, t; s) = \limsup_{\tau \to t, \lambda \searrow 0} \frac{F(x, \tau + \lambda s) - F(x, \tau)}{\lambda}, \tag{2.32}$$

for every $t, s \in \mathbb{R}$ and a.e. $x \in \mathbb{R}^N$.

Remark 2.4 The following two propositions remain true under the growth condition (f_1), but we observe that it is enough to consider only the case when the function f has the growth $|f(x, s)| \leq c|s|^{p-1}$ for a.e. $x \in \mathbb{R}^N$, $\forall s \in \mathbb{R}$. For the sake of simplicity we denote by $h(s) = c|s|^{p-1}$ and in the sequel we basically use only the fact that the function h is convex, $h(0) = 0$, and monotone increasing on $(0, \infty)$.

Proposition 2.15 *The function* $\Phi : X \to \mathbb{R}$, *defined by* $\Phi(u) := \int_{\mathbb{R}^N} F(x, u(x))dx$ *is locally Lipschitz on bounded sets of* X.

Proof For every $u, v \in X$, with $\|u\|$, $\|v\| < r$, we have

$$|\Phi(u) - \Phi(v)| \leq \int_{\mathbb{R}^N} |F(x, u(x)) - F(x, v(x))|dx$$

$$\leq c_1 \int_{\mathbb{R}^N} |u(x) - v(x)|[h(|u(x)|) + h(|v(x)|)]dx$$

$$\leq c_2 \left(\int_{\mathbb{R}^N} |u(x) - v(x)|^p dx\right)^{\frac{1}{p}} \left[\left(\int_{\mathbb{R}^N} (h(|u(x)|)^{p'} dx\right)^{\frac{1}{p'}} \right.$$

$$\left. + \left(\int_{\mathbb{R}^N} (h(|v(x)|)^{p'} dx\right)^{\frac{1}{p'}} dx\right]$$

$$\leq c_2 \|u - v\|_p [\|h(|u|)\|_{p'} + \|h(|v|)\|_{p'})$$

$$\leq C(u, v)\|u - v\|,$$

where $\frac{1}{p} + \frac{1}{p'} = 1$ and we used the Hölder inequality, the subadditivity of the norm $\|\cdot\|_{p'}$ and the fact that the inclusion $X \hookrightarrow L^p(\mathbb{R}^N)$ is continuous. $C(u, v)$ is a constant which depends only on u and v. □

Proposition 2.16 *If the condition* (f_1) *holds, then for every* $u, v \in X$, *we have*

$$\Phi^0(u; v) \leq \int_{\mathbb{R}^N} F_2^0(x, u(x); v(x)) \, dx. \tag{2.33}$$

Proof Due to Remark 2.4, it suffices to prove the proposition for a such function f which satisfies only the growth condition $|f(x, s)| \leq c|s|^{p-1}$. Let us fix the elements $u, v \in X$. The function $F(x, \cdot)$ is locally Lipschitz and therefore continuous. Thus $F_2^0(x, u(x); v(x))$ can be expressed as the upper limit of

$$\frac{F(x, y + tv(x)) - F(x, y)}{t},$$

where $t \searrow 0$ takes rational values and $y \to u(x)$ takes values in a countable subset of $\mathbb{R}$. Therefore, the map $x \mapsto F_2^0(x, u(x); v(x))$ is measurable as the "countable limsup" of measurable functions in x. From condition (f_1) we get that the function $x \mapsto F_2^0(x, u(x); v(x))$ is from $L^1(\mathbb{R}^N)$.

Using the fact that the Banach space X is separable, there exists a sequence $w_n \in X$ with $\|w_n - u\| \to 0$ and a real number sequence $t_n \to 0^+$, such that

$$\Phi^0(u; v) = \lim_{n \to \infty} \frac{\Phi(w_n + t_n v) - \Phi(w_n)}{t_n}. \tag{2.34}$$

Since the inclusion $X \hookrightarrow L^p(\mathbb{R}^N)$ is continuous, we get $\|w_n - u\|_p \to 0$. In particular, there exists a subsequence of $\{w_n\}$, denoted in the same way, such that $w_n(x) \to u(x)$ a.e. $x \in \mathbb{R}^N$. Now, let $\varphi_n : \mathbb{R}^N \to \mathbb{R} \cup \{+\infty\}$ be the function defined by

$$\varphi_n(x) = -\frac{F(x, w_n(x) + t_n v(x)) - F(x, w_n(x))}{t_n}$$

$$+ c_1 |v(x)| [h(|w_n(x) + t_n v(x)|) + h(|w_n(x)|)].$$

We see that the functions φ_n are measurable and non-negative. If we apply Fatou's lemma, we get

$$\int_{\mathbb{R}^N} \liminf_{n \to \infty} \varphi_n(x) \, dx \leq \liminf_{n \to \infty} \int_{\mathbb{R}^N} \varphi_n(x) \, dx.$$

This inequality is equivalent with

$$\int_{\mathbb{R}^N} \limsup_{n \to \infty} [-\varphi_n(x)] \, dx \geq \limsup_{n \to \infty} \int_{\mathbb{R}^N} [-\varphi_n(x)] \, dx. \tag{2.35}$$

For the sake of simplicity we introduce the following notations:

(i) $\varphi_n^1(x) := \frac{F(x, w_n(x) + t_n v(x)) - F(x, w_n(x))}{t_n}$;

(ii) $\varphi_n^2(x) = c_1 |v(x)| [h(|w_n(x) + t_n v(x)|) + h(|w_n(x)|)]$.

With these notations, we have $\varphi_n(x) = -\varphi_n^1(x) + \varphi_n^2(x)$. Now we prove the existence of the limit $b = \lim_{n \to \infty} \int_{\mathbb{R}^N} \varphi_n^2(x) \, dx$. Since $\|w_n - u\|_p \to 0$, in particular, there exists a positive function $g \in L^p(\mathbb{R}^N)$, such that $|w_n(x)| \leq g(x)$ a.e. $x \in \mathbb{R}^N$. Considering that the function h is monotone increasing on positive numbers, we get

$$|\varphi_n^2(x)| \leq c_1 |v(x)| [h(g(x) + |v(x)|) + h(g(x))], \quad \text{a.e. } x \in \mathbb{R}^N.$$

Moreover, $\varphi_n^2(x) \to 2c_1|v(x)|h(|u(x)|)$ for a.e. $x \in \mathbb{R}^N$. Thus, using the Lebesgue dominated convergence theorem, we have

$$b = \lim_{n\to\infty} \int_{\mathbb{R}^N} \varphi_n^2(x)\,dx = \int_{\mathbb{R}^N} 2c_1|v(x)|h(|u(x)|)\,dx. \tag{2.36}$$

If we denote by $I_1 = \limsup_{n\to\infty} \int_{\mathbb{R}^N} [-\varphi_n(x)]\,dx$, then using (2.34) and (2.36), we have

$$I_1 = \limsup_{n\to\infty} \int_{\mathbb{R}^N} [-\varphi_n(x)]\,dx = \Phi^0(u;v) - b. \tag{2.37}$$

In the next we estimate the expression $I_2 = \int_{\mathbb{R}^N} \limsup_{n\to\infty}[-\varphi_n(x)]\,dx$. We have the following inequality:

$$\int_{\mathbb{R}^N} \limsup_{n\to\infty} \varphi_n^1(x)\,dx - \int_{\mathbb{R}^N} \lim_{n\to\infty} \varphi_n^2(x)\,dx \geq I_2. \tag{2.38}$$

Using the fact that $w_n(x) \to u(x)$ a.e. $x \in \mathbb{R}^N$ and $t_n \to 0^+$, we get

$$\int_{\mathbb{R}^N} \lim_{n\to\infty} \varphi_n^2(x)\,dx = 2c_1 \int_{\mathbb{R}^N} |v(x)|h(|u(x)|)\,dx.$$

On the other hand, we have

$$\int_{\mathbb{R}^N} \limsup_{n\to\infty} \varphi_n^1(x)\,dx \leq \int_{\mathbb{R}^N} \limsup_{y\to u(x),\, t\to 0^+} \frac{F(x, y + tv(x)) - F(x, y)}{t}\,dx$$

$$= \int_{\mathbb{R}^N} F_2^0(x, u(x); v(x))\,dx.$$

Using the relations (2.35), (2.37), (2.38) and the above estimations we obtain the desired result.   □

References

1. J.P. Aubin, F.H. Clarke, Shadow prices and duality for a class of optimal control problems. SIAM J. Control Optim. **17**, 567–586 (1979)
2. D. Azagra, J. Ferrera, F. López-Mesas, Nonsmooth analysis and Hamilton-Jacobi equations on Riemannian manifolds. J. Funct. Anal. **220**, 304–361 (2005)
3. J.M. Borwein, Q.J. Zhu, *Techniques of Variational Analysis* (Springer, Berlin, 2005)
4. S. Carl, V.K. Le, D. Motreanu, *Nonsmooth Variational Problems and Their Inequalities. Comparison Principles and Applications* (Springer, Berlin, 2007)
5. K.-C. Chang, Variational methods for non-differentiable functionals and their applications to partial differential equations. J. Math. Anal. Appl. **80**, 102–129 (1981)

6. F.H. Clarke, Optimization and nonsmooth analysis, in *Classics in Applied Mathematics, Society for Industrial and Applied Mathematics* (1990)
7. F.H. Clarke, Y.S. Ledyaev, R.J. Stern, P.R. Wolenski, *Nonsmooth Analysis and Control Theory*, vol. 178. Graduate Texts in Mathematics (Springer, New York, 1998)
8. P. Clément, B. de Pagter, G. Sweers, F. de Thélin, Existence of solutions to a semilinear elliptic system through Orlicz-Sobolev spaces. Mediterr. J. Math. **3**, 241–267 (2004)
9. N. Costea, G. Moroşanu, C. Varga, Weak solvability for Dirichlet partial differential inclusions in Orlicz-Sobolev spaces. Adv. Diff. Equ. **23**, 523–554 (2018)
10. Z. Denkowski, S. Migorski, N.S. Papageorgiou, *An Introduction to Nonlinear Analysis: Theory* (Kluwer Academic Publishers, Berlin, 2003)
11. M.P. do Carmo, *Riemannian Geometry*. Mathematics: Theory & Applications (Birkhäuser Boston Inc., Boston, 1992). Translated from the second Portuguese edition by Francis Flaherty
12. A.D. Ioffe, V.L. Levin, Subdifferentials of Convex Functions. Trans. Moscow Math. Soc. **26**, 1–72 (1972)
13. W.P.A. Klingenberg, *Riemannian Geometry*, vol. 1. De Gruyter Studies in Mathematics, 2nd edn. (Walter de Gruyter & Co., Berlin, 1995)
14. A. Kristály, Infinitely many radial and non-radial solutions for a class of hemivariational inequalities. Rocky Mount. J. Math. **35**, 1173–1190 (2005)
15. A. Kristály, Nash-type equilibria on Riemannian manifolds: a variational approach. J. Math. Pures Appl. (9) **101**, 660–688 (2014)
16. G. Lebourg, Valeur moyenne pour gradient généralisé. C. R. Math. Acad. Sci. Paris **281**, 795–797 (1975)
17. Y.S. Ledyaev, Q.J. Zhu, Nonsmooth analysis on smooth manifolds. Trans. Amer. Math. Soc. **359**, 3687–3732 (2007)
18. C. Li, G. López, V. Martín-Márquez, Monotone vector fields and the proximal point algorithm on Hadamard manifolds. J. Lond. Math. Soc. (2) **79**, 663–683 (2009)
19. D. Motreanu, P.D. Panagiotopoulos, *Minimax Theorems and Qualitative Properties of the Solutions of Hemivariational Inequalities and Applications*, vol. 29. Nonconvex Optimization and its Applications (Kluwer Academic Publishers, Boston, 1999)
20. Z. Naniewicz, P.D. Panagiotopoulos, *Mathematical Theory of Hemivariational Inequalities and Applications* (Marcel Dekker, New York, 1995)

Critical Points, Compactness Conditions and Symmetric Criticality

3.1 Locally Lipschitz Functionals

In 1981, Chang [1] used the properties of the Clarke subdifferential to develop a *critical point theory* for locally Lipschitz functionals that are not necessarily differentiable. The main details and notions are given below.

Proposition 3.1 ([1]) *Let* $f : X \to \mathbb{R}$ *be a locally Lipschitz function. Then the function* $\lambda_f : X \to \mathbb{R}$ *defined by* $\lambda_f(u) := \inf_{\zeta \in \partial_C f(u)} \|\zeta\|_*$, *is well defined and it is lower semicontinuous.*

Proof Since $\partial_C f(u)$ is a nonempty, convex and weak* compact subset of X^* and the function $\zeta \mapsto \|\zeta\|_*$ is weakly lower semicontinuous and bounded below, it follows that for every $u_0 \in X$ there exists $\zeta_0 \in \partial_C f(u_0)$ such that

$$\|\zeta_0\|_* = \inf_{\zeta \in \partial_C f(u_0)} \|\zeta\|_*.$$

Now, we prove that the function $u \mapsto \lambda_f(u)$ is lower semicontinuous. Fix $u_0 \in X$, arbitrary. Arguing by contradiction, there exist sequences $\{u_n\} \subset X$ and $\{\zeta_n\} \subset X^*$ such that

$$u_n \to u_0, \ \liminf_{n \to \infty} \lambda_f(u_n) < \lambda_f(u_0), \ \zeta_n \in \partial_C f(u_n) \text{ and } \|\zeta_n\|_* = \lambda_f(u_n).$$

© The Author(s), under exclusive license to Springer Nature Switzerland AG 2021
N. Costea et al., *Variational and Monotonicity Methods in Nonsmooth Analysis*,
Frontiers in Mathematics, https://doi.org/10.1007/978-3-030-81671-1_3

Using the fact that the set valued map $u \mapsto \partial_C f(u)$ is weak*-upper semicontinuous we choose a subsequence $\{\zeta_{n_k}\}$ such that $\zeta_{n_k} \rightharpoonup \zeta_0 \in \partial_C f(u_0)$. But

$$\liminf_{k \to \infty} \|\zeta_{n_k}\|_* \geq \|\zeta_0\|_* \geq \lambda_f(u_0),$$

which is a contradiction.   □

Definition 3.1 ([1]) Let $f : X \to \mathbb{R}$ be a locally Lipschitz function. We say that $u \in X$ is a *critical point* of f, if $\lambda_f(u) = 0$, or equivalently $0 \in \partial_C f(u)$.

Proposition 3.2 *If $u \in X$ is a local minimum or maximum of the locally Lipschitz function $f : X \to \mathbb{R}$, then u is a critical point of f.*

Proof Using Proposition 2.5-(i) for $\lambda = -1$ we see that it suffices to consider the case when the point $u \in X$ is a local minimum. Then, for sufficiently small $t > 0$, $f(u + tv) \geq f(u)$. Thus

$$f^0(u; v) \geq \limsup_{t \searrow 0} \frac{f(u + tv) - f(u)}{t} \geq 0,$$

which ensures that $0 \in \partial_C f(u)$.   □

A sequence $\{u_n\} \subset X$ is called *Palais-Smale sequence* for f if $\lambda_f(u_n) \to 0$ as $n \to \infty$. So a Palais-Smale sequence is a sequence of "almost critical points" and it is readily seen that any accumulation point of such a sequence is a critical point f. It is well-known that Palais-Smale sequences do not necessarily lead to critical points, therefore the following compactness condition is usually imposed.

Definition 3.2 Let $f : X \to \mathbb{R}$ be a locally Lipschitz functional. We say f satisfies the *Palais-Smale condition* if any Palais-Smale sequence $\{u_n\} \subset X$ such that $\{f(u_n)\}$ is bounded possesses a (strongly) convergent subsequence.

Sometimes it is useful to work with weaker compactness conditions, such as the Cerami condition, given below.

Definition 3.3 Let $f : X \to \mathbb{R}$ be a locally Lipschitz functional. We say f satisfies the *Cerami condition at level $c \in \mathbb{R}$* (in short, $(C)_c$-condition) if any sequence $\{u_n\} \subset \mathbb{R}$ satisfying

(i) $f(u_n) \to c$ as $n \to \infty$;
(ii) $(1 + \|u_n\|)\lambda_f(u_n) \to 0$ as $n \to \infty$,

possesses a (strongly) convergent subsequence. If this holds for every level $c \in \mathbb{R}$, then we simply say that f satisfies the *Cerami condition* (in short, (C)-condition).

Remark 3.1 If $f \in C^1(X, \mathbb{R})$, then it is readily seen that f is locally Lipschitz and $\lambda_f(u) = \|f'(u)\|_*$. Then u is a critical point of f if and only if $f'(u) = 0$, i.e., the critical point in the sense of Chang reduces to the usual notion of critical point. Also, the $(PS)_c$ and $(C)_c$ compactness conditions reduce to their counterparts from smooth analysis.

3.2 Szulkin Functionals

Let X be a real Banach space and I a functional on X satisfying the structure hypothesis (see Szulkin [10]):

(H) $I := \varphi + \psi$, where $\varphi \in C^1(X, \mathbb{R})$ and $\psi : X \to (-\infty, +\infty]$ is proper, l.s.c. and convex.

Definition 3.4 ([10]) A point $u \in X$ is said to be *critical point* of I if $u \in D(\psi)$ and if it satisfies the inequality

$$\langle \varphi'(u), v - u \rangle + \psi(v) - \psi(u) \geq 0, \quad \forall v \in X. \tag{3.1}$$

Note that X can be replaced by $D(\psi)$ in (3.1). Now, we recall some basic facts on the functionals which verify the structure hypothesis (H). Here and hereafter such functionals will be called *Szulkin functionals*.

Remark 3.2 The inequality (3.1) can be formulated equivalently as

$$-\varphi'(u) \in \partial \psi(u).$$

A number $c \in \mathbb{R}$ such that $I^{-1}(c)$ contains a critical point will be called *a critical value*. We shall use the following notations:

$$K = \{u \in X : u \text{ is critical point of } I\};$$

$$I_c = \{u \in X : I(u) \leq c\}, \quad K_c = \{u \in K : I(u) = c\}.$$

Proposition 3.3 *If I satisfies (H), each local minimum is a critical point of I.*

Proof Let u be a local minimum of I. Using convexity of ψ it follows that for small $t > 0$,

$$0 \leq I((1-t)u + tv) - I(u) = \varphi(u + t(v-u)) - \varphi(u) + \psi((1-t)u + tv) - \psi(u)$$

$$\leq \varphi(u + t(v-u)) - \varphi(u) + t(\psi(v) - \psi(u)).$$

Dividing by t and letting $t \to 0$ we obtain (3.1). □

Definition 3.5 ([10]) We say that I satisfies the *Palais-Smale compactness condition at level c*, denoted $(PS)_c$, if any sequence $\{u_n\} \subset X$ satisfying

(i) $I(u_n) \to c \in \mathbb{R}$;
(ii) there exists $\varepsilon_n \subset \mathbb{R}$, $\varepsilon_n \searrow 0$ such that

$$\langle \varphi'(u_n), v - u_n \rangle + \psi(v) - \psi(u_n) \geq -\varepsilon_n \|v - u_n\|, \quad \forall v \in X; \tag{3.2}$$

possesses a (strongly) convergent subsequence.

As before, a sequence satisfying (i) and (ii) will be called *Palais-Smale sequence*. If $(PS)_c$ holds for every $c \in \mathbb{R}$ we say that I satisfies the *Palais-Smale condition*, denoted by (PS).

It will be proved in the sequel that condition $(PS)_c$ can be also formulated as follows: $(PS)_c'$: Any sequence $\{u_n\} \subset X$ satisfying:

(*i*) $I(u_n) \to c \in \mathbb{R}$;
(*ii*) there exists $\zeta_n \in X^*$, such that $\zeta_n \to 0$ in X^* and

$$\langle \varphi'(u_n), v - u_n \rangle + \psi(v) - \psi(u_n) \geq \langle \zeta_n, v - u_n \rangle; \tag{3.3}$$

possesses a convergent subsequence.

Lemma 3.1 *Let X be a real Banach space and $\chi : X \to (-\infty, +\infty]$ a l.s.c. convex functional with $\chi(0) = 0$. If*

$$\chi(u) \geq -\|u\|, \quad \forall u \in X,$$

then there exists $\zeta \in X^*$ such that $\|\zeta\| \leq 1$ and

$$\chi(u) \geq \langle \zeta, u \rangle, \quad \forall u \in X.$$

Proposition 3.4 *Conditions $(PS)_c$ and $(PS)_c'$ are equivalent.*

Proof It suffices to prove that (3.2) and (3.3) are equivalent; it is clear that (3.3) implies (3.2), so suppose that (3.2) is satisfied.

Let $u := v - u_n$ and

$$\chi(u) := \frac{1}{\varepsilon_n}\left[\langle \varphi'(u_n), u \rangle + \psi(u + u_n) - \psi(u_n)\right].$$

Then (3.2) is actually $\chi(u) \geq -\|u\|$ for all $u \in X$. According to Lemma 3.1 there is a $\zeta_n \in X^*$ with $\|\zeta_n\| \leq 1$ and $\chi(u) \geq \langle \zeta_n, u \rangle$. Choosing $\zeta_n = \varepsilon_n \zeta_n$ one has

$$\frac{\langle \varphi'(u_n), v - u_n \rangle + \psi(v) - \psi(u_n)}{\varepsilon_n} \geq \left\langle \frac{\zeta_n}{\varepsilon_n}, v - u_n \right\rangle.$$

Hence (3.3) is satisfied and $\zeta_n \to 0$ because $\varepsilon_n \to 0$. $\qquad\qquad\square$

Proposition 3.5 *Suppose that I satisfies (H) and $(PS)_c$ and let $\{u_n\}$ Palais-Smale sequence. If u is an accumulation point of $\{u_n\}$, then $u \in K_c$. In particular, K_c is a compact set.*

Proof We may assume that $u_n \to u$. Passing to the limit in (3.2) and using the fact that $\lim_{n\to\infty} \psi(u_n) \geq \psi(u)$, we obtain (3.1). Hence $u \in K$. Moreover, since the inequality (3.1) cannot be strict for $v = u$, $\lim_{n\to\infty} \psi(u_n) = \psi(u)$. Consequently, $I(u_n) \to I(u) = c$ and $u \in K_c$. If $\{u_n\} \subset K_c$, then $I(u_n) = c$ and (3.2) is satisfied with $\varepsilon_n = 0$. It follows that a subsequence of $\{u_n\}$ converges to some $u \in X$. By the first part of the proposition, $u \in K_c$. Hence K_c is compact. $\qquad\qquad\square$

3.3   Motreanu-Panagiotopoulos Functionals

In this subsection we present some results from the critical point theory for *Motreanu-Panagiotopoulos functionals* (see [7]), i.e., functionals satisfying the structure hypothesis: (H') $I := h + \psi$, with $h : X \to \mathbb{R}$ locally Lipschitz and $\psi : X \to (-\infty, +\infty]$ convex, proper and l.s.c.

Definition 3.6 ([7]) An element $u \in X$ is said to be a *critical point of $I := h + \psi$*, if

$$h^0(u; v - u) + \psi(v) - \psi(u) \geq 0, \forall v \in X.$$

In this case, $I(u)$ is a *critical value of I*.

We have the following result, see Gasinski and Papageorgiou [2, Remark 2.3.1].

Proposition 3.6 *An element* $u \in X$ *is a critical point of* $I := h + \psi$, *if and only if* $0 \in \partial_C h(u) + \partial \psi(u)$.

Definition 3.7 The functional $I := h + \psi$ is said to satisfy the *Palais-Smale condition at level* $c \in \mathbb{R}$ (shortly, $(PS)_c$), if every sequence $\{u_n\} \subset X$ satisfying

(i) $I(u_n) \to c$ as $n \to \infty$;
(ii) there exists $\{\varepsilon_n\} \subset \mathbb{R}$ such that $\varepsilon_n \searrow 0$ and

$$h^0(u_n; v - u_n) + \psi(v) - \psi(u_n) \geq -\varepsilon_n \|v - u_n\|, \forall v \in X,$$

possesses a convergent subsequence.

If $(PS)_c$ is verified for all $c \in \mathbb{R}$, then I is said to satisfy the *Palais-Smale condition* (shortly, (PS)).

Remark 3.3 The Motreanu-Panagiotopoulos critical point theory contains as particular cases both critical the point theory in the sense of Chang as well as in the sense of Szulkin. More precisely, we have the following:

(i) If $\psi \equiv 0$ in (H'), then Definition 3.6 reduces to Definition 3.1 and Definition 3.7 reduces to Definition 3.2;
(ii) If $h \in C^1(X; \mathbb{R})$ in (H'), then Definition 3.6 reduces to Definition 3.4 and Definition 3.7 reduces to Definition 3.5.

3.4 Principle of Symmetric Criticality

Let G be a group and let π a representation of G over X, that is $\pi(g) \in \mathcal{L}(X)$ for each $g \in G$ (where $\mathcal{L}(X)$ denotes the set of the linear and bounded operator from X into X), and

- $\pi(e)u = u, \forall u \in X$;
- $\pi(g_1 g_2)u = \pi(g_1)(\pi(g_2)u), \quad \forall g_1, g_2 \in G \forall u \in X,$

where e is the identity element of G.

The representation π_* of G over X^* is naturally induced by π through the relation

$$\langle \pi_*(g)\zeta, u \rangle := \left\langle \zeta, \pi(g^{-1})u \right\rangle, \forall g \in G, \zeta \in X^* \text{ and } u \in X. \tag{3.4}$$

For simplicity, we shall often write gu or $g\zeta$ instead of $\pi(g)u$ or $\pi_*(g)\zeta$, respectively. A function $h : X \to \mathbb{R}$ (or $h : X^* \to \mathbb{R}$) is called G-invariant if $h(gu) = h(u)$ $(h(g\zeta) =$

$h(\zeta)$) for every $u \in X$ ($\zeta \in X^*$) and $g \in G$. A subset M of X is called G-invariant (or M^* of X^*) if

$$gM = \{gu : u \in M\} \subseteq M \quad (\text{or } gM^* \subseteq M^*). \quad \forall g \in G.$$

The fixed point sets of the group action G on X and X^* (other authors call them G-symmetric points) are defined as

$$\Sigma = X^G := \{u \in X : gu = u, \ \forall g \in G\},$$

$$\Sigma_* = (X^*)^G := \{\zeta \in X^* : g\zeta = \zeta, \ \forall g \in G\}.$$

Hence, by (3.4), we can see that $\zeta \in X^*$ is symmetric if and only if ζ is a G-invariant functional. The sets Σ and Σ_* are closed linear subspaces of X and X^*, respectively. So Σ and Σ_* are regarded as Banach spaces with their induced topologies. We introduce the following notations:

- $C_G^1(X) = \{\varphi \in C^1(X, \mathbb{R}) : \varphi \text{ is } G\text{-invariant}\}$;
- $\mathcal{L}_G(X) = \{\varphi \in \text{Lip}_{\text{loc}}(X, \mathbb{R}) : \varphi \text{ is } G\text{-invariant}\}$.

The principle of symmetric criticality for C^1–functionals reads like this:
  (PSC) :If $\varphi \in C_G^1(X)$ and $(\varphi|_\Sigma)'(u) = 0$, then $\varphi'(u) = 0$.

Theorem 3.1 (Palais [8]) *The principle (PSC) is valid if and only if $\Sigma_* \cap \Sigma^\perp = \{0\}$, where $\Sigma^\perp := \{\zeta \in X^* : \langle \zeta, u \rangle = 0, \ \forall u \in \Sigma\}$.*

Proof "$\Leftarrow$" Suppose that $\Sigma_* \cap \Sigma^\perp = \{0\}$ and let $u_0 \in \Sigma$ be a critical point of $\varphi|_\Sigma$. We must show $\varphi'(u_0) = 0$. Because $\varphi(u_0) = \varphi|_\Sigma(u_0)$ and $\varphi(u_0 + v) = \varphi|_\Sigma(u_0 + v)$ for all $v \in \Sigma$, we obtain

$$\langle \varphi'(u_0), v \rangle_{X^*, X} = \langle (\varphi|_\Sigma)'(u_0), v \rangle_{\Sigma^*, \Sigma},$$

for every $v \in \Sigma$, where $\langle \cdot, \cdot \rangle_{\Sigma^*, \Sigma}$ denotes the duality pairing between Σ and its dual Σ^*. This implies $\varphi'(u_0) \in \Sigma^\perp$. On the other hand, from the G-invariance of φ follows that

$$\langle \varphi'(gu), v \rangle = \lim_{t \to 0} \frac{\varphi(gu + tv) - \varphi(gu)}{t} = \lim_{t \to 0} \frac{\varphi(u + tg^{-1}v) - \varphi(u)}{t}$$

$$= \langle \varphi'(u), g^{-1}v \rangle = \langle g\varphi'(u), v \rangle$$

for all $g \in G$ and $u, v \in X$. This means that φ' is G-equivariant, i.e.

$$\varphi'(gu) = g\varphi'(u), \tag{3.5}$$

for every $g \in G$ and $u \in X$. Since $u_0 \in \Sigma$, we obtain $g\varphi'(u_0) = \varphi'(u_0)$ for all $g \in G$, i.e. $\varphi'(u_0) \in \Sigma_*$.

Thus we conclude $\varphi'(u_0) \in \Sigma_* \cap \Sigma^\perp = \{0\}$, that is, $\varphi'(u_0) = 0$.

"$\Rightarrow$" Suppose that there exists a non-zero element $\zeta \in \Sigma_* \cap \Sigma^\perp$ and define $\varphi_*(\cdot)$ by $\varphi_*(u) := \langle \zeta, u \rangle$. It is clear that $\varphi_* \in C_G^1(X)$ and $(\varphi_*)'(\cdot) = \zeta \neq 0$, so φ_* has no critical point in X.

On the other hand $\zeta \in \Sigma^\perp$ implies $\zeta|_\Sigma = 0$, therefore $(\varphi_*|_\Sigma)'(u) = 0$ for every $u \in \Sigma$. This contradicts the principle (PSC). Therefore the condition $\Sigma_* \cap \Sigma^\perp = \{0\}$ is necessary for the principle (PSC). $\square$

We assume next that the following condition holds:

(A_1) G is a compact topological group and the representation π of G over X is continuous, i.e., $(g, u) \to gu$ is a continuous function $G \times X$ into X.

In this situation for each $u \in X$, there exists a unique element $Au \in X$ such that

$$\langle \zeta, Au \rangle = \int_G \langle \zeta, gu \rangle dg, \quad \forall \zeta \in X^*, \tag{3.6}$$

where dg is the normalized Haar measure on G. The mapping A is called the averaging operator on G. The averaging operator A has the following important properties:

- $A : X \to \Sigma$ is a continuous linear projection;
- If $K \subset X$ is a G-invariant closed convex, then $A(K) \subset K$.

Moreover, if we denote by $\Gamma_G(X^*)$ the set of G-invariant weakly*-closed convex subsets of X^*, we have

Lemma 3.2 *The adjoint operator A^* is a mapping from X^* to Σ_*. If $K \in \Gamma_G(X^*)$, then $A^*(K) \subset K$.*

Proof We first prove that for all $\zeta \in X^*$ implies $A^*\zeta \in \Sigma_*$. By the right invariance of the Haar measure, we get

$$Agu = Au, \quad \forall g \in G, \forall u \in X.$$

Therefore for every $g \in G$ and $u \in X$ we have

$$\langle g A^* \zeta, u \rangle = \langle \zeta, A g^{-1} u \rangle = \langle \zeta, Au \rangle = \langle A^* \zeta, u \rangle,$$

that is $A^* \zeta \in \Sigma_*$.

In the sequel we prove $A^*(K) \subset K$. Suppose that there exists an element $\zeta \in K$ such that $A^* \zeta \notin K$. We apply the Hahn-Banach theorem in X^* with the weak* topology τ_{w^*}. Then there exists $u \in X$, $c \in \mathbb{R}$ and $\varepsilon > 0$ such that for every $w^* \in K$ we have

$$\langle A^* \zeta, u \rangle \leq c - \varepsilon < c \leq \langle \xi, u \rangle.$$

By putting $\xi := g^{-1} \zeta \in K$ for all $g \in G$, we get

$$\langle \zeta, Au \rangle \leq c - \varepsilon < c \leq \langle \zeta, gu \rangle,$$

which contradicts (3.6). □

We have the following result due to Palais [8].

Theorem 3.2 *If (A1) is satisfied, then (PSC) is valid.*

Proof We verify the condition $\Sigma_* \cap \Sigma^\perp = \{0\}$. Let $\zeta \in \Sigma_* \cap \Sigma^\perp$ fixed and suppose that $\zeta \neq 0$. Because $\zeta \in \Sigma_*$, the hyperplane $H = \{u \in X : \langle \zeta, u \rangle = 1\}$ becomes a nonempty G-invariant closed convex subset of X. Thus, for any $u \in H$, we have $Au \in H \cap \Sigma$ and because $\zeta \in \Sigma^\perp$ we have $\langle \zeta, Au \rangle = 0$. This contradicts the fact that $Au \in H$. □

We present next a version of the principle of symmetric criticality for locally Lipschitz functions due to Krawcewicz and Marzantowicz [5]. Let $\varphi : X \to \mathbb{R}$ be a G-invariant locally Lipschitz function. Using the Chain Rule we obtain that $g \partial_C \varphi(u) = \partial_C \varphi(gu)$, i.e. the set $\partial \varphi(u)$ is G-invariant.

We consider the following principle:

$(PSCL)$: If $\varphi \in \mathcal{L}_G^1(X)$ and $0 \in \partial_C(\varphi|_\Sigma)(u)$ then $0 \in \partial_C \varphi(u)$.

Theorem 3.3 *If (A1) is satisfied then (PSCL) is valid.*

Proof Without loss of generality we may suppose that $u = 0$ is a critical point of $\varphi|_\Sigma$. Let $A : X \to \Sigma$ be the averaging operator over G. Since, $\varphi^0(0; \cdot)$ is a continuous convex function, then

$$\varphi^0(0; Av) \leq \int_G \varphi^0(0; gv) \mathrm{d}g = \int_G (\varphi \circ g)^0(0; v) \mathrm{d}g = \int_G \varphi^0(0; v) \mathrm{d}g = \varphi^0(0; v).$$

Let us remark that for $v \in \Sigma$ we have $(\varphi|_\Sigma)^0 (0; v) \leq \varphi^0(0; v)$ and $A^*X^* = \Sigma_* = (\Sigma)^*$. Thus

$$\partial_C(\varphi|_\Sigma)(0) = \left\{ w \in \Sigma_* : \langle w, v \rangle \leq (\varphi|_\Sigma)^0(0; v), \ \forall v \in \Sigma \right\}$$

$$\subseteq \left\{ w \in \Sigma_* : \langle w, v \rangle \leq \varphi^0(0; v), \ \forall v \in \Sigma \right\}$$

$$= \left\{ w \in A^*X^* : \langle w, v \rangle = \langle w, Av \rangle \leq \varphi^0(0; Av) \leq \varphi^0(0; v), \ \forall v \in X \right\}$$

$$\subseteq A^*(\partial_C\varphi(0)).$$

Therefore, if $0 \in \partial_C(\varphi|_\Sigma)(0)$ then $0 \in A^*(\partial_C\varphi(0))$ and, since $A^*(\partial_C\varphi(0)) \subseteq \partial_C\varphi(0)$, this implies that $0 \in \partial_C\varphi(0)$ and the $(PSCL)$ is satisfied. □

Now, we suppose that G is a compact Lie group and let M be a Banach G-manifold modelled on the Banach space E. Let us recall that, for each $g \in G$, there is a diffeomorphism $g : M \to M$ defined by $g(x) = g \cdot x, x \in M$. The G-action on $T^*(M)$ is defined as follows: if $(x, w) \in T^*(M)$, i.e. $w \in T_x^*(M)$, then $g \cdot (x, w) = (gx, w')$, where $w' = (T_{g(x)}g^{-1})^*w$.

Suppose that $\varphi : M \to \mathbb{R}$ is a G-invariant locally Lipschitz functional. It follows that

$$g \cdot \partial_C\varphi(x) = (T^*g^{-1})(\partial_C\varphi(x)) = (\partial_C(\varphi \circ g)^{-1})(gx) = \partial_C\varphi(gx), \tag{3.7}$$

i.e.,

$$g \cdot \partial_C\varphi(x) = \partial_C\varphi(gx), \tag{3.8}$$

for every $g \in G, x \in M$.

This means that the generalized gradient $\partial_C\varphi : M \to T^*M$ of a G-invariant functional φ is G-invariant. We denote by M^G the fixed point set (symmetric point set) of the action G over M. Now, let x be a symmetric point of M. There is a natural linear representation of G on T_xM given by $g \to Dg(x)$. The action G is called *linearizable* at x if there exists an open $U \in \mathcal{V}_M(x)$ G-invariant neighborhood of x and a diffeomorphism $\varphi : U \to \varphi(U)$, where $\varphi(U) \subset T_x(M)$ is open and G-equivariant such that the map $\varphi \circ g \circ \varphi^{-1} : \varphi(U) \to T_xM$ is the restriction to $\varphi(U)$ of the linear map $Dg(x)$. We observe that if (U, ψ) is a chart at x such that U is invariant and $\psi(x) = 0$, where $\psi : U \to E$ and $E \cong T_xM$, by identifying E with T_xM, we can define

$$\varphi(y) = \int_G (Dg(y) \cdot \psi)(g^{-1}y)dg, \quad y \in U.$$

The map φ linearizes the action of G about of x. Thus we have

Proposition 3.7 *Any action of a compact Lie group G by diffeomorphisms on a Banach manifold is linearizable at symmetric points.*

Since E^G is a closed linear subspace of E and $\varphi(U \cap M^G) = \varphi(U) \cap E^G$ we conclude that M^G is a submanifold of M.

Using Proposition 3.7 we see that the (PSCL) remains true for G-Banach manifold M, when G is a compact Lie group.

Theorem 3.4 *Let $x \in M^G$ and $\varphi : M \to \mathbb{R}$ be a locally Lipschitz G-invariant function. Then x is a critical point of φ if and only if x is a critical point of $\varphi^G := \varphi|_{M^G} : M^G \to \mathbb{R}$.*

In the next we give a direct application of Theorem 3.4 which is very useful in the study of eigenvalue problems. For this we consider a Hilbert space $(H, \langle \cdot, \cdot \rangle)$, a locally Lipschitz function $f : H \to \mathbb{R}$ and $h : H \to \mathbb{R}$ a C^1-function such that $a \in \mathbb{R}$ is a regular value of h, i.e. $h'(x) \neq 0$, if $h(x) = a$. Then $S = h^{-1}(0)$ is a C^1-manifold of H whose tangent space $T_u S$ at any $u \in S$ is expressed by

$$T_u S = h'(u)^{-1}(0) = \{x \in H : \langle h'(u), x \rangle = 0\}.$$

The generalized gradient $\partial_C(f|_S)(u)$ of $f|_S$ at any $u \in S$ is given by

$$\partial_C(f|_S)(u) = \left\{ z - \frac{\langle z, \nabla h(u) \rangle}{\| \nabla h(u) \|^2} h'(u) : z \in \partial_C f(u) \right\},$$

where $\nabla h(u)$ means the gradient of h at u, that is the element $\nabla h(u) \in H$ satisfying

$$\langle h'(u), v \rangle = \langle \nabla h(u), v \rangle, \quad \forall v \in H.$$

Now, let G be a compact Lie group which acts linearly and isometrically on H. We suppose that the functions f and h are G-invariant. We introduce the notations:

$$\Sigma := \{u \in H : gu = u, \forall g \in G\} \text{ and } S^G := S \cap \Sigma.$$

As above, we get that S^G is a submanifold of S and $T_u S^G = \{w \in \Sigma : (dh)_u(w) = 0\}$ for every $u \in S^G$. As a direct consequence of Theorem 3.4 we have

Corollary 3.1 $0 \in \partial_C(f|_{S^G})(u) \Leftrightarrow 0 \in \partial_C(f|_S)(u)$.

We denote by $\Phi(X)$ the set of functions $\varphi : X \to (-\infty, +\infty]$ which are convex, proper and lower semicontinuous. Recall that $\Gamma_G(X^*)$ is the set of G-invariant weakly*-closed convex subset of X^* and let $\Phi_G(X)$ denote the set of all G-invariant functionals belonging to $\Phi(X)$.

We first consider the following principle.

$(PSCI)$: Forevery $\varphi \in \Phi_G(X)$ and all $K \in \Gamma_G(X^*)$ it holds that:

$$\partial(\varphi|_\Sigma)(u) \cap K|_\Sigma \neq \varnothing \Rightarrow \partial\varphi(u) \cap K \neq \varnothing,$$

where $K|_\Sigma = \{\zeta|_\Sigma : \zeta \in K\}$ with $\langle \zeta|_\Sigma, u \rangle_{\Sigma, \Sigma^*} = \langle \zeta, u \rangle_{X, X^*}$ and $u \in \Sigma$.

Remark 3.4 The principle $(PSCI)$ is a generalization of the classical (PSC), of the $(PSCL)$, and includes in particular the principle for Szulkin type functionals.

To conclude this we observe that for every $J \in C^1$ and $u \in \Sigma$ we have

$$(J|_\Sigma)'(u) = \big(J'(u)\big)|_\Sigma. \tag{3.9}$$

Indeed, for every $h \in X$, $J'(u)$ satisfies:

$$J(u + h) = J(u) + \langle J'(u), h \rangle + o(h)$$

and $\big(J'(u)\big)|_\Sigma$ satisfies

$$J(u + h) = \varphi(u) + \langle (J'(u))|_\Sigma, h \rangle_\Sigma + o(h)$$

for every $h \in \Sigma$. Noticing that $u, u + h \in \Sigma$ imply $J(u + h) = J|_\Sigma(u + h)$ and $J(u) = J|_\Sigma(u)$, we get that $(J|_\Sigma)'(u) = \big(J'(u)\big)|_\Sigma$.

Now, let $J \in C_G^1(X)$ and put $K = \{-J'(u)\}$ with $u \in \Sigma$, then by virtue of (3.5), we get $K \in \Gamma_G(X^*)$. Therefore, in view of (3.9), we find that $(PSCI)$ yields

$(PSCI)'$: For all $\varphi \in \Phi_G(X)$ and all $J \in C_G^1(X)$, it holds that

$$\partial\,(\varphi|_\Sigma)\,(u) + (J|_\Sigma)'\,(u) \ni 0 \Rightarrow \partial\varphi(u) + J'(u) \ni 0.$$

Remark 3.5 Principle $(PSCI)'$ corresponds exactly to the Szulkin type functions, see Remark 3.2. Moreover, in particular, take $\varphi \equiv 0$, then $\partial\,(\varphi|_\Sigma)\,(u) = \partial\varphi(u) = 0$. Thus, $(PSCI)'$ with $\varphi \equiv 0$ gives the classical principle of symmetric criticality (PSC). Finally, let $\varphi : X \to \mathbb{R}$ be a G-invariant locally Lipschitz function. For $u \in \Sigma$, let us choose $K = \partial\varphi(u)$ and $\psi \equiv 0$. Then $(PSCI)$ reduces to $(PSCL)$ since we obviously have $\partial(\psi|_\Sigma)(u) \subseteq \partial\psi(u)|_\Sigma$. By a mild modification of the above arguments, the principle of symmetric criticality has been extended to Motreanu-Panagiotopoulos functionals by Kristály et al. [6] and to continuous functions by Squassina [9].

Proposition 3.8 *For all $\varphi \in \Phi_G(X)$, the subdifferential $\partial\varphi$ of φ is G-equivariant, i.e. for every $g \in G$ and $u \in X$ we have $\partial\varphi(gu) = g\partial\varphi(u)$.*

Proof First, we prove that $\partial\varphi(gu) \subset g\partial\varphi(u)$. Let $\zeta \in \partial\varphi(gu)$. Then we have

$$\varphi(v) - \varphi(u) = \varphi(gv) - \varphi(gu) \geq \langle \zeta, gv - gu \rangle = \langle g^{-1}\zeta, v - u \rangle,$$

for all $v \in X$. This implies $g^{-1}\zeta \in \partial\varphi(u)$ and hence $\zeta \in g\partial\varphi(u)$.
Moreover, the above relation with g replaced by g^{-1} gives

$$g\partial\varphi(u) = g\partial\varphi(g^{-1}gu) \subset gg^{-1}\partial\varphi(gu) = \partial\varphi(gu),$$

which completes the proof. $\square$

If we take $K := -J'(u) + \partial\psi(u)$ with $J \in C_G^1(X)$, $\psi \in \Phi_G(X)$ and $u \in \Sigma$, then (3.5) and Proposition 3.8 assure that $K \in \Gamma_G(X^*)$. Then $(PSCI)$ can be reformulated in the following way:
   $(PSCI)''$: For all $\varphi, \psi \in \Phi_G(X)$ and all $J \in C_G^1(X)$, it holds that

$$\partial\left(\varphi|_\Sigma\right)(u) + \left(J|_\Sigma\right)'(u) - \partial\left(\psi|_\Sigma\right)(u) \ni 0 \Rightarrow \partial\varphi(u) + J'(u) - \partial\psi(u) \ni 0,$$

provide that $\partial\left(\psi|_\Sigma\right) = \left(\partial\psi(u)\right)|_\Sigma$.
   We consider the following further hypotheses:

(B_1) X is reflexive and the norms of X and X^* are strictly convex;
(B_2) The action of G over X is isometric, i.e., $\|gu\| = \|u\|$, for all $g \in G$ and $u \in X$.

   One can prove the following results.

Theorem 3.5 *Assume that (B_1) and (B_2) are satisfied. Then the principle $(PSCI)$ is valid.*

Theorem 3.6 *Assume that (A_1) is satisfied and $\partial\varphi + \partial I_\Sigma$ is maximal monotone. Then the principle $(PSCI)$ is valid.*

   The proofs of these results are fairly technical, so we will omit them. However, an interested reader can consult Kobayashi [3], and Kobayashi and Otani [4].

References

1. K.-C. Chang, Variational methods for non-differentiable functionals and their applications to partial differential equations. J. Math. Anal. Appl. **80**, 102–129 (1981)
2. L. Gasinski, N. Papageorgiou, *Nonsmooth Critical Point Theory and Nonlinear Boundary Value Problems*. Mathematical Analysis and Applications (CRC Press, Boca Raton, 2004)
3. J. Kobayashi, A principle of symmetric criticality for variational inequalities, in *Mathematical Aspects of Modelling Structure Formation Phenomena*, vol. 17. Gakuto International Series Mathematical Sciences and Applications (2001), pp. 94–106
4. J. Kobayashi, M. Ôtani, The principle of symmetric criticality for non-differentiable mappings. J. Funct. Anal. **214**, 428–449 (2004)
5. W. Krawciewicz, W. Marzantovicz, Some remarks on the Lusternik-Schnirelmann method for non-differentiable functionals invariant with respect to a finite group action. Rocky Mount. J. Math. **20**, 1041–1049 (1990)
6. A. Kristály, C. Varga, V. Varga, A nonsmooth principle of symmetric criticality and variational-hemivariational inequalities. J. Math. Anal. Appl. **325**, 975–986 (2007)
7. D. Motreanu, P.D. Panagiotopoulos, *Minimax Theorems and Qualitative Properties of the Solutions of Hemivariational Inequalities and Applications*, vol. 29. Nonconvex Optimization and its Applications (Kluwer Academic Publishers, Boston, 1999)
8. R.S. Palais, Principle of symmetric criticality. Comm. Math. Phys. **69**, 19–31 (1979)
9. M. Squassina, On Palais' principle for non-smooth functionals. Nonlin. Anal. **74**, 3786–3804 (2011)
10. A. Szulkin, Minimax principles for lower semicontinuous functions and applications to nonlinear boundary problems. Ann. Inst. H. Poincaré Anal. Non. Linéaire **3**, 77–109 (1986)

Part II

Variational Techniques in Nonsmooth Analysis and Applications

Deformation Results

4

4.1 Deformations Using a Cerami-Type Compactness Condition

In this section we present two deformation results for locally Lipschitz functions defined on Banach spaces. These results extend those of Chang [2] and Kourogenis-Papageorgiou [7].

Let us consider $f : X \to \mathbb{R}$ to be a locally Lipschitz function. Our approach is based on using a general compactness condition which contains as particular cases both the Palais-Smale and Cerami compactness conditions. More precisely, we consider a globally Lipschitz functional $\varphi : X \to \mathbb{R}$ such that $\varphi(u) \geq 1, \forall u \in X$ (or, $\varphi(u) \geq \alpha$, for some $\alpha > 0$).

Definition 4.1 We say that the function f satisfies the $(\varphi - C)$-*condition at level c* (in short, $(\varphi - C)_c$) if every sequence $\{u_n\} \subset X$ such that $f(u_n) \to c$ and $\varphi(u_n)\lambda_f(u_n) \to 0$ has a (strongly) convergent subsequence.

As pointed out before, the $(\varphi - C)_c$-condition contains the $(PS)_c$ and $(C)_c$ compactness conditions, respectively. Indeed if $\varphi \equiv 1$ we get the $(PS)_c$-condition and if $\varphi(u) := 1 + \|u\|$ we have the $(C)_c$-condition.

Throughout in this chapter we use the following notations for the locally Lipschitz function $f : X \to \mathbb{R}$ and a number $c \in \mathbb{R}$:

$$f^c := \{u \in X : f(u) \leq c\}, \quad f_c := \{u \in X : f(u) \geq c\},$$

$$K_c := \{u \in X : \lambda_f(u) = 0, f(u) = c\}, \quad (K_c)_\delta := \{u \in X : d(u, K_c) < \delta\},$$

$$(K_c)_\delta^c := X \setminus (K_c)_\delta.$$

© The Author(s), under exclusive license to Springer Nature Switzerland AG 2021
N. Costea et al., *Variational and Monotonicity Methods in Nonsmooth Analysis*,
Frontiers in Mathematics, https://doi.org/10.1007/978-3-030-81671-1_4

We need the following result in order to obtain the existence of a suitable locally Lipschitz vector field.

Lemma 4.1 *Let X be a Banach space and let $f : X \to \mathbb{R}$ be a locally Lipschitz function satisfying the $(\varphi - C)_c$-condition with $\varphi : X \to \mathbb{R}$ a globally Lipschitz function such that $\varphi(u) \geq 1$, $\forall u \in X$. Then for each $\delta > 0$ there exist constants $\gamma, \varepsilon > 0$ and a locally Lipschitz vector field $\Lambda : f^{-1}([c - \varepsilon, c + \varepsilon]) \cap (K_c)_\delta^c \to X$ such that for each $u \in f^{-1}([c - \varepsilon, c + \varepsilon]) \cap (K_c)_\delta^c$ one has*

$$\|\Lambda(u)\| \leq \varphi(u) \tag{4.1}$$

and

$$\langle \zeta, \Lambda(u) \rangle \geq \frac{\gamma}{2}, \quad \forall \zeta \in \partial_C f(x). \tag{4.2}$$

Proof From the $(\varphi - C)_c$-condition we get $\gamma, \varepsilon > 0$ such that

$$\varphi(u)\lambda_f(u) \geq \gamma, \tag{4.3}$$

for each $u \in (K_c)_\delta^c$ and $c - \varepsilon \leq f(u) \leq c + \varepsilon$. Assuming by contradiction this not the case, we could find a sequence $\{u_n\} \subset (K_c)_\delta^c$ such that $f(u_n) \to c$ and $\varphi(u_n)\lambda_f(u_n) \to 0$. Using the $(\varphi - C)_c$-condition we obtain a convergent subsequence of $\{u_n\}$ (denoted again by $\{u_n\}$), say $u_n \to u_0 \in (K_c)_\delta^c$. Since f is continuous and λ_f is lower semicontinuous we obtain that $f(u_0) = c$ and $\varphi(u_0)\lambda_f(u_0) = 0$. This implies $u_0 \in K_c$, which is a contradiction. Thus (4.3) holds.

Let $u_0 \in f^{-1}([c - \varepsilon, c + \varepsilon]) \cap (K_c)_\delta^c$ and $\zeta_0 \in \partial_C f(u_0)$ be such that $\lambda_f(u_0) = \|\zeta_0\|$. Then we have $B_{\|\zeta_0\|} \cap \partial_C f(u_0) = \varnothing$, where $B_r := \{\xi \in X^* : \|\xi\| < r\}, r > 0$. Using the separation theorem in X^* endowed with the weak*-topology we obtain that there exists some $h_0 \in X$ such that $\|h_0\| = 1$ and $\langle \xi, h_0 \rangle \leq \langle \zeta_0, h_0 \rangle \leq \langle \zeta, h_0 \rangle$ for each $\xi \in B_{\|\zeta_0\|}$ and $\zeta \in \partial_C f(x_0)$. From Corollary A.2 and (4.3) we get

$$\sup_{\xi \in B_{\|\zeta_0\|}} \langle \xi, h_0 \rangle = \|\zeta_0\| > \frac{\gamma}{2\varphi(u_0)}.$$

Therefore $\langle \zeta, h_0 \rangle \geq \|\zeta_0\| > \frac{\gamma}{2\varphi(x_0)}$, for every $\zeta \in \partial_C f(u_0)$. As the set-valued map $u \mapsto \partial_C f(u)$ is weakly*-upper semicontinuous, for each $u \in f^{-1}([c - \varepsilon, c + \varepsilon]) \cap (K_c)_\delta^c$ there exists $r_0 > 0$ and $h_0 \in X$ such that for every $v \in B(u_0, r_0)$ and every $\zeta \in \partial_C f(v)$ we have $\langle \zeta, h_0 \rangle > \frac{\gamma}{2\varphi(v)}$. The set of all such balls $\{B(u_0, r_0)\}$ covers $f^{-1}([c - \varepsilon, c + \varepsilon]) \cap (K_c)_\delta^c$. By paracompactness there is a locally finite covering $\{V_i\}_{i \in I}$ subordinated to it. If we consider the functions $\rho_i : X \to \mathbb{R}$ defined by $\rho_i(u) := \text{dist}(u, X \setminus V_i)$ for all $u \in X$, then the functions ρ_i are Lipschitz continuous and $\rho_i |_{X \setminus V_i} = 0$.

For every $u \in \bigcup_{i \in I} V_i$, let $\beta_i(u) := \frac{\rho_i(u)}{\sum_{j \in I} \rho_j(u)}$ and $\Lambda(u) := \varphi(u) \sum_{i \in I} \beta_i(x) h_i$, where h_i plays the same role for u_i as h_0 for u_0. It follows that the function $\Lambda : f^{-1}([c - \varepsilon, c + \varepsilon]) \cap (K_c)^c_\delta \to X$ is locally Lipschitz and for every $u \in f^{-1}([c - \varepsilon, c + \varepsilon])$ we have

$$\|\Lambda(u)\| \le \varphi(u) \sum_{i \in I} \beta_i(x) \|h_i\| = \varphi(u)$$

and

$$\langle \zeta, \Lambda(u) \rangle = \varphi(u) \sum_{i \in I} \beta_i(u) \langle \zeta, h_i \rangle > \frac{\gamma}{2}, \quad \forall \zeta \in \partial_C f(x).$$

Thus properties (4.1) and (4.2) are satisfied.   □

The next result can be proved in the same way as the above; thus we will omit it.

Lemma 4.2 *Let X be a Banach space and let $f : X \to \mathbb{R}$ be a locally Lipschitz function and $S \subset X$ a subset. Suppose that the numbers $c \in \mathbb{R}$, $\varepsilon, \delta > 0$ are such that*

$$\lambda_f(u) \ge \frac{4\varepsilon}{\delta}, \quad \forall u \in f^{-1}([c - 2\varepsilon, c + 2\varepsilon]) \cap S_{2\delta}. \tag{4.4}$$

Then there exists a locally Lipschitz vector field $\Lambda : f^{-1}([c - 2\varepsilon, c + 2\varepsilon]) \cap S_{2\delta} \to X$ such that:

(a) $\|\Lambda(u)\| \le 1$;
(b) for every $\zeta \in \partial_C f(x)$ we have $\langle \zeta, \Lambda(u) \rangle > \frac{2\varepsilon}{\delta}$.

The next result is a quantitative deformation lemma for locally Lipschitz functionals and it appears in the paper of Varga and Varga [15].

Theorem 4.1 *Let X be a Banach space and let $f : X \to \mathbb{R}$ be a locally Lipschitz function and S a subset of X. Let $c \in \mathbb{R}$ and $\varepsilon, \delta > 0$ be numbers such that (4.4) holds. Then there exists a continuous function $\eta : [0, 1] \times X \to X$ with the properties:*

(i) $\eta(0, u) = u$, for every $u \in X$;
(ii) $\eta(t, \cdot) : X \to X$ is homeomorphism for every $t \in [0, 1]$;
(iii) $\eta(t, u) = u$, for every $u \notin f^{-1}([c - 2\varepsilon, c + 2\varepsilon]) \cap S_{2\delta}$ and $t \in [0, 1]$;
(iv) $\|\eta(t, u) - u\| \le \delta$, for all $u \in X$ and $t \in [0, 1]$;
(v) $f(u) - f(\eta(t, u)) \ge 2\varepsilon t$, for $t \in [0, 1]$ with $\eta(t, u) \in f^{-1}([c - 2\varepsilon, c + 2\varepsilon]) \cap S_{2\delta}$;
(vi) $\eta(1, f^{c+\varepsilon} \cap S) \subset f^{c-\varepsilon}$.

Proof We introduce the sets

$$A := S_{2\delta} \cap f^{-1}([c - 2\varepsilon, c + 2\varepsilon]) \text{ and } B := S_{\delta} \cap f^{-1}([c - \varepsilon, c + \varepsilon]),$$

and define $\psi : X \to \mathbb{R}$ by

$$\psi(u) := \frac{\text{dist}(u, X \setminus A)}{\text{dist}(u, X \setminus A) + \text{dist}(u, B)}.$$

The function ψ is locally Lipschitz. Using Lemma 4.2 we get a locally Lipschitz vector field $\Lambda : A \to X$ such that conditions (*a*) and (*b*) hold.

Let $V : X \to X$ be the vector field given by

$$V(u) := \begin{cases} -\psi(u)\Lambda(u), & \text{if } u \in A \\ 0, & \text{otherwise.} \end{cases} \tag{4.5}$$

The vector field V is locally Lipschitz and $\|V(u)\| \leq 1$, for every $u \in X$, hence the corresponding ODE

$$\begin{cases} \dot{\sigma}(t, u) = V(\sigma(t, u)); \\ \sigma(0, u) = u, \end{cases} \tag{4.6}$$

has a unique global solution $\sigma(\cdot, u)$ for every $u \in X$. Let $\eta : [0, 1] \times X \to X$ be the function given by $\eta(t, u) := \sigma(\delta t, u)$. For each $t \in [0, 1]$ the function $\eta(t, \cdot) : X \to X$ is a homeomorphism and $\eta(0, u) = u$ for every $u \in X$. Thus (*i*) and (*ii*) hold.

From (4.6) it results that $\eta(t, u) = u$ for every $u \notin f^{-1}([c - 2\varepsilon, c + 2\varepsilon]) \cap S_{2\delta}$ and $t \in [0, 1]$. Therefore (*iii*) is true. In order to prove (*iv*), note that

$$\frac{\text{d}}{\text{d}t}\eta(t, u) = \frac{\text{d}}{\text{d}t}\sigma(\delta t, u) = \delta\dot{\sigma}(\delta t, u) = \delta V(\sigma(\delta t, u)),$$

so

$$\int_0^t \frac{\text{d}}{\text{d}s}\eta(s, u)ds = \delta \int_0^t V(\sigma(\delta s, u))ds.$$

Thus

$$\|\eta(t, u) - \eta(0, u)\| \leq \delta \int_0^1 \|V(\sigma(\delta s, u))ds\| \leq \delta,$$

which proves (*iv*).

For every $u \in X$ we consider the function $h : \mathbb{R} \to \mathbb{R}$ given by $h(t) := f(\eta(t, u))$. Using Proposition 2.7 we obtain

$$h'(s) \leq \max_{\zeta \in \partial_C f(\eta(s,u))} \left\langle \zeta, \frac{d}{ds}\eta(s, u) \right\rangle = \max_{\zeta \in \partial_C f(\eta(s,u))} \langle \zeta, \delta\dot{\sigma}(\delta s, u) \rangle$$

$$= \delta \max_{\zeta \in \partial_C f(\eta(s,u))} \langle \zeta, V(\sigma(\delta s, u)) \rangle = -\delta \min_{\zeta \in \partial_C f(\eta(s,u))} \{\langle \zeta, \psi(\sigma(\delta s, u)) \Lambda(\sigma(\delta s, u)) \rangle$$

$$\leq \begin{cases} -2\varepsilon, & \text{if } \eta(s, u) \in A \\ 0, & \text{if } \eta(s, u) \in X \setminus A. \end{cases}$$

From this we obtain that if $\eta(t, u) \in A$, then

$$f(u) - f(\eta(t, u)) = h(0) - h(u) = -\int_0^t h'(s)ds \geq 2\varepsilon t,$$

with $\eta(t, u) \in A$ and $t \in [0, 1]$. Therefore the function f is decreasing along the path $\eta(\cdot, u)$.

Now let $u \notin A$ a fixed element. Then $\psi(u) = 0$, hence $V(u) = 0$. Using the Cauchy problem (4.6) we obtain $\eta(t, u) = u$ for every $t \in [0, 1]$. Thus (v) is proved.

In order to prove (vi) fix $u \in f^{c+\varepsilon} \cap S$. We shall prove that $f(\eta(1, u)) \leq c - \varepsilon$. Therefore we can suppose that $u \in (f^{c+\varepsilon} \setminus f^{c-\varepsilon}) \cap S$, i.e. $f(u) \leq c+\varepsilon$, $f(u) \geq c-\varepsilon$ and $u \in S$. If we assume by contradiction that $f(\eta(1, u)) > c-\varepsilon$, then $f(u)-f(\eta(1, u)) < 2\varepsilon$. On the other hand, if $t \in [0, 1]$ and $\eta(t, u) \in A$ then $f(u) - f(\eta(t, u)) \geq 2\varepsilon t$ and the contradiction completes the proof.   □

In the sequel we prove a very general deformation result which unifies several results of this kind; it appears in the paper of Kristály et al. [8].

Theorem 4.2 *Let $f : X \to \mathbb{R}$ be a locally Lipschitz function on the Banach space X satisfying the $(\varphi - C)_c$-condition, with $c \in \mathbb{R}$ and a globally Lipschitz function $\varphi : X \to \mathbb{R}$ with Lipschitz constant $L > 0$ and $\varphi(u) \geq 1$, $\forall u \in X$. Then for every $\varepsilon_0 > 0$ and every neighborhood U of K_c (if $K_c = \emptyset$, then we choose $U = \emptyset$) there exist a number $0 < \varepsilon < \varepsilon_0$ and a continuous function $\eta : X \times [0, 1] \to X$, such that for every $(u, t) \in X \times [0, 1]$ we have:*

(a) $\|\eta(u, t) - u\| \leq \varphi(u)te^{Lt}$;
(b) $\eta(u, t) = u$, *for every* $u \notin f^{-1}([c - \varepsilon_0, c + \varepsilon_0])$ *and* $t \in [0, 1]$;
(c) $f(\eta(u, t)) \leq f(u)$;
(d) $\eta(u, t) \neq u \Rightarrow f(\eta(u, t)) < f(u)$.
(e) $\eta(f^{c+\varepsilon}, 1) \subset f^{c-\varepsilon} \cup U$;
(f) $\eta(f^{c+\varepsilon} \setminus U, 1) \subset f^{c-\varepsilon}$.

Proof Fix $\varepsilon_0 > 0$ and a neighborhood U of K_c. From the compactness of K_c we can find $\delta > 0$ such that $(K_c)_{3\delta} \subseteq U$. Moreover, the proof of Lemma 4.1 guarantees the existence of $\gamma > 0$ and $0 < \bar{\varepsilon} < \varepsilon_0$ such that $\varphi(u)\lambda_f(u) \geq \gamma$ for all $u \in f^{-1}([c - \bar{\varepsilon}, c + \bar{\varepsilon}]) \cap (K_c)_\delta^c$. We consider the following two closed sets:

$$A := \{u \in X : |f(u) - c| \geq \bar{\varepsilon}\} \cup \overline{(K_c)_\delta} \tag{4.7}$$

$$B := \left\{u \in X : |f(u) - c| \leq \frac{\bar{\varepsilon}}{2}\right\} \cap (K_c)_{2\delta}^c. \tag{4.8}$$

Because $A \cap B = \varnothing$ there exists a locally Lipschitz function $\psi : X \to [0, 1]$ such that $\psi = 0$ on a closed neighborhood of A, say $\tilde{A}$, disjoint of B, $\psi|_B = 1$ and $0 \leq \psi \leq 1$. For instance, we can take $\psi(u) := \frac{d(u,\tilde{A})}{d(u,\tilde{A})+d(u,B)}$, $\forall u \in X$.

Let $V : X \to X$ be defined by

$$V(u) := \begin{cases} -\psi(x) \cdot \Lambda(u), & u \in f^{-1}([c - \bar{\varepsilon}, c + \bar{\varepsilon}]) \cap (K_c)_\delta^c; \\ 0, & \text{otherwise}, \end{cases} \tag{4.9}$$

where $\Lambda(u)$ is constructed in Lemma 4.1. The vector field V is locally Lipschitz and by the same lemma, for $u \in f^{-1}([c - \bar{\varepsilon}, c + \bar{\varepsilon}]) \cap (K_c)_\delta^c$ we have

$$\|V(u)\| = \psi(u)\|\Lambda(u)\| \leq \varphi(u) \tag{4.10}$$

and

$$\langle \zeta, V(u) \rangle = -\psi(u)\langle \zeta, \Lambda(u) \rangle \leq -\psi(u)\frac{\gamma}{2}, \ \forall \zeta \in \partial_C f(u). \tag{4.11}$$

Since V is locally Lipschitz and $\|V(u)\| \leq \varphi(0) + L\|u\|$, the following Cauchy problem:

$$\begin{cases} \dot{\eta}(u, t) = V(\eta(u, t)) \text{ a.e. on } [0, 1] \\ \eta(u, 0) = u \end{cases} \tag{4.12}$$

has a unique solution $\eta(u, \cdot)$ on $\mathbb{R}$, for each $u \in X$. By (4.10) we have

$$\|\eta(u, t) - u\| \leq \int_0^t \|V(\eta(u, s))\| ds \leq \int_0^t \varphi(\eta(x, s)) ds = \int_0^t [\varphi(\eta(u, s)) - \varphi(u)] ds$$

$$+ \int_0^t \varphi(u) ds \leq L \int_0^t \|\eta(u, s) - u\| ds + \varphi(u)t.$$

Using Gronwall's inequality we get $\|\eta(u, t) - u\| \leq \varphi(u)te^{Lt}$, therefore the assertion (a) is proved.

If $u \notin f^{-1}([c - \bar{\varepsilon}, c + \bar{\varepsilon}])$, then $u \in A$, so $\psi(u) = 0$. By (4.9) it follows that $V(u) = 0$ and from (4.12) we obtain that $\eta(u, t) = u$, for each $t \in [0, 1]$. This yields (b).

Next, for a fixed $u \in X$, let us consider the function $h_u : [0, 1] \to \mathbb{R}$ given by $h_u(t) := f(\eta(u, t))$. Using the chain rule we have

$$\frac{d}{dt} h_u(t) \leq \max_{\zeta \in \partial_C f(\eta(x,t))} \left\langle \zeta, \frac{d}{dt} \eta(x, t) \right\rangle = \max_{\zeta \in \partial_C f(\eta(x,t))} \langle \zeta, V(\eta(x, t)) \rangle$$

a.e. on $[0, 1]$. Therefore, taking into account (4.11), we infer

$$\frac{d}{dt} h_u(t) \leq -\psi(\eta(u, t)) \frac{\gamma}{2} \leq 0 \text{ if } \eta(u, t) \in f^{-1}([c - \bar{\varepsilon}, c + \bar{\varepsilon}]) \cap (K_c)_\delta^c, \tag{4.13}$$

and clearly, by (4.9)

$$\frac{d}{dt} h_u(t) \leq 0, \text{ if } \eta(u, t) \notin f^{-1}([c - \bar{\varepsilon}, c + \bar{\varepsilon}]) \cap (K_c)_\delta^c.$$

Hence property (c) holds true.

In order to prove property (d), suppose that $\eta(u, t) \neq u$. First, we show that

$$\eta(u, s) \in f^{-1}([c - \bar{\varepsilon}, c + \bar{\varepsilon}]) \cap (K_c)_\delta^c, \ \forall s \in [0, t]. \tag{4.14}$$

On the contrary, there would exist $s_0 \in [0, t]$ such that $\eta(u, s_0) \in A$. This implies that $V(\eta(u, s_0)) = 0$. Using the uniqueness of solution to the Cauchy problem formed by the equation in (4.12) and the initial condition with the initial value $\eta(u, s_0)$, we see that

$$\eta(u, \tau + s_0) = \eta(u, s_0), \ \forall \tau \in \mathbb{R}.$$

Letting $\tau := t - s_0$ and $\tau := -s_0$ one obtains $\eta(u, t) = u$, which contradicts our assumption. Thus the claim in (4.14) is true.

Using (4.13) and (4.14) it follows that

$$f(u) - f(\eta(u, t)) = -\int_0^t \frac{d}{ds} h_u(s) ds \geq \frac{\gamma}{2} \int_0^t \psi(\eta(u, s)) ds. \tag{4.15}$$

We show that there is $s \in [0, t]$ such that

$$\psi(\eta(u, s)) \neq 0, \tag{4.16}$$

for otherwise, if $\psi(\eta(u, s)) = 0$, $\forall s \in [0, t]$, then $V(\eta(u, s)) = 0$, $\forall s \in [0, t]$. By (4.12), we get that $\eta(u, \cdot)$ is constant on $[0, t]$, which contradicts $\eta(u, t) \neq u$. It results that (4.16) is valid. Since $\psi \geq 0$, from (4.15) and (4.16) we infer that $f(\eta(u,t)) < f(u)$, which proves assertion (d).

Let us prove next assertion (e). Let $\rho > 0$ be such that $\overline{(K_c)}_{3\delta} \subset B(0; \rho)$. We choose

$$0 < \varepsilon \leq \min\left\{ \frac{\bar{\varepsilon}}{2}, \frac{\gamma}{4}, \frac{\delta\gamma}{8} e^{-L} (\varphi(0) + L\rho)^{-1} \right\}, \tag{4.17}$$

and proceed by contradiction. Let $u \in f^{c+\varepsilon}$ be such that $f(\eta(u,1)) > c - \varepsilon$ and $\eta(u,1) \notin U$. Since, by (c), $f(\eta(u,t)) \leq f(u) \leq c + \varepsilon$ and $f(\eta(u,t)) \geq f(\eta(u,1))$ for each $t \in [0,1]$, we get

$$c - \varepsilon < f(\eta(u,t)) \leq c + \varepsilon, \quad \forall t \in [0,1]. \tag{4.18}$$

We claim that

$$\eta(\{u\} \times [0,1]) \cap (K_c)_{2\delta} \neq \emptyset. \tag{4.19}$$

Suppose that (4.19) does not hold, i.e,

$$\eta(\{u\} \times [0,1]) \cap (K_c)_{2\delta} = \emptyset. \tag{4.20}$$

First, we show that

$$\eta(u,t) \in B, \quad \forall t \in [0,1]. \tag{4.21}$$

The fact that $\eta(u,t) \in f^{-1}\left(\left[c - \frac{\bar{\varepsilon}}{2}, c + \frac{\bar{\varepsilon}}{2}\right]\right)$ follows from (4.17) and (4.18). By (4.20) one has that $\eta(u,t) \in (K_c)_{2\delta}^c$. Consequently, from (4.8) we conclude that (4.21) is established. On the basis of (4.21) and (4.13) we may write

$$f(u) - f(\eta(u,1)) = h_u(0) - h_u(1) = -\int_0^1 \frac{d}{dt} h_u(t) dt \geq \int_0^1 \frac{\gamma}{2} \psi(\eta(u,t)) dt.$$

Then, combining (4.21) and the definition of ψ it is clear that

$$f(u) - f(\eta(u,1)) \geq \frac{\gamma}{2}. \tag{4.22}$$

On the other hand, from (4.18) we obtain that

$$f(u) - f(\eta(u,1)) < 2\varepsilon. \tag{4.23}$$

From (4.22) and (4.23) we get $\frac{\gamma}{2} < 2\varepsilon$, which contradicts (4.17). This justifies (4.19).

The next step in the proof is to show that there exist $0 \le t_1 < t_2 \le 1$ such that

$$\text{dist}(\eta(u, t_1), K_c) = 2\delta, \quad \text{dist}(\eta(u, t_2), K_c) = 3\delta \tag{4.24}$$

and

$$2\delta < \text{dist}(\eta(u, t), K_c) < 3\delta, \quad \forall t_1 < t < t_2. \tag{4.25}$$

Denote by $g(t) := \text{dist}(\eta(u, t), K_c)$, $\forall t \in [0, 1]$. In view of (4.19) we have that $\{t \in [0, 1] : g(t) \le 2\delta\} \ne \emptyset$. Thus it is permitted to consider

$$t_1 := \sup\{t \in [0, 1] : g(t) \le 2\delta\}.$$

Since it is known that $(K_c)_{3\delta} \subset U$ and $\eta(u, 1) \notin U$, we derive that $\eta(u, 1) \notin (K_c)_{3\delta}$. This means that $g(1) \ge 3\delta$. Since $g(t_1) \le 2\delta$ it is necessary to have $t_1 < 1$. The definition of t_1 implies $g(t) > 2\delta$ for all $t \in (t_1, 1]$ (which is the first inequality in (4.25)). Letting $t \downarrow t_1$ we deduce that $g(t_1) \ge 2\delta$. We obtain that $g(t_1) = 2\delta$, so the first part in (4.24) is proved. Taking into account that $g(1) \ge 3\delta$, we see that $\{t \in [t_1, 1] : g(t) \ge 3\delta\}$ is nonempty. Then we can define

$$t_2 := \inf\{t \in [t_1, 1] : g(t) \ge 3\delta\}.$$

Since $g(t_2) \ge 3\delta$ and $g(t_1) = 2\delta$ it is clear that $t_1 < t_2$. By the definition of t_2 we have that $g(t) < 3\delta$ for all $t_1 \le t < t_2$, so (4.25) holds. In addition, letting $t \uparrow t_2$, we get $g(t_2) = 3\delta$, so (4.24) holds, too.

Let us show that

$$t_2 - t_1 < \frac{4\varepsilon}{\gamma}. \tag{4.26}$$

From (4.25) it follows that $\eta(u, t) \notin (K_c)_{2\delta}$, $\forall t \in [t_1, t_2]$, while (4.18) and (4.17) imply $\eta(u, t) \in f^{-1}\left(\left[c - \frac{\bar{\varepsilon}}{2}, c + \frac{\bar{\varepsilon}}{2}\right]\right)$, $\forall t \in [t_1, t_2]$. The definition of the set B in (4.8) yields

$$\eta(u, t) \in B, \quad \forall t \in [t_1, t_2].$$

Using the definition of ψ, (4.13) and (4.18) we see that

$$\frac{\gamma}{2}(t_2 - t_1) = \frac{\gamma}{2} \int_{t_1}^{t_2} \psi(\eta(u, t)) dt \le -\int_{t_1}^{t_2} \frac{d}{dt} h_u(t) dt = h_u(t_1) - h_x(t_2)$$

$$= f(\eta(u, t_1)) - f(\eta(u, t_2)) < 2\varepsilon.$$

Thus (4.26) is proved.

We need the following inequality

$$\|\eta(u, t_2) - \eta(u, t_1)\| \geq \delta. \tag{4.27}$$

To check (4.27) consider a point $v \in K_c$ so that

$$\operatorname{dist}(\eta(u, t_1), K_c) = \|\eta(u, t_1) - v\| = 2\delta.$$

Here the compactness of K_c and the first part in (4.24) have been used. Then, on the basis of the second part in (4.24) we can write

$$\|\eta(u, t_2) - \eta(u, t_1)\| \geq \|\eta(u, t_2) - v\| - \|\eta(u, t_1) - v\| \geq 3\delta - 2\delta = \delta.$$

Therefore (4.27) holds.

Using (4.12), (4.10) and the Lipschtz property of φ we can write

$$\|\eta(u, t_2) - \eta(u, t_1)\| \leq \int_{t_1}^{t_2} \|V(\eta(u, s))\| ds \leq \int_{t_1}^{t_2} \varphi(\eta(u, s)) ds$$

$$= \int_{t_1}^{t_2} [\varphi(\eta(u, s)) - \varphi(\eta(u, t_1))] ds + \varphi(\eta(u, t_1))(t_2 - t_1)$$

$$\leq \int_{t_1}^{t_2} L\|\eta(u, s) - \eta(u, t_1)\| ds + \varphi(\eta(u, t_1))(t_2 - t_1). \tag{4.28}$$

By (4.28) and Gronwall's inequality we get

$$\|\eta(u, t_2) - \eta(u, t_1)\| \leq \varphi(\eta(u, t_1))(t_2 - t_1) e^{L(t_2 - t_1)}. \tag{4.29}$$

From (4.27), (4.29), (4.26) and the Lipschitz property of φ we deduce that

$$\delta \leq \|\eta(u, t_2) - \eta(u, t_1)\| < \frac{4\varepsilon}{\gamma} e^L \varphi(\eta(u, t_1))$$

$$\leq \frac{4\varepsilon}{\gamma} e^L (\varphi(0) + L\|\eta(u, t_1)\|). \tag{4.30}$$

In view of (4.24) and the choice of ρ to satisfy $\overline{(K_c)_{3\delta}} \subset B(0; \rho)$ we have $\eta(u, t_1) \in (K_c)_{3\delta} \subset B(0; \rho)$. This property and (4.17) yield from (4.30) that

$$\delta \leq \frac{4\varepsilon}{\gamma} e^L (\varphi(0) + L\rho) \leq \frac{\delta}{2},$$

which is a contradiction and the proof of (e) is now complete.

In order to show (f), since $(K_c)_{3\delta} \subset U$ it is enough to prove that

$$\eta(f^{c+\varepsilon} \setminus (K_c)_{3\delta}, 1) \subset f^{c-\varepsilon}. \tag{4.31}$$

Let us denote

$$C := (f^{c+\varepsilon} \setminus f^{c-\varepsilon}) \cap (K_c)^c_{3\delta}.$$

To check (4.31), we note that it is sufficient to verify that

$$\eta(u, 1) \in f^{c-\varepsilon}, \quad \forall u \in C, \tag{4.32}$$

because for $u \in f^{c-\varepsilon}$ we have $f(\eta(u, 1)) \le f(u) \le c - \varepsilon$, due to the nondecreasing monotonicity of $f(\eta(u, \cdot))$.

To show (4.32), denote by

$$D := (f^{c+\varepsilon} \setminus f^{c-\varepsilon}) \cap (K_c)^c_{\frac{5}{2}\delta}.$$

First, we verify that

$$\forall u \in C, \ \exists t_u \in \left(0, \frac{4\varepsilon}{\gamma}\right] \text{ such that } \eta(u, t_u) \notin D. \tag{4.33}$$

To this end, we prove the following inclusion

$$\{t > 0 : \eta(u, \tau) \in D, \ \forall \tau \in [0, t]\} \subset \left(0, \frac{4\varepsilon}{\gamma}\right), \quad \forall u \in C. \tag{4.34}$$

Indeed, if $\eta(u, \tau)$ is in $D \subset B$, $\forall \tau \in [0, t]$, we have $\psi(\eta(u, \tau)) = 1$, $\forall \tau \in [0, t]$. Therefore, by (4.13), we have $\frac{d}{d\tau} h_u(\tau) \le -\frac{\gamma}{2}$, $\forall \tau \in [0, t]$. From this and (4.18) we obtain

$$2\varepsilon > h_u(0) - h_u(t) = -\int_0^t \frac{d}{d\tau} h_u(\tau) d\tau \ge \frac{\gamma}{2} t,$$

so $t < \frac{4\varepsilon}{\gamma}$. Thus (4.34) is satisfied.

We are now in the position to prove (4.33). We proceed by contradiction. Assuming that there exist $u \in C$ such that $\eta(u, t) \in D$, $\forall t \in \left(0, \frac{4\varepsilon}{\gamma}\right]$, by (4.34), we arrive at the contradiction

$$\frac{4\varepsilon}{\gamma} \in \{t > 0 : \eta(u, \tau) \in D, \ \forall \tau \in [0, t]\} \subset \left(0, \frac{4\varepsilon}{\gamma}\right),$$

which proves (4.33).

Let us show that for every $u \in C$, it is true that

$$\eta(\{u\} \times [0, 1]) \cap (K_c)_{\frac{5\delta}{2}} \neq \varnothing \Rightarrow \exists t_0 \in (0, t_3] \text{ such that } \eta(u, t_0) \in f^{c-\varepsilon}, \qquad (4.35)$$

with

$$t_3 := \inf\left\{t \in [0, 1] : \operatorname{dist}(\eta(u, t), K_c) \leq \frac{5\delta}{2}\right\},$$

where the set $\left\{t \in [0, 1] : \operatorname{dist}(\eta(u, t), K_c) \leq \frac{5\delta}{2}\right\}$ is nonempty in view of (4.25). If (4.35) were not true it would exist $u \in C$ with $\eta(\{u\} \times [0, 1]) \cap (K_c)_{\frac{5\delta}{2}} \neq \varnothing$ and $f(\eta(u, t)) > c - \varepsilon, \forall t \in [0, t_3]$. Hence $\eta(u, t) \in D, \forall t \in [0, t_3]$. This follows from the definition of t_3 and since $u \in C$. The inclusion in (4.34) implies that

$$t_3 < \frac{4\varepsilon}{\gamma}. \qquad (4.36)$$

Introduce

$$t_4 := \sup\{t \in [0, t_3] : \operatorname{dist}(\eta(u, t), K_c) \geq 3\delta\}.$$

Since $u \in C$, then $u \in (K_c)_{3\delta}^c$, thus the set $\{t \in [0, t_3] : \operatorname{dist}(\eta(u, t), K_c) \geq 3\delta\}$ is nonempty. By the definitions of t_3 and t_4 it follows that

$$\eta(u, t) \in \left(f^{c+\varepsilon} \setminus f^{c-\varepsilon}\right) \cap \left((K_c)_{3\delta} \setminus (K_c)_{\frac{5\delta}{2}}\right), \quad \forall t \in [t_4, t_3].$$

Note that

$$\|\eta(u, t_3) - \eta(u, t_4)\| \geq \frac{\delta}{2}. \qquad (4.37)$$

Indeed, by the definition of t_4 we have

$$\|\eta(u, t_3) - \eta(u, t_4)\| \geq \|\eta(u, t_4) - v\| - \|\eta(u, t_3) - v\|$$
$$\geq 3\delta - \|\eta(u, t_3) - v\|, \quad \forall v \in K_c.$$

This leads to

$$\|\eta(u, t_3) - \eta(u, t_4)\| \geq 3\delta - \operatorname{dist}(\eta(u, t_3), K_c) = 3\delta - \frac{5\delta}{2} = \frac{\delta}{2},$$

so (4.37) is verified.

Using (4.12), (4.10) and the Lipschitz property of φ we can write

$$\|\eta(u, t_3) - \eta(u, t_4)\| \leq \int_{t_4}^{t_3} \|V(\eta(u, s))\| ds \leq \int_{t_4}^{t_3} \varphi(\eta(u, s)) ds$$

$$= \int_{t_4}^{t_3} [\varphi(\eta(u, s)) - \varphi(\eta(u, t_4))] ds + \varphi(\eta(u, t_4))(t_3 - t_4)$$

$$\leq \int_{t_4}^{t_3} L \|\eta(u, s) - \eta(u, t_4)\| ds + \varphi(\eta(u, t_4))(t_3 - t_4).$$

By Gronwall's inequality we get

$$\|\eta(u, t_3) - \eta(u, t_4)\| \leq \varphi(\eta(u, t_4))(t_3 - t_4)e^{L(t_3 - t_4)}. \tag{4.38}$$

Using (4.37), (4.38), the Lipschitz property of φ, the inclusion $\overline{(K_c)}_{3\delta} \subset B(0; \rho)$ and (4.36), we have that

$$\frac{\delta}{2} \leq \|\eta(u, t_3) - \eta(u, t_4)\| \leq e^{L(t_3 - t_4)} \varphi(\eta(u, t_4))(t_3 - t_4)$$

$$\leq e^L(\varphi(0) + L\|\eta(u, t_4)\|)t_3 < e^L(\varphi(0) + L\rho)\frac{4\varepsilon}{\gamma}.$$

This contradicts the choice of ε in (4.17), therefore (4.35) is true.

In order to complete the proof of (f), let $u \in C$. From (4.33), there exists $t_u \in (0, \frac{4\varepsilon}{\gamma}]$ such that $\eta(u, t_u) \notin D$. This means that

$$\eta(u, t_u) \in (X \setminus f^{c+\varepsilon}) \cup f^{c-\varepsilon} \cup (K_c)_{\frac{5\delta}{2}}.$$

On the other hand, $\eta(u, t_u) \in f^{c+\varepsilon}$ since, as $u \in C$, $f(\eta(u, t_u)) \leq f(u) \leq c + \varepsilon$. Consequently, we deduce that $\eta(u, t_u) \in f^{c-\varepsilon} \cup (K_c)_{\frac{5\delta}{2}}$. Two cases arise:

(1) $\eta(u, t_u) \in f^{c-\varepsilon}$;
(2) $\eta(u, t_u) \in (K_c)_{\frac{5\delta}{2}}$.

In case (1) we have directly that

$$f(\eta(u, 1)) \leq f(\eta(u, t_u)) \leq c - \varepsilon,$$

which ensures the desired conclusion.

Should (2) occur, we make use of property (4.35). Therefore, we find $t_0 \in (0, t_3]$ such that $\eta(u, t_0) \in f^{c-\varepsilon}$. Thus we may write $f(\eta(u, 1)) \leq f(\eta(u, t_0)) \leq c - \varepsilon$.   □

Remark 4.1 If we choose $\varphi \equiv 1$ or $\varphi(u) := 1 + \|u\|$ then we obtain the deformation lemmas of Chang [2] and Kourogenis-Papageorgiou [7], respectively.

4.2   Deformations with Compactness Condition of Ghoussoub-Preiss Type

The following variant of Palais-Smale condition is an extension to the locally Lipschitz case of the one introduced by Ghoussoub and Preiss [6] for C^1−functionals. Let $f : X \to \mathbb{R}$ be a locally Lipschitz functional, $c \in \mathbb{R}$ a real number and $B \subseteq X$.

Definition 4.2 We say that the locally Lipschitz function f satisfies the *Palais-Smale condition around B at level c* (shortly, $(PS)_{B,c}$), if every sequence $\{u_n\} \subset X$ with $f(u_n) \to c$, $\mathrm{dist}(u_n, B) \to 0$ and $\lambda_f(u_n) \to 0$ as $n \to \infty$, contains a (strongly) convergent subsequence in X.

In particular, we write $(PS)_c$ instead of $(PS)_{X,c}$ and simply (PS) if $(PS)_c$ holds for every $c \in \mathbb{R}$.

For a fixed $B \subseteq X$ and a fixed number $\delta > 0$, we denote the closed δ-neighborhood of B by $N_\delta(B)$, that is,

$$N_\delta(B) := \{u \in X : \mathrm{dist}(u, B) \leq \delta\}.$$

Definition 4.3 A *generalized normalized pseudo-gradient vector field* of the locally Lipschitz $f : X \to \mathbb{R}$ with respect to a subset $B \subset X$ and a number $c \in \mathbb{R}$ is a locally Lipschitz mapping $\Lambda : N_\delta(B) \cap f^{-1}[c - \delta, c + \delta] \to X$ with some $\delta > 0$, such that $\|\Lambda(u)\| \leq 1$ and

$$\langle \zeta, \Lambda(u) \rangle > \frac{1}{2} \inf_{u \in \mathrm{dom}(\Lambda)} \lambda_f(u) > 0$$

for all $\zeta \in \partial_C f(u)$ and $u \in \mathrm{dom}(\Lambda) := N_\delta(B) \cap f^{-1}[c - \delta, c + \delta]$.

The existence of a generalized normalized pseudo-gradient vector field in the sense of Definition 4.3 is given by the result below.

Lemma 4.3 *Let $f : X \to \mathbb{R}$ be a locally Lipschitz function, $c \in \mathbb{R}$ and a closed subset B of X, such that $(PS)_{B,c}$ is satisfied together with $B \cap K_c(f) = \varnothing$ and $B \subset f^c$. Then there exists $\delta > 0$ and a generalized normalized pseudo-gradient vector field $\Lambda : N_\delta(B) \cap f^{-1}[c - \delta, c + \delta] \to X$ of f with respect to B and c.*

Proof Let us show that there exists a number $\delta > 0$ such that

$$\lambda_f(u) \geq \sigma > 0, \quad \forall u \in N_\delta(B) \cap f^{-1}[c - \delta, c + \delta], \tag{4.39}$$

with

$$\sigma := \inf \left\{ \lambda_f(u) : u \in N_\delta(B) \cap f^{-1}[c - \delta, c + \delta] \right\}.$$

Indeed, arguing by contradiction we assume that there exists a sequence $\{u_n\} \subset X$ with $\lambda_f(u_n) \to 0$, $\mathrm{dist}(u_n, B) \to 0$ and $f(u_n) \to c$. By $(PS)_{B,c}$ we derive the existence of a convergent subsequence of $\{u_n\}$, denoted again by $\{u_n\}$, such that $u_n \to u$ in X as $n \to \infty$. The lower semicontinuity of the function λ_f, yields $\lambda_f(u) \leq \liminf_{n \to \infty} \lambda_f(u_n) = 0$. We deduce that $u \in K_c(f)$ which contradicts the condition $B \cap K_c(I) = \varnothing$. The claim in (4.39) is verified.

Along the line of the proof of Lemma 4.1 and the property (4.39), we construct a locally Lipschitz map

$$\Lambda : N_\delta(B) \cap f^{c-\delta} \cap f_{c+\delta} \to X$$

such that

$$\|\Lambda(u)\| \leq 1 \tag{4.40}$$

and

$$\langle \zeta, \Lambda(u) \rangle > \frac{1}{2}\sigma, \quad \forall \zeta \in \partial_C f(u), \ u \in N_\delta(B) \cap f^{c-\delta} \cap f_{c+\delta}. \tag{4.41}$$

Now, it remains to make use of the usual partition of unity argument.   □

The following deformation result has been proved by Motreanu and Varga [10].

Theorem 4.3 *Let* $f : X \to \mathbb{R}$ *be a locally Lipschitz functional,* $c \in \mathbb{R}$ *and a closed subset* B *of* X *provided one has* $(PS)_{B,c}$, $B \cap K_c(f) = \varnothing$ *and* $B \subset f^c$. *Let* Λ *be a generalized normalized pseudo-gradient vector field of* f *with respect to* B *and* c. *Then for every* $\bar{\varepsilon} > 0$ *there exist an* $\varepsilon \in (0, \bar{\varepsilon})$ *and a number* $\delta < c$ *such that for each closed subset* A *of* X *with* $A \cap B = \varnothing$ *and* $A \subset f_{c-\varepsilon_A}$, *where*

$$\varepsilon_A := \min(\varepsilon, \varepsilon d(A, B)), \tag{4.42}$$

there exists a continuous mapping $\eta_A : \mathbb{R} \times X \to X$ with the following properties

(i) *$\eta_A(\cdot, u)$ is the solution of the vector field $V_A := -\varphi_A \Lambda$ with the initial condition $u \in X$ for some locally Lipschitz function $\varphi_A : X \to [0, 1]$ whose support is contained in the set $(X \setminus A)$;*

(ii) *$\eta_A(t, u) = u$, for all $t \in \mathbb{R}$ and $u \in A \cup f^{c-\bar{\varepsilon}} \cup f_{c+\bar{\varepsilon}}$;*

(iii) *for every $\delta \leq d \leq c$ one has $\eta_A(1, B \cap f^d) \subset f^{d-\varepsilon}$.*

Proof Note that the existence of a normalized generalized pseudo-gradient vector field $\Lambda : N_{3\delta_1}(B) \cap f^{-1}[c - 3\varepsilon_1, c + 3\varepsilon_1] \to X$ of f with respect to B and c is assured by Lemma 4.3, for some constants $\delta_1 > 0$ and $\varepsilon_1 > 0$. Consequently, a constant, $\sigma_1 > 0$ can be found such that

$$\langle \zeta, \Lambda(u) \rangle > \frac{1}{2}\sigma_1, \ \forall \zeta \in \partial_C f(u), \ u \in N_{3\delta_1}(B) \cap f_{c-3\varepsilon_1} \cap f^{c+3\varepsilon_1}. \tag{4.43}$$

We claim that the result of Theorem 4.3 holds for every $\varepsilon > 0$ with

$$\varepsilon < \min\left\{\bar{\varepsilon}, \varepsilon_1, \frac{1}{2}\sigma_1, \frac{1}{2}\sigma_1\delta_1\right\}. \tag{4.44}$$

In order to check the claim in (4.44) let us fix two locally Lipschitz functions $\varphi, \ \psi : X \to [0, 1]$ satisfying

$$\varphi = 1 \text{ on } N_{\delta_1}(B) \cap f^{c+\varepsilon_1} \cap f_{c-\varepsilon_1}; \ \varphi = 0 \text{ on } X \setminus (N_{2\delta_1}(B) \cap f^{c+2\varepsilon_1} \cap f_{c-2\varepsilon_1});$$

$$\psi = 0 \text{ on } f^{c-\bar{\varepsilon}} \cup f_{c+\bar{\varepsilon}}; \ \psi = 1 \text{ on } f^{c+\varepsilon_0} \cap f_{c-\varepsilon_0},$$

for some ε_0 with

$$\varepsilon < \varepsilon_0 < \min\{\bar{\varepsilon}, \varepsilon_1\}. \tag{4.45}$$

Then we are able to construct the locally Lipschitz vector field $V : X \to X$ by setting

$$V(u) := \begin{cases} -\delta_1\varphi(u)\psi(u)\Lambda(u), & \forall u \in N_{3\delta_1}(B) \cap f_{c-3\varepsilon_1} \cap f^{c+3\varepsilon_1}, \\ 0, & \text{otherwise.} \end{cases} \tag{4.46}$$

Using (4.46) we see that the vector field V is locally Lipschitz and bounded, namely

$$\|V(u)\| \leq \delta_1, \quad \forall u \in X. \tag{4.47}$$

From (4.43), (4.46) and (4.47) we derive

$$- \langle \zeta, V(u) \rangle \geq \frac{1}{2} \delta_1 \sigma_1, \ \forall u \in N_{\delta_1}(B) \cap f_{c-\varepsilon_0} \cap f^{c+\varepsilon_0}, \ \forall \zeta \in \partial_C f(u). \tag{4.48}$$

In view of (4.47) we may consider the *global flow* $\gamma : \mathbb{R} \times X \to X$ of V defined by (4.46), i.e.

$$\frac{d\gamma}{dt}(t, u) = V(\gamma(t, u)), \ \forall (t, u) \in \mathbb{R} \times X,$$

$$\gamma(0, u) = u, \ \forall u \in X.$$

In the next we set

$$B_1 := \gamma([0, 1]) \times B). \tag{4.49}$$

We notice that B_1 in (4.49) is a closed subset of X. To see this let $v_n := \gamma(t_n, u_n) \in B_1$ be a sequence with $t_n \in [0, 1]$, $u_n \in B$ and $v_n \to v$ in X. Passing to a subsequence we can suppose that $t_n \to t \in [0, 1]$ in $\mathbb{R}$. Putting $w_n := \gamma(t, u_n)$ we get

$$\|w_n - v_n\| = \|\gamma(t, u_n) - \gamma(t_n, u_n)\| = \left\| \int_{t_n}^t \frac{d}{dt} \gamma(\tau, u_n) d\tau \right\| \leq \delta_1 |t_n - t|,$$

where (4.47) has been used. Since $w_n \to v$ in X, it turns out that $u_n \to \gamma(-t, u) \in B$. Finally, we obtain $u = \gamma(t, \gamma(-t, u)) \in B_1$ which establishes that B_1 is indeed closed.

The next step is to justify that $f(\gamma(t, u))$ is a decreasing function of $t \in \mathbb{R}$, for each $u \in X$. Toward this, by applying Lebourg's mean value theorem and the chain rule for generalized gradients we infer for arbitrary real numbers $t > t_0$ the following inclusions

$$f(t, u) - f(t_0, u) \in \partial_C^1 (f(\gamma(t, u))) \Big|_{t=\tau} \subset \partial_C f(\gamma(\tau, u)) \frac{d\gamma}{dt}(\tau, u)(t - t_0)$$

$$= \partial_C f(\gamma(\tau, u)) V(\gamma(\tau, u))(t - t_0)$$

with some $\tau \in (t_0, t)$. By (4.43) and (4.46) we derive that $f(t, u) \leq f(t_0, u)$. Now we prove the relation

$$A \cap B_1 = \emptyset. \tag{4.50}$$

To check (4.50), we admit by contradiction that there exist $u_0 \in B$ and $t_0 \in [0, 1]$ provided $\gamma(t_0, u_0) \in A$. Since A and B are disjoint we have necessarily that $t_0 > 0$.

From the relations $A \subset f_{c-\varepsilon_A}$ and $B \subset f^c$ we deduce

$$c - \varepsilon_A \le f(\gamma(t_0, u_0)) \le f(\gamma(t, u_0)) \le f(u_0) \le c, \ \forall t \in [0, t_0]. \tag{4.51}$$

It turns out that

$$\gamma(t, u_0) \in N_{\delta_1}(B) \cap f^c \cap f_{c-\varepsilon_A}, \ \forall t \in [0, t_0].$$

On the other hand from (4.47) we infer the estimate

$$d(A, B) \le \|\gamma(t_0, u_0) - u_0\| = \left\| \int_0^{t_0} V(\gamma(s, u_0))ds \right\| \le \delta_1 t_0.$$

If we denote $h(t) := f(\gamma(t, u_0))$, then h is a locally Lipschitz function, and (4.46), (4.48) allow to write

$$h'(s) \le \max_{\zeta \in \partial_C f(\gamma(s,u))} \langle \zeta, \frac{d\gamma}{ds}(s, u) \rangle = \max_{\zeta \in \partial_C f(\gamma(s,u))} \langle \zeta, V(\gamma(s, u)) \rangle \le -\frac{1}{2}\delta_1 \sigma_1,$$

for a.e. $s \in [0, t_0]$. Therefore, by virtue of (4.44), we have the following estimate

$$f(\gamma(t_0, u_0)) - f(u_0) = h(t_0) - h(0) = \int_0^{t_0} h'(s)ds \le -\frac{1}{2}\delta_1 \sigma_1 t_0 < -\delta_1 \varepsilon t_0$$

$$\le -\varepsilon d(A, B) \le -\varepsilon_A. \tag{4.52}$$

The contradiction between (4.51) and (4.52) shows that the property (4.50) is actually true. Taking into account (4.50) there exists a locally Lipschitz function $\psi_A : X \to \mathbb{R}$ verifying $\psi_A = 0$ on a neighborhood of A and $\psi_A = 1$ on B_1. Then we define the homotopy $\eta_A : \mathbb{R} \times X \to X$ as being the global flow of the vector field $V_A = \psi_A V$. The assertion (i) is clear from the construction of η_A because one can take $\varphi_A = -\delta_1 \psi_A \varphi \psi$. Assertion (ii) follows easily because $V_A = 0$ on $A \cup f^{c-\bar{\varepsilon}} \cup f_{c+\bar{\varepsilon}}$. We show that (iii) is valid for $\delta = c + \varepsilon - \varepsilon_0$ with ε described in (4.44) and ε_0 in (4.45). To this end we argue by contradiction. Suppose that for some $d \in [\delta, c]$ there exists $u \in B \cap f^d$ such that

$$f(\eta_A(1, u)) > d - \varepsilon. \tag{4.53}$$

Using the fact that $\psi_A = 1$ on B_1 we deduce

$$\eta_A(t, u) = \gamma(t, u) \in N_{\delta_1}(B) \cap f^d \cap f_{d-\varepsilon}, \ \forall t \in [0, 1].$$

Then a reasoning similar to the one in (4.52) can be carried out to write

$$f(\eta_A(1, u)) - f(u) \leq -\frac{1}{2}\delta_1\sigma_1 < -\varepsilon.$$

This contradicts the relation (4.53) because $f(u) \leq d$.   □

Remark 4.2 Theorem 4.3 unifies different deformation results as for instance those in Chang [2], Du [5], Motreanu [9], Pucci and Serrin [12].

Corollary 4.1 (Chang [2]) *Let $f : X \to \mathbb{R}$ be a locally Lipschitz function which satisfies the (PS) condition. If c is not a critical value of f, i.e. $K_c(f) = \varnothing$, then given any $\bar{\varepsilon} > 0$ there exist an $\varepsilon \in (0, \bar{\varepsilon})$ and a homeomorphism $\eta : X \to X$ such that*

(i) $\eta(u) = u$ for all $u \in f^{c-\bar{\varepsilon}} \cup f_{c-\bar{\varepsilon}}$;
(ii) $\eta(f^{c+\varepsilon}) \subset f^{c-\varepsilon}$.

Proof Let us fix a positive number $\bar{a} < \bar{\varepsilon}$ such that the interval $[c - \bar{a}, c + \bar{a}]$ be without critical values of f. We apply Theorem 4.3 for $B_a := f^{c+a} \cap f_{c-a}$ and $c + a$ in place of B and c, respectively, for each $a \in (0, \bar{a}]$. Theorem 4.3 provides $\varepsilon_a > 0, \delta_a < c + a$ and, with $A := f_{c+\bar{\varepsilon}}$, the homotopy $\eta_a \in C(\mathbb{R} \times X, X)$ satisfying the requirements (i)-(iii) for $\varepsilon, \delta, \eta_A$ replaced by $\varepsilon_a, \delta_a, \eta_a$, respectively. Note that this claim holds because $f_{c+\bar{\varepsilon}} \subset f_{c-\varepsilon_{a,A}}$, where $\varepsilon_{a,A} := \min\{\varepsilon_a, \varepsilon_a d(A, B_a)\}$. Then $1°$ follows from (ii) of Theorem 4.3. The relations (4.43) and (4.44) show that $\varepsilon_{a,A}$ is bounded away from zero, say $\varepsilon_{a,A} \geq \bar{\varepsilon} > 0$ for $a \in (0, \bar{a}]$. Set $d := c + \frac{\min\{a, \bar{\varepsilon}\}}{2}$. We observe that if $a > 0$ is small enough, d can be used in (iii) of Theorem 4.3 relative to η_a, that is $\delta_a \leq d \leq c + a$, because ε_0 in (4.45) can be chosen independently of $a \in (0, \bar{a}]$. Then $2°$ is checked with $\eta(x) := \eta_a(1, u)$ for all $u \in X$ and $\varepsilon = \frac{\min\{a, \bar{\varepsilon}\}}{2}$ by means of property (iii) in Theorem 4.3 for B_a and $c + a$ in place B and c, respectively, with $a > 0$ sufficiently small. This occurs in view of the relations $c + \varepsilon = d$ and $d - \varepsilon_a \leq c_{\varepsilon}$, so one can conclude.   □

The following result extends Lemma 1.1 in Du [5] to the case of locally Lipschitz functions (see again Motreanu and Varga [11]).

Corollary 4.2 *Let $f : X \to \mathbb{R}$ be a locally Lipschitz function, let A and B be two closed disjoint subset of X and let $c \in \mathbb{R}$ such that $B \cap K_c(f) = \varnothing$, $B \subset f^c$, $A \subset f_c$ and f satisfies the $(PS)_{B,c}$ condition. Then there exist a number $\varepsilon > 0$ and a homeomorphism η of X such that*

(i) $f(\eta(u)) \leq f(u)$, $\forall u \in X$;
(ii) $\eta(u) = u$, $\forall u \in A$;
(iii) $\eta(B) \subset f^{c-\varepsilon}$.

Proof Apply Theorem 4.3 for the set B and the number c. One obtains an $\varepsilon > 0$ and $\eta := \eta_A(1, \cdot) \in C(X, X)$ corresponding to $A \subset f_c \subset f_{c-\varepsilon_A}$. It is obvious that the conclusion of Corollary 4.2 follows from Theorem 4.3, where (iii) is deduced for $c = d$.

$\square$

4.3 Deformations Without a Compactness Condition

In this section we establish a deformation result for locally Lipschitz functionals defined on B_R, which will be used the following to derive minimax theorems in nonsmooth critical point theory. In this section, unless otherwise stated, we always assume that

(H_0) X *is a smooth reflexive Banach space.*

If there is no danger of confusion we shall simply write B_R and S_R instead of $B_X(0, R)$ and $\partial B_X(0, R)$. Sometimes we shall denote $(-\infty, 0]$ $([0, \infty), (-\infty, 0), (0, \infty))$ by $\mathbb{R}_-$ $(\mathbb{R}_+, \mathbb{R}_-^*, \mathbb{R}_+^*)$, while $\mathbb{R}_-\xi := \{\alpha\xi : \alpha \in \mathbb{R}_-\}$. The closed convex hull of a set $A \subset X$ is denoted by $\overline{A}^{\pm co}$.

Let $\phi : [0, \infty) \to [0, \infty)$ be a given normalization function and denote by J_ϕ the corresponding duality mapping.

Note that the reflexivity of X implies that the weak- and weak*-topology on X coincide. Theorem C.1 and Corollaries C.1 and C.3 imply that J_ϕ is single-valued and the norm is Gâteaux differentiable on $X \setminus \{0\}$ and X^* is strictly convex. We also point out the fact that assumption (H_0) is not very restrictive as for any reflexive Banach space X with norm $\|\cdot\|$ there exists an equivalent norm $\|\cdot\|_0$ on X such that $(X, \|\cdot\|_0)$ and $(X^*, \|\cdot\|_{0*})$ are strictly convex (see e.g. Asplund [1]).

The following propositions will turn out useful in the subsequent sections.

Proposition 4.1 *Let* $f : X \to \mathbb{R}$ *be a locally Lipschitz functional. If* $u \in X$, $\{u_n\} \subset X$ *and* $\{\zeta_n\} \subset X^*$ *are such that* $u_n \to u$ *and* $\zeta_n \in \partial_C f(u_n)$, *for all* $n \in \mathbb{N}$, *then there exist* $\zeta \in \partial_C f(u)$ *and a subsequence* $\{\zeta_{n_k}\}$ *of* $\{\zeta_n\}$ *such that* $\zeta_{n_k} \rightharpoonup \zeta$ *in* X^*.

Proof The upper semicontinuity of $\partial_C f$ together with Proposition 2.4 ensures that there exists $n_0 \in \mathbb{N}$ such that

$$\partial_C f(u_n) \subset B_{X^*}(0, 2L_u), \quad \forall n \geq n_0,$$

with $L_u > 0$ the Lipschitz constant near u. Therefore $\{\zeta_n\}$ is a bounded sequence in X^*. Since X is reflexive, then X^* is also reflexive, hence the Eberlein-Šmulian theorem ensures that $\{\zeta_n\}$ possesses subsequence $\{\zeta_{n_k}\}$ such that $\zeta_{n_k} \rightharpoonup \zeta$, for some $\zeta \in X^*$. It follows at once that $\zeta \in \partial_C f(u)$ since $\partial_C f$ is weakly closed.

$\square$

Proposition 4.2 *Let* $\gamma : [0, \infty) \to X \setminus \{0\}$ *be a* C^1-*curve and* $\Phi(t) := \int_0^t \phi(s) \mathrm{d}s$. *Then*

$$\frac{\mathrm{d}}{\mathrm{d}t} \Phi(\|\gamma(t)\|) = \langle J_\phi \gamma(t), \gamma'(t) \rangle.$$

Proof Clearly, for all $t, s > 0$ the following relations hold

$$\langle J_\phi \gamma(t), \gamma(t) \rangle = \phi(\|\gamma(t)\|) \|\gamma(t)\|,$$

and

$$\langle J_\phi \gamma(t), \gamma(s) \rangle \le \phi(\|\gamma(t)\|) \|\gamma(s)\|,$$

hence by substraction we get

$$\langle J_\phi \gamma(t), \gamma(s) - \gamma(t) \rangle \le \phi(\|\gamma(t)\|) \left[\|\gamma(s)\| - \|\gamma(t)\| \right].$$

If $s > t$, then

$$\left\langle J_\phi \gamma(t), \frac{\gamma(s) - \gamma(t)}{s - t} \right\rangle \le \phi(\|\gamma(t)\|) \frac{\|\gamma(s)\| - \|\gamma(t)\|}{s - t},$$

and letting $s \downarrow t$ we get

$$\langle J_\phi \gamma(t), \gamma'(t) \rangle \le \phi(\|\gamma(t)\|) \frac{\mathrm{d}}{\mathrm{d}t} \|\gamma(t)\|.$$

For $s < t$ we get the converse inequality, hence

$$\langle J_\phi \gamma(t), \gamma'(t) \rangle = \phi(\|\gamma(t)\|) \frac{\mathrm{d}}{\mathrm{d}t} \|\gamma(t)\| = \frac{\mathrm{d}}{\mathrm{d}t} \Phi(\|\gamma(t)\|).$$

□

The following lemma ensures the existence of a locally Lipschitz vector field which plays the role of a pseudo-gradient field in the smooth case and will be used in the sequel.

Lemma 4.4 *Let* $f : X \to \mathbb{R}$ *be a locally Lipschitz functional and let* $F_0 \subset F \subset X$ *be such that*

(A) *there exists* $\gamma > 0$ *such that* $\lambda_f(u) \ge \gamma$, *for all* $u \in F$;
(B) *there exists* $\theta \in (0, 1)$ *such that*

$$0 \notin C(u, \theta), \quad \textit{for all } u \in F_0,$$

where $C(u, \theta) := \overline{[\partial_C f]_\theta(u) \cup J_\phi u}^{co}$ *and* $[\partial_C f]_\theta(u) := \partial_C f(u) + \theta \lambda_f(u) \overline{B}_{X^*}(0, 1)$.

Then there exists a locally Lipschitz vector field $\Lambda : F \to X$ such that

($P1$) $\|\Lambda(u)\| \leq 1$, *for all* $u \in F$;
($P2$) $\langle \zeta, \Lambda(u) \rangle > \theta\gamma/2$, *for all* $u \in F$ *and all* $\zeta \in \partial_C f(u)$;
($P3$) $\langle J_\phi u, \Lambda(u) \rangle > 0$, *for all* $u \in F_0$.

Proof Let $u \in F_0$ be fixed. The Krein-Šmulian theorem (see, e.g., Conway [3, V.13.4]) implies that the convex set $C(u, \theta)$ is weakly compact. Using the weak lower semicontinuity of the norm and assumption (B) we deduce that there exists $r_0 > 0$ such that $r_0 := \inf_{\xi \in C(u,\theta)} \|\xi\|$. Since $B_{X^*}(0, r_0) \cap C(u, \theta) = \varnothing$, the Hahn-Banach weak separation theorem implies that there exists $w_u \in \partial B_X(0, 1)$ and $\alpha \in \mathbb{R}$ such that

$$\langle \eta, w_u \rangle \leq \alpha \leq \langle \xi, w_u \rangle, \quad \forall \eta \in B_{X^*}(0, r_0), \ \forall \xi \in C(u, \theta).$$

Taking supremum with respect to η, we get

$$0 < r_0 \leq \langle \xi, w_u \rangle, \quad \forall \xi \in C(u, \theta). \tag{4.54}$$

In particular,

$$\langle J_\phi u, w_u \rangle > 0. \tag{4.55}$$

We claim that

$$\langle \zeta, w_u \rangle > \theta\gamma/2, \quad \forall \zeta \in \partial_C f(u). \tag{4.56}$$

Recall that $\langle \zeta, w_u \rangle = d(\zeta, \ker w_u)$ (see, e.g., Costara and Popa [4, p. 87]), where by $\ker w_u$ we have denoted the following subset of X^*

$$\ker w_u := \left\{ \xi \in X^* : \langle \xi, w_u \rangle = 0 \right\}.$$

Therefore, it suffices to prove that $d(\partial_C f(u), \ker w_u) > \theta\gamma/2$. Let $\eta \in \ker w_u$ be fixed. Obviously $\eta \notin [\partial_C f]_\theta(u)$, otherwise η would belong to $C(u, \theta)$ and (4.54) would be violated. By the definition of $[\partial_C f]_\theta(u)$, we have $d(\eta, \partial_C f(u)) \geq \theta\lambda_f(u)$. Since η was arbitrarily chosen it follows that

$$d(\partial_C f(u), \ker w_u) \geq \theta\lambda_f(u) > \theta\gamma/2.$$

We prove next that there exists $r_u > 0$ such that

$$\langle \zeta, w_u \rangle > \theta\gamma/2, \quad \forall v \in B_X(u, r_u) \cap F, \ \forall \zeta \in \partial_C f(v), \tag{4.57}$$

and

$$\langle J_\phi v, w_u \rangle > 0, \quad \forall v \in B_X(u, r_u). \tag{4.58}$$

Arguing by contradiction, assume that (4.57) does not hold, i.e. for each $r > 0$ there exist $v \in B_X(u, r) \cap F$ and $\zeta \in \partial_C f(v)$ such that

$$\langle \zeta, w_u \rangle \leq \theta \gamma / 2.$$

Taking $r = 1/n$ we obtain the existence of two sequences $\{v_n\} \subset X$ and $\{\zeta_n\} \subset X^*$ such that

$$v_n \to u, \ \zeta_n \in \partial_C f(v_n) \text{ and } \langle \zeta_n, w_u \rangle \leq \theta \gamma / 2.$$

According to Proposition 4.1 there exists $\zeta_0 \in \partial_C f(u)$ such that, up to a subsequence,

$$\zeta_n \rightharpoonup \zeta_0, \text{ in } X^*.$$

Letting $n \to \infty$ we get $\langle \zeta_0, w_u \rangle \leq \theta \gamma / 2$ which contradicts (4.56). Relation (4.58) may be proved in a similar manner by using Proposition C.7 which asserts that J_ϕ is demicontinuous on reflexive Banach spaces.

If $u \in F \setminus F_0$, we can employ a similar argument as above with $\partial_C f(u)$ instead of $C(u, \theta)$ to get the existence of an element $w_u \in \partial B_X(0, 1)$ such that (4.56) holds.

Thus, the family $\{B_X(u, r_u)\}_{u \in F}$ is an open covering of F and it is paracompact, hence it possesses a locally finite refinement say $\{U_\alpha\}_{\alpha \in I}$. Standard arguments ensure the existence of a locally Lipschitz partition of unity, denoted $\{\rho_\alpha\}_{\alpha \in I}$, subordinated to the covering $\{U_\alpha\}_{\alpha \in I}$. The required locally Lipschitz vector field $\Lambda : F \to X$ can now be defined by

$$\Lambda(u) := \sum_{\alpha \in I} \rho_\alpha(u) w_u.$$

Simple computations show Λ satisfies the required conditions.   □

The following proposition provides an equivalent form of condition (B) in the previous lemma, which will be useful the following sections.

Proposition 4.3 *Let $u \in X \setminus \{0\}$ and $\theta \in (0, 1)$ be fixed. Then the following statements are equivalent:*

(*i*) $0 \notin C(u, \theta)$;
(*ii*) $\mathbb{R}_- J_\phi u \cap [\partial_C f]_\theta(u) = \varnothing$.

Proof $(i) \Rightarrow (ii)$ Arguing by contradiction, assume there exist $\alpha \in \mathbb{R}_-$ and $\xi \in [\partial_C f]_\theta(u)$ such that $\xi = \alpha J_\phi u$. Then, for $t := \frac{1}{1-\alpha} \in (0, 1]$ we get

$$0 = \frac{1}{1-\alpha}(-\alpha J_\phi u + \xi) = (1-t)J_\phi u + t\xi,$$

which shows that $0 \in C(u, \theta)$, contradicting (i).

$(ii) \Rightarrow (i)$ Assume by contradiction that $0 \in C(u, \theta)$. Then there exist $t_n \in [0, 1]$ and $\xi_n \in [\partial_C f]_\theta(u)$ such that

$$\rho_n := (1-t_n)J_\phi u + t_n \xi_n \to 0, \text{ as } n \to \infty.$$

Since $\{t_n\}$ is a bounded sequence in $\mathbb{R}$, it follows that it possesses a subsequence $\{t_{n_k}\}$ such that

$$t_{n_k} \to t \in [0, 1].$$

Obviously the set $[\partial_C f]_\theta(u)$ is bounded, hence if $t = 0$, then $t_{n_k}\xi_{n_k} \to 0$. Thus $\rho_{n_k} \to J_\phi u$ and the uniqueness of the limit leads to $J_\phi u = 0$ which is a contradiction, as $u \neq 0$. If $t \in (0, 1]$, then

$$\xi_{n_k} = \frac{1}{t_{n_k}}\rho_{n_k} + \frac{t_{n_k}-1}{t_{n_k}}J_\phi u \to \frac{t-1}{t}J_\phi u \in \mathbb{R}_- J_\phi u.$$

Since $[\partial_C f]_\theta(u)$ is also closed, it follows that $\frac{t-1}{t}J_\phi u \in [\partial_C f]_\theta(u)$, but this contradicts (ii).   □

We are now in position to prove the main result of this section which is given by the following deformation theorem. The set $Z \subset \overline{B}_R$ in the statement may be regarded as a "restriction" set that allows us to control the deformation. The reader may think of $Z = \overline{B}_R$ as the "unrestricted" case. Here and hereafter in this section, if $f : \overline{B}_R \to \mathbb{R}$ is a functional and Z is a subset of $\overline{B}_R$, we adopt the following notations

$$f^a := \left\{ u \in \overline{B}_R : f(u) \le a \right\},$$

and

$$Z_b := \left\{ u \in \overline{B}_R : d(u, Z) \le b \right\}.$$

Theorem 4.4 *Let $f : \overline{B}_R \to \mathbb{R}$ be a locally Lipschitz and $Z \subset \overline{B}_R$. Assume that there exist $c, \rho \in \mathbb{R}$, $\delta > 0$ and $\theta \in (0, 1)$ such that the following conditions hold:*

(H_1) $\lambda_f(u) \geq \frac{4\delta}{\rho\theta^2}$, on $\left\{u \in \overline{B}_R : |f(u) - c| \leq 3\delta\right\} \cap Z_{3\rho}$;
(H_2) $0 \notin C(u, \theta)$, on $\{u \in S_R : |f(u) - c| \leq 3\delta\} \cap Z_{3\rho}$.

Then there exists a continuous deformation $\sigma : [0, 1] \times \overline{B}_R \to \overline{B}_R$ such that:

   (i) $\sigma(0, \cdot) = id$;
  (ii) $\sigma(t, \cdot) : \overline{B}_R \to \overline{B}_R$ *is a homeomorphism for all $t \in [0, 1]$*;
 (iii) $\sigma(t, u) = id$, *for all $u \in \overline{B}_R \setminus \left\{u \in \overline{B}_R : d(u, Z) \leq 2\rho, |f(u) - c| \leq 2\delta\right\}$*;
 (iv) *The function $f(\sigma(\cdot, u))$ is nonincreasing for all $u \in \overline{B}_R$. Moreover, $f(\sigma(t, u)) < f(u)$, whenever $\sigma(t, u) \neq u$*;
  (v) $\|\sigma(t_1, u) - \sigma(t_2, u)\| \leq \rho\theta|t_1 - t_2|$ *for all $t_1, t_2 \in [0, 1]$*;
 (vi) $\sigma\left(1, f^{c+\delta} \cap Z\right) \subseteq f^{c-\delta} \cap Z_\rho$.

Proof Let us define the following subsets of $\overline{B}_R$ as follows

$$F := \left\{u \in \overline{B}_R : \lambda_f(u) \geq \frac{4\delta}{\rho\theta^2}\right\}, \quad F_0 := \{u \in S_R : d(u, Z) \leq 3\rho, |f(u) - c| \leq 3\delta\},$$

$$F_1 := \left\{u \in \overline{B}_R : d(u, Z) \leq 2\rho, |f(u) - c| \leq 2\delta\right\},$$

$$F_2 := \left\{u \in \overline{B}_R : d(u, Z) \leq \rho, |f(u) - c| \leq \delta\right\},$$

and consider the locally Lipschitz function $\chi : \overline{B}_R \to \mathbb{R}$ defined as

$$\chi(u) := \frac{d(u, \overline{B}_R \setminus F_1)}{d(u, \overline{B}_R \setminus F_1) + d(u, F_2)}.$$

Obviously $\chi \equiv 0$ on $\overline{B}_R \setminus F_1$, whereas $\chi \equiv 1$ on F_2 and $0 < \chi < 1$ in-between. Applying Lemma 4.4 with F and F_0 defined as above, we get the existence of a locally Lipschitz vector field $\Lambda : F \to X$ having the properties $(P1)$–$(P3)$. Using the cutoff function we define $V : \overline{B}_R \to X$ to be given by

$$V(u) := \begin{cases} -\chi(u)\Lambda(u), & \text{if } u \in F, \\ 0, & \text{otherwise.} \end{cases}$$

Then V can be extended to a locally Lipschitz and globally bounded map defined on the whole X by setting

$$V(u) = V\left(\frac{R}{\|u\|}u\right), \quad \text{whenever } \|u\| > R.$$

By an extended version of the Picard–Lindelöf existence theorem for Banach spaces (see, e.g., [13, Lemma 2.11.1]) the initial value problem

$$\begin{cases} \dfrac{d}{dt}\eta(t, u) = V(\eta(t, u)), \\[2mm] \eta(0, u) = u. \end{cases}$$

possesses a unique maximal solution $\eta : \mathbb{R} \times X \to X$. We define the required deformation via time dilation,

$$\sigma(t, \cdot) := \eta(\rho\theta t, \cdot), \quad \forall t \in \mathbb{R}.$$

The initial value ensures that $\sigma(0, \cdot) = id$, thus establishing (i). It follows from the aforementioned result that $\sigma(t, \cdot) : X \to X$ is a homeomorphism (with inverse $\sigma(t, \cdot)^{-1} = \sigma(-t, \cdot)$). For convenience, we denote by $\sigma_u : X \to X$, the *orbit* defined by $\sigma_u(t) := \sigma(t, u)$, for all $(t, u) \in \mathbb{R} \times X$.

We claim that, for each $u \in \overline{B}_R$, the orbit $\{\sigma_u(t)\}_{t \geq 0}$ lies entirely in $\overline{B}_R$. In order to check this, assume that $T_0 \geq 0$ is such that

$$u_1 := \sigma_u(T_0) \in S_R,$$

and

$$\|\sigma_u(t)\| \leq R, \quad \forall t \in [0, T_0).$$

By Proposition 4.2 we have

$$\frac{d}{dt}\Phi\left(\|\sigma_u(t)\|\right) = \rho\theta\left\langle J_\phi\sigma_u(t), V(\sigma_u(t))\right\rangle, \tag{4.59}$$

and

$$\left\langle J_\phi\sigma_u(t), V(\sigma_u(t))\right\rangle = \begin{cases} -\chi(\sigma_u(t))\langle J_\phi\sigma_u(t), \Lambda(\sigma_u(t))\rangle, & \text{if } \sigma_u(t) \in F, \\ 0, & \text{otherwise}, \end{cases} \tag{4.60}$$

whenever $\sigma_u(t) \neq 0$.

If $u_1 \in F_0$, then $\langle J_\phi u_1, \Lambda(u_1) \rangle > 0$, hence there exists a neighborhood U of u_1 such that

$$\langle J_\phi v, \Lambda(v) \rangle > 0, \quad \forall v \in U \cap F. \tag{4.61}$$

The continuity of $\sigma_u(\cdot)$ and relations (4.59)–(4.61) ensure that

$$\frac{\mathrm{d}}{\mathrm{d}t} \Phi(\|\sigma_u(t)\|) \leq 0,$$

holds in a neighborhood $[T_0, T_0 + s)$ of T_0.

If $u_1 \notin F_0$, then V vanishes in a neighborhood of u_1 and by a similar reasoning we obtain

$$\frac{\mathrm{d}}{\mathrm{d}t} \Phi(\|\sigma_u(t)\|) = 0, \quad \forall t \in [T_0, T_0 + s).$$

Thus $\Phi(\|\sigma_u(\cdot)\|)$ is nonincreasing in $[T_0, T_0 + s)$, while $\Phi(\cdot)$ is strictly increasing on $\mathbb{R}_+$, hence $\|\sigma_u(t)\| \leq R$ for all $t \in [T_0, T_0 + s)$. The argument can be repeated whenever $\{\sigma_u(t)\}_{t \geq 0}$ reaches S_R.

Henceforth we restrict σ to $[0, 1] \times \overline{B}_R$, without changing the notation. It is clear from above that $\sigma(t, \cdot)$ is a homeomorphism for all $t \in [0, 1]$ and $\chi \equiv 0$ on $\overline{B}_R \setminus F_1$, therefore (ii) and (iii) hold.

In order to prove (iv), fix $u \in \overline{B}_R$ and define $h : [0, 1] \to \mathbb{R}$ by $h(t) := f(\sigma_u(t))$. Then, by Proposition 2.7 h is differentiable almost everywhere and for a.e. $s \in [0, 1]$ we have

$$h'(s) \leq \max_{\zeta \in \partial_C f(\sigma_u(s))} \langle \zeta, \sigma'_u(s) \rangle = \max_{\zeta \in \partial_C f(\sigma_u(s))} \rho\theta \langle \zeta, V(\sigma_u(s)) \rangle.$$

Since Λ satisfies property $(P2)$ and χ vanishes on $\overline{B}_R \setminus F_1$, we get $h'(s) \leq 0$ if $\sigma_u(s) \in \overline{B}_R \setminus F_1$ and

$$h'(s) \leq -\rho\theta\chi(\sigma_u(s))\langle \zeta, \Lambda(\sigma_u(s)) \rangle \leq -\rho\theta\chi(\sigma_u(s))\frac{\theta}{2}\frac{4\delta}{\rho\theta^2} = -2\delta\chi(\sigma_u(s)),$$

otherwise. This shows that $f(\sigma_u(\cdot))$ is nonincreasing.

If $\sigma_u(t) \neq u$, then $t > 0$ and $\sigma_u(t) \notin \overline{B}_R \setminus F_1$. Therefore there exists $\epsilon > 0$ such that $\sigma_u(s) \notin \overline{B}_R \setminus F_1$ for all $s \in (t - \epsilon, t + \epsilon)$. Thus $\chi(\sigma_u(s)) > 0$ for all $s \in (t - \epsilon, t)$ and

$$f(\sigma_u(t)) - f(u) = f(\sigma_u(t)) - f(\sigma_u(0)) = \int_0^t h'(s)\mathrm{d}s < 0.$$

For a fixed $u \in \overline{B}_R$ and $0 \le t_1 < t_2 \le 1$ we have

$$\|\sigma_u(t_2) - \sigma_u(t_1)\| = \left\| \int_{t_1}^{t_2} \sigma_u'(s)ds \right\| \le \rho\theta \int_{t_1}^{t_2} \|V(\sigma_u(s))\|ds \le \rho\theta(t_2 - t_1),$$

which shows that (v) holds. Moreover, if $u \in Z$, then $\|\sigma_u(t) - u\| \le \rho\theta t < \rho$, hence $\sigma_u(t) \in Z_\rho$, for all $t \in [0, 1]$.

Finally, in order to complete the proof it suffices to show that for any $u \in Z \subset \overline{B}_R$ such that $f(u) \le c + \delta$ we have $f(\sigma_u(1)) \le c - \delta$. We distinguish two cases:

(a) $f(u) \le c - \delta$. Then

$$f(\sigma_u(1)) \le f(\sigma_u(0)) = f(u) \le c - \delta.$$

(b) $c - \delta < f(u) \le c + \delta$. Then $u \in F_2$. Let $t_{max} \in [0, 1]$ be the maximal time for which the $\sigma_u(\cdot)$ does not exit F_2, i.e.,

$$\sigma_u(t) \in F_2 \text{ for } t \in [0, t_{max}].$$

If $t_{max} = 1$, then $\chi(\sigma_u(s)) = 1$ for all $s \in [0, 1]$ and

$$f(\sigma_u(1)) - f(u) = \int_0^1 h'(s)ds \le \int_0^1 -2\delta\chi(\sigma_u(s))ds = -2\delta,$$

which leads to

$$f(\sigma_u(1)) \le f(u) - 2\delta \le c + \delta - 2\delta = c - \delta.$$

If $t_{max} < 1$, then there exists $t_0 \in (t_{max}, 1]$ such that $\sigma_u(t_0) \notin F_2$. Since $\sigma_u(t_0) \in Z_\rho$, it follows that either $f(\sigma_u(t_0)) < c - \delta$, or $f(\sigma_u(t_0)) > c + \delta$. The latter cannot occur due to (i) and (iv).   □

4.4   A Deformation Lemma for Szulkin Functionals

In this section we present a deformation result for Szulkin type functionals, see [14].

As in the Sect. 3.2, let X be a real Banach space and I a function on X satisfying the following structure hypothesis:

(H) $f := \varphi + \psi$, where $\varphi \in C^1(X, \mathbb{R})$ and $\psi : X \to (-\infty, +\infty]$ is convex, proper and l.s.c.

Lemma 4.5 *Suppose that f satisfies (H) and $(PS)_c$ and let N be a neighbourhood of K_c. Then for each $\overline{\varepsilon} > 0$ there exists an $\varepsilon \in (0, \overline{\varepsilon})$ such that if $u_0 \notin N$ and $c - \varepsilon \leq f(u_0) \leq c + \varepsilon$, then*

$$\langle \varphi'(u_0), v_0 - u_0 \rangle + \psi(v_0) - \psi(u_0) \leq -3\varepsilon \|v_0 - u_0\| \tag{4.62}$$

for some $v_0 \in X$.

Proof If the conclusion is false, there exists a sequence $\{u_n\} \subset X \setminus N$ such that $f(u_n) \to c$ and

$$\langle \varphi'(u_n), v - u_n \rangle + \psi(v) - \psi(u_n) \geq -\frac{1}{n}\|v - u_n\|, \quad \forall v \in X.$$

Thus, by $(PS)_c$ and Proposition 3.5, a subsequence of $\{u_n\}$ converges to $u \in K_c$. This contradicts the fact that $u_n \notin N$ for every $n \in \mathbb{N}$ and N is a neighborhood of K_c. $\qquad\square$

Lemma 4.6 *Suppose that f satisfies (H) and (PS). Let N be a neighborhood of K_c. Let $\varepsilon > 0$ be such that the assertion of Lemma 4.5 is satisfied. Then for every $u_0 \in f^{c+\varepsilon} \setminus N$, there exists $v_0 \in X$ and an open neighborhood U_0 of v_0 such that*

$$\langle \varphi'(u), v_0 - u \rangle + \psi(v_0) - \psi(u) \leq \|v_0 - u\| \tag{4.63}$$

for all $u \in U_0$,

$$\langle \varphi'(u), v_0 - u \rangle + \psi(v_0) - \psi(u) \leq -3\varepsilon \|v_0 - u\| \tag{4.64}$$

for all $u \in U_0 \cap f_{c-\varepsilon}$. Moreover, if $u_0 \in K$ we can take $v_0 := u_0$ and if $u_0 \notin K$, v_0, U_0 and a number $\delta_0 > 0$ can be chosen so that $v_0 \notin \overline{U_0}$ and

$$\langle \varphi'(u), v_0 - u \rangle + \psi(v_0) - \psi(u) \leq -\delta_0 \|v_0 - u\|, \quad \forall u \in U_0. \tag{4.65}$$

Proof We distinguish two cases: (i) $u_0 \in K$ and (ii) $u_0 \notin K$.

(i) From the definition of the critical point of the function $f = \varphi + \psi$ follows that

$$\langle \varphi'(u_0), u - u_0 \rangle + \psi(u) - \psi(u_0) \geq 0$$

for all $u \in X$. We now choose a small neighbourhood U_0 of u_0 such that

$$\langle \varphi'(u), u_0 - u \rangle + \psi(u_0) - \psi(u) \leq \langle \varphi'(u) - \varphi'(u_0), u_0 - u \rangle + \psi(u) - \psi(u_0)$$

$$\leq \|\varphi'(u) - \varphi'(u_0)\| \cdot \|u_0 - u\| \leq \|u_0 - u\|$$

for all $u \in U_0$. This show that (4.63) is satisfied with $v_0 := u_0$. We now observe that if $c - \varepsilon \leq f(u) \leq c + \varepsilon$ and $u \in K$, then by Lemma 4.5, $u \in N$. Since $u_0 \notin N$, we must have $f(u_0) < c - \varepsilon$. If $f(u) < c - \varepsilon$ in some neighborhood of u_0, we may choose U_0 contained in this neighborhood. Therefore the condition (4.64) is empty. If every neighborhood of u_0 contains a point u at which $f(u) \geq c - \varepsilon$ we easily check, using the continuity of φ, that $\psi(u) - \psi(u_0) \geq d > 0$ for some constant d and u sufficiently close to u_0 and satisfying $f(u) \geq c - \varepsilon$. This means that if U_0 is sufficiently small neighborhood of u_0, then

$$\langle \varphi'(u), u_0 - u \rangle + \psi(u_0) - \psi(u) \leq \|\varphi'(u)\| \cdot \|u_0 - u\| - d \leq -3\varepsilon \|u_0 - u\|$$

for all $u \in U_0$ such that $f(u) \geq c - \varepsilon$.

(ii) First we suppose that $f(u_0) < c - \varepsilon$. Since u_0 is not a critical point of f, there exists $v_0 \in X$ such that

$$\langle \varphi'(u_0), v_0 - u_0 \rangle + \psi(v_0) - \psi(u_0) < 0. \tag{4.66}$$

Letting $w_0 = tv_0 + (1 - t)u_0, 0 < t < 1$, we get by the convexity of ψ that

$$\langle \varphi'(u_0), w_0 - u_0 \rangle + \psi(w_0) - \psi(u_0) \leq$$

$$\leq t \left(\langle \varphi'(u_0), v_0 - u_0 \rangle + \psi(v_0) - \psi(u_0) \right) < 0.$$

Hence we may assume that v_0 is close to u_0. As in the Case (i) we show that there exists $d > 0$ such that $\psi(u) - \psi(u_0) \geq d > 0$ for all u sufficiently close to u_0 and such that $f(u) \geq c - \varepsilon$. It then follows from (4.66) that if U_0 and $\|v_0 - u_0\|$ are sufficiently small then

$$(\psi(v_0) - \psi(u_0)) + (\psi(u_0) - \psi(u)) \leq \frac{d}{2} - d = -\frac{d}{2}$$

and

$$\langle \varphi'(u), v_0 - u \rangle + \psi(v_0) - \psi(u) \leq \|\varphi'(u)\| \cdot \|v_0 - u\| - \frac{d}{2} \leq -3\varepsilon \|v_0 - u\|$$

for all $u \in U_0$ with $f(u) \geq c - \varepsilon$. This means that (4.64) holds. Since $v_0 \neq u_0$, we may assume that $v_0 \notin \overline{U}_0$ and moreover U_0 can be chosen smaller, if necessary, to ensure that the inequality (4.66) remains true in U_0, that is

$$\langle \varphi'(u), v_0 - u \rangle + \psi(v_0) - \psi(u) \leq -\delta_0 \|v_0 - u\|$$

for all $u \in U_0$, thus (4.65) is satisfied. It now remains to consider the case $f(u_0) \geq c - \varepsilon$. We can apply Lemma 4.5 in order to obtain the existence of v_0 such that the inequality from this lemma is satisfied. By the continuity of φ' and the lower semicontinuity of ψ we can extend this inequality to a suitably small neighborhood of U_0 of u_0, with $v_0 \notin \overline{U}_0$, that is

$$\langle \varphi'(u), v_0 - u \rangle + \psi(v_0) - \psi(u) \leq -3\varepsilon \|v_0 - u\|$$

for all $u \in U_0$. This shows that (4.65) holds.   □

A family of mappings $\alpha(\cdot, s) \equiv \alpha_s : W \to X, 0 \leq s \leq s_0, s_0 > 0$, is said to be a *deformation* if $\alpha \in C(W \times [0, s_0], X)$ and $\alpha_0 = id$ (*id* identity on W).

Lemma 4.7 (Szulkin Deformation Lemma) *Suppose that f satisfies (H) and the (PS) condition and let N be a neighborhood of K_c. Then for each $\overline{\varepsilon} > 0$ there exists $\varepsilon \in (0, \overline{\varepsilon})$ such that for each compact subset A of $X \setminus N$ satisfying*

$$c \leq \sup_{u \in A} f(u) \leq c + \varepsilon,$$

we can find a closed set W, with $A \subset \operatorname{int}(W)$ and a deformation $\alpha_s : W \to X, 0 \leq s \leq s_0$, having the following properties

$$\|u - \alpha_s(u)\| \leq s, \ \forall u \in W, \tag{4.67}$$

$$f(\alpha_s(u)) - f(u) \leq 2s, \ \forall u \in W, \tag{4.68}$$

$$f(\alpha_s(u)) - f(u) \leq -2\varepsilon s, \ \forall u \in W, \ f(u) \geq c - \varepsilon \tag{4.69}$$

and

$$\sup_{u \in A} f(\alpha_s(u)) - \sup_{u \in A} f(u) \leq -2\varepsilon s. \tag{4.70}$$

Moreover, if W_0 is a closed set such that $W_0 \cap K = \varnothing$, then W and α_s can be chosen so that

$$f(\alpha_s(u)) - f(u) \leq 0, \quad \forall u \in W \cap W_0. \tag{4.71}$$

Proof By Lemma 4.6 there exists $\varepsilon \in (0, \overline{\varepsilon})$ such that for each $u_0 \in A$ there correspond a neighborhood of U_0 satisfying conditions stated in that lemma. If $u_0 \in K$ we may always assume that $U_0 \cap W_0 = \varnothing$. The sets U_0 corresponding to $u_0 \in A$ form a covering of a compact A. Let $\{U_i\}_{i \in J}$ be a finite subcovering. Let $\{u_i\}$ and $\{v_i\}$ be points corresponding to U_i from Lemma 4.6. By taking a suitable refinement, if necessary, we may always assume that if a $i_0 \in J$ and $u_{i_0} \in K$, then $\text{dist}(u_{i_0}, U_i) > 0$ for each $i \neq i_0$. Let ρ_i be a continuous function such that $\rho_i(u) > 0$ for $u \in U_i$ and $\rho_i(u) = 0$ for $u \notin U_i$. We set

$$\sigma_i(u) := \frac{\rho_i(u)}{\sum_{j \in J} \rho_j(u)}$$

$u \in V = \cup_{j \in J} U_j$ and define a deformation mapping α_s as follows: if $u_{i_0} \in A \cap K$, then

$$\alpha_s(u) := \begin{cases} u + s\frac{u_{i_0} - u}{\|u_{i_0} - u\|}, & \text{for } 0 \leq s < \|u_{i_0} - u\|, \ u \in U_{i_0} \setminus \cup_{i \neq i_0} U_i \\ u_{i_0}, & \text{for } s \geq \|u_{i_0} - u\|, \ u \in U_{i_0} \setminus \cup_{i \neq i_0} U_i \end{cases}$$

and in all other cases

$$\alpha_s(u) := u + s \sum_{i \in J} \sigma_i(u) \frac{v_i - u}{\|v_i - u\|}.$$

It is easy to check that α_s is well defined and continuous for sufficiently small s. It is clear that $\alpha_0 = id$. To check the remaining properties of α_s we write

$$f(\alpha_s(u)) = f(u + sw) = \varphi(u) + s\langle \varphi'(u), w \rangle + r(s) + \psi(u + sw), \qquad (4.72)$$

where

$$\alpha_s(u) = u + sw \quad \text{and} \quad |r(s)| \leq s \sup_{0 \leq t \leq s} \|\varphi'(u + tw) - \varphi'(w)\|.$$

We choose $\delta > 0$ such that

$$0 < 3\delta < \min\{1, \varepsilon, \delta_i\},$$

where $\delta_i > 0$ corresponds to U_i from the relation (4.65). Since A is compact, there exists a closed set W, with $A \subset \text{int}(W)$ and $\bar{s} > 0$ such that $|r(s)| \leq \delta s$ for all $0 < s \leq \bar{s}$, $u \in W$ and $w \in X$ with $\|w\| \leq 1$. If $u \notin U_{i_0} \setminus \cup_{i \neq i_0} U_i$, then

$$\alpha_s(u) = u + sw = \left(1 - s\sum_{i \in J} \sigma_i(u)\|v_i - u\|^{-1}\right) u + s\sum_{i \in J} \sigma_i(u)\|v_i - u\|^{-1} v_i.$$

For s sufficiently small we have

$$0 \leq s\sum_{i \in J} \sigma_i(u)\|v_i - u\|^{-1} \leq 1$$

and using the convexity of ψ we deduce from (4.72) that

$$f(\alpha_s(u)) \leq \varphi(u) + s \sum_{i \in J} \frac{\sigma_i(u)}{\|v_i - u\|} \langle \varphi'(u), v_i - u \rangle + \delta s +$$

$$+ \left(1 - s \sum_{i \in J} \frac{\sigma_i(u)}{\|v_i - u\|}\right) \psi(u) + s \sum_{i \in J} \frac{\sigma_i(u)}{\|v_i - u\|} \psi(v_i) =$$

$$= f(u) + s \sum_{i \in J} \frac{\sigma_i(u)}{\|v_i - u\|} \left(\langle \varphi'(u), v_i - u \rangle + \psi(v_i) - \psi(u) \right) + \delta s.$$

According to (4.63) each term in the last summation is less than or equal to $\sigma_i(u)$, hence

$$f(\alpha_s(u)) \leq f(u) + s + \delta s \qquad (4.73)$$

and (4.68) holds. In a similar manner we show, using (4.64) and (4.65), that

$$f(\alpha_s(u)) \leq f(u) - 3\varepsilon s + \delta s \qquad (4.74)$$

for all $u \in W$ with $f(u) \geq c - \varepsilon$ and

$$f(\alpha_s(u)) \leq f(u) - 3\delta s + \delta s \qquad (4.75)$$

for all $u \in W \cap W_0$. Suppose that $u \in U_{i_0} \setminus \cup_{i \neq i_0} U_i$. We have

$$\alpha_s(u) = u + sw = \left(1 - s\|u_{i_0} - u\|^{-1}\right) u + s\|u_{i_0} - u\|^{-1} u_{i_0}$$

for $s < \|u_{i_0} - u\| = \bar{s}$. In this case we repeat the previous part of the proof to show (4.68) and (4.69). On the other hand, if $s \geq \bar{s}$, then

$$f(\alpha_s(u)) = f(\alpha_{\bar{s}}(u)) \leq f(u) + \bar{s} + \delta \bar{s} \leq f(u) + 2s$$

and $f(\alpha_s(u)) = f(u_{i_0}) < c - \varepsilon$. This means that (4.68) and (4.69) hold for small s. The inequality (4.71) follows from (4.75) if $u \notin U_{i_0} \setminus \cup_{i \neq i_0} U_i$. If $u \in U_{i_0} \setminus \cup_{i \neq i_0} U_i$, then $u \in U_{i_0}$ and $u_{i_0} \in K$. Hence $U_{i_0} \cap W_0 = \emptyset$ and $u \notin W \cap W_0$. Finally, to show (4.70) let us first assume that $\sup_{u \in A} f(\alpha_s(u)) \leq c - \frac{\varepsilon}{2}$. Then taking $s \leq \frac{1}{4}$ we get (4.70) since $\sup_{u \in A} f(u) \geq c$. On the other hand if $\sup_{u \in A} f(\alpha_s(u)) > c - \frac{\varepsilon}{2}$, we set

$$B := \{u \in A : f(u) > c - \varepsilon\}.$$

It follows from (4.68) that

$$\sup_{u \in A} f(\alpha_s(u)) = \sup_{u \in B} f(\alpha_s(u))$$

for s small, say $s \le \frac{\varepsilon}{4}$. This combined with (4.69) implies that

$$\sup_{u \in A} f(\alpha_s(u)) - \sup_{u \in A} f(u) = \sup_{u \in B} f(\alpha_s(u)) - \sup_{u \in B} f(u) \tag{4.76}$$

$$\le \sup_{u \in B}(f(\alpha_s(u)) - f(u)) \le -2\varepsilon s.$$

□

Corollary 4.3 *Suppose that φ and ψ are even and that A is symmetric. Then α_s is odd.*

Proof We may assume that W is symmetric. We define

$$\beta_s(u) := \frac{1}{2}(\alpha_s(u) - \alpha_s(-u)),$$

then β_s is odd and satisfies (4.67). Writing $\alpha_s(u) = u + h_s(u)$, we have by Taylor's formula

$$f(\beta_s(u)) = \varphi(u) + \frac{1}{2}\langle \varphi'(u), h_s(u) - h_s(-u) \rangle + r_1(s) + \psi\left(\frac{(u + h_s(u)) + (u - h_s(-u))}{2}\right).$$

From this we deduce that

$$f(\beta_s(u)) \le \frac{1}{2}\left[\varphi(u) + \langle \varphi'(u), h_s(u) \rangle + \psi(u + h_s(u))\right]$$

$$+ \frac{1}{2}\left[\varphi(-u) + \langle \varphi'(-u), h_s(-u) \rangle + \psi(-u + h_s(-u))\right] + \delta s.$$

Applying Taylor's formula again we get

$$f(\beta_s(u)) \le \frac{1}{2}f(\alpha_s(u)) + \frac{1}{2}f(\alpha_s(-u)) + 2\delta s.$$

This combined with (4.73) gives

$$f(\beta_s(u)) \le f(u) + s + 3\delta s \le f(u) + 2s$$

for s small and (4.68) holds. Similarly, using (4.74) and (4.75) we show that β_s satisfies (4.69) and (4.71). Finally, β_s satisfies (4.70) since (4.76) continues to hold for u satisfying (4.68) and (4.69).

□

References

1. E. Asplund, Average norms. Israel J. Math. **5**, 227–233 (1967)
2. K.-C. Chang, Variational methods for non-differentiable functionals and their applications to partial differential equations. J. Math. Anal. Appl. **80**, 102–129 (1981)
3. J. Conway, *A Course in Functional Analysis*. Graduate Texts in Mathematics (Springer, Berlin, 1990)
4. C. Costara, D. Popa, *Exercises in Functional Analysis* (Springer, Berlin, 2003)
5. Y. Du, A deformation lemma and some critical point theorems. Bull. Aust. Math. Soc. **43**, 161–168 (1991)
6. N. Ghoussoub, D. Preiss, A general mountain pass principle for locating and classifying critical points. Ann. Inst. H. Poincaré Anal. Non Linéaire **6**, 321–330 (1989)
7. N. Kourogenis, N. Papageorgiou, Nonsmooth critical point theory and nonlinear elliptic equations at resonance. J. Austral. Math. Soc. Ser. A **69**, 245–271 (2000)
8. A. Kristály, V.V. Motreanu, C. Varga, A minimax principle with a general Palais-Smale condition. Commun. Appl. Anal. **9**, 285–299 (2005)
9. D. Motreanu, A multiple linking minimax principle. Bull. Aust. Math. Soc. **53**, 39–49 (1996)
10. D. Motreanu, C. Varga, Some critical point results for locally Lipschitz functionals. Comm. Appl. Nonlin. Anal. **4**, 17–33 (1997)
11. D. Motreanu, C. Varga, A nonsmooth equivariant minimax principle. Commun. Appl. Anal. **3**, 115–130 (1999)
12. P. Pucci, J. Serrin, Extensions of the mountain pass theorem. J. Funct. Anal. **59**, 185–210 (1984)
13. M. Schechter, *Linking Methods in Critical Point Theory* (Birkhäuser, Basel, 1999)
14. A. Szulkin, Ljusternik-Schnirelmann theory on C^1-manifolds. Ann. Inst. H. Poincaré Anal. Non Linéaire **5**, 119–139 (1988)
15. C. Varga, V. Varga, A note on the Palais-Smale condition for non-differentiable functionals, in *Proceedings of the 23rd Conference on Geometry and Topology*, Cluj-Napoca (1993), pp. 209–214

Minimax and Multiplicity Results

5

5.1   Minimax Results with Weakened Compactness Condition

Throughout this section we use the following notion of "linking sets". For more details and examples check out Appendix E.

Definition 5.1 Let X be a Banach space and $A, C \subseteq X$ two subsets. We say that C *links* A, if $A \cap C = \varnothing$, and C is not contractible in $X \setminus A$.

Remark 5.1 It is well known that if X is a finite dimensional and U is an open bounded neighborhood of an element $u \in X$, then the boundary ∂U (the boundary of U) is not contractible in $X \setminus \{u\}$.

Theorem 5.1 *If $A, C \subseteq X$ are nonempty, A is closed, C links A, Γ_C is the set of all contractions of C, and $f : X \to \mathbb{R}$ is a locally Lipschitz which satisfies the $(\varphi - C)_c$-condition with*

$$c := \inf_{h \in \Gamma_C} \sup_{[0,1] \times C} f \circ h < \infty \text{ and } \sup_{u \in C} f(u) \leq \inf_{u \in A} f(u),$$

then $c \geq \inf_{u \in A} f(u)$ and c is a critical value of f. Moreover, if $c = \inf_{u \in A} f(u)$, then there exists $u \in A$ such that $u \in K_c$.

Proof Since by hypothesis C links A, for every $h \in \Gamma_C$ we have $h([0, 1] \times C) \neq \varnothing$. So we infer that $c \geq \inf_{u \in A} f(u)$.

© The Author(s), under exclusive license to Springer Nature Switzerland AG 2021
N. Costea et al., *Variational and Monotonicity Methods in Nonsmooth Analysis,*
Frontiers in Mathematics, https://doi.org/10.1007/978-3-030-81671-1_5

First we assume that $\inf_{u \in A} f(u) < c$. Suppose that $K_c = \varnothing$. Let $U = \varnothing$ and let $\varepsilon > 0$ and $\eta : [0, 1] \times X \to X$ be as in Theorem 4.2. Also from the definition of c, we can find $h \in \Gamma_C$ such that

$$f(h(t, u)) \le c + \varepsilon, \quad \forall (t, u) \in [0, 1] \times C. \tag{5.1}$$

Let $H : [0, 1] \times C \to X$ defined by

$$H(t, x) := \begin{cases} \eta(2t, u), & \text{if } 0 \le t \le \frac{1}{2}, \\ \eta(1, h(2t - 1, u)), & \text{if } \frac{1}{2} \le t \le 1. \end{cases}$$

It is easy to check that $H \in \Gamma_C$ and from (d) and (c) of Theorem 4.2 we obtain that for every $u \in C$ we have

$$f(H(t, u)) = f(\eta(2t, u)) \le f(u) \le \sup_{u \in C} f(u) < c, \text{ if } t \in [0, 1/2]$$

$$f(H(t, u)) = f(\eta(1, h(2t - 1, u))) \le c - \varepsilon < c, \text{ if } t \in [1/2, 1]$$

and from (5.1) we get

$$h(t, u) \in f^{c+\varepsilon} \text{ for every } t \in [0, 1].$$

So we have contradicted the definition of c. This proves that $K_c \neq \varnothing$, when $c > \inf_{u \in A} f(u)$.

Next assume that $c = \inf_{u \in A} f(u)$. We need to show that $K_c \cap A \neq \varnothing$. Suppose the contrary and let U be a neighborhood of K_c with $U \cap A = \varnothing$. Let $\varepsilon > 0$ and $\eta : [0, 1] \times X \to X$ be as in Theorem 4.2. As before let $h \in \Gamma_C$ such that $f(h(t, u)) \le c + \varepsilon$ for all $(t, u) \in [0, 1] \times C$. Then we define $H : [0, 1] \times C \to X$ by

$$H(t, u) := \begin{cases} \eta(2t, u), & \text{if } 0 \le t \le \frac{1}{2} \\ \eta(1, h(2t - 1, u)), & \text{if } \frac{1}{2} \le t \le 1. \end{cases}$$

Again, we have $H \in \Gamma_C$. From Theorem 4.2 follows that for all $0 \le t \le \frac{1}{2}$ and all $u \in C$, we have

$$\eta(2t, u) = u \text{ or } f(\eta(2t, u)) < f(u) \le \inf_{u \in A} f(u) = c$$

which implies

$$\eta(2t, u) \notin A, \quad \forall (t, u) \in [0, 1/2] \times C.$$

For all $t \in [1/2, 1]$ and all $u \in C$, we have from (d) of Theorem 4.2

$$\eta(1, h(2t - 1, u)) \subseteq f^{c-\varepsilon} \cup U,$$

while $(f^{c-\varepsilon} \cup U) \cap A = \emptyset$. So H is a contraction of C in $X \setminus A$ and this contradiction completes the proof. □

Theorem 5.2 (Mountain Pass Theorem) *Let X be a Banach space, $f : X \to \mathbb{R}$ be a locally Lipschitz function and $\varphi : X \to \mathbb{R}$ a globally Lipschitz function such that $\varphi(u) \geq 1$, $\forall u \in X$. Suppose that there exist $u_1 \in X$ and $r > 0$ such that $\|u_1\| > r$ and*

(i) $\max\{f(0), f(u_1)\} \leq \inf\{f(u) : \|u\| = r\}$;
(ii) the function f satisfies the $(\varphi - C)_c$-condition $(c \in \mathbb{R})$,

where

$$c := \inf_{\gamma \in \Gamma} \max_{t \in [0,1]} f(\gamma(t)),$$

with $\Gamma := \{\gamma \in C([0, 1], X) : \gamma(0) = 0, \ \gamma(1) = u_1\}$. Then the minimax value c in (ii) is a critical value of f. Moreover, if $c = \inf\{f(u) : \|u\| = r\}$, there exist a critical point u_2 of f with $f(u_2) = c$ and $\|u_2\| = r$.

Proof We will apply Theorem 5.1 with $A := \{u \in X : \|u\| = r\}$ and $C := \{0, u_1\}$. Clearly C links A and $c < \infty$. Let $\gamma \in \Gamma$ and define

$$h(t, u) := \begin{cases} \gamma(t), & \text{if } u = 0 \\ u_1, & \text{if } u = u_1 \end{cases}$$

Then $h \in \Gamma_C$. Therefore

$$\inf_{\overline{h} \in \Gamma_C} \sup_{[0,1] \times C} f(\overline{h}(t, u)) \leq f(h(t, u)) \leq c. \tag{5.2}$$

On the other hand, if $h \in \Gamma_C$, then

$$\gamma(t) := \begin{cases} h(2t, 0), & \text{if } t \in [0, 1/2] \\ h(2 - 2t, x_1), & \text{if } t \in [1/2, 1] \end{cases}$$

belongs to Γ and so

$$\inf_{h \in \Gamma_C} \sup_{[0,1] \times C} f(h(t, u)) \geq c. \tag{5.3}$$

By (5.2) and (5.3) we have

$$c = \inf_{h \in \Gamma_C} \sup_{[0,1] \times C} f(h(t, u))$$

and so we can apply Theorem 5.1 and finish the proof.   □

Theorem 5.3 (Saddle Point Theorem) *Let X be a Banach space and $f : X \to \mathbb{R}$ be a locally Lipschitz function. Suppose that $X := Y \oplus V$, with $\dim Y < \infty$, and there exists $r > 0$ such that:*

(i) $\max\{f(u) : u \in Y, \|u\| = r\} \leq \inf\{f(u) : u \in V\}$;
(ii) *the function f satisfies the $(\varphi - C)_c$-condition where*

$$c := \inf_{\gamma \in \Gamma} \max_{u \in E} f(\gamma(u))$$

with $\Gamma := \{\gamma \in C(E, X) : \gamma|_{\partial E} = id\}$, $E := \{u \in Y : \|u\| \leq r\}$ and $\partial E = \{u \in Y : \|u\| = r\}$.
    Then $c \geq \inf_V f$ and c is a critical value of f. Moreover, if $c = \inf_V f$, then $V \cap K_c \neq \emptyset$.

Proof We will apply Theorem 5.1 with $A := V$ and $C := \partial E$. Clearly from the compactness of E, we have that $c < \infty$. Let $P : X \to Y$ be the projection. We show that C links A. Suppose not and let h be a contraction of C in $X \setminus V$. Let $H(t, u) := Ph(t, u)$, which is a contraction of C in $Y \setminus \{0\}$. This contradicts the Remark 5.1.
    Next let $\gamma \in \Gamma$ and define $h(t, u) := \gamma((1 - t)u)$. Clearly $h \in \Gamma_C$. So, we have

$$\inf_{h \in \Gamma_C} \sup_{[0,1] \times C} f(h(t, u)) \leq f(h(t, u)) \leq c. \tag{5.4}$$

Also if $h \in \Gamma_C$ and $h(1, u) = z_1$ for all $u \in C$, then we define

$$\xi(t, u) = \begin{cases} h(t, u), & \text{if } (t, u) \in [0, 1] \times C \\ z_1, & \text{if } (t, x) \in \{1\} \times E \end{cases}$$

which is continuous from $([0, 1] \times C) \cup (\{1\} \times E)$ into X.
    Let $Q : E \to ([0, 1] \times C) \cup (\{1\} \times E)$ be a homeomorphism such that $Q(C) = \{0\} \times C$. Then we see that $\xi \circ Q \in \Gamma$, so

$$c \leq \inf_{h \in \Gamma_C} \sup_{[0,1] \times C} f(h(t, u)). \tag{5.5}$$

By (5.4) and (5.5) it follows that

$$c = \inf_{h \in \Gamma_C} \sup_{[0,1] \times C} f(h(t,u))$$

and so we can apply Theorem 5.1 and complete the proof.   □

Theorem 5.4 (Linking Theorem)   *Let X be a Banach space and $f : X \to \mathbb{R}$ be a locally Lipschitz function. Let $X := Y \oplus V$ be with $\dim Y < \infty$ and $0 < r < R$, $e \in V$ with $\|e\| = 1$. We consider the set*

$$Q := \{u = v + te : v \in Y, \ t \geq 0, \ \|u\| \leq R\}$$

and ∂Q its boundary in $Y \oplus \mathbb{R}e$. We suppose that

(i) $\max\{f(u) : u \in \partial Q\} \leq \inf\{f(u) : u \in \partial B(0,r) \cap V\}$;
(ii) the function f satisfies the $(\varphi - C)_c$-condition, where $c := \inf\limits_{\gamma \in \Gamma} \max\limits_{u \in Q} f(\gamma(u))$ with
  $\Gamma = \{\gamma \in C(Q,X) : \gamma|_{\partial Q} = id\}$.

Then $c \geq \inf\{f(u) : u \in \partial B(0,r) \cap V\}$ and c is a critical value of f. Moreover, if $c = \inf\{f(u) : u \in \partial B(0,r) \cap V\}$, then $K_c \cap (\partial B(0,r) \cap V) \neq \emptyset$.

Proof   Because Q is compact, it is clear that $c < \infty$. Let $P_1 : X \to Y$ and $P_2 : X \to V$ be the projection operators on Y and V, respectively and let $A := \partial B(0,r) \cap V$ and $C := \partial Q$. If $h(t,u)$ is a contraction of C in $X \setminus A$, then $H(t,u) := P_1 h(t,u) + \|P_2 h(t,u)\|e$ is a contraction of C in $(V \oplus Re) \setminus \{re\}$ which contradicts Remark 5.1.

As in Theorem 5.1, we can verify that $c = \inf\limits_{h \in \Gamma_C} \sup\limits_{[0,1] \times C} f \circ h$. Therefore we apply Theorem 5.1 and conclude the proof.   □

5.2   A General Minimax Principle: The "Zero Altitude" Case

In this section we present a general minimax principle for locally Lipschitz functionals that appears in the paper of Motreanu and Varga [11].

Theorem 5.5   *Let $f : X \to \mathbb{R}$ be a locally Lipschitz functional and $B \subseteq X$ a closed set such that $c := \inf_B f > -\infty$ and f satisfies $(PS)_{B,c}$. Let M be a nonempty family of subsets M of X such that*

$$c := \inf_{M \in \mathcal{M}} \sup_{u \in M} f(u). \tag{5.6}$$

Assume that for a generalized normalized pseudo-gradient vector field Λ of f with respect to B and c the following hypothesis holds

(H) for each set $M \in \mathcal{M}$ and each number $\varepsilon > 0$ with $f|_M < c + \varepsilon$ there exists a closed subset A of X with $f|_A \leq c + \varepsilon_A$ (see (4.42)), and $A \cap B = \emptyset$ such that for each locally Lipschitz function $\varphi_A : X \to [0, 1]$ with $\text{supp}\, \varphi_A \subset (X \setminus A) \cap \text{supp}\, \Lambda$ the global flow ξ_A of $\varphi_A \Lambda$ satisfies $\xi_A(1, M) \cap B \neq \emptyset$.

Then the following assertions are true

(i) $c = \inf_B f$ is attained;
(ii) $K_c(f) \setminus A \neq \emptyset$ for each set A entering (H);
(iii) $K_c(f) \cap B \neq \emptyset$.

Proof The assertions (i) and (ii) are direct consequences of the property (iii). The proof of (iii) is achieved arguing by contradiction. Accordingly, we suppose $K_{-c}(-f) \cap B = \emptyset$. By hypothesis we know that $B \subset (-f)_{-c}$, so Theorem 4.3 can be applied for $-f$ and $-c$ (in place of f and c, respectively). Thus Theorem 4.3 yields an $\varepsilon > 0$ with the properties there stated. Then from the minimax description of c, by means of $\mathcal{M}$, we obtain the existence of a set $M \in \mathcal{M}$ satisfying $f|_M < c + \varepsilon$. Corresponding to M, assumption (H) allows to find a closed set $A \subset X \setminus B$ which satisfies $A \subset (-f)^{-c - \varepsilon_A}$ and the linking property formulated in (H). Theorem 4.3 gives rise to the deformation $\eta_A \in C(\mathbb{R} \times X, X)$ which verifies $\eta_A(1, B \cap (-f)_{-c}) \subset (-f)_{-c-\varepsilon}$. This reads as

$$\eta_A(1, B) \subset f^{c+\varepsilon}. \tag{5.7}$$

By Theorem 4.3 and assumption (H) it is seen that

$$\xi_A(t, u) = \eta_A(-t, u), \tag{5.8}$$

for all $(t, u) \in \mathbb{R} \times X$. As shown in (H) one has the intersection property

$$\xi(1, M) \cap B \neq \emptyset.$$

Combining with (5.8) it turns out

$$\eta_A(1, B) \cap M \neq \emptyset.$$

Taking into account (5.6) we obtain the existence of some point $u_0 \in M$ with $f(u_0) \geq c + \varepsilon$. This contradicts the choice of the set M. $\square$

Corollary 5.1 *Let $f : X \rightarrow \mathbb{R}$ be a locally Lipschitz functional satisfying (PS) and let a family $\mathcal{M}$ of subsets M of X be such that c defined by (5.6) is a real number. Assume that the hypothesis below holds*

(H') *for each $M \in \mathcal{M}$ there exists a closed set A in X with $f|_A < c$ such that for every homeomorphism h of X with $h|_A = id_A$ one has $h(M) \cap f^c \neq \varnothing$.*

Then c in (5.6) is a critical value of f and $K_c(f) \cap A = \varnothing$ for every A in (H').

Proof We consider the global flow ξ_A (see (5.7)) and we apply Theorem 5.5 with $B := f^c$. It is clear that (H') implies (H) because $A \subset M \setminus B$ and $\xi_A(1, \cdot)$ is a homeomorphism of X with $\xi_A(1, \cdot) = id$ on A. Then Theorem 5.5 concludes the proof. □

Remark 5.2 The minimax principle in Corollary 5.1 includes and extends to the locally Lipschitz functionals many classic minimax results, e.g. those in Ambrosetti and Rabinowitz [1], Chang [2], Du [5], Ghoussoub and Preiss [6], Motreanu [9], Motreanu and Varga [10]).

Theorem 5.5 is useful in locating the critical points. We illustrate this aspect by deriving from Theorem 5.5 an extension for locally Lipschitz functionals of a result due to Ghoussoub & Preiss [6].

Corollary 5.2 *Let $f : X \rightarrow \mathbb{R}$ be a locally Lipschitz functional and for the points u, $v \in X$ let the number*

$$c := \inf_{g \in \Gamma} \max_{0 \leq t \leq 1} f(g(t)),$$

where Γ is the set of paths $g \in C([0, 1], X)$ joining u and v. Suppose F is a closed subset of X such that $F \cap f^c$ separates u and v, i.e. u, v belong to disjoint connected components of $X \setminus F \cap f^c$, and condition $(PS)_{F,c}$ is verified. Then there exists a critical point of f in F with critical value c.

Proof Set $\mathcal{M} := \{g([0, 1]) : g \in \Gamma\}$, $B := F \cap f^c$ and $A := \{u, v\}$. Applying Theorem 5.5 we see that $\xi_A(1, M) \in \mathcal{M}$ whenever $M \in \mathcal{M}$. Thus hypothesis (H) is verified. Theorem 5.5 implies the conclusion of corollary. □

Theorem 5.5 is suitable for applications to multiple linking problems.

Definition 5.2 Let Q, Q_0 be closed subsets of X, with $Q_0 \neq \varnothing$, $Q_0 \subset Q$, and let S be a subset of X such that $Q_0 \cap S = \varnothing$. We say that the pair (Q, Q_0) links with S if for each mapping $g \in C(Q, X)$ with $g|_{Q_0} = id|_{Q_0}$ one has $g(Q) \cap S \neq \varnothing$.

A common situation of linking is presented in the following result given in Motreanu and Varga [11] (it unifies the minimax principles in Chang [2] and Du [5]).

Corollary 5.3 *Given the subsets* Q, Q_0, S *of the real Banach space* X *we assume that* (Q, Q_0) *links with* S *in* X *in the sense above. Let* $f : X \to \mathbb{R}$ *be a locally Lipschitz functional such that* $\sup_Q f < \infty$ *and, for some number* $\alpha \in \mathbb{R}_+$,

$$Q_0 \subset f_\alpha, \quad S \subset f^\alpha.$$

Then assuming that for the minimax value

$$c := \inf_{g \in \Gamma} \sup_{u \in Q} f(g(u)),$$

where

$$\Gamma := \{ g \in C(Q, X) : g|_{Q_0} = id|_{Q_0} \},$$

$(PS)_{S,c}$ *is satisfied, the following properties hold*

(i) $c \geq \alpha$;
(ii) $K_c(f) \setminus Q_0 \neq \varnothing$;
(iii) $K_c(f) \cap S \neq \varnothing$ *if* $c = \alpha$.

Proof Since the case $\alpha < c$ follows immediately we discuss only the situation where $\alpha = c$. The conclusion is readily obtained from Theorem 5.5 by choosing $\mathcal{M} := \{ g(Q) : g \in \Gamma \}$ and $B := S$. $\square$

A direct consequence of this corollary is the following.

Corollary 5.4 (Zero Altitude Mountain Pass Theorem) *Let* $f : X \to \mathbb{R}$ *be a locally Lipschitz function on a Banach space satisfying* $(PS)_c$ *for every* $c \in \mathbb{R}$ *and the conditions:*

(i) $f(u) \geq \alpha \geq f(0)$ *for all* $\|u\| = \rho$ *where* α *and* $\rho > 0$ *are constants;*
(ii) *there is* $e \in X$ *with* $\|e\| > \rho$ *and* $f(e) \leq \alpha$.

Then the number

$$c := \inf_{g \in \Gamma} \max_{u \in [0,e]} f(g(u)),$$

where $[0, e]$ is the closed line segment in X joining 0 and e and

$$\Gamma := \{g \in C([0, e], X) : \ g(0) = 0, \ \ g(e) = e\},$$

is a critical value of f with $c \geq \alpha$.

Proof It is sufficient to take in Corollary 5.3 the following choices $Q := [0, e]$, $Q_0 := \{0, e\}$ and $S := \{u \in X : \ ||u|| = \rho \}$. □

Corollary 5.5 (Zero Altitute Linking Theorem) *Let X be a real Banach space, $f :$ $X \to \mathbb{R}$ be a locally Lipschitz function which satisfies the $(PS)_c$ condition for every $c \in \mathbb{R}$. We suppose that that the following conditions are fulfilled:*

(i) $X := X_1 \oplus X_2$ with $\dim X_1 < \infty$;
(ii) for some constant $\alpha \in \mathbb{R}$ and a closed neighbourhood N of 0 in X whose boundary is ∂N we have $f|_{\partial N} \leq \alpha \leq f|_{X_2}$.

Then the number

$$c := \inf_{g \in \Gamma} \max_{u \in N} f(g(u)),$$

where

$$\Gamma = \{g \in C(N, X) : \ g|_{\partial N} = id|_{\partial N} \},$$

is a critical value of f with $c \geq \alpha$.

Proof We choose $Q := N$, $Q_0 := \partial N$ and $S := X_2$ in Corollary 5.3. □

5.3 $\mathbb{Z}_2$–Symmetric Mountain Pass Theorem

In this section we present a $\mathbb{Z}_2$-version of the Mountain Pass theorem for locally Lipschitz functions, which satisfy the generalized $(\varphi - C)_c$ condition. This result is an extension of Theorem 9.12 of Rabinowitz [12]. Since we proved a deformation results for locally Lipschitz functions which satisfy the $(\varphi - C)_c$ condition, the proof is similar as in the above mentioned result of Rabinowitz. For the sake of completeness we give this proof.

First of all, we recall some basic facts on the simplest index theory, see Rabinowitz [12]. Let E be a real Banach space and $\mathcal{E}$ denote the family of sets $A \subset E \setminus \{0\}$ such that A is closed in E and symmetric, i.e. $A = -A$.

Definition 5.3 We say that the positive integer n is the genus of $A \in \mathcal{E}$, if there exists an odd map $\varphi \in C(A, \mathbb{R}^n \setminus \{0\})$ and n is the smallest integer with this property. The genus of the set A is denoted by $\gamma(A) = n$. When there does not exist a finite such n, set $\gamma(A) = \infty$. Finally set $\gamma(\varnothing) = 0$.

Example 5.1 Suppose $B \subset E$ is closed and $B \cap (-B) = \varnothing$. Let $A = B \cup (-B)$. Then $\gamma(A) = 1$ since the function $\varphi(u) = 1$ for $u \in B$ and $\varphi(u) = -1$ for $u \in -B$ is odd and lies in $C(A, \mathbb{R} \setminus \{0\})$.

Remark 5.3 If $A \in \mathcal{E}$ and $\gamma(A) > 1$, then A contains infinitely many distinct points. Indeed, if A were finite we could write $A = B \cup (-B)$ with B as in Example 5.1. But then $\gamma(A) = 1$.

Example 5.2 If $n \geq 1$ and A is homeomorphic to S^n by an odd map, then $\gamma(A) > 1$. Otherwise there is a mapping $\varphi \in C(A, \mathbb{R} \setminus \{0\})$ with φ odd. Choose any $u \in A$ such that $\varphi(u) > 0$. Then $\varphi(-u) < 0$ and by Intermediate Value Theorem, φ must vanish somewhere on any path in A joining u and $-u$, a contradiction.

For $A \in \mathcal{E}$ and $\delta > 0$ we denote by $N_\delta(A)$ the uniform δ-neighborhood of A, i.e. $N_\delta(A) := \{u \in E : \operatorname{dist}(u, A) \leq \delta\}$. The genus has the following properties.

Proposition 5.1 *Let* $A, B \in \mathcal{E}$. *Then*

$1°$. *Normalization: If* $u \neq 0$, $\gamma(\{u\} \cup \{-u\}) = 1$;
$2°$. *Mapping property: If there exists an odd map* $f \in C(A, B)$, *then* $\gamma(A) \leq \gamma(B)$;
$3°$. *Monotonicity property: If* $A \subset B$, $\gamma(A) \leq \gamma(B)$;
$4°$. *Subadditivity:* $\gamma(A \cup B) \leq \gamma(A) + \gamma(B)$;
$5°$. *Continuity property: If* A *is compact then* $\gamma(A) < \infty$, *and there is a* $\delta > 0$ *such that* $N_\delta(A) \in \mathcal{E}$ *then* $\gamma(N_\delta(A)) = \gamma(A)$.

Proof

$1°$. follows from the Example 5.1.
$2°$. Here and hereafter, we assume that $\gamma(A), \gamma(B) < \infty$; the remaining cases are trivial.
    We suppose $\gamma(B) = n$. Then there exists a function φ belonging to $C(B, \mathbb{R}^n \setminus \{0\})$. Consequently $\varphi \circ f$ is odd and $\varphi \circ f \in C(A, \mathbb{R}^n \setminus \{0\})$. Therefore $\gamma(A) \leq n = \gamma(B)$.
$3°$. Choosing $f := id$ in $2°$ we get the assertion.
$4°$. Suppose that $\gamma(A) = m$ and $\gamma(B) = n$. Then there exist mapping $\varphi \in C(A, \mathbb{R}^m \setminus \{0\})$ and $\psi \in C(B, \mathbb{R}^n \setminus \{0\})$, both odd. By the Tietze Extension Theorem, there are mappings $\widehat{\varphi} \in C(E, \mathbb{R}^m)$ and $\widehat{\psi} \in C(E, \mathbb{R}^n)$ such that $\widehat{\varphi}|_A = \varphi$ and $\widehat{\psi}|_B = \psi$. Replacing $\widehat{\varphi}, \widehat{\psi}$ by their odd parts, we can assume $\widehat{\varphi}, \widehat{\psi}$ are odd. Set $f = (\widehat{\varphi}, \widehat{\psi})$. Then $f \in C(A \cup B, \mathbb{R}^{m+n} \setminus \{0\})$ and is odd. Therefore $\gamma(A \cup B) \leq m + n = \gamma(A) + \gamma(B)$.

5°. For each $u \in A$, set $r(u) \equiv 1/2\|u\| = r(-u)$ and $T_u := B_{r(u)}(u) \cup B_{r(u)}(-u)$. Then $\gamma(\overline{T_u}) = 1$ by Example 5.1. Certainly $A \subset \bigcup_{u \in A} T_u$ and by the compactness of A, $A \subset \bigcup_{i=1}^{k} T_{u_i}$ for some finite set of points $u_1, \ldots, u_k$. Therefore $\gamma(A) < \infty$ via 4°. If $\gamma(A) = n$, there is a mapping $\varphi \in C(A, \mathbb{R}^n \setminus \{0\})$ with φ odd. Extend φ to an odd function $\widehat{\varphi}$ as in 4°. Since A is compact, there is a $\delta > 0$ such that $\widehat{\varphi} \neq 0$ on $N_\delta(A)$. Therefore $\gamma(N_\delta(A)) \leq n = \gamma(A)$. But by 3°, $\gamma(A) \leq \gamma(N_\delta(A))$ so we have equality. $\square$

Remark 5.4 For later arguments it is useful to observe that if $\gamma(B) < \infty$, $\gamma(\overline{A \setminus B}) \geq \gamma(A) - \gamma(B)$. Indeed $A \subset \overline{A \setminus B} \cup B$ so the inequality follows from 3° – 4° of Proposition 5.1.

Next we will calculate the genus of an important class of sets.

Proposition 5.2 *If $A \subset E$, Ω is a bounded neighborhood of 0 in $\mathbb{R}^k$, and there exists a mapping $h \in C(A, \partial\Omega)$ with h an odd homeomorphism, then $\gamma(A) = k$.*

Proof Plainly $\gamma(A) \leq k$. If $\gamma(A) = j < k$, there is a $\varphi \in C(A, \mathbb{R}^j \setminus \{0\})$ with φ odd. Then $\varphi \circ h^{-1}$ is odd and belongs to $C(\partial\Omega, \mathbb{R}^j \setminus \{0\})$. But this is contrary to the Borsuk-Ulam Theorem since $k > j$. Therefore $\gamma(A) = k$. $\square$

Proposition 5.3 *Let X be a subspace of E of codimension k and $A \in \mathcal{E}$ with $\gamma(A) > k$. Then $A \cap X \neq \varnothing$.*

Proof Writing $E = V \oplus X$ with V a dimensional complement of X, let P denote the projector of E onto V. If $A \cap X = \varnothing$, $P \in C(A, V \setminus \{0\})$. Moreover P is odd. Hence by 2° of Proposition 5.1, $\gamma(A) \leq \gamma(PA)$. The radial projection of PA into $\partial B_1 \cap V$ is another continuous odd map. Hence $\gamma(A) \leq \gamma(\partial B_1 \cap V) = k$ via Proposition 5.2, contrary to hypothesis. $\square$

The main result of this section is the following, which represents an extension to non-smooth case of the multiplicity result Theorem 9.12 of Rabinowitz [12].

Theorem 5.6 *Let E be an infinite dimensional Banach space and let $f : E \to \mathbb{R}$ be an even locally Lipschitz function which satisfies the $(\varphi - C)_c$ condition for every $c \in \mathbb{R}$, and $f(0) = 0$. If $E := V \oplus X$, where V is finite dimensional, and f satisfies*

(f_1') *there are constants $\rho, \alpha > 0$ such that $f|_{\partial B_\rho \cap X} \geq \alpha$;*
(f_2') *for each finite dimensional subspace $\widetilde{E} \subset E$, there is an $R = R(\widetilde{E})$ such that $f \leq 0$ on $\widetilde{E} \setminus B_{R(\widetilde{E})}$,*

then f possesses an unbounded sequence of critical values.

Proof The proof is given by in more steps. First we define sequence of families of sets Γ_m and we associate the corresponding sequence $\{c_m\}$ of critical values of f, which are obtained by taking the minimax of f over each Γ_m. A separate argument then shows $\{c_m\}$ is unbounded.

Suppose V is k dimensional and $V := \text{span}\{e_1, \ldots, e_k\}$. For $m \geq k$, inductively choose $e_{m+1} \notin \text{span}\{e_1, \ldots, e_m\} \equiv E_m$. Set $R_m \equiv R(E_m)$ and $D_m \equiv B_{R_m} \cap E_m$. Let

$$G_m \equiv \{ h \in C(D_m, E) : h \text{ is odd and } h = id \text{ on } \partial B_{R_m} \cap E_m \}. \tag{5.9}$$

Note that $id \in G_m$ for all $m \in \mathbb{N}$ so $G_m \neq \varnothing$. Set

$$\Gamma_j \equiv \{ h(\overline{D_m \setminus Y}) : h \in G_m, \ m \geq j, \ Y \in \mathcal{E}, \text{ and } \gamma(Y) \leq m - j \}. \tag{5.10}$$

Proposition 5.4 *The sets Γ_j possess the following properties:*

$1°$ $\Gamma_j \neq \varnothing$ for all $j \in \mathbb{N}$;
$2°$ *(Monotonicity)* $\Gamma_{j+1} \subset \Gamma_j$;
$3°$ *(Invariance)* If $\varphi \in C(E, E)$ is odd, and $\varphi = id$ on $\partial B_{R_m} \cap E_m$ for all $m \geq j$, then $\varphi : \Gamma_j \to \Gamma_j$;
$4°$ *(Excision)* If $B \in \Gamma_j$, $Z \in \mathcal{E}$, and $\gamma(Z) \leq s < j$, then $\overline{B \setminus Z} \in \Gamma_{j-s}$.   □

Proof

$1°$ Since $id \in G_m$ for all $m \in \mathbb{N}$, it follows that $\Gamma_j \neq \varnothing$ for all $j \in \mathbb{N}$.
$2°$ If $B = h(\overline{D_m \setminus Y}) \in \Gamma_{j+1}$, then $m \geq j + 1 \geq j$, $h \in G_m$, $Y \in \mathcal{E}$, and $\gamma(Y) \leq m - (j+1) \leq m - j$. Therefore $B \in \Gamma_j$.
$3°$ Suppose $B = h(\overline{D_m \setminus Y}) \in \Gamma_j$ and φ is as above. Then $\varphi \circ h$ is odd, belongs to $C(D_m, E)$, and $\varphi \circ h = id$ in $\partial B_{R_m} \cap E_m$. Therefore $\varphi \circ h \in G_m$ and $\varphi \circ h(\overline{D_m \setminus Y}) = \varphi(B) \in \Gamma_j$.
$4°$ Again let $B = h(\overline{D_m \setminus Y}) \in \Gamma_j$ and $Z \in \mathcal{E}$ with $\gamma(Z) \leq s < j$. We claim

$$\overline{B \setminus Z} = h(\overline{D_m \setminus (Y \cup h^{-1}(Z))}). \tag{5.11}$$

Assuming (5.11), note that since h is odd and continuous and $Z \in \mathcal{E}$, $h^{-1}(Z) \in \mathcal{E}$. Therefore $Y \cup h^{-1}(Z) \in \mathcal{E}$ and by $4°$ and $2°$ of Proposition 5.1,

$$\gamma(Y \cup h^{-1}(Z)) \leq \gamma(Y) + \gamma(h^{-1}(Z)) \leq \gamma(Y) + \gamma(Z)$$

$$\leq m - j + s = m - (j - s).$$

Hence $\overline{B \setminus Z} \in \Gamma_{j-s}$.

In order to prove (5.11), suppose $b \in h(D_m \setminus (Y \cup h^{-1}(Z)))$. Then $b \in h(D_m \setminus Y) \setminus Z \subset B \setminus Z \subset \overline{B \setminus Z}$. Therefore

$$h(D_m \setminus (Y \cup h^{-1}(Z))) \subset \overline{B \setminus Z}. \tag{5.12}$$

On the other hand if $b \in B \setminus Z$, then $b = h(w)$ where

$$w \in \overline{D_m \setminus Y} \setminus h^{-1}(Z) \subset \overline{D_m \setminus (Y \cup h^{-1}(Z))}.$$

Thus

$$B \setminus Z \subset h(\overline{D_m \setminus (Y \cup h^{-1}(Z))}). \tag{5.13}$$

Comparing (5.12)–(5.13) yields (5.11) since h is continuous.   □

Now a sequence of minimax values of f can be defined. Set

$$c_j = \inf_{B \in \Gamma_j} \max_{u \in B} f(u), \quad j \in \mathbb{N}. \tag{5.14}$$

It will soon be seen that if $j > k = \dim V$, c_j is a critical value of f. The following intersection theorem is needed to provide a key estimate.

Proposition 5.5 *If $j > k$ and $B \in \Gamma_j$, then*

$$B \cap \partial B_\rho \cap X \neq \varnothing. \tag{5.15}$$

Proof Set $B = h(\overline{D_m \setminus Y})$ where $m \geq j$ and $\gamma(Y) \leq m - j$. Let $\widehat{O} = \{u \in D_m \mid h(u) \in B_\rho\}$. Since h is odd, $0 \in \widehat{O}$. Let O denote the component of $\widehat{O}$ containing 0. Since D_m is bounded, O is a symmetric (with respect to 0) bounded neighborhood of 0 in E_m. Therefore by Proposition 5.2, $\gamma(\partial O) = m$.
   We claim

$$h(\partial O) \subset \partial B_\rho. \tag{5.16}$$

Assuming (5.16) for the moment, set $W \equiv \{u \in D_m : h(u) \in \partial B_\rho\}$. Therefore (5.16) implies $W \supset \partial O$. Hence by 3° of Proposition 5.1, $\gamma(W) = m$ and by Remark 5.4, $\gamma(\overline{W \setminus Y}) \geq m - (m - j) = j > k$. Thus by 2° of Proposition 5.1, $\gamma(h(\overline{W \setminus Y})) > k$. Since $codim X = k$, $h(\overline{W \setminus Y}) \cap X \neq \varnothing$ by Proposition 5.3. But $h(\overline{W \setminus Y}) \subset (B \cap \partial B_\rho)$. Consequently (5.15) holds.

It remains to prove (5.16). Note first that by the choice of R_m,

$$f \leq 0 \text{ on } E_m \setminus B_{R_m}. \tag{5.17}$$

Since $m > k$, $\partial B_\rho \cap X \cap E_m \neq \varnothing$. Hence by (f_1'),

$$f|_{\partial B_\rho \cap X \cap E_m} \geq \alpha > 0. \tag{5.18}$$

Comparing (5.17) and (5.18) shows $R_m > \rho$. Now to verify (5.16), suppose $u \in \partial O$ and $h(u) \in B_\rho$. If $x \in D_m$ there is a neighborhood N of X such that $h(N) \subset B_\rho$. But then $u \notin \partial O$. Thus $u \in \partial D_m$ (with ∂ relative to E_m). But on ∂D_m, $h = id$. Consequently if $u \in \partial D_m$ and $h(u) \in B_\rho$, $\|h(u)\| = \|u\| = R_m < \rho$ contrary to what we just proved. Thus (5.16) must hold.

Remark 5.5 A closer inspection of the above proof shows that

$$\gamma(B \cap \partial B_\rho \cap X) \geq j - k.$$

Corollary 5.6 *If $j > k$, $c_j \geq \alpha > 0$.*   □

Proof If $j > k$ and $B \in \Gamma_j$, by (5.15) and (f_1'), $\max_{u \in B} f(u) \geq \alpha$. Therefore by (5.14), $c_j \geq \alpha$.   □

The next proposition both shows c_j is a critical value of f for $j > k$ and makes an appropriate multiplicity statement about degenerate critical values.

Proposition 5.6 *If $j > k$, and $c_j = \cdots = c_{j+p} \equiv c$, then $\gamma(K_c) \geq p + 1$.*   □

Proof Since $f(0) = 0$ while $c \geq \alpha > 0$ via Corollary 5.6, $0 \notin K_c$. Therefore $K_c \in \mathcal{E}$ and by the compactness condition, K_c is compact. If $\gamma(K_c) \leq p$, by 5° of Proposition 5.1, there is a $\delta > 0$ such that $\gamma(N_\delta(K_c)) \leq p$. Invoking (f) of Theorem 4.2 with $U = O = N_\delta(K_c)$ and $\varepsilon_0 = \alpha/2$, there is an $\varepsilon \in (0, \varepsilon_0)$ and $\eta \in C([0, 1] \times E, E)$ such that $\eta(1, \cdot)$ is odd and

$$\eta(1, f^{c+\varepsilon} \setminus O) \subset f^{c-\varepsilon}. \tag{5.19}$$

Choose $B \in \Gamma_{j+p}$ such that

$$\max_{u \in B} f(u) \leq c + \varepsilon. \tag{5.20}$$

By 4° of Proposition 5.4, $\overline{B \setminus O} \in \Gamma_j$. The definition of R_m shows $f(u) \leq 0$ for $u \in \partial B_{r_m} \cap E_m$ for any $m \in \mathbb{N}$. Hence (b) of Theorem 4.2 and our choice of ε_0 imply $\eta(1, \cdot) = id$ on $\partial B_{R_m} \cap E_m$ for each $m \in \mathbb{N}$. Consequently $\eta(1, \overline{B \setminus O}) \in \Gamma_j$ by 3° of Proposition 5.4. The definition of c_j and (5.19)–(5.20) then imply

$$\max_{u \in \eta(1, \overline{B \setminus O})} f(u) \leq c - \varepsilon,$$

a contradiction.   □

Proposition 5.7 $c_j \to \infty$ as $j \to \infty$.   □

Proof By 2° of Proposition 5.4 and (5.14), $c_{j+1} \geq c_j$. Suppose the sequence (c_j) is bounded. Then $c_j \to \overline{c} < \infty$ as $j \to \infty$. If $c_j = \overline{c}$ for all large j, Proposition 5.6 implies $\gamma(K_{\overline{c}}) = \infty$. But condition $(\varphi - C)_{\overline{c}}$ implies $K_{\overline{c}}$ is compact so $\gamma(K_{\overline{c}}) < \infty$ via 5° of Proposition 5.1. Thus $\overline{c} > c_j$ for all $j \in \mathbb{N}$. Set

$$\mathcal{K} \equiv \{u \in E : c_{k+1} \leq f(u) \leq \overline{c} \text{ and } f'(u) = 0\}.$$

By condition $(\varphi - C)$ we have $\mathcal{K}$ is compact and 5° of Proposition 5.1 implies $\gamma(\mathcal{K}) < \infty$ and there is a $\delta > 0$ such that $\gamma(N_\delta(\mathcal{K})) = \gamma(\mathcal{K}) \equiv q$. Let $s = \max(q, k+1)$. The deformation Theorem 4.2 with $c = \overline{c}$, $\varepsilon_0 = \overline{c} - c_s$, and $U = O = N_\delta(\mathcal{K})$ yields an ε and η as usual such that

$$\eta(1, f^{\overline{c}+\varepsilon} \setminus O) \subset f^{\overline{c}-\varepsilon}. \tag{5.21}$$

Choose $j \in \mathbb{N}$ such that $c_j > \overline{c} - \varepsilon$ and $B \in \Gamma_{j+s}$ such that

$$\max_B f \leq \overline{c} + \varepsilon. \tag{5.22}$$

Arguing as in the proof of Proposition 5.4 shows $\overline{B \setminus O}$ is in Γ_j as is $\eta(1, \overline{B \setminus O})$ provided that $\eta(1, \cdot) = id$ on $\partial B_{R_m} \cap E_m$ for all $m \geq j$. But $f \leq 0$ on $\partial B_{R_m} \cap E_m$ for all $m \in \mathbb{N}$ while $\overline{c} - \varepsilon_0 = c_s \geq c_{k+1} \geq \alpha > 0$ via Corollary 5.6. Consequently $\eta(1, \overline{B \setminus O}) \in \Gamma_j$ and by (5.21)–(5.22) and the choice of c_j,

$$c_j \leq \max_{\eta(1, \overline{B \setminus O})} f \leq \overline{c} - \varepsilon < c_j,$$

a contradiction. The proof is complete. The above proposition completes the proof of Theorem 5.6.   □

5.4   Bounded Saddle Point Methods for Locally Lipschitz Functionals

Using the Schechter type deformation result from the Sect. 4.3 we prove results regarding the existence Palais-Smale sequences in a ball for a given locally Lipschitz function f. More precisely, we show that if there exists $\theta \in (0, 1)$ such that $0 \notin C(u, \theta)$ holds in a certain region of S_R, then f possesses a Palais-Smale sequence. This boundary condition actually replaces the compactness condition (be it $(PS)_c$, $(C)_c$, or $(\varphi - C)_c$) required in the previous sections. If the boundary condition is dropped, then an alternative is obtained: either f possesses a Palais-Smale sequence in the ball, or a sequence leading to a negative eigenvalue exists on the sphere. Finally, if we impose a mild compactness condition, namely the Schechter-Palais-Smale compactness condition, then existence and multiplicity results regarding the critical points of f are established.

We start with the case when f is bounded below on B_R.

Theorem 5.7 *Let $f : \overline{B}_R \to \mathbb{R}$ be a locally Lipschitz function such that*

$$m_R := \inf_{B_R} f > -\infty. \tag{5.23}$$

Suppose that there exist $\theta \in (0, 1)$ and $\varepsilon > 0$ such that

$$0 \notin C(u, \theta), \ \ on \ \{u \in S_R : \ |f(u) - m_R| \le \varepsilon\}.$$

Then there exists a sequence $\{u_n\} \subset \overline{B}_R$ such that

$$f(u_n) \to m_R \ \ and \ \ \lambda_f(u_n) \to 0.$$

Proof Arguing by contradiction, assume that such a sequence does not exist. Then there exist $\gamma, \ \delta > 0$ such that

$$\lambda_f(u) \ge \gamma, \ \ on \ \{u \in \overline{B}_R : \ |f(u) - m_R| \le 3\delta\}.$$

Shrinking δ if necessary, we may assume that $3\delta \le \varepsilon$. Applying Theorem 4.4 with $Z := \overline{B}_R$ and $c := m_R$ and $\rho := \frac{4\delta}{\gamma\theta^2}$ we get the existence of a continuous deformation $\sigma : [0, 1] \times \overline{B}_R \to \overline{B}_R$ which satisfies

$$\sigma\left(1, f^{m_R+\delta}\right) \subseteq f^{m_R-\delta}. \tag{5.24}$$

Due to (5.23), the set in the left-hand side is nonempty, while the set in the right-hand side is empty, thus (5.24) yields a contradiction.   □

In the next we shall work with Schechter's definition of linking for the ball $\overline{B}_R$ (see Definition E.5). The following linking-type theorem says that if A and B are linked, i.e. cannot be pulled apart without intersecting and the energy over A is dominated by the energy over B, then there is a *bounded* sequence whose energy is converging to a minimax level—given that a certain boundary condition holds on S_R.

For later convenience we introduce the following notation for the above mentioned condition on A, B and f,

$$(LC)_{A,B,f} : \begin{cases} \overline{B}_R \supset A \text{ links } B \subset \overline{B}_R \text{ w.r.t } \Phi; \\ \sup_A f := a_0 \le b_0 := \inf_B f; \\ c_R := \inf_{\Gamma \in \Phi} \sup_{\substack{t \in [0,1] \\ u \in A}} f(\Gamma(t,u)) < +\infty. \end{cases}$$

The following is a direct generalization of Schechter's result [14, Theorem 5.2.1].

Theorem 5.8 *Let* $f : \overline{B}_R \to \mathbb{R}$ *be a locally Lipschitz functional such that* $(LC)_{A,B,f}$ *holds for some* $A, B \subset \overline{B}_R$. *Suppose that there exist* $\theta \in (0,1)$ *and* $\varepsilon > 0$ *such that*

$$0 \notin C(u, \theta), \text{ on } \{u \in S_R : |f(u) - c_R| \le \varepsilon\}. \tag{5.25}$$

Then there exists a sequence $\{u_n\} \subset \overline{B}_R$ *such that*

$$f(u_n) \to c_R \text{ and } \lambda_f(u_n) \to 0.$$

Furthermore, if $c_R = b_0$, *then* $d(u_n, B) \to 0$ *also holds.*

Proof Clearly, $b_0 \le c_R$. We distinguish two cases.

Case 1. $b_0 < c_R$.

Assume by contradiction that a sequence satisfying the required properties does not exist. Then one can find γ, $\delta > 0$ such that

$$\lambda_f(u) \ge \gamma, \text{ on } \{u \in \overline{B}_R : |f(u) - c_R| \le 3\delta\}.$$

Without loss of generality we may assume that $\delta < \min\{\varepsilon/3, c_R - b_0\}$. For $Z := \overline{B}_R$ and $c := c_R$ and $\rho := \frac{4\delta}{\gamma\theta^2}$, Theorem 4.4 ensures that there exists a continuous deformation $\sigma : [0, 1] \times \overline{B}_R \to \overline{B}_R$ such that (i)–(vi) hold. We reach contradiction by constructing a deformation $\overline{\Gamma} \in \Phi$ for which the "sup" in the definition of c_R is actually lower than c_R. By the definition of c_R, there exists $\Gamma \in \Phi$ such that

$$\sup_{t \in [0,1], \, u \in A} f(\Gamma(t,u)) \le c_R + \delta.$$

In other words

$$\Gamma(t, A) \subseteq f^{c_R + \delta}, \text{ for all } t \in [0, 1]. \tag{5.26}$$

Now let $\overline{\Gamma} : [0, 1] \times \overline{B}_R \to \overline{B}_R$ to be defined by

$$\overline{\Gamma}(t, u) := \begin{cases} \sigma(4t/3, u), & \text{if } t \in [0, 3/4], \\ \sigma(1, \Gamma(4t - 3, u)), & \text{if } t \in (3/4, 1]. \end{cases} \tag{5.27}$$

We claim that $\overline{\Gamma} \in \Phi$. Obviously (Φ_1) and (Φ_2) follow directly from the deformation theorem. In order to check (Φ_3), let $u_\Gamma \in \overline{B}_R$ be the element for which Γ satisfies (Φ_3), then $u_{\overline{\Gamma}} = \sigma(1, u_\Gamma)$ is suitable for $\overline{\Gamma}$.

Furthermore, we claim that

$$\overline{\Gamma}(t, A) \subseteq f^{c_R - \delta}, \text{ for all } t \in [0, 1].$$

Indeed, if $t \in [0, 3/4]$, then

$$f\left(\overline{\Gamma}(t, u)\right) = f\left(\sigma(4t/3, u)\right) \le f(u) \le a_0 \le b_0 < c_R - \delta,$$

for all $u \in A$. On the other hand, if $t \in (3/4, 1]$ then

$$f\left(\overline{\Gamma}(t, u)\right) = f(\sigma(1, \Gamma(4t - 3, u))) \le c - \delta,$$

for all $u \in A$.

In conclusion we constructed $\overline{\Gamma} \in \Phi$ such that

$$f\left(\overline{\Gamma}(t, u)\right) \le c_R - \delta, \text{ for all } u \in A \text{ and all } t \in [0, 1],$$

which contradicts the definition of c_R.

Case 2. $b_0 = c_R$.

We point out the fact that it suffices to prove that for any γ, $\delta > 0$ there exists $u \in \overline{B}_R$ such that

$$|f(u) - c_R| \le 3\delta, \ d(u, B) \le \frac{16\delta}{\gamma\theta^2} \text{ and } \lambda_f(u) < \gamma, \tag{5.28}$$

as we can set $\delta := 1/n^2$ and $\gamma := 1/n$ to get the desired sequence.

Assume by contradiction that (5.28) does not hold, i.e. there exist γ, $\delta > 0$ such that

$$\lambda_f(u) \geq \gamma, \ \text{on} \ \left\{ u \in \overline{B}_R : \ |f(u) - c_R| \leq 3\delta, \ d(u, B) \leq \frac{16\delta}{\gamma\theta^2} \right\}$$

and let $\sigma : [0, 1] \times \overline{B}_R \to \overline{B}_R$ be the deformation given by Theorem 4.4 with $c := c_R$, $\rho := \frac{4\delta}{\gamma\theta^2}$ and $Z := \{u \in \overline{B}_R : d(u, B) \leq \rho\}$.

We claim that

$$\sigma\left(1, f^{c_R+\delta}\right) \cap B = \emptyset, \tag{5.29}$$

and

$$\sigma(t, A) \cap B = \emptyset, \ \text{for all} \ t \in (0, 1]. \tag{5.30}$$

If there exists $u \in f^{c_R+\delta}$ such that $\sigma(1, u) \in B$, then

$$\|\sigma(1, u) - u\| = \|\sigma(1, u) - \sigma(0, u)\| \leq \rho\theta < \rho,$$

hence $u \in Z$. Property (vi) implies that

$$f(\sigma(1, u)) \leq c_R - \delta = b_0 - \delta,$$

which violates the definition of b_0.

In order to show that (5.30) holds, assume by contradiction that there exists $(t, u) \in (0, 1] \times A$ such that $\sigma(t, u) \in B$. If $\sigma(t, u) = u$, then $u \in A \cap B$, which contradicts the fact that A links B. If $\sigma(t, u) \neq u$, then

$$f(\sigma(t, u)) < f(u) \leq a_0 \leq b_0,$$

and this contradicts the definition of b_0.

Define $\overline{\Gamma} : [0, 1] \times \overline{B}_R \to \overline{B}_R$ formally as in (5.27). Clearly, $\overline{\Gamma} \in \Phi$, but (5.26), (5.29) and (5.30) imply that $\overline{\Gamma}(t, A) \cap B = \emptyset$ for all $t \in (0, 1]$ which contradicts the fact that A links B.   □

In the sequel we suppose that the boundary condition is dropped. Of course, one cannot expect to get the existence of a bounded Palais-Smale sequence in this case. However, we are able to prove that the following alternative holds:

either f possesses a Palais-Smale sequence in $\overline{B}_R$,

or,

there exist $\{u_n\} \subset S_R$ *and* $\zeta_n \in \partial_C f(u_n)$ *such that*

$$\zeta_n - \frac{\langle \zeta_n, u_n \rangle}{R\phi(R)} J_\phi u_n \to 0, \ \text{as } n \to \infty.$$

Before stating the result, for each $u \in \overline{B}_R$ we define projection $\pi_u : X^* \to \ker u$ as follows

$$\pi_u(\xi) := \begin{cases} \xi - \frac{\langle \xi, u \rangle}{\|u\|\phi(\|u\|)} J_\phi u, & \text{if } u \neq 0, \\ \xi, & \text{if } u = 0. \end{cases}$$

Obviously,

$$\|\pi_u(\xi)\| = \left\| \xi - \frac{\langle \xi, u \rangle}{\|u\|\phi(\|u\|)} J_\phi u \right\| \leq \|\xi\| + \frac{|\langle \xi, u \rangle|}{\|u\|} \leq 2\|\xi\|, \ \text{for all } \xi \in X^*.$$

For $u \neq 0$ and $\alpha \in \mathbb{R}$ and $\xi \in X^*$ we have the following estimates

$$\|\xi - \alpha J_\phi u\| = \left\| \pi_u(\xi) + \left(\frac{\langle \xi, u \rangle}{\|u\|\phi(\|u\|)} - \alpha \right) J_\phi u \right\|$$

$$\leq \|\pi_u(\xi)\| + \left| \frac{\langle \xi, u \rangle}{\|u\|\phi(\|u\|)} - \alpha \right| \phi(\|u\|),$$

and

$$\|\pi_u(\xi)\| = \|\pi_u(\xi - \alpha J_\phi u)\| \leq 2\|\xi - \alpha J_\phi u\|.$$

Taking the infimum as $\alpha \in \mathbb{R}$ we get

$$d(\xi, \mathbb{R} J_\phi u) \leq \|\pi_u(\xi)\| \leq 2d(\xi, \mathbb{R} J_\phi u), \ \text{for all } \xi \in X^*. \tag{5.31}$$

Moreover, restricting the infimum to $\mathbb{R}_-$ or $\mathbb{R}_+^*$ we also have

$$\langle \xi, u \rangle \leq 0 \Rightarrow d(\xi, \mathbb{R}_- J_\phi u) \leq \|\pi_u(\xi)\| \leq 2d(\xi, \mathbb{R}_- J_\phi u), \tag{5.32}$$

and

$$\langle \xi, u \rangle > 0 \Rightarrow d(\xi, \mathbb{R}_+^* J_\phi u) \leq \|\pi_u(\xi)\| \leq 2d(\xi, \mathbb{R}_+^* J_\phi u). \tag{5.33}$$

We will also make use of the following decomposition of $\partial_C f(u)$

$$\partial_C f^-(u) := \{\zeta \in \partial_C f(u) : \langle \zeta, u \rangle \leq 0\}, \quad \partial_C f^+(u) := \{\zeta \in \partial_C f(u) : \langle \zeta, u \rangle > 0\}.$$

Theorem 5.9 ([4]) *Let $f : \overline{B}_R \to \mathbb{R}$ be a locally Lipschitz functional and let $A, B \subset \overline{B}_R$ be such that $(LC)_{A,B,f}$ holds. Assume in addition that there exists $\Lambda_R > 0$ such that*

$$|\langle \zeta, u \rangle| \leq \Lambda_R, \text{ for all } u \in S_R \text{ and all } \zeta \in \partial_C f(u). \tag{5.34}$$

Then the following alternative holds:

(A_1) *there exists $\{u_n\} \subset \overline{B}_R$ such that $f(u_n) \to c_R$ and $\lambda_f(u_n) \to 0$. Furthermore, if $c_R = b_0$, then $d(u_n, B \cup S_R) \to 0$;*
(A_2) *there exist $\{u_n\} \subset S_R$ and $\{\zeta_n\} \subset X^*$ with $\zeta_n \in \partial_C f(u_n)$ such that*

$$f(u_n) \to c_R, \ \|\pi_{u_n}(\zeta_n)\| \to 0 \text{ and } \langle \zeta_n, u_n \rangle \leq 0.$$

Proof Assume option (A_2) does not hold. Then there exist $\gamma, \delta > 0$ such that

$$\|\pi_u(\zeta)\| \geq \gamma, \tag{5.35}$$

whenever $u \in S_R$ and $\zeta \in \partial_C f(u)$ satisfy

$$|f(u) - c_R| \leq \delta \text{ and } \langle \zeta, u \rangle \leq 0. \tag{5.36}$$

Obviously if there exist $\theta \in (0, 1)$ and $\varepsilon > 0$ such that

$$0 \notin C(u, \theta), \text{ on } \{u \in S_R : |f(u) - c_R| \leq \varepsilon\},$$

then (A_1) is obtained via Theorem 5.8.

If this is not the case, then for each $n \in \mathbb{N}$ there exists $u_n \in S_R$ such that

$$|f(u_n) - c_R| \leq \frac{1}{n} \text{ and } 0 \in C\left(u_n, \frac{1}{n}\right).$$

Proposition 4.3 implies that $\mathbb{R}_- J_\phi u_n \cap [\partial_C f]_{\theta_n}(u_n) \neq \varnothing$, that is, there exist $\zeta_n \in \partial_C f(u_n)$, $\eta_n \in \overline{B}_{X^*}(0, 1)$ and $\xi_n \in \mathbb{R}_- J_\phi u_n$ such that

$$\zeta_n + \frac{1}{n}\lambda_f(u_n)\eta_n = \xi_n,$$

hence

$$d(\zeta_n, \mathbb{R}_- J_\phi u_n) \leq \|\zeta_n - \xi_n\| \leq \frac{1}{n}\lambda_f(u_n) \leq \frac{1}{n}\|\zeta_n\| \leq \frac{1}{n}\|\pi_{u_n}(\zeta_n)\| + \frac{1}{n}\frac{|\langle \zeta_n, u_n \rangle|}{R}$$

$$\leq \frac{2}{n}d(\zeta_n, \mathbb{R}J_\phi u_n) + \frac{\Lambda_R}{nR} \leq \frac{2}{n}d(\zeta_n, \mathbb{R}_- J_\phi u_n) + \frac{\Lambda_R}{nR},$$

which leads to

$$d(\zeta_n, \mathbb{R}_- J_\phi u_n) \to 0, \text{ as } n \to \infty. \tag{5.37}$$

Conditions (5.32), (5.35) and (5.36) ensure that there exists $n_0 \in \mathbb{N}$ such that

$$\langle \zeta_n, u_n \rangle > 0, \text{ for all } n \geq n_0. \tag{5.38}$$

From (5.37) and (5.38) we deduce that

$$d(\partial_C f^+(u_n), \mathbb{R}_- J_\phi u_n) \to 0, \text{ as } n \to \infty. \tag{5.39}$$

On the other hand, taking the infimum as $\zeta \in \partial_C f^+(u_n)$ in (5.33) and keeping in mind (5.31) we get

$$d(\partial_C f^+(u_n), \mathbb{R}^*_+ J_\phi u_n) \leq \inf_{\zeta \in \partial_C f^+(u_n)} \|\pi_{u_n}(\zeta)\| \leq 2 \inf_{\zeta \in \partial_C f^+(u_n)} d(\zeta, \mathbb{R}J_\phi u_n)$$

$$\leq 2d(\zeta_n, \mathbb{R}J_\phi u_n) \leq 2d(\zeta_n, \mathbb{R}_- J_\phi u_n),$$

hence

$$d(\partial_C f^+(u_n), \mathbb{R}^*_+ J_\phi u_n) \to 0, \text{ as } n \to \infty. \tag{5.40}$$

Relations (5.39) and (5.40) ensure that for sufficiently large $n \in \mathbb{N}$ there exist $\alpha_n \in \mathbb{R}_-$, $\beta_n \in \mathbb{R}^*_+$ and $\zeta'_n, \zeta''_n \in \partial_C f^+(u_n)$ such that

$$\max\{\|\zeta'_n - \alpha_n J_\phi u_n\|, \|\zeta''_n - \beta_n J_\phi u_n\|\} \to 0, \text{ as } n \to \infty.$$

Define $t_n := \frac{\beta_n}{\beta_n - \alpha_n} \in (0, 1]$ and $\bar{\zeta}_n := t_n \zeta'_n + (1 - t_n)\zeta''_n$. Since $\partial_C f^+(u_n)$ is convex it follows that $\bar{\zeta}_n \in \partial_C f^+(u_n)$. Then

$$\|\bar{\zeta}_n\| = \|t_n \zeta'_n + (1 - t_n)\zeta''_n\| = \|t_n(\zeta'_n - \alpha_n J_\phi u_n) + (1 - t_n)(\zeta''_n - \beta_n J_\phi u_n)\|$$

$$\leq t_n \|\zeta'_n - \alpha_n J_\phi u_n\| + (1 - t_n)\|\zeta''_n - \beta_n J_\phi u_n\|$$

$$\leq \max\{\|\zeta'_n - \alpha_n J_\phi u_n\|, \|\zeta''_n - \beta_n J_\phi u_n\|\}.$$

We have proved thus that there exists $\{u_n\} \subseteq S_R$ and such that $|f(u_n) - c_R| \leq \frac{1}{n}$ and

$$\lambda_f(u_n) \leq \|\bar{\zeta}_n\| \to 0, \text{ as } n \to \infty,$$

that is, (A_1) holds.   □

Corollary 5.7 *Assume the hypotheses of Theorem 5.9 are fulfilled. Then there exists* $\{u_n\} \subset \overline{B}_R$, $\{\zeta_n\} \subset X^*$ *with* $\zeta_n \in \partial_C f(u_n)$ *and* $v \in \mathbb{R}_-$ *such that*

$$f(u_n) \to c_R, \quad \|\pi_{u_n}(\zeta_n)\| \to 0 \quad and \quad \langle \zeta_n, u_n \rangle \to v.$$

Furthermore, if $c_R = b_0$*, then* $d(u_n, B \cup S_R) \to 0$.

Proof Suppose that (A_1) of the alternative theorem holds, i.e. $f(u_n) \to c_R$ and $\lambda_f(u_n) \to 0$ and let $\zeta_n \in \partial_C f(u_n)$ be such that $\|\zeta_n\| = \lambda_f(u_n)$. Then

$$\|\pi_{u_n}(\zeta_n)\| \leq 2\|\zeta_n\| \to 0, \text{ as } n \to \infty,$$

and

$$|\langle \zeta_n, u_n \rangle| \leq \|\zeta_n\|\|u_n\| \leq R\|\zeta_n\| \to 0, \text{ as } n \to \infty,$$

hence we can choose $v := 0$ in this case.

On the other hand, if (A_2) holds, then condition (5.34) implies that the sequence $v_n := \langle \zeta_n, u_n \rangle \leq 0$ is bounded in $\mathbb{R}$ hence possesses a convergent subsequence.

Finally, if $c_R = b_0$, then (A_1) implies $d(u_n, B \cup S_R) \to 0$, while (A_2) ensures that $d(u_n, S_R) = 0$, hence the proof is complete.   □

Remark 5.6 If $f : \overline{B}_R \to \mathbb{R}$ is a C^1-functional, then the conclusion of the previous corollary reads as follows: there exists $\{u_n\} \subset \overline{B}_R$ such that

$$f(u_n) \to c_R, \quad \left\| f'(u_n) - \frac{\langle f'(u_n), u_n \rangle}{\|u_n\|\phi(\|u_n\|)} J_\phi u_n \right\| \to 0, \langle f'(u_n), u_n \rangle \to v \leq 0,$$

If, in addition, X is a Hilbert space, then this reduces to Schechter's conclusion (see e.g. [14, Corollary 5.3.2.]).

If f is bounded below, then following similar steps as in the proofs of Theorems 5.7, 5.9 and Corollary 5.7 one can prove the following result.

Theorem 5.10 *Let* $f : \overline{B}_R \to \mathbb{R}$ *be a locally Lipschitz satisfying (5.23) and (5.34). Then there exist* $\{u_n\} \subset \overline{B}_R$, $\{\zeta_n\} \subset X^*$ *with* $\zeta_n \in \partial_C f(u_n)$ *and* $v \in \mathbb{R}_-$ *such that*

$$f(u_n) \to m_R, \ \|\pi_{u_n}(\zeta_n)\| \to 0 \text{ and } \langle \zeta_n, u_n \rangle \to v.$$

Definition 5.4 We say that a locally Lipschitz functional $f : X \to \mathbb{R}$ satisfies the *Schechter-Palais-Smale condition at level* c *in* $\overline{B}_R$, $(SPS)_c$ for short, if any sequence $\{u_n\} \subset \overline{B}_R$ satisfying:

(SPS_1) $f(u_n) \to c$, as $n \to \infty$;
(SPS_2) there exist $\zeta_n \in \partial_C f(u_n)$ and $v \leq 0$ s.t. $\|\pi_{u_n}(\zeta_n)\| \to 0$ and $\langle \zeta_n, u_n \rangle \to v$,

possesses a (strongly) convergent subsequence.

Theorem 5.11 *Let* $f : \overline{B}_R \to \mathbb{R}$ *be a locally Lipschitz functional such that the* $(LC)_{A,B,f}$ *holds for some* $A, B \subset \overline{B}_R$. *Assume in addition that (5.34) and* $(SPS)_{c_R}$ *hold. Then the following alternative holds:*

(A'_1) *there exists* $u \in \overline{B}_R$ *such that* $f(u) = c_R$ *and* $0 \in \partial_C f(u)$. *Furthermore, if* $c_R = b_0$, *then* $u \in \overline{B} \cup S_R$;
(A'_2) *there exist* $u \in S_R$ *and* $\lambda < 0$ *such that* $f(u) = c_R$ *and* $\lambda J_\phi u \in \partial_C f(u)$.

Proof If case (A_1) of Theorem 5.9 holds, then there exists $\{u_n\} \subset \overline{B}_R$ such that

$$f(u_n) \to c_R, \ \text{and} \ \lambda_f(u_n) \to 0.$$

Let $\zeta_n \in \partial_C f(u_n)$ be such that $\|\zeta_n\| = \lambda_f(u_n)$. Then

$$\|\pi_{u_n}(\zeta_n)\| \leq 2\|\zeta_n\| \to 0, \ \text{as} \ n \to \infty,$$

and

$$|\langle \zeta_n, u_n \rangle| \leq R\|\zeta_n\| \to v = 0, \ \text{as} \ n \to \infty.$$

The $(SPS)_{c_R}$ condition there exists a subsequence $\{u_{n_k}\}$ of $\{u_n\}$ and $u \in \overline{B}_R$ such that $u_{n_k} \to u$ in X. Moreover, $\zeta_{n_k} \in \partial_C f(u_{n_k})$ and $\zeta_{n_k} \to 0$, thus Proposition 2.3 ensures that $0 \in \partial_C f(u)$. If $c_R = b_0$, then $d(u_{n_k}, B \cup S_R) \to 0$, hence $u \in \overline{B} \cup S_R$.

On the other hand, if case (A_2) of Theorem 5.9 holds, then there exist $\{u_n\} \in S_R$, $\zeta_n \in \partial_C f(u_n)$ and $v \leq 0$ such that

$$f(u_n) \to c_R, \ \|\pi_{u_n}(\zeta_n)\| \to 0 \text{ and } \langle \zeta_n, u_n \rangle \to v.$$

The $(SPS)_{c_R}$ condition and Proposition 4.1 show that there exist $u \in S_R$, $\zeta \in \partial_C f(u)$ and two subsequences $\{u_{n_k}\}$, $\{\zeta_{n_k}\}$ of $\{u_n\}$ and $\{\zeta_n\}$, respectively, such that

$$u_{n_k} \to u \text{ and } \zeta_{n_k} \rightharpoonup \zeta.$$

But J_ϕ is demicontinuous, hence

$$\pi_{u_{n_k}}(\zeta_{n_k}) = \zeta_{n_k} - \frac{\langle \zeta_{n_k}, u_{n_k} \rangle}{R\phi(R)} J_\phi u_{n_k} \rightharpoonup \zeta - \frac{v}{R\phi(R)} J_\phi u,$$

which together with $\pi_{u_{n_k}}(\zeta_{n_k}) \to 0$ gives

$$\zeta = \frac{v}{R\phi(R)} J_\phi u \in \partial_C f(u).$$

If $v = 0$, then (A'_1) holds, while $v < 0$ implies that (A'_2) holds for $\lambda := \frac{v}{R\phi(R)}$. $\square$

The next result follows directly from Theorem 5.10 and the (SPS)-condition.

Theorem 5.12 *Assume the hypotheses of Theorem 5.10 are fulfilled and assume* $(SPS)_{m_R}$ *also holds. Then there exist* $u \in \overline{B}_R$ *and* $\lambda \leq 0$ *such that*

$$f(u) = m_R \text{ and } \lambda J_\phi u \in \partial_C f(u).$$

Furthermore, $\lambda \neq 0 \implies u \in S_R$.

Assuming the hypotheses of Theorems 5.11 and 5.12 are simultaneously satisfied, one can obtain multiplicity results of the following type.

Theorem 5.13 *Let* $f : \overline{B}_R \to \mathbb{R}$ *be a locally Lipschitz functional such that (5.23) and (5.34) hold. Suppose there exist two subsets* A, B *of* $\overline{B}_R$ *such that* $(LC)_{A,B,f}$ *holds and condition* $(SPS)_c$ *is satisfied for* $c \in \{c_R, m_R\}$. *Then there exist* $u_1, u_2 \in \overline{B}_R$ *and* $\lambda_1, \lambda_2 \leq 0$ *such that* $u_1 \neq u_2$ *and*

$$\lambda_k J_\phi u_k \in \partial_C f(u_k), \quad k = 1, 2. \tag{5.41}$$

Furthermore, if $\lambda_k < 0$, *then* $u_k \in S_R$. *Also, if there exist* $v_0, v_1 \in A \cap B_R$ *distinct such that* $f(v_1) \leq f(v_0)$ *and* $v_0 \notin B$, *then* u_1 *and* u_2 *can be chosen in such a way that* $v_0 \notin \{u_1, u_2\}$.

Proof It follows from Theorems 5.11 and 5.12 that there exist $u_1, u_2 \in \overline{B}_R$ and $\lambda_1, \lambda_2 \leq 0$ such that

$$f(u_1) = m_R \leq c_R = f(u_2), \text{ and } \lambda_k J_\phi u_k \in \partial_C f(u_k), \ k = 1, 2.$$

The fact that $\lambda_k < 0 \Rightarrow u_k \in S_R$, follows directly from Theorems 5.11 and 5.12, respectively. In order to complete the proof we consider the following cases:

(i) $m_R \leq b_0 < c_R$.
    Then

$$f(u_1) = m_R \leq f(v_1) \leq f(v_0) \leq a_0 \leq b_0 < c_R = f(u_2),$$

hence $u_1 \neq u_2$ and $v_0 \neq u_2$. If $u_1 = v_0$, then $f(v_1) = m_R$, that is v_1 is a global minimum point of f on B_R. As any extremum point of a locally Lipschitz functional is in fact a critical point, we conclude that $0 \in \partial_C f(v_1)$, which shows that v_1, u_2 satisfy the conclusion of the theorem.

(ii) $m_R < b_0 = c_R$.
    Then

$$f(u_1) = m_R < b_0 = c_R = f(u_2),$$

hence $u_1 \neq u_2$. Moreover, $u_2 \in \overline{B} \cup S_R$ which shows that $v_0 \neq u_2$. Again, if $u_1 = v_0$, then we can replace u_1 with v_1.

(iii) $m_R = b_0 = c_R$.
    Then each point of A is a solution of (5.41). Note that A must have at least two points in order to link B. It is readily seen that we only need to discuss the case $A = \{v_0, v_1\} \subset B_R$ and $v_1 \in \overline{B}$. Let $\rho \in (0, \|v_1 - v_0\|)$ be such that $S_\rho(v_0) \subset B_R$. Then A links $S_\rho(v_0)$ (see Example E.1 and Remark E.1 in Appendix E) and

$$m_R \leq \inf_{S_\rho(v_0)} f \leq \inf_{\Gamma \in \Phi} \sup_{t \in [0,1], \, u \in A} f(\Gamma(t, u)) = m_R.$$

Theorem 5.11 ensures that (5.41) possesses a solution $u_* \in S_\rho(v_0) \cup S_R$, hence $u_* \neq v_0$. $\square$

5.5 Minimax Results for Szulkin Functionals

In this section we suppose that X is a real Banach space and f a function on X satisfying the hypothesis:

(H) $f := \varphi + \psi$, where $\varphi \in C^1(X, \mathbb{R})$ and $\psi : X \to (-\infty, +\infty]$ is convex and proper and l.s.c.

In this section we prove the Mountain Pass Theorem for functionals satisfying (H).

Theorem 5.14 (Mountain Pass Theorem, Szulkin [15]) *Suppose that* $f : X \to$ $(-\infty, \infty]$ *satisfies* (H) *and the* (PS) *condition. Moreover, assume that*

(i) $f(0) = 0$ *and there exist constants* $\rho > 0$ *and* $\alpha > 0$ *such that*

$$f(u) \geq \alpha \text{ for } u \in S(0, \rho);$$

(ii) $f(e) \leq 0$ *for some* $e \notin \overline{B(0, \rho)}$.

Then

$$\alpha \leq c := \inf_{g \in \Gamma} \sup_{t \in [0,1]} f(g(t)),$$

and c *is a critical value of* f, *where*

$$\Gamma := \{g \in C([0, 1], X) : g(0) = 0, g(1) = e\}.$$

Proof Since $g([0, 1]) \cap S(0, \rho) \neq \emptyset$ for all $g \in \Gamma$, then $c \geq \alpha$. Suppose that c is not a critical value. We now apply Lemma 4.7 with $N := \emptyset$ and $\bar{\varepsilon} := c$ and $\varepsilon < \bar{\varepsilon}$ be a positive constant from that lemma. It follows from the definition of c that $f^{c-\frac{\varepsilon}{4}}$ is not path connected and let us denote by W_0 and W_e components containing 0 and e, respectively. We now replace Γ by a collection of paths Γ_1 with "loose ends" defined by

$$\Gamma_1 := \left\{ g \in C([0, 1], X); \ g(0) \in W_0 \cap f^{c-\frac{\varepsilon}{2}}, \ g(1) \in W_e \cap f^{c-\frac{\varepsilon}{2}} \right\}$$

and set

$$c_1 := \inf_{g \in \Gamma_1} \sup_{t \in [0,1]} f(g(t)).$$

We now show that $c_1 = c$. Since $\Gamma \subset \Gamma_1$, $c_1 \leq c$. If $c_1 < c$, then there exists $g \in \Gamma_1$ such that $\sup_{t \in [0,1]} f(g(t)) < c$. Since $g(0)$ can be joined to 0 and $g(1)$ to e by paths lying in $f^{c-\frac{\varepsilon}{4}}$, we see that there exists a path $g \in \Gamma$ such that $\sup_{t \in [0,1]} f(g(t)) < c$, which is impossible. Since φ is continuous and ψ is convex it is routine to show that Γ_1 is a closed subspace of Γ and consequently Γ_1 is a complete metric space. Since Γ_1 is a complete metric space it is easy to show that a functional $\Pi : \Gamma_1 \to (-\infty, \infty]$ defined by

$$\Pi(g) := \sup_t f(g(t))$$

is lower semicontinuous. We now apply Ekeland's variational principle with $\varepsilon > 0$ and $\lambda := 1$ to obtain a path $\gamma \in \Gamma_1$ such that $\Pi(\gamma) \leq c + \varepsilon$ and

$$\Pi(g) - \Pi(\gamma) \geq -\varepsilon d(\gamma, g) \tag{5.42}$$

for all $g \in \Gamma_1$. Let $A := \gamma([0, 1])$ and let α_s be deformation mapping from Lemma 4.7 corresponding to A and satisfying (4.67)–(4.71). Setting $g := \alpha_s \circ \gamma$ we check that $g \in \Gamma_1$ for s sufficiently small. Indeed, if $f(\gamma(0)) \in \left(c - \varepsilon, c - \frac{\varepsilon}{2}\right]$, then by (4.68) $f(g(0)) \leq f(\gamma(0)) \leq c - \frac{\varepsilon}{2}$ and if $f(\gamma(0)) \leq c - \varepsilon$, then by (4.70) $f(g(0)) \leq f(\gamma(0)) + 2s < c - \frac{\varepsilon}{2}$. In both cases $g \in W_0 \cap f^{c - \frac{\varepsilon}{2}}$. Similarly, we show that $g(1) \in W_e \cap f^{c - \frac{\varepsilon}{2}}$. Therefore, $g \in \Gamma_1$. According to (4.67) $d(\gamma, g) \leq s$, it then follows from (4.70) and (5.42) that

$$-\varepsilon s \leq -\varepsilon d(\gamma, g) \leq \Pi(g) - \Pi(\gamma) \leq -2\varepsilon s$$

and we arrived at a contradiction.   □

Corollary 5.8 *Suppose that f satisfies (H) and the condition (PS). If 0 is a local minimum of f and if $f(e) \leq f(0)$ for some $e \neq 0$, then f has a critical point u distinct from 0 and e. In particular, if f has two local minima, then it has at least three critical points.*

Proof Without loss of generality we may always assume that $f(0) = 0$. If there exist constants $\alpha > 0$ and $\rho > 0$ such that $e \notin B(0, \rho)$ and $f(u) \geq \alpha$ for all $u \in S(0, \rho)$, then the existence of a critical point distinct from 0 and e is a consequence of Theorem 5.14. If such constants do not exist we choose $r < \|e\|$ so that $f(u) \geq 0$ for $u \in \overline{B(0, r)}$. We now apply Ekeland's variational principle with $\varepsilon := \frac{1}{m^2}$ and $\lambda := m$ and f restricted to $\overline{B(0, r)}$. Let $0 < \rho < r$. Since $\inf_{u \in S(0, \rho)} f(u) = 0$, there exists $w_m \in S(0, \rho)$ and $u_m \in \overline{B(0, \rho)}$ such that

$$0 \leq f(u_m) \leq f(w_m) \leq \frac{1}{m^2}, \quad \|u_m - w_m\| \leq \frac{1}{m}$$

and

$$f(z) - f(u_m) \geq -\frac{1}{m}\|z - w_m\|$$

for all $z \in \overline{B(0, r)}$. For m sufficiently large $u_m \in B(0, r)$. Let $v \in X$ and let $z = (1 - t)u_m + tv$. If $t > 0$ is sufficiently small then $z \in \overline{B(0, r)}$. We deduce from the last inequality and the convexity of ψ that

$$-\frac{t}{m}\|v - u_m\| \leq f((1 - t)u_m + tv) - f(u_m)$$

$$\leq \varphi(u_m + t(v - u_m)) - \varphi(u_m) + t(\psi(v) - \psi(u_m)).$$

Dividing by t and letting $t \to 0$ we get

$$\langle \varphi'(u_m), v - u_m \rangle + \psi(v) - \psi(u_m) \geq -\frac{1}{m} \|v - u_m\|$$

for all $v \in X$. According to the (PS) condition we may assume that $u_m \to u$, with $u \in S(0, \rho)$ and that u is a critical point which is distinct from 0 and e. Finally, if f has two local minima u_0 and u_1, we may always assume that $u_0 = 0$ and $f(u_1) \leq f(u_0) = 0$. By the previous part of the proof there exists a critical point u distinct from u_0 and u_1 and this completes the proof.   □

Remark 5.7 It is easy to observe that the "Saddle Point", "Linking" and "$\mathbb{Z}_2$-symmetric version of Mountain Pass" theorems remain true for functionals which satisfy the structure condition (H).

5.6   Ricceri-Type Multiplicity Results for Locally Lipschitz Functions

In this section we establish some multiplicity results for locally Lipschitz functionals depending on a real parameter.

For every $\tau \geq 0$, we introduce the following class of functions:

$$\mathcal{G}_\tau := \left\{ g \in C^1(\mathbb{R}, \mathbb{R}) : g \text{ is bounded and } g(t) = t, \forall t \in [-\tau, \tau] \right\}$$

The first result represents the main result from the paper of [7] and reads like this.

Theorem 5.15 Let $(X, \| \cdot \|)$ be a real reflexive Banach space and $\tilde{X}_i$ $(i = 1, 2)$ be two Banach spaces such that the embeddings $X \hookrightarrow \tilde{X}_i$ are compact. Let Λ be a real interval, $h : [0, \infty) \to [0, \infty)$ be a non-decreasing convex function, and let $\Phi_i : \tilde{X}_i \to \mathbb{R}$ $(i = 1, 2)$ be two locally Lipschitz functions such that $E_{\lambda,\mu} := h(\| \cdot \|) + \lambda \Phi_1 + \mu g \circ \Phi_2$ restricted to X satisfies the $(PS)_c$-condition for every $c \in \mathbb{R}$, $\lambda \in \Lambda$, $\mu \in [0, |\lambda| + 1]$ and $g \in \mathcal{G}_\tau$, $\tau \geq 0$. Assume that $h(\| \cdot \|) + \lambda \Phi_1$ is coercive on X for all $\lambda \in \Lambda$ and that there exists $\rho \in \mathbb{R}$ such that

$$\sup_{\lambda \in \Lambda} \inf_{x \in X} [h(\|x\|) + \lambda(\Phi_1(x) + \rho)] < \inf_{x \in X} \sup_{\lambda \in \Lambda} [h(\|x\|) + \lambda(\Phi_1(x) + \rho)]. \tag{5.43}$$

Then, there exist a non-empty open set $A \subset \Lambda$ and $r > 0$ with the property that for every $\lambda \in A$ there exists $\mu_0 \in]0, |\lambda| + 1]$ such that, for each $\mu \in [0, \mu_0]$ the functional $\mathcal{E}_{\lambda,\mu} := h(\| \cdot \|) + \lambda \Phi_1 + \mu \Phi_2$ has at least three critical points in X whose norms are less than r.

Proof Since h is a non-decreasing convex function, $X \ni x \mapsto h(\|u\|)$ is also convex; thus, $h(\| \cdot \|)$ is sequentially weakly lower semicontinuous on X. From the fact that the embeddings $X \hookrightarrow \tilde{X}_i$ $(i = 1, 2)$ are compact and $\Phi_i : \tilde{X}_i \to \mathbb{R}$ $(i = 1, 2)$ are locally Lipschitz functions, it follows that the function $E_{\lambda,\mu}$ as well as $\varphi : X \times \Lambda \to \mathbb{R}$ (in the first variable) given by

$$\varphi(u, \lambda) := h(\|u\|) + \lambda(\Phi_1(u) + \rho)$$

are sequentially weakly lower semicontinuous on X.

The function φ satisfies the hypotheses of Theorem D.10. Fix $\sigma > \sup_\Lambda \inf_X \varphi$ and consider a nonempty open set Λ_0 with the property expressed in Theorem D.10. Let $A := [a, b] \subset \Lambda_0$.

Fix $\lambda \in [a, b]$; then, for every $\tau \geq 0$ and $g_\tau \in \mathcal{G}_\tau$, there exists $\mu_\tau > 0$ such that, for any $\mu \in]0, \mu_\tau[$, the functional $E_{\lambda,\mu}^\tau = h(\| \cdot \|) + \lambda\Phi_1 + \mu g_\tau \circ \Phi_2$ restricted to X has two local minima, say u_1^τ, u_2^τ, lying in the set $\{u \in X : \varphi(u, \lambda) < \sigma\}$.

Note that

$$\cup_{\lambda \in [a,b]} \{u \in X : \varphi(u, \lambda) < \sigma\} \subset \{u \in X : h(\|u\|) + a\Phi_1(u) < \sigma - a\rho\}$$

$$\cup \{u \in X : h(\|u\|) + b\Phi_1(u) < \sigma - b\rho\}.$$

Because the function $h(\| \cdot \|) + \lambda\Phi_1$ is coercive on X, the set on the right-side is bounded. Consequently, there is some $\eta > 0$, such that

$$\cup_{\lambda \in [a,b]} \{u \in X : \varphi(u, \lambda) < \sigma\} \subset B_\eta, \tag{5.44}$$

where $B_\eta := \{u \in X : \|u\| < \eta\}$. Therefore, $u_1^\tau, u_2^\tau \in B_\eta$. Now, set $c^* := \sup_{t \in [0,\eta]} h(t) + \max\{|a|, |b|\} \sup_{B_\eta} |\Phi_1|$ and fix $r > \eta$ large enough such that for any $\lambda \in [a, b]$ to have

$$\{u \in X : h(\|u\|) + \lambda\Phi_1(u) \leq c^* + 2\} \subset B_r. \tag{5.45}$$

Let $r^* := \sup_{B_r} |\Phi_2|$ and correspondingly, fix a function $g = g_{r^*} \in \mathcal{G}_{r^*}$. Let us define $\mu_0 := \min\left\{|\lambda| + 1, \frac{1}{1+\sup|g|}\right\}$. Since the functional $E_{\lambda,\mu} := E_{\lambda,\mu}^{r^*} = h(\| \cdot \|) + \lambda\Phi_1 + \mu g_{r^*} \circ \Phi_2$ restricted to X satisfies the $(PS)_c$ condition for every $c \in \mathbb{R}$, $\mu \in [0, \mu_0]$, and $u_1 = u_1^{r^*}, u_2 = u_2^{r^*}$ are local minima of $E_{\lambda,\mu}$, we may apply Corollary 5.4, obtaining that

$$c_{\lambda,\mu} = \inf_{\gamma \in \Gamma} \max_{s \in [0,1]} E_{\lambda,\mu}(\gamma(s)) \geq \max\{E_{\lambda,\mu}(u_1), E_{\lambda,\mu}(u_2)\} \tag{5.46}$$

is a critical value for $E_{\lambda,\mu}$, where Γ is the family of continuous paths $\gamma : [0, 1] \to X$ joining u_1 and u_2. Therefore, there exists $u_3 \in X$ such that

$$c_{\lambda,\mu} = E_{\lambda,\mu}(u_3) \text{ and } 0 \in \partial_C E_{\lambda,\mu}(u_3).$$

If we consider the path $\gamma \in \Gamma$ given by $\gamma(s) := u_1 + s(u_2 - u_1) \subset B_\eta$ we have

$$h(\|u_3\|) + \lambda\Phi_1(u_3) = E_{\lambda,\mu}(u_3) - \mu g(\Phi_2(u_3)) = c_{\lambda,\mu} - \mu g(\Phi_2(u_3))$$

$$\leq \sup_{s\in[0,1]} (h(\|\gamma(s)\|) + \lambda\Phi_1(\gamma(s)) + \mu g(\Phi_2(\gamma(s)))) - \mu g(\Phi_2(u_3))$$

$$\leq \sup_{t\in[0,\eta]} h(t) + \max\{|a|, |b|\} \sup_{B_\eta} |\Phi_1| + 2\mu_0 \sup |g|$$

$$\leq c^* + 2.$$

From (5.45) it follows that $u_3 \in B_r$. Therefore, u_i, $i = 1, 2, 3$ are critical points for $E_{\lambda,\mu}$, all belonging to the ball B_r. It remains to prove that these elements are critical points not only for $E_{\lambda,\mu}$ but also for $\mathcal{E}_{\lambda,\mu} = h(\|\cdot\|) + \lambda\Phi_1 + \mu\Phi_2$. Let $u = u_i$, $i \in \{1, 2, 3\}$. Since $u \in B_r$, we have that $|\Phi_2(u)| \leq r^*$. Note that $g(t) = t$ on $[-r^*, r^*]$; thus, $g(\Phi_2(u)) = \Phi_2(u)$. Consequently, on the open set B_r the functionals $E_{\lambda,\mu}$ and $\mathcal{E}_{\lambda,\mu}$ coincide, which completes the proof. $\qquad\square$

We present next the main theoretical result from the paper of Costea & Varga [3]. For this first we describe the framework.

Let X be a real reflexive Banach space and Y, Z two Banach spaces such that there exist $T : X \to Y$ and $S : X \to Z$ linear and compact. Let $L : X \to \mathbb{R}$ be a sequentially weakly lower semicontinuous C^1 functional such that $L' : X \to X^*$ has the $(S)_+$ property, i.e. if $u_n \rightharpoonup u$ in X and $\limsup_{n\to\infty}\langle L'(u_n), u_n - u\rangle \leq 0$, then $u_n \to u$. Assume in addition that $J_1 : Y \to \mathbb{R}$, $J_2 : Z \to \mathbb{R}$ are two locally Lipschitz functionals.

We are interested in studying the existence of critical points for functionals $\mathcal{E}_\lambda : X \to \mathbb{R}$ of the following type

$$\mathcal{E}_\lambda(u) := L(u) - (J_1 \circ T)(u) - \lambda(J_2 \circ S)(u), \qquad (5.47)$$

where $\lambda > 0$ is a real parameter.

We point out the fact that it makes sense to talk about critical points for the functional defined in (5.47) as $\mathcal{E}_\lambda$ is locally Lipschitz. We also point out the fact that the functional $\mathcal{E}_\lambda$ is sequentially weakly lower semicontinuous since we assumed L to be sequentially weakly lower semicontinuous and T, S to be compact operators.

We assume the following conditions are fulfilled:

$(\mathcal{H}_1)$ there exists $u_0 \in X$ such that u_0 is a strict local minimum for L and $L(u_0) = (J_1 \circ T)(u_0) = (J_2 \circ S)(u_0) = 0$;

$(\mathcal{H}_2)$ for each $\lambda > 0$ the functional $\mathcal{E}_\lambda$ is coercive and we can determine $u_\lambda^0 \in X$ such that $\mathcal{E}_\lambda(u_\lambda^0) < 0$;

($\mathcal{H}_3$) there exists $R_0 > 0$ such that

$$(J_1 \circ T)(u) \leq L(u) \quad \text{and} \quad (J_2 \circ S)(u) \leq 0, \quad \forall u \in \bar{B}(u_0; R_0) \setminus \{u_0\};$$

($\mathcal{H}_4$) there exists $\rho \in \mathbb{R}$ such that

$$\sup_{\lambda > 0} \inf_{u \in X} \{\lambda\,[L(u) - (J_1 \circ T)(u) + \rho] - (J_2 \circ S)(u)\} <$$

$$\inf_{u \in X} \sup_{\lambda > 0} \{\lambda\,[L(u) - (J_1 \circ T)(u) + \rho] - (J_2 \circ S)(u)\}.$$

Theorem 5.16 *Assume that conditions ($\mathcal{H}_1$)–($\mathcal{H}_3$) are fulfilled. Then for each $\lambda > 0$ the functional $\mathcal{E}_\lambda$ defined in (5.47) has at least three critical points. If in addition ($\mathcal{H}_4$) holds, then there exists $\lambda^* > 0$ such that $\mathcal{E}_{\lambda^*}$ has at least four critical points.*

Proof The proof will be carried out in four steps an relies essentially on the zero altitude mountain pass theorem (see Corollary 5.4) combined with a technique of finding global minima for parametrized functions developed by Ricceri (see Theorem D.11). Let us first fix $\lambda > 0$ and assume that ($\mathcal{H}_1$)–($\mathcal{H}_3$) are fulfilled.

Step 1. *u_0 is a critical point for $\mathcal{E}_\lambda$.*
Since $u_0 \in X$ is a strict local minimum for L there exists $R_1 > 0$ such that

$$L(u) > 0, \quad \forall u \in \bar{B}(u_0; R_1) \setminus \{u_0\}. \tag{5.48}$$

From ($\mathcal{H}_3$) we deduce that

$$\frac{(J_1 \circ T)(u) + \lambda(J_2 \circ S)(u)}{L(u)} \leq 1, \quad \forall u \in \bar{B}(u_0; R_0) \setminus \{u_0\}. \tag{5.49}$$

Taking $R_2 = \min\{R_0, R_1\}$ from (5.48) and (5.49) we have

$$\mathcal{E}_\lambda(u) = L(u) - (J_1 \circ T)(u) - \lambda(J_2 \circ S)(u) \geq 0,\, \forall u \in \bar{B}(u_0; R_2) \setminus \{u_0\} \tag{5.50}$$

We have proved thus that $u_0 \in X$ is a local minimum for $\mathcal{E}_\lambda$, therefore it is a critical point for this functional.

Step 2. *The functional $\mathcal{E}_\lambda$ admits a global minimum point $u_1 \in X \setminus \{u_0\}$.*
Indeed, such a point exists since the functional $\mathcal{E}_\lambda$ is sequentially weakly lower semicontinuous and coercive, therefore it admits a global minimizer denoted u_1. Moreover, from ($\mathcal{H}_2$) we deduce that $\mathcal{E}_\lambda(u_1) < 0$, hence $u_1 \neq u_0$.

Step 3. *There exists $u_2 \in X \setminus \{u_0, u_1\}$ such that u_2 is a critical point for $\mathcal{E}_\lambda$.*

We shall prove first that the functional $\mathcal{E}_\lambda$ satisfies the (PS)-condition. Let $c \in \mathbb{R}$ be fixed and $\{u_n\} \subset X$ be a sequence such that

- $\mathcal{E}_\lambda(u_n) \to c$;
- there exists $\{\varepsilon_n\} \subset \mathbb{R}$, $\varepsilon_n \downarrow 0$ s.t. $\mathcal{E}_\lambda^0(u_n; v - u_n) + \varepsilon_n \|v - u_n\|_X \geq 0, \forall v \in X$.

Obviously $\{u_n\}$ is bounded due to the fact that $\mathcal{E}_\lambda$ is coercive. Then there exists $u \in X$ such that, up to a subsequence, $u_n \rightharpoonup u$ in X. Taking into account that T, S are compact operators we infer that $Tu_n \to Tu$ in Y and $Su_n \to Su$ in Z. For $v = u$ we have

$$0 \leq \varepsilon_n \|u - u_n\|_X + \mathcal{E}_\lambda^0(u_n; u - u_n) = \varepsilon_n \|u - u_n\|_X + (L - J_1 \circ T - \lambda J_2 \circ S)^0(u_n; u - u_n)$$

$$\leq \varepsilon_n \|u - u_n\|_X + L^0(u_n; u - u_n) + J_1^0(Tu_n; Tu_n - Tu) + (\lambda J_2)^0(Su_n; Su_n - Su).$$

But, $L \in C^1(X; \mathbb{R})$ and thus $L^0(u_n; u - u_n) = \langle L'(u_n), u - u_n \rangle$. On the other hand $\varepsilon_n \downarrow 0$ and $\{u_n\}$ is bounded hence $\limsup_{n \to \infty} \varepsilon_n \|u - u_n\|_X = 0$. Taking into account Proposition 2.3 we deduce that

$$\limsup_{n \to \infty} J_1^0(Tu_n; Tu_n - Tu) \leq J_1^0(Tu; 0) = 0$$

and

$$\limsup_{n \to \infty} (\lambda J_2)^0(Su_n; Su_n - Su) \leq (\lambda J_2)^0(Su; 0) = 0.$$

Gathering the above information we have

$$\limsup_{n \to \infty} \langle L'(u_n), u_n - u \rangle \leq \limsup_{n \to \infty} \varepsilon_n \|u - u_n\|_X + \limsup_{n \to \infty} J_1^0(Tu_n; Tu_n - Tu)$$

$$+ \limsup_{n \to \infty} (\lambda J_2)^0(Su_n; Su_n - Su) \leq 0.$$

But, $L' : X \to X^*$ has the $(S)_+$ property, and this allows us to conclude that $\{u_n\}$ has a convergent subsequence, therefore $\mathcal{E}_\lambda$ satisfies the (PS)-condition. According to Step 2 there exists $u_1 \in X$ such that $\mathcal{E}_\lambda(u_1) < 0$. On the other hand, $\mathcal{E}_\lambda(u_0) = 0$ and we can choose $0 < r < \min\{R_2, \|u_1 - u_0\|_X\}$ such that

$$\mathcal{E}_\lambda(u) \geq \max\{\mathcal{E}_\lambda(u_0), \mathcal{E}_\lambda(u_1)\} = 0, \quad \forall u \in \partial \bar{B}(u_0; r).$$

Applying Corollary 5.4 we obtain that there exists a critical point $u_2 \in X \setminus \{u_0, u_1\}$ for $\mathcal{E}_\lambda$ and $\mathcal{E}_\lambda(u_1) \geq 0$. This completes the proof of the first part of the theorem, i.e. the functional $\mathcal{E}_\lambda$ has at least three distinct critical points.

Step 4. *If in addition* $(\mathcal{H}_4)$ *holds, then there exists* $\lambda^* > 0$ *such that* $\mathcal{E}_{\lambda^*}$ *has two global minima.*

Let us consider the functional $f : X \times (0, \infty) \to \mathbb{R}$ defined by

$$f(u, \mu) := \mu [L(u) - (J_1 \circ T)(u) + \rho] - (J_2 \circ S)(u) = \mu \mathcal{E}_{1/\mu}(u) + \mu \rho,$$

where $\rho \in \mathbb{R}$ is the number from $(\mathcal{H}_4)$.

We observe that for each $u \in X$ the functional $\mu \mapsto f(u, \mu)$ is affine, therefore it is quasi-concave. We also note that for each $\mu > 0$ the mapping $u \mapsto f(u, \mu)$ is sequentially weakly lower semicontinuous. Therefore for each $\mu > 0$, the sub-level sets of $u \mapsto f(u, \mu)$ are sequentially weakly closed.

Let us consider now the set $S^\mu(c) := \{u \in X : f(u, \mu) \leq c\}$ for some $c \in \mathbb{R}$ and a sequence $\{u_n\} \subset S^\mu(c)$. Obviously $\{u_n\}$ is bounded due to the fact that the functional $u \mapsto f(u, \mu)$ is coercive, which is clear since $f(u, \mu) = \mu \mathcal{E}_{1/\mu}(u) + \mu \rho$, $\mathcal{E}_{1/\mu}$ is coercive and $\mu > 0$. According to the Eberlein-Smulyan Theorem $\{u_n\}$ admits a subsequence, still denoted $\{u_n\}$, which converges weakly to some $u \in X$. Keeping in mind that $u_n \in S^\mu(c)$ for $n > 0$ we deduce that

$$\mathcal{E}_{1/\mu}(u_n) \leq \frac{c - \mu \rho}{\mu}, \qquad \text{for all } n > 0.$$

Combining the above relation with the fact that $\mathcal{E}_{1/\mu}$ is sequentially weakly lower semicontinuous we get

$$\mathcal{E}_{1/\mu}(u) \leq \liminf_{n \to \infty} \mathcal{E}_{1/\mu}(u_n) \leq \frac{c - \mu \rho}{\mu},$$

which shows that $f(u, \mu) \leq c$, therefore the set $S^\mu(c)$ is a sequentially weakly compact subset of X. We have proved thus that, for each $\mu > 0$, the sub-level sets of $u \mapsto f(u, \mu)$ are sequentially weakly compact. Taking into account Remark 1 in [13] which states that we can replace "closed and compact" by "sequentially closed and sequentially compact" in Theorem D.11 and using condition $(\mathcal{H}_4)$ we can apply this theorem for the weak topology of X and conclude that there exists $\mu^* > 0$ for which $f(u, \mu^*) = \mu^* \mathcal{E}_{1/\mu^*}(u) + \mu^* \rho$ has two global minima. It is easy to check that any global minimum point of $f(u, \mu^*)$ is also a global minimum point for $\mathcal{E}_{1/\mu^*}$, and thus we get the existence of a point $u_3 \in X \setminus \{u_1\}$ such that

$$\mathcal{E}_{1/\mu^*}(u_1) = \mathcal{E}_{1/\mu^*}(u_3) \leq \mathcal{E}_{1/\mu^*}(u^0_{1/\mu^*}) < 0 = \mathcal{E}_{1/\mu^*}(u_0) \leq \mathcal{E}_{1/\mu^*}(u_2),$$

showing that $u_3 \in X \setminus \{u_0, u_1, u_2\}$. Taking $\lambda^* = 1/\mu^*$ the proof is now complete.

$\square$

We conclude this section by a nonsmooth form of a Ricceri-type alternative result, extended to locally Lipschitz functions by Marano & Motreanu [8].

Theorem 5.17 ([8]) *Let $(X, \| \cdot \|)$ be a reflexive real Banach space, and $\tilde{X}$ another real Banach spaces such that X is compactly embedded into $\tilde{X}$. Let $\Phi : \tilde{X} \to \mathbb{R}$ and $\Psi : X \to \mathbb{R}$ be two locally Lipschitz functions, such that Ψ is weakly sequentially lower semicontinuous and coercive. For every $\rho > \inf_X \Psi$, put*

$$\varphi(\rho) = \inf_{u \in \Psi^{-1}(]-\infty,\rho[)} \frac{\Phi(u) - \inf_{v \in \overline{(\Psi^{-1}(]-\infty,\rho[))}_w} \Phi(v)}{\rho - \Psi(u)}, \qquad (5.51)$$

where $\overline{(\Psi^{-1}(]-\infty, \rho[))}_w$ is the closure of $\Psi^{-1}(]-\infty, \rho[)$ in the weak topology. Furthermore, set

$$\gamma := \liminf_{\rho \to +\infty} \varphi(\rho), \qquad \delta := \liminf_{\rho \to (\inf_X \Psi)^+} \varphi(\rho). \qquad (5.52)$$

Then, the following conclusions hold.

(A) *If $\gamma < +\infty$ then, for every $\lambda > \gamma$, either*
    (A1) *$\Phi + \lambda\Psi$ possesses a global minimum, or*
    (A2) *there is a sequence $\{u_n\}$ of critical points of $\Phi + \lambda\Psi$ such that $\lim_{n \to +\infty} \Psi(u_n) = +\infty$.*
(B) *If $\delta < +\infty$ then, for every $\lambda > \delta$, either*
    (B1) *$\Phi + \lambda\Psi$ possesses a local minimum, which is also a global minimum of Ψ, or*
    (B2) *there is a sequence $\{u_n\}$ of pairwise distinct critical points of $\Phi + \lambda\Psi$, with $\lim_{n \to +\infty} \Psi(u_n) = \inf_X \Psi$, weakly converging to a global minimum of Ψ.*

Proof One can observe that for every $\rho > \inf_X \Psi$ and $\lambda > \varphi(\rho)$ the function $\Phi + \lambda\Psi$ has a local minimum belonging to $\Psi^{-1}(] - \infty, \rho[)$.

(A) Let us fix $\lambda > \gamma$ and choose a sequence $\{\rho_n\} \subset I =] \inf_X \Psi, +\infty[$ such that

$$\lim_{n \to \infty} \rho_n = +\infty, \quad \varphi(\rho_n) < \lambda, \quad n \in \mathbb{N}. \qquad (5.53)$$

For every $n \in \mathbb{N}$ there exists a point u_n with the property that

$$\Phi(u_n) + \lambda\Psi(u_n) = \min_{v \in \Psi^{-1}(]-\infty,\rho_n[)} (\Phi(v) + \lambda\Psi(v)). \qquad (5.54)$$

On one hand, if $\lim_{n\to\infty} \Psi(u_n) = +\infty$ then (A2) holds, since u_n is a critical point of $\Phi + \lambda\Psi$. On the other hand, if $\liminf_{n\to\infty} \Psi(u_n) < +\infty$, let us fix

$$\rho > \max\left\{\inf_X \Psi, \liminf_{n\to\infty} \Psi(u_n)\right\}.$$

Since $\rho \in I$, one can assume (up to a subsequence) that $\{u_n\}$ weakly converges to $u \in X$. Let $v \in X$ be a fixed element. By the weak sequential lower semicontinuity of $\Phi + \lambda\Psi$ and relations (5.53) and (5.54), we obtain that

$$\Phi(u) + \lambda\Psi(u) \leq \Phi(v) + \lambda\Psi(v).$$

Since $v \in X$ is arbitrary, then $u \in X$ is a global minimum of $\Phi + \lambda\Psi$, which proves the assertion (A1).

In order to prove (B), let us fix $\lambda > \delta$ and choose a sequence $\{\rho_n\} \subset I$ such that

$$\lim_{n\to\infty} \rho_n = \inf_X \Psi, \quad \varphi(\rho_n) < \lambda, \quad n \in \mathbb{N}. \tag{5.55}$$

A similar argument as above shows the existence of a sequence $\{u_n\}$ of critical points of $\Phi + \lambda\Psi$ verifying relation (5.54). If $\rho \geq \max_{n\in\mathbb{N}} \rho_n$ then it turns out that $u_n \in \Psi^{-1}(] -\infty, \rho[)$ and $\{u_n\}$ weakly converges to $u \in X$ (up to a subsequence). It follows that $u \in X$ is a global minimum of Ψ; indeed, by the weak sequential lower semicontinuity of Ψ one has

$$\Psi(u) \leq \liminf_{n\to\infty} \Psi(u_n) \leq \liminf_{n\to\infty} \rho_n = \inf_X \Psi.$$

In particular, by taking a subsequence if necessary, it follows that

$$\lim_{n\to\infty} \Psi(u_n) = \Psi(u) = \inf_X \Psi.$$

The latter relation easily concludes the alternative in (B). $\qquad\qquad\qquad\square$

References

1. A. Ambrosetti, P. Rabinowitz, Dual variational methods in critical point theory and applications J. Funct. Anal. **14**, 349–381 (1973)
2. K.-C. Chang, Variational methods for non-differentiable functionals and their applications to partial differential equations. J. Math. Anal. Appl. **80**, 102–129 (1981)
3. N. Costea, C. Varga, Multiple critical points for non-differentiable parametrized functionals and applications to differential inclusions. J. Global Optim. **56**, 399–416 (2013)
4. N. Costea, M. Csirik, C. Varga, Linking-type results in nonsmooth critical point theory and applications. Set-Valued Var. Anal. **25**, 333–356 (2017)

5. Y. Du, A deformation lemma and some critical point theorems. Bull. Aust. Math. Soc. **43**, 161–168 (1991)

6. N. Ghoussoub, D. Preiss, A general mountain pass principle for locating and classifying critical points. Ann. Inst. H. Poincaré Anal. Non Linéaire **6**, 321–330 (1989)

7. A. Kristály, W. Marzantowicz, C. Varga, A non-smooth three critical points theorem with applications in differential inclusions. J. Global Optim. **46**, 49–62 (2010)

8. S. Marano, D. Motreanu, Infinitely many critical points of non-differentiable functions and applications to a Neumann-type problem involving the p-Laplacian. J. Diff. Equ. **182**, 108–120 (2002)

9. D. Motreanu, A multiple linking minimax principle. Bull. Aust. Math. Soc. **53**, 39–49 (1996)

10. D. Motreanu, C. Varga, Some critical point results for locally Lipschitz functionals. Comm. Appl. Nonlin. Anal. **4**, 17–33 (1997)

11. D. Motreanu, C. Varga, A nonsmooth equivariant minimax principle. Commun. Appl. Anal. **3**, 115–130 (1999)

12. P.H. Rabinowitz, Minimax methods in critical point theory with applications to differential equations, in *Regional Conference Series in Mathematics*, vol. 65. Conference Series in Mathematics, CBMS, Providence (1986)

13. B. Ricceri, Multiplicity of global minima for parametrized functions. Atti Accad. Naz. Lincei Cl. Sci. Fis. Mat. Natur. Rend. Lincei (9) Mat. Appl. **21**, 47–57 (2010)

14. M. Schechter, *Linking Methods in Critical Point Theory* (Birkhäuser, Basel, 1999)

15. A. Szulkin, Ljusternik-Schnirelmann theory on C^1-manifolds. Ann. Inst. H. Poincaré Anal. Non Linéaire **5**, 119–139 (1988)

Existence and Multiplicity Results for Differential Inclusions on Bounded Domains

6

6.1   Boundary Value Problems with Discontinuous Nonlinearities

Let Ω be a bounded open domain in $\mathbb{R}^N$ whose boundary $\partial\Omega$ is of class C^1 and consider the following elliptic boundary problem

$$\begin{cases} -\Delta u = g(x, u), & \text{in } \Omega, \\ u = 0, & \text{on } \partial\Omega, \end{cases} \tag{DP}$$

with $g : \Omega \times \mathbb{R} \to \mathbb{R}$ a prescribed function. If $t \mapsto g(x, t)$ is continuous, then *a weak solution* $u \in H_0^1(\Omega)$ of problem (DP) is defined to satisfy the following *variational equality*

$$\int_\Omega \nabla u \cdot \nabla v \, dx = \int_\Omega g(x, u) v \, dx, \quad \forall v \in H_0^1(\Omega).$$

The *energy functional* corresponding to our problem $E : H_0^1(\Omega) \to \mathbb{R}$ is defined by

$$E(u) := \int_\Omega |\nabla u|^2 dx - \int_\Omega f(x, u(x)) dx, \tag{6.1}$$

with $f(x, t) := \int_0^t g(x, s) ds$. Standard arguments show that $E \in C^1(H_0^1(\Omega, \mathbb{R}))$ and any critical point of E, i.e, $E'(u) = 0$ is also a weak solution for (DP).

© The Author(s), under exclusive license to Springer Nature Switzerland AG 2021
N. Costea et al., *Variational and Monotonicity Methods in Nonsmooth Analysis*,
Frontiers in Mathematics, https://doi.org/10.1007/978-3-030-81671-1_6

However, if $t \mapsto g(x, t)$ is not continuous, but only locally bounded, then it is well known that (DP) need not have a solution. In order to overcome this difficulty, Chang [5] had the idea to "fill in the gaps" at the discontinuity points of $g(x, \cdot)$, thus obtaining a *multivalued equation* that approximates the initial problem

$$\begin{cases} -\Delta u \in [g_-(x, u(x)), g_+(x, u(x))], & \text{in } \Omega, \\ u = 0, & \text{on } \partial\Omega, \end{cases} \quad (ME)$$

with

$$g_-(x, t) := \lim_{\varepsilon \to 0} \text{ess} \inf_{|s-t|<\varepsilon} g(x, s) \text{ and } g_+(x, t) := \lim_{\varepsilon \to 0} \text{ess} \sup_{|s-t|<\varepsilon} g(x, s).$$

Using the subdifferential calculus developed by Clarke [6], Chang showed that $t \mapsto f(x, t) := \int_0^t g(x, s)\mathrm{d}s$ is locally Lipschitz and

$$\partial_C^2 f(x, t) = [g_-(x, u(x)), g_+(x, u(x))],$$

and thus (ME) can be equivalently written as a *differential inclusion*

$$\begin{cases} -\Delta u \in \partial_C^2 f(x, u(x)), & \text{in } \Omega, \\ u = 0, & \text{on } \partial\Omega, \end{cases} \quad (DI)$$

As before, we define $u \in H_0^1(\Omega)$ to be a *weak solution* of (DI) if there exists $\zeta \in L^2(\Omega)$ such that $\zeta(x) \in \partial_C^2 f(x, u(x))$ and

$$\int_\Omega \nabla u \cdot \nabla v \mathrm{d}x = \int_\Omega \zeta(x)v(x)\mathrm{d}x, \quad \forall v \in H_0^1(\Omega).$$

Now, by the definition of the Clarke subdifferential, one can define a weak solution of (DI) to satisfy not a variational equality, but a *hemivariational inequality* of the type

$$\int_\Omega \nabla u \cdot \nabla v \mathrm{d}x \leq \int_\Omega f^0(x, u(x); v(x))\mathrm{d}x, \quad \forall v \in H_0^1(\Omega).$$

One can also easily prove that the energy functional E defined by (6.1), corresponding to (DI), is no longer differentiable, but only locally Lipschitz and any critical point of E is a weak solution of (DI) in the sense that it satisfies the above hemivariational inequality.

Remark 6.1 As this argument can be repeated whenever necessary, in the sequel we shall work with boundary value problems with discontinuous nonlinearities expressed as a differential inclusions of the type (DI).

We prove next an existence result for (DI) provided the following conditions hold.

(H_0) $f : \overline{\Omega} \times \mathbb{R} \to \mathbb{R}$ is a Carathéodory function such that:
  (i) $f(\cdot, t)$ is measurable for all $t \in \mathbb{R}$;
  (ii) $f(x, \cdot)$ is locally Lipschitz for all $x \in \overline{\Omega}$;
  (iii) $f(x, 0) = 0$ for every $x \in \Omega$.

(H_1) $|\zeta| \leq a_1 + a_2 |t|^s$, $\forall (x, t) \in \Omega \times \mathbb{R}$, $\forall \zeta \in \partial_C^2 f(x, t)$ with constants $a_1, a_2 \geq 0$, $0 \leq s < \frac{N+2}{N-2}$ if $N \geq 3$;

(H_2) $\sup_{\|v\|_{H_0^1} = \rho} \int_\Omega f(x, v) dx \leq \frac{1}{2} \rho^2$, for some $\rho > 0$;

(H_3) $t\zeta - \mu^{-1} f(x, \zeta) \geq -b_1 |t|^\sigma - b_2$, $\forall (x, t) \in \Omega \times \mathbb{R}$ and $\zeta \in \partial_C^2 f(x, t)$ with constants $\mu > 2$, $1 \leq \sigma < 2$ and $b_1, b_2 \geq 0$;

(H_4) $\limsup_{n \to \infty} \frac{1}{n^\sigma} \int_\Omega f(x, n v_0) dx = +\infty$, for some $v_0 \in H_0^1(\Omega)$.

Theorem 6.1 ([32]) *Assume that conditions* $(H_0) - (H_4)$ *are verified. Then problem* (DI) *possesses a nontrivial weak solution* $u \in H_0^1(\Omega)$.

Proof In view of (H_1) the energy functional $E : H_0^1(\Omega) \to \mathbb{R}$ defined by (6.1) is well defined and locally Lipschitz. Theorem 2.6 ensures that $\partial_C E(u)$ at any $u \in H_0^1(\Omega)$ satisfies the relation

$$\partial_C E(u) \subset I'(u) - \partial_C F(u), \text{ in } H^{-1}(\Omega),$$

where

$$I(u) := \int_\Omega |\nabla u|^2 dx,$$

and

$$F(u) := \int_\Omega f(x, u(x)) dx.$$

Consequently, it suffices to show that the functional E admits a nontrivial critical point $u \in H_0^1(\Omega)$, i.e., $0 \in \partial_C E(u)$. To this end we shall apply Corollary 5.4. Notice that $E(0) = 0$. We check that E satisfies the condition $(PS)_c$ for every $c \in \mathbb{R}$. Let $\{u_n\}$ be a sequence in $H_0^1(\Omega)$ such that $E(u_n) \to c$ as $n \to \infty$ and there exists $\zeta_n \in L^{\frac{s+1}{s}}(\Omega)$ provided

$$\nabla u_n - \zeta_n \to 0 \text{ in } H^{-1}(\Omega) \text{ as } n \to \infty \tag{6.2}$$

and

$$\zeta_n(x) \in \partial_C^2 f(x, u_n(x)) \text{ a.e. } x \in \Omega. \tag{6.3}$$

Then, for any n sufficiently large we can write

$$c + 1 + \frac{1}{\mu}\|u_n\|_{H_0^1} \geq \left(\frac{1}{2} - \frac{1}{\mu}\right)\|u_n\|_{H_0^1}^2 + \frac{1}{\mu}\int_\Omega \left(\zeta_n u_n - \frac{1}{\mu}f(x, u_n)\right)dx$$

$$\geq \left(\frac{1}{2} - \frac{1}{\mu}\right)\|u_n\|_{H_0^1}^2 - b_1\|u_n\|_{L^\sigma}^\sigma - b_2\text{meas}(\Omega)$$

$$\geq \left(\frac{1}{2} - \frac{1}{\mu}\right)\|u_n\|_{H_0^1}^2 - b\|u_n\|_{H_0^1}^\sigma - b_2\text{meas}(\Omega)$$

with a new constant $b > 0$. Above we used assumption (H_3). Since $\mu > 2$ and $\sigma < 2$ we conclude that $\{u_n\}$ is bounded in $H_0^1(\Omega)$. Taking into account that the embedding $H_0^1(\Omega) \hookrightarrow L^{s+1}(\Omega)$ is compact, relation (6.3) ensures that, up to a subsequence, $\{\zeta_n\}$ converges in $H^{-1}(\Omega)$. Thus from (6.2) we derive that $\{u_n\}$ contains a convergent subsequence in $H_0^1(\Omega)$, i.e., condition $(PS)_c$ is verified. Now we justify condition (i) of Corollary 5.4 with $\alpha := 0$. Indeed, by (H_2) it is seen that

$$E(v) \geq \frac{1}{2}\|v\|_{H_0^1}^2 - \sup_{\|v\|_{H_0^1}=\rho}\int_\Omega f(x, v)dx \geq 0$$

for all $v \in H_0^1(\Omega)$ with $\|v\|_{H_0^1} = \rho$. The final step is to check condition (ii) of Corollary 5.4. Due to (H_3) we have

$$\partial_C^t(|ty|^{-\mu}f(x, ty)) = \mu|y|^{-\mu}t^{-1-t}(m^{-1}ty\partial f(x, ty)) - f(x, ty)$$

$$\geq -\mu|y|^{-\mu}t^{-1-\mu}(b_1 t^\sigma|y|^\sigma + b_2)$$

for every $y \in \mathbb{R} \setminus \{0\}$ and $t > 0$.

By Lebourg's mean value theorem and assumption (H_3) we infer that

$$\frac{f(x, (n+1)y)}{(n+1)^\mu|y|^\mu} - \frac{f(x, ny)}{n^\mu|y|^\mu} \geq \min_{n \leq t \leq n+1} \partial_C^t(|ty|^{-\mu}f(x, ty))$$

$$\geq -\mu|y|^{-\mu}(b_1 n^{\sigma-\mu-1}|y|^\sigma + b_2 n^{-\mu-1})$$

for all $y \in \mathbb{R} \setminus \{0\}$ and positive integers $n \geq 1$. Taking the sum of the inequalities above where n is replaced by $1, \cdots, n-1$ we get

$$f(x, ny) \geq n^{\mu} \left[f(x, y) - \mu b_1 |y|^{\sigma} \sum_{i \geq 1} \left(\frac{1}{i} \right)^{\mu+1-\sigma} + b_2 \sum_{i \geq 1} \left(\frac{1}{i} \right)^{\mu+1} \right]$$

for all $y \in \mathbb{R}$ and $n \geq 1$. Therefore, with the new constants $c_1, c_2 \geq 0$ one has

$$f(x, ny) \geq n^{\mu} (f(x, y) - c_1 |y|^{\mu} - c_2), \quad \forall y \in \mathbb{R}, \forall n \geq 1.$$

One obtains

$$E(nv) \leq \frac{1}{2} n^2 \|v\|_{H_0^1}^2 - n^{\mu} \left(\int_{\Omega} f(x, v) dx - c_1 \|v\|_{L^{\sigma}}^{\sigma} - c_2 meas(\Omega) \right) \tag{6.4}$$

for all $v \in H_0^1(\Omega)$ and all $n \geq 1$.
   By (H_4) we can find $n_0 \geq 1$ such that

$$\frac{1}{n_0^{\sigma}} \left[\int_{\Omega} f(x, n_0 v_0(x) x) dx + c_2 meas(\Omega) \right] \geq c_1 \|v\|_{L^{\sigma}}^{\sigma},$$

therefore

$$c_0 := \left[\int_{\Omega} f(x, n_0 v_0(x)) dx - c_1 \|v\|_{L^{\sigma}}^{\sigma} n_0^{\sigma} + c_2 \, meas(\Omega) \right] > 0. \tag{6.5}$$

Combining (6.4), (6.5) we get

$$E(nn_0 v_0) \leq \frac{1}{2} n^2 n_0^2 \|v_0\|_{H_0^1}^2 - c_0 n^{\mu} n_0^{\mu}, \quad \forall n \geq 1. \tag{6.6}$$

If we pass to the limit in (6.6) as $n \to \infty$ it is clear that

$$\lim_{n \to \infty} I(nn_0 v_0) = -\infty$$

because $\mu > 2$ and $c_0 > 0$ as shown in (6.5). Corollary 5.4 with $\alpha := E(0) = 0$ completes the proof of theorem.   □

6.2   Parametric Problems with Locally Lipschitz Energy Functional

Let Ω be a non-empty, bounded, open subset of the real Euclidian space $\mathbb{R}^N$, $N \geq 3$, having a smooth boundary $\partial\Omega$ and let $W^{1,2}(\Omega)$ be the closure of $C^\infty(\Omega)$ with the respect to the norm

$$\|u\| := \left(\int_\Omega |\nabla u(x)|^2 + \int_\Omega u^2(x) \right)^{1/2}.$$

Denote by $2^\star := \dfrac{2N}{N-2}$ and $\overline{2}^\star := \dfrac{2(N-1)}{N-2}$ the critical Sobolev exponent for the embedding $W^{1,2}(\Omega) \hookrightarrow L^p(\Omega)$ and for the trace mapping $W^{1,2}(\Omega) \hookrightarrow L^q(\partial\Omega)$, respectively. If $p \in [1, 2^\star]$ then the embedding $W^{1,2}(\Omega) \hookrightarrow L^p(\Omega)$ is continuous while if $p \in [1, 2^\star)$, it is compact. In the same way for $q \in \left[1, \overline{2}^\star\right]$, $W^{1,2}(\Omega) \hookrightarrow L^q(\partial\Omega)$ is continuous, and for $q \in \left[1, \overline{2}^\star\right)$ it is compact. Therefore, there exist constants $c_p, \overline{c}_q > 0$ such that

$$\|u\|_{L^p(\Omega)} \leq c_p \|u\|, \quad \text{and} \quad \|u\|_{L^q(\partial\Omega)} \leq \overline{c}_q \|u\|, \quad \forall u \in W^{1,2}(\Omega).$$

Now, we consider a locally Lipschitz function $F : \mathbb{R} \to \mathbb{R}$ which satisfies the following conditions:

(F_1) $F(0) = 0$ and there exists $C_1 > 0$ and $p \in [1, 2^\star)$ such that

$$|\xi| \leq C_1(1 + |t|^{p-1}), \quad \forall \xi \in \partial_C F(t), \forall t \in \mathbb{R}; \tag{6.7}$$

(F_2) $\lim\limits_{t \to 0} \dfrac{\max\{|\xi| : \xi \in \partial_C F(t)\}}{t} = 0;$

(F_3) $\limsup\limits_{|t| \to +\infty} \dfrac{F(t)}{t^2} \leq 0;$

(F_4) There exists $\tilde{t} \in \mathbb{R}$ such that $F(\tilde{t}) > 0$.

Example 6.1 Let $p \in (1, 2]$ and $F : \mathbb{R} \to \mathbb{R}$ be defined by $F(t) := \min\{|t|^{p+1}, \arctan(t_+)\}$, where $t_+ := \max\{t, 0\}$. The function F enjoys properties $(F1) - (F4)$.

Let also $G : \mathbb{R} \to \mathbb{R}$ be another locally Lipschitz function satisfying the following condition:

(G_0) There exists $C_2 > 0$ and $q \in \left[1, \overline{2}^\star\right)$ such that

$$|\xi| \leq C_2(1 + |t|^{q-1}), \quad \forall \xi \in \partial_C G(t), \forall t \in \mathbb{R}. \tag{6.8}$$

For $\lambda, \mu > 0$, we consider the following differential inclusion problem, with inhomogeneous Neumann condition:

$$\begin{cases} -\Delta u + u \in \lambda \partial_C F(u(x)), & \text{in } \Omega; \\ \dfrac{\partial u}{\partial n} \in \mu \partial_C G(u(x)), & \text{on } \partial\Omega. \end{cases} \qquad (P_{\lambda,\mu})$$

Definition 6.1 We say that $u \in W^{1,2}(\Omega)$ is a solution of the problem $(P_{\lambda,\mu})$, if there exist $\xi_F(x) \in \partial_C F(u(x))$ and $\xi_G(x) \in \partial_C G(u(x))$ for a.e. $x \in \Omega$ such that for all $v \in W^{1,2}(\Omega)$ we have

$$\int_\Omega (-\Delta u + u)v\,dx = \lambda \int_\Omega \xi_F v\,dx \ \text{ and } \ \int_{\partial\Omega} \frac{\partial u}{\partial n} v\,d\sigma = \mu \int_{\partial\Omega} \xi_G v\,d\sigma.$$

The main result of this section reads as follows.

Theorem 6.2 ([23, Theorem 3.1]) *Let $F, G : \mathbb{R} \to \mathbb{R}$ be two locally Lipschitz functions satisfying the conditions $(F_1)-(F_4)$ and (G_0). Then there exists a non-degenerate compact interval $[a, b] \subset (0, +\infty)$ and a number $r > 0$, such that for every $\lambda \in [a, b]$ there exists $\mu_0 \in (0, \lambda + 1]$ such that for each $\mu \in [0, \mu_0]$, the problem $(P_{\lambda,\mu})$ has at least three distinct solutions with $W^{1,2}$-norms less than r.*

In the sequel, we are going to prove Theorem 6.2, assuming from now on that its assumptions are verified.

Since F, G are locally Lipschitz, it follows through (6.7) and (6.8) in a standard way that $\Phi_1 : L^p(\Omega) \to \mathbb{R}$ ($p \in [1, 2^\star]$) and $\Phi_2 : L^q(\partial\Omega) \to \mathbb{R}$ ($q \in [1, \overrightarrow{2}^\star]$) defined by

$$\Phi_1(u) := -\int_\Omega F(u(x))dx \ \ (u \in L^p(\Omega)),$$

and

$$\Phi_2(u) := -\int_{\partial\Omega} G(u(x))d\sigma \ \ (u \in L^q(\partial\Omega))$$

are well-defined, locally Lipschitz functionals and due to Theorem 2.6, we have

$$\partial_C \Phi_1(u) \subseteq -\int_\Omega \partial_C F(u(x))dx \ \ (u \in L^p(\Omega)),$$

and

$$\partial_C \Phi_2(u) \subseteq -\int_{\partial\Omega} \partial_C G(u(x))d\sigma \ \ (u \in L^q(\partial\Omega)).$$

We introduce the energy functional $\mathcal{E}_{\lambda,\mu} : W^{1,2}(\Omega) \to \mathbb{R}$ associated to the problem $(P_{\lambda,\mu})$, given by

$$\mathcal{E}_{\lambda,\mu}(u) := \frac{1}{2}\|u\|^2 + \lambda\Phi_1(u) + \mu\Phi_2(u), \quad u \in W^{1,2}(\Omega).$$

Using the latter inclusions and the Green formula, the critical points of the functional $\mathcal{E}_{\lambda,\mu}$ are solutions of the problem $(P_{\lambda,\mu})$ in the sense of Definition 6.1. Before proving Theorem 6.2, we need the following auxiliary result.

Proposition 6.1 $\displaystyle\lim_{t\to 0^+} \frac{\inf\{\Phi_1(u) : u \in W^{1,2}(\Omega), \|u\|^2 < 2t\}}{t} = 0.$

Proof Fix $\tilde{p} \in (\max\{2, p\}, 2^*)$. Applying Lebourg's mean value theorem and using (F_1) and (F_2), for any $\varepsilon > 0$, there exists $K(\varepsilon) > 0$ such that

$$|F(t)| \leq \varepsilon t^2 + K(\varepsilon)|t|^{\tilde{p}}, \quad \forall t \in \mathbb{R}. \tag{6.9}$$

Taking into account (6.9) and the continuous embedding $W^{1,2}(\Omega) \hookrightarrow L^{\tilde{p}}(\Omega)$ we have

$$\Phi_1(u) \geq -\varepsilon c_2^2\|u\|^2 - K(\varepsilon)c_{\tilde{p}}^{\tilde{p}}\|u\|^{\tilde{p}}, \quad u \in W^{1,2}(\Omega). \tag{6.10}$$

For $t > 0$ define the set $S_t := \{u \in W^{1,2}(\Omega) : \|u\|^2 < 2t\}$. Using (6.10) we have

$$0 \geq \frac{\inf_{u \in S_t} \Phi_1(u)}{t} \geq -2c_2^2\varepsilon - 2^{\tilde{p}/2}K(\varepsilon)c_{\tilde{p}}^{\tilde{p}}t^{\frac{\tilde{p}}{2}-1}.$$

Since $\varepsilon > 0$ is arbitrary and since $t \to 0^+$, we get the desired limit. $\qquad\square$

Proof of Theorem 6.2 Let us define the function for every $t > 0$ by

$$\beta(t) := \inf\left\{\Phi_1(u) : u \in W^{1,2}(\Omega), \frac{\|u\|^2}{2} < t\right\}.$$

We have that $\beta(t) \leq 0$, for $t > 0$, and Proposition 6.1 yields that

$$\lim_{t\to 0^+} \frac{\beta(t)}{t} = 0. \tag{6.11}$$

We consider the constant function $u_0 \in W^{1,2}(\Omega)$ by $u_0(x) := \tilde{t}$ for every $x \in \Omega$, $\tilde{t}$ being from (F_4). Note that $\tilde{t} \neq 0$ (since $F(0) = 0$), so $\Phi_1(u_0) < 0$. Therefore it is possible to choose a number $\eta > 0$ such that

$$0 < \eta < -\Phi_1(u_0) \left[\frac{\|u_0\|^2}{2} \right]^{-1}.$$

By (6.11) we get the existence of a number $t_0 \in \left(0, \frac{\|u_0\|^2}{2} \right)$ such that $-\beta(t_0) < \eta t_0$. Thus

$$\beta(t_0) > \left[\frac{\|u_0\|^2}{2} \right]^{-1} \Phi_1(u_0)t_0. \tag{6.12}$$

Due to the choice of t_0 and using (6.12), we conclude that there exists $\rho_0 > 0$ such that

$$- \beta(t_0) < \rho_0 < -\Phi_1(u_0) \left[\frac{\|u_0\|^2}{2} \right]^{-1} t_0 < -\Phi_1(u_0). \tag{6.13}$$

Define now the function $\varphi : W^{1,2}(\Omega) \times \mathbb{I} \to \mathbb{R}$ by

$$\varphi(u, \lambda) := \frac{\|u\|^2}{2} + \lambda\Phi_1(u) + \lambda\rho_0,$$

where $\mathbb{I} := [0, +\infty)$. We prove that the function φ satisfies the inequality

$$\sup_{\lambda \in \mathbb{I}} \inf_{u \in W^{1,2}(\Omega)} \varphi(u, \lambda) < \inf_{u \in W^{1,2}(\Omega)} \sup_{\lambda \in \mathbb{I}} \varphi(u, \lambda). \tag{6.14}$$

The function

$$\mathbb{I} \ni \lambda \mapsto \inf_{u \in W^{1,2}(\Omega)} \left[\frac{\|u\|^2}{2} + \lambda(\rho_0 + \Phi_1(u)) \right]$$

is obviously upper semicontinuous on $\mathbb{I}$. It follows from (6.13) that

$$\lim_{\lambda \to +\infty} \inf_{u \in W^{1,2}(\Omega)} \varphi(u, \lambda) \leq \lim_{\lambda \to +\infty} \left[\frac{\|u_0\|^2}{2} + \lambda(\rho_0 + \Phi_1(u_0)) \right] = -\infty.$$

Thus we find an element $\bar{\lambda} \in \mathbb{I}$ such that

$$\sup_{\lambda \in \mathbb{I}} \inf_{u \in W^{1,2}(\Omega)} \varphi(u, \lambda) = \inf_{u \in W^{1,2}(\Omega)} \left[\frac{\|u\|^2}{2} + \bar{\lambda}(\rho_0 + \Phi_1(u)) \right]. \tag{6.15}$$

Since $-\beta(t_0) < \rho_0$, it follows from the definition of β that for all $u \in W^{1,2}(\Omega)$ with $\frac{\|u\|^2}{2} < t_0$ we have $-\Phi_1(u) < \rho_0$. Hence

$$t_0 \leq \inf\left\{\frac{\|u\|^2}{2} : u \in W^{1,2}(\Omega), \ -\Phi_1(u) \geq \rho_0\right\}. \tag{6.16}$$

On the other hand,

$$\inf_{u \in W^{1,2}(\Omega)} \sup_{\lambda \in \mathbb{I}} \varphi(u, \lambda) = \inf_{u \in W^{1,2}(\Omega)} \left[\frac{\|u\|^2}{2} + \sup_{\lambda \in \mathbb{I}} (\lambda(\rho_0 + \Phi_1(u)))\right]$$

$$= \inf_{u \in W^{1,2}(\Omega)} \left\{\frac{\|u\|^2}{2} : -\Phi_1(u) \geq \rho_0\right\}.$$

Thus inequality (6.16) is equivalent to

$$t_0 \leq \inf_{u \in W^{1,2}(\Omega)} \sup_{\lambda \in \mathbb{I}} \varphi(u, \lambda). \tag{6.17}$$

We consider two cases. First, when $0 \leq \overline{\lambda} < \frac{t_0}{\rho_0}$, then we have that

$$\inf_{u \in W^{1,2}(\Omega)} \left[\frac{\|u\|^2}{2} + \overline{\lambda}(\rho_0 + \Phi_1(u))\right] \leq \varphi(0, \overline{\lambda}) = \overline{\lambda}\rho_0 < t_0.$$

Combining this inequality with (6.15) and (6.17) we obtain (6.14).

Now, if $\frac{t_0}{\rho_0} \leq \overline{\lambda}$, then from (6.12) and (6.13), it follows that

$$\inf_{u \in W^{1,2}(\Omega)} \left[\frac{\|u\|^2}{2} + \overline{\lambda}(\rho_0 + \Phi_1(u))\right] \leq \frac{\|u_0\|^2}{2} + \overline{\lambda}(\rho_0 + \Phi_1(u_0))$$

$$\leq \frac{\|u_0\|^2}{2} + \frac{t_0}{\rho_0}(\rho_0 + \Phi_1(u_0)) < t_0.$$

It remains to apply again (6.15) and (6.17), which concludes the proof of (6.14).

Now, we are in the position to apply Theorem 5.15; we choose $X := W^{1,2}(\Omega)$, $\tilde{X}_1 := L^p(\Omega)$ with $p \in [1, 2^*)$, $\tilde{X}_2 := L^q(\partial\Omega)$ with $q \in [1, \overline{2^*})$, $\Lambda := \mathbb{I} = [0, +\infty)$, $h(t) := t^2/2$, $t \geq 0$.

Now, we fix $g \in \mathcal{G}_\tau$ ($\tau \geq 0$), $\lambda \in \Lambda$, $\mu \in [0, \lambda + 1]$, and $c \in \mathbb{R}$. We shall prove that the functional $E_{\lambda,\mu} : W^{1,2}(\Omega) \to \mathbb{R}$ given by

$$E_{\lambda,\mu}(u) := \frac{1}{2}\|u\|^2 + \lambda\Phi_1(u) + \mu(g \circ \Phi_2)(u), \ u \in W^{1,2}(\Omega),$$

satisfies the $(PS)_c$. Note that due to Proposition 2.3, we have for every $u, v \in W^{1,2}(\Omega)$ that

$$E^\circ_{\lambda,\mu}(u; v) \leq \langle u, v \rangle_{W^{1,2}} + \lambda \Phi^\circ_1(u; v) + \mu(g \circ \Phi_2)^\circ(u; v). \tag{6.18}$$

First of all, let us observe that $\frac{1}{2}\|\cdot\|^2 + \lambda \Phi_1$ is coercive on $W^{1,2}(\Omega)$, due to (F_3); thus, the functional $E_{\lambda,\mu}$ is also coercive on $W^{1,2}(\Omega)$. Consequently, it is enough to consider a bounded sequence $\{u_n\} \subset W^{1,2}(\Omega)$ such that

$$E^\circ_{\lambda,\mu}(u_n; v - u_n) \geq -\varepsilon_n \|v - u_n\| \text{ for all } v \in W^{1,2}(\Omega), \tag{6.19}$$

where $\{\varepsilon_n\}$ is a positive sequence such that $\varepsilon_n \to 0$. Because the sequence $\{u_n\}$ is bounded, there exists an element $u \in W^{1,2}(\Omega)$ such that $u_n \rightharpoonup u$ weakly in $W^{1,2}(\Omega)$, $u_n \to u$ strongly in $L^p(\Omega)$, $p \in [1, 2^*)$ (since $W^{1,2}(\Omega) \hookrightarrow L^p(\Omega)$ is compact), and $u_n \to u$ strongly in $L^q(\partial\Omega)$, $q \in \left[1, \overline{2}^*\right)$ (since $W^{1,2}(\Omega) \hookrightarrow L^q(\partial\Omega)$ is compact). Using (6.19) with $v := u$ and apply relation (6.18) for the pairs $(u_n, u - u_n)$ and $(u, u_n - u)$, we have that

$$\|u - u_n\|^2 \leq \varepsilon_n \|u - u_n\| - E^\circ_{\lambda,\mu}(u; u_n - u) + \lambda[\Phi^\circ_1(u_n; u - u_n) + \Phi^\circ_1(u; u_n - u)]$$
$$+ \mu[(g \circ \Phi_2)^\circ(u_n; u - u_n) + (g \circ \Phi_2)^\circ(u; u_n - u)].$$

Since $\{u_n\}$ is bounded in $W^{1,2}(\Omega)$, we clearly have that $\lim_{n\to\infty} \varepsilon_n \|u - u_n\| = 0$. Now, fix $\zeta \in \partial_C E_{\lambda,\mu}(u)$; in particular, we have $\langle \zeta, u_n - u \rangle_{W^{1,2}} \leq E^\circ_{\lambda,\mu}(u; u_n - u)$. Since $u_n \rightharpoonup u$ weakly in $W^{1,2}(\Omega)$, we have that $\liminf_{n\to\infty} E^\circ_{\lambda,\mu}(u; u_n - u) \geq 0$. Now, for the remaining four terms in the above estimation we use the fact that $\Phi^\circ_1(\cdot; \cdot)$ and $(g \circ \Phi_2)^\circ(\cdot; \cdot)$ are upper semicontinuous functions on $L^p(\Omega)$ and $L^q(\partial\Omega)$, respectively. Since $u_n \to u$ strongly in $L^p(\Omega)$, we have for instance $\limsup_{n\to\infty} \Phi^\circ_1(u_n; u - u_n) \leq \Phi^\circ_1(u; 0) = 0$; the remaining terms are similar. Combining the above outcomes, we obtain finally that $\limsup_{n\to\infty} \|u - u_n\|^2 \leq 0$, i.e., $u_n \to u$ strongly in $W^{1,2}(\Omega)$. It remains to apply Theorem 5.15 in order to obtain the conclusion.   □

6.3   Multiplicity Alternative for Parametric Differential Inclusions Driven by the p−Laplacian

In this section we use the theoretical results obtained in the Sect. 5.4 to study differential inclusions involving the p-Laplace operator. More exactly we prove that either the problem

$$(P) : \begin{cases} -\Delta_p u \in \partial^2_C f(x, u(x)), & \text{in } \Omega, \\ u = 0, & \text{on } \partial\Omega, \end{cases}$$

possesses at least two nontrivial weak solutions, or the corresponding eigenvalue problem

$$(P_\lambda): \begin{cases} -\Delta_p u \in \lambda \partial_C^2 f(x, u(x)), & \text{in } \Omega, \\ u = 0, & \text{on } \partial\Omega, \end{cases}$$

has a rich family of eigenfunctions corresponding to eigenvalues located in the interval $(0, 1)$.

Here, $\Delta_p u := \text{div}(|\nabla u|^{p-2}\nabla u)$, $1 < p < \infty$, is the p-Laplacian, $\Omega \subset \mathbb{R}^N$ ($N \geq 2$) is a bounded domain with $C^{1,\alpha}$ boundary, $f : \Omega \times \mathbb{R} \to \mathbb{R}$ is a locally Lipschitz function with respect to the second variable and $\partial_C^2 f(x, t)$ denotes the Clarke subdifferential of the map $t \mapsto f(x, t)$. As usual, we consider the Sobolev space

$$W^{1,p}(\Omega) := \left\{ u \in L^p(\Omega) : \frac{\partial u}{\partial x_i} \in L^p(\Omega), i = 1, \ldots, N \right\}$$

endowed with the norm $\|u\|_{1,p} := \|u\|_p + \|\nabla u\|_p$, with $\|\cdot\|_p$ being the usual norm on $L^p(\Omega)$. Since we work with Dirichlet boundary condition, the natural space to seek weak solution of problem (P) is the Sobolev space

$$W_0^{1,p}(\Omega) = \overline{C_0^\infty(\Omega)}^{\|\cdot\|_{1,p}} = \left\{ u \in W^{1,p}(\Omega) : u = 0 \text{ on } \partial\Omega \right\},$$

with the value of u on $\partial\Omega$ understood in the sense of traces.

Definition 6.2 A function $u \in W_0^{1,p}(\Omega)$ is a *weak solution* of problem (P) if, there exists $\xi \in W^{-1,p'}(\Omega)$ such that $\xi(x) \in \partial_C^2 f(x, u(x))$ for a.e. $x \in \Omega$ and

$$\int_\Omega |\nabla u|^{p-2}\nabla u \cdot \nabla v dx = \int_\Omega \xi(x) v(x) dx, \quad \forall v \in W_0^{1,p}(\Omega),$$

Definition 6.3 A real number λ is said to be an *eigenvalue* of (P_λ) if there exist $u_\lambda \in W_0^{1,p}(\Omega) \setminus \{0\}$ and $\xi_\lambda \in W^{-1,p'}(\Omega)$ such that $\xi_\lambda(x) \in \partial_C^2 f(x, u_\lambda(x))$ for a.e. $x \in \Omega$ and

$$\int_\Omega |\nabla u|^{p-2}\nabla u \cdot \nabla v dx = \lambda \int_\Omega \xi_\lambda(x) v(x) dx, \quad \forall v \in W_0^{1,p}(\Omega).$$

The function u_λ satisfying the above relation is called an *eigenfunction* corresponding to λ.

Following a well-known idea of Lions [26] (see Brezis [4] also), we may regard $-\Delta_p$ as an operator acting from $W_0^{1,p}(\Omega)$ into its dual $W^{-1,p'}(\Omega)$ by

$$\langle -\Delta_p u, v \rangle := \int_\Omega |\nabla u|^{p-2}\nabla u \cdot \nabla v dx, \quad \forall u, v \in W_0^{1,p}(\Omega).$$

Henceforth we consider $W_0^{1,p}(\Omega)$ to be endowed with the norm $|u|_{1,p} := \|\nabla u\|_p$, which is equivalent to $\|u\|_{1,p}$ due to the Poincaré inequality. Then the duality mapping corresponding to the normalization function $\phi(t) := t^{p-1}$, i.e., $J_\phi : W_0^{1,p}(\Omega) \to W^{-1,p'}(\Omega)$ satisfies

$$J_\phi(u) = -\Delta_p u. \tag{6.20}$$

It is also known that $-\Delta_p$ is a potential operator in the sense that

$$\Phi'(u) = -\Delta_p u,$$

with $\Phi : W_0^{1,p}(\Omega) \to \mathbb{R}$ being the C^1-functional defined as follows

$$\Phi(u) := \frac{1}{p}|u|_{1,p}^p = \frac{1}{p}\int_\Omega |\nabla u|^p \mathrm{d}x.$$

Finally, we note that $X := W_0^{1,p}(\Omega)$ is separable and uniformly convex (see, e.g., [1, Theorem 3.6]), therefore the theory developed in the preceding chapters is applicable. Here and hereafter, we denote by p^* the critical Sobolev exponent, that is,

$$p^* := \begin{cases} \frac{Np}{N-p}, & \text{if } p < N, \\ \infty, & \text{otherwise.} \end{cases}$$

Assumption 1 The function $f : \Omega \times \mathbb{R} \to \mathbb{R}$ satisfies:

(f_1) For all $t \in \mathbb{R}$ the map $x \mapsto f(x,t)$ is measurable and $f(x,0) = 0$;
(f_2) For almost all $x \in \Omega$, the map $t \mapsto f(x,t)$ is locally Lipschitz;
(f_3) There exists $C > 0$ and $q \in (p, p^*)$ such that $|\xi| \leq C|t|^{q-1}$, for a.e. $x \in \Omega$, all $t \in \mathbb{R}$ and all $\xi \in \partial_C^2 f(x,t)$.

Assumption 2 There exists $u_0 \in W_0^{1,p}(\Omega) \setminus \{0\}$ such that $|u_0|_{1,p}^p \leq p \int_\Omega f(x, u_0(x))\mathrm{d}x$.

Theorem 6.3 ([9]) *Suppose that Assumptions 1–2 hold. Then the following alternative holds:*
Either

(A_1) *Problem (P) possesses at least two nontrivial weak solutions;*
        or
(A_2) *For each $R \in (|u_0|_{1,p}, \infty)$ problem (P_λ) possesses an eigenvalue $\lambda \in (0, 1)$ with the corresponding eigenfunction satisfying $|u_\lambda|_{1,p} = R$.*

Proof Assumption 1 ensures that we can apply the Aubin-Clarke theorem to conclude that the function $F : L^q(\Omega) \to \mathbb{R}$ defined by

$$F(w) := \int_\Omega f(x, w(x))dx,$$

is Lipschitz continuous on bounded domains and

$$\partial_C F(w) \subseteq \int_\Omega \partial_C^2 f(x, w(x))dx, \quad \forall w \in L^q(\Omega),$$

in the sense that for each $\zeta \in \partial_C F(w)$, there exists $\xi \in L^{q'}(\Omega)$ such that $\xi(x) \in \partial_C^2 f(x, w(x))$ for a.e. $x \in \Omega$ and

$$\langle \zeta, w \rangle = \int_\Omega \xi(x)w(x)dx.$$

Define now the energy functional $E : W_0^{1,p}(\Omega) \to \mathbb{R}$ as follows

$$E(u) := \frac{1}{p}|u|_{1,p}^p - F(u).$$

It follows from the Rellich-Kondrachov theorem (see, e.g., [1, Theorem 6.3]) that the inclusion $W_0^{1,p}(\Omega) \hookrightarrow L^q(\Omega)$ is compact, hence E is well defined. Moreover,

$$\partial_C E(u) \subset -\Delta_p u - \partial_C F(u).$$

In conclusion, if $\mu \leq 0$ and $u \in W_0^{1,p}(\Omega)$ are such that

$$\mu J_\phi u \in \partial_C E(u),$$

then there exists $\xi \in L^{q'}(\Omega) \subset W^{-1,p'}(\Omega)$ such that $\xi(x) \in \partial_C^2 f(x, u(x))$ for almost all $x \in \Omega$ and

$$\mu \int_\Omega |\nabla u|^{p-2}\nabla u \cdot \nabla v dx = \int_\Omega |\nabla u|^{p-2}\nabla u \cdot \nabla v dx - \int_\Omega \xi(x)v(x)dx.$$

Moreover, if $\mu = 0$, then u is a weak solution of (P), while $\mu < 0$ implies that $\lambda := \frac{1}{1-\mu} \in (0, 1)$ is an eigenvalue of (P_λ), provided that $u \neq 0$.

Fix $R \in (|u_0|_{1,p}, \infty)$. We prove next that $E|_{\overline{B}_R}$ satisfies the hypotheses of Theorem 5.13.

STEP 1.   *The functional E maps bounded sets into bounded sets.*
Fix $u \in W_0^{1,p}(\Omega)$ and $M > 0$ such that $|u|_{1,p} \leq M$. According to Lebourg's mean value theorem there exist $t \in (0, 1)$ and $\bar{\xi}(x) \in \partial_C^2 f(x, tu(x))$ such that

$$f(x, u(x)) = f(x, u(x)) - f(x, 0) = \bar{\xi}(x)u(x), \text{ for a.e. } x \in \Omega.$$

Therefore,

$$|F(u)| \leq \int_\Omega |f(x, u(x))|dx \leq \int_\Omega |\bar{\xi}(x)||u(x)|dx \leq \int_\Omega C|t|^{q-1}|u(x)|^q dx \leq C\|u\|_q^q.$$

Then

$$|E(u)| \leq \frac{1}{p}M^p + CC_q^q M^q,$$

with $C_q > 0$ being given by the compact embedding $W_0^{1,p}(\Omega) \hookrightarrow L^q(\Omega)$.
STEP 2.   *There exists $\rho \in \left(0, |u_0|_{1,p}\right)$ such that $E(u) \geq 0$ for all $u \in S_\rho$.*
By Assumption 2 and STEP 1 we have

$$\frac{1}{p}|u_0|_{1,p}^p \leq F(u_0) \leq CC_q^q|u_0|_{1,p}^q.$$

Pick $\rho := \frac{1}{2}\left(\frac{1}{pc_0}\right)^{\frac{1}{q-p}}$, with $c_0 := CC_q^q$. Then $\rho < \left(\frac{1}{pc_0}\right)^{\frac{1}{q-p}} \leq |u_0|_{1,p}$ and for all $u \in W_0^{1,p}(\Omega)$ satisfying $|u|_{1,p} = \rho$ we have

$$E(u) = \frac{1}{p}|u|_{1,p}^p - F(u) \geq \frac{1}{p}|u|_{1,p}^p - c_0|u|_{1,p}^q = \left(\frac{1}{p}\right)^{\frac{q}{q-p}}\left(\frac{1}{c_0}\right)^{\frac{p}{q-p}}\left(\frac{1}{2^p} - \frac{1}{2^q}\right) \geq 0.$$

STEP 3.   *The functional E satisfies $(SPS)_c$ in $\overline{B}_R$ for all $c \in \mathbb{R}$.*
Let $c \in \mathbb{R}$, $\{u_n\} \subset \overline{B}_R$ be s.t. $E(u_n) \to c$ and assume there exists $\{\zeta_n\} \subset W^{-1,p'}(\Omega)$ satisfies

$$\zeta_n \in \partial_C E(u_n), \quad \|\pi_{u_n}(\zeta_n)\| \to 0, \quad \langle \zeta_n, u_n \rangle \to v \leq 0. \tag{6.21}$$

Since $\{u_n\}$ is bounded and $W_0^{1,p}(\Omega)$ is reflexive, it follows from the Eberlein-Šmulian theorem (see Theorem A.8) that there exist $u \in W_0^{1,p}(\Omega)$ and a subsequence of $\{u_n\}$, still denoted $\{u_n\}$, such that

$$u_n \rightharpoonup u, \text{ in } W_0^{1,p}(\Omega).$$

We may assume that $|u_n|_{1,p} \to r$. If $r = 0$, then $u_n \to 0$ in $W_0^{1,p}(\Omega)$. Assume now that $r > 0$. Then the compactness of the embedding $W_0^{1,p}(\Omega) \hookrightarrow L^q(\Omega)$ implies

$$u_n \to u, \text{ in } L^q(\Omega).$$

Since $\partial_C E(u_n) \subset -\Delta_p u_n - \partial_C F(u_n)$, it follows that there exists $\eta_n \in \partial_C F(u_n)$ such that

$$\zeta_n = -\Delta_p u_n - \eta_n.$$

Since $u_n \to u$ in $L^q(\Omega)$, it follows from Proposition 4.1 that there exists $\eta \in \partial_C F(u)$ such that

$$\eta_n \rightharpoonup \eta, \text{ in } L^{q'}(\Omega).$$

But $L^{q'}(\Omega)$ is compactly embedded into $W^{-1,p'}(\Omega)$ which means

$$\eta_n \to \eta, \text{ in } W^{-1,p'}(\Omega).$$

It follows that

$$-\zeta_n - \Delta_p u_n \to \eta, \text{ in } W^{-1,p'}(\Omega). \tag{6.22}$$

On the other hand, the second relation of (6.21) implies

$$\zeta_n + \frac{\langle \zeta_n, u_n \rangle}{|u_n|_{1,p}^p} \Delta_p u_n \to 0 \text{ in } W^{-1,p'}. \tag{6.23}$$

Adding (6.22) and (6.23) we get

$$\left(1 - \frac{\langle \zeta_n, u_n \rangle}{|u_n|_{1,p}^p}\right)(-\Delta_p u_n) \to \eta, \text{ in } W^{-1,p'}(\Omega).$$

Consequently,

$$\lim_{n\to\infty} \left(1 - \frac{\langle \zeta_n, u_n \rangle}{|u_n|_{1,p}^p}\right)\langle -\Delta_p u_n, u_n - u \rangle = 0.$$

But, $\lim_{n\to\infty}(1 - \langle \zeta_n, u_n \rangle / |u_n|_{1,p}^p) = 1 - v/r^p \geq 1$, which combined with the above relation gives

$$\lim_{n\to\infty} \langle -\Delta_p u_n, u_n - u \rangle = 0.$$

It follows that $u_n \to u$ in $W_0^{1,p}(\Omega)$ due to the fact that $-\Delta_p$ satisfies the (S_+) condition (see Proposition C.6).

STEP 4. *There exists $\Lambda_R > 0$ s.t. $|\langle \zeta, u \rangle| \le \Lambda_R$, for all $u \in S_R$ and all $\zeta \in \partial_C E(u)$.*

Fix $u \in S_R$ and $\zeta \in \partial_C E(u)$. Then there exists $\xi \in W^{-1,p'}(\Omega)$ satisfying $\xi(x) \in \partial_C^2 f(x, u(x))$ such that

$$|\langle \zeta, u \rangle| = \left| \langle -\Delta_p u, u \rangle - \int_\Omega \xi(x) u(x) \mathrm{d}x \right| \le |\langle -\Delta_p u, u \rangle| + \int_\Omega |\xi(x)| |u(x)| \mathrm{d}x$$

$$\le R^p + C \|u\|_q^q \le R^p + C C_q^q R^q := \Lambda_R.$$

Applying Theorem 5.13 with $A := \{0, u_0\}$, $B := S_\rho$ (with $\rho > 0$ given by STEP 2), $v_0 := 0$, $v_1 := u_0$ we get the desired conclusion. □

6.4 Differential Inclusions Involving the $p(\cdot)$–Laplacian and Steklov-Type Boundary Conditions

In this section we are concerned with the study of a differential inclusion of the type

$$\begin{cases} -\mathrm{div}(|\nabla u|^{p(x)-2} \nabla u) + |u|^{p(x)-2} u \in \partial_C^2 \phi(x, u) & \text{in } \Omega, \\ \dfrac{\partial u}{\partial n_{p(x)}} \in \lambda \partial_C^2 \psi(x, u) & \text{on } \partial\Omega, \end{cases} \qquad (P_\lambda)$$

where $\Omega \subset \mathbb{R}^N$ ($N \ge 3$) is a bounded domain with smooth boundary, $\lambda > 0$ is a real parameter, $p : \overline{\Omega} \to \mathbb{R}$ is a continuous function such that $\inf_{x \in \overline{\Omega}} p(x) > N$, $\phi : \Omega \times \mathbb{R} \to \mathbb{R}$ and $\psi : \partial\Omega \times \mathbb{R} \to \mathbb{R}$ are locally Lipschitz functionals with respect to the second variable and $\dfrac{\partial u}{\partial n_{p(x)}} := |\nabla u|^{p(x)-2} \nabla u \cdot n$, n being the unit outward normal on $\partial\Omega$.

In the case when $p(x) \equiv p$, $\phi(x, t) \equiv 0$ and $\psi(x, t) := \frac{1}{q}|t|^q$ the problem (P_λ) becomes

$$\begin{cases} \Delta_p u = |u|^{p-2} u & \text{in } \Omega, \\ |\nabla u|^{p-2} \dfrac{\partial u}{\partial n} = \lambda |u|^{q-2} u & \text{on } \partial\Omega, \end{cases} \qquad (P)$$

and it was studied by Fernández-Bonder and Rossi [19] in the case $1 < q < p^* = \frac{p(N-1)}{N-p}$ by using variational arguments combined with the Sobolev trace inequality. In [19] it is also proved that if $p = q$ then problem (P) admits a sequence of eigenvalues $\{\lambda_n\}$, such that $\lambda_n \to \infty$ as $n \to \infty$. Furthermore, Martinez and Rossi [28] proved that the first eigenvalue λ_1 of problem (P) (that is, $\lambda_1 \le \lambda$ for any other eigenvalue) when $p = q$ is isolated and simple. In the linear case, that is $p = q = 2$, problem (P) is known in the literature as the *Steklov* problem (see, e.g., Babuška and Osborn [3]).

Let us present next some basic notions and results from the theory of Lebesgue-Sobolev spaces with variable exponent. For more details one can consult the book by Musielak [33] and the papers by Edmunds et al. [12–14], O. Kováčik and J. Rákosník [21], Fan et al. [16, 18], M. Mihăilescu and V. Rădulescu [29].

Set

$$C_+(\overline{\Omega}) := \{\varphi \in C(\overline{\Omega}) : \varphi(x) > 1, \forall x \in \overline{\Omega}\},$$

and for $\varphi \in C_+(\overline{\Omega})$ we denote

$$\varphi^- := \inf_{x \in \overline{\Omega}} \varphi(x) \quad \text{and} \quad \varphi^+ := \sup_{x \in \overline{\Omega}} \varphi(x).$$

For a function $p \in C_+(\overline{\Omega})$ we define the *variable exponent Lebesgue space*

$$L^{p(\cdot)}(\Omega) := \left\{u : u \text{ is a real valued-function and } \int_\Omega |u(x)|^{p(x)} dx < \infty\right\}$$

which can be endowed with the so-called *Luxemburg norm* given by the formula

$$|u|_{L^{p(\cdot)}(\Omega)} := \inf\left\{\zeta > 0 : \int_\Omega \left|\frac{u(x)}{\zeta}\right|^{p(x)} dx \le 1\right\}.$$

We recall that $\left(L^{p(\cdot)}(\Omega), |\cdot|_{L^{p(\cdot)}(\Omega)}\right)$ is a separable and reflexive Banach space. If $0 < \text{meas}(\Omega) < \infty$ and p, q are variable exponents such that $p(x) \le q(x)$ in Ω, then the embedding $L^{q(\cdot)}(\Omega) \hookrightarrow L^{p(\cdot)}(\Omega)$ is continuous. We also remember that the following *Hölder type inequality* holds

$$\int_\Omega |u(x)v(x)| dx \le \left(\frac{1}{p^-} + \frac{1}{p'^-}\right) |u|_{L^{p(\cdot)}(\Omega)} |v|_{L^{p'(\cdot)}(\Omega)},$$

for all $u \in L^{p(\cdot)}(\Omega)$ and all $v \in L^{p'(\cdot)}(\Omega)$, where by $p'(x)$ we have denoted the conjugated exponent of $p(x)$, that is, $p'(x) := \frac{p(x)}{p(x)-1}$.

We recall that $\left(W^{1,p(\cdot)}(\Omega), \|\cdot\|\right)$ is a separable and reflexive Banach space. If we set

$$I(u) := \int_\Omega \left(|\nabla u(x)|^{p(x)} + |u(x)|^{p(x)}\right) dx$$

then for $u \in W^{1,p(\cdot)}(\Omega)$ the following relations hold true

$$\|u\| > 1 \Longrightarrow \|u\|^{p^-} \le I(u) \le \|u\|^{p^+}, \tag{6.24}$$

$$\|u\| < 1 \Longrightarrow \|u\|^{p^+} \le I(u) \le \|u\|^{p^-}. \tag{6.25}$$

Remark 6.2 If $N < p^- \le p(x)$ for any $x \in \overline{\Omega}$, then Fan and Zhao [17, Theorem 2.2] proved that the space $W^{1,p(\cdot)}(\Omega)$ is continuously embedded in $W^{1,p^-}(\Omega)$, and, since $N < p^-$ it follows that $W^{1,p(\cdot)}(\Omega)$ is compactly embedded in $C(\overline{\Omega})$. Therefore, there exists a positive constant $c_\infty > 0$ such that

$$\|u\|_\infty \le c_\infty\|u\|, \quad \forall u \in W^{1,p(\cdot)}(\Omega), \tag{6.26}$$

where by $\|\cdot\|_\infty$ we have denoted the usual norm on $C(\overline{\Omega})$, that is, $\|u\|_\infty = \sup_{x \in \overline{\Omega}}|u(x)|$.

Definition 6.4 We say that $u \in W^{1,p(\cdot)}(\Omega)$ is a *weak solution* of problem (P_λ) if there exist $\xi, \zeta \in \left(W^{1,p(\cdot)}(\Omega)\right)^*$ such that $\xi(x) \in \partial_C^2\phi(x, u(x))$, $\zeta(x) \in \partial_C^2\psi(x, u(x))$ for almost every $x \in \overline{\Omega}$ and for all $v \in W^{1,p(\cdot)}(\Omega)$ we have

$$\int_\Omega \left(-\mathrm{div}(|\nabla u(x)|^{p(x)-2}\nabla u(x)) + |u(x)|^{p(x)-2}u(x)\right) v(x)\mathrm{d}x = \int_\Omega \xi(x)v(x)\mathrm{d}x$$

and

$$\int_{\partial\Omega} \frac{\partial u}{\partial n_{p(\cdot)}} v(x)\mathrm{d}\sigma = \lambda \int_{\partial\Omega} \zeta(x)v(x)\mathrm{d}\sigma.$$

Using the Green formula and the definition of the Clarke subdifferential one has that a weak solution $u \in W^{1,p(\cdot)}(\Omega)$ needs to satisfy the following hemivariational inequality

$$\int_\Omega \left(|\nabla u|^{p(x)-2}\nabla u \cdot \nabla v + |u|^{p(x)-2}uv\right) \mathrm{d}x \le \int_\Omega \phi_{,2}^0(x, u(x); v(x))\mathrm{d}x$$

$$+ \lambda \int_{\partial\Omega} \psi_{,2}^0(x, u(x); v(x))\mathrm{d}\sigma \tag{6.27}$$

Here, and hereafter we shall assume the following hypotheses hold:

(H_1) $\phi : \Omega \times \mathbb{R} \to \mathbb{R}$ is a functional such that
    (i) $\phi(x, 0) = 0$ for a.e. $x \in \Omega$;
    (ii) the function $x \mapsto \phi(x, t)$ is measurable for every $t \in \mathbb{R}$;
    (iii) the function $t \mapsto \phi(x, t)$ is locally Lipschitz for a.e. $x \in \Omega$;
    (iv) there exist $c_\phi > 0$ and $q \in C(\overline{\Omega})$ with $1 < q(x) \le q^+ < p^-$ s.t.

$$|\xi(x)| \le c_\phi|t|^{q(x)-1}$$

    for a.e. $x \in \Omega$, every $t \in \mathbb{R}$ and every $\xi(x) \in \partial_C^2\phi(x, t)$.
    (v) there exists $\delta_1 > 0$ s.t. $\phi(x, t) \le 0$ when $0 < |t| \le \delta_1$, for a.e. $x \in \Omega$.

(H_2) $\psi : \partial\Omega \times \mathbb{R} \to \mathbb{R}$ is a functional such that

    (i) $\psi(x, 0) = 0$ for a.e. $x \in \partial\Omega$;

    (ii) the function $x \mapsto \psi(x, t)$ is measurable for every $t \in \mathbb{R}$;

    (iii) the function $t \mapsto \psi(x, t)$ is locally Lipschitz for a.e. $x \in \partial\Omega$;

    (iv) there exist $c_\psi > 0$ and $r \in C(\partial\Omega)$ with $1 < r(x) \le r^+ < p^-$ s.t.

$$|\zeta(x)| \le c_\psi |t|^{r(x)-1}$$

for a.e. $x \in \partial\Omega$, every $t \in \mathbb{R}$ and every $\zeta(x)^2_C \in \partial\psi(x, t)$;

    (v) there exists $\delta_2 > 0$ s.t. $\psi(x, t) \le 0$ when $0 < |t| \le \delta_2$, for a.e. $x \in \partial\Omega$.

(H_3) There exist $\eta > \max\{\delta_1, \delta_2\}$ s.t. $\eta^{p(x)} \le p(x)\phi(x, \eta)$ for a.e. $x \in \Omega$ and $\psi(x, \eta) > 0$ for a.e. $x \in \partial\Omega$.

(H_4) There exists $m \in L^1(\Omega)$ s.t. $\phi(x, t) \le m(x)$ for all $t \in \mathbb{R}$ and a.e. $x \in \Omega$.

(H_5) There exists $\mu > \max\left\{c_\infty(p^+ \|m\|_{L^1(\Omega)})^{1/p^-}; c_\infty(p^+ \|m\|_{L^1(\Omega)})^{1/p^+}\right\}$ s.t.

$$\sup_{|t|\le\mu} \psi(x, t) \le \psi(x, \eta) < \sup_{t\in\mathbb{R}} \psi(x, t).$$

Theorem 6.4 ([8]) *Assume that (H^1)–(H^3) are fulfilled. Then for each $\lambda > 0$ problem (P_λ) admits at least two non-zero solutions. If in addition (H^4) and (H^5) hold, then there exists $\lambda^* > 0$ such that problem (P_{λ^*}) admits at least three non-zero solutions.*

Proof Let us denote $X := W^{1,p(\cdot)}(\Omega)$, $Y = Z := C(\overline{\Omega})$ and consider $T : X \to Y$, $S : X \to Z$ to be the embedding operators. It is clear that T, S are compact operators and for the sake of simplicity, everywhere below, we will omit to write Tu and Su to denote the above operators, writing u instead of Tu or Su. We introduce next $L : X \to \mathbb{R}$, $J_1 : Y \to \mathbb{R}$ and $J_2 : Z \to \mathbb{R}$ as follows

$$L(u) := \int_\Omega \frac{1}{p(x)}\left[|\nabla u(x)|^{p(x)} + |u(x)|^{p(x)}\right] dx, \quad \text{for } u \in X,$$

$$J_1(y) := \int_\Omega \phi(x, y(x))dx, \quad \text{for } y \in Y,$$

and

$$J_2(z) := \int_{\partial\Omega} \psi(x, z(x))d\sigma, \quad \text{for } z \in Z.$$

We point out the fact that L is sequentially weakly lower semicontinuous and $L' : X \to X^*$,

$$\langle L'(u), v \rangle = \int_\Omega \left(|\nabla u|^{p(x)-2} \nabla u \cdot \nabla v + |u|^{p(x)-2} uv \right) dx$$

has the $(S)_+$ property according to Fan and Zhang [15, Theorem 3.1].

The idea is to prove that the functional $\mathcal{E}_\lambda : X \to \mathbb{R}$ defined by

$$\mathcal{E}_\lambda(u) := L(u) - J_1(u) - \lambda J_2(u)$$

satisfies the conditions of Theorem 5.16 and each critical point of this functional is a solution of problem (P_λ) in the sense of Definition 6.4. With this end in view we divide the proof in several steps.

STEP 1. *The functionals J_1 and J_2 defined above are locally Lipschitz.*

Let $y \in Y$, $R > 0$ and $y_1, y_2 \in B_Y(y; R)$ be fixed. According to Lebourg's mean value theorem there exists $\bar{y} := t_0 y_1 + (1 - t_0) y_2$ and $\xi^*(x) \in \partial_C^2 \phi(x, \bar{y}(x))$, for some $t_0 \in (0, 1)$, such that

$$\phi(x, y_1(x)) - \phi(x, y_2(x)) = \xi^*(x)(y_1(x) - y_2(x)).$$

Thus,

$$|J_1(y_1) - J_1(y_2)| = \left| \int_\Omega \phi(x, y_1(x)) - \phi(x, y_2(x)) dx \right| \leq \int_\Omega |\phi(x, y_1(x)) - \phi(x, y_2(x))| dx$$

$$= \int_\Omega |\xi^*(x)||y_1(x) - y_2(x)| dx \leq \int_\Omega c_\phi |\bar{y}(x)|^{q(x)-1} |y_1(x) - y_2(x)| dx \leq \tilde{c}_0 \|y_1 - y_2\|_\infty,$$

where $\tilde{c}_0 = \tilde{c}_0(y, R)$ is a suitable constant. In a similar way we prove that J_2 is a locally Lipschitz functional.

STEP 2. $u_0 := 0$ *satisfies hypothesis $(\mathcal{H}_1)$ of Theorem 5.16.*

Indeed, $L(0) = J_1(0) = J_2(0) = 0$ and for each $R > 0$ we have

$$L(u) > 0, \quad \forall u \in B_X(0; R) \setminus \{0\},$$

which shows that $u_0 = 0$ is a strict minimum point for L.

STEP 3. $\mathcal{E}_\lambda$ *is coercive.*

Let $u \in X$ be fixed. According to Lebourg's mean value theorem there exist $s_0, s_1 \in (0, 1)$ and $\xi^*(x) \in \partial_C^2 \phi(x, s_0 u(x))$, $\zeta^*(x) \in \partial_C^2 \psi(x, s_1 u(x))$ such that

$$\phi(x, u(x)) - \phi(x, 0) = \xi^*(x)u(x) \text{ and } \psi(x, u(x)) - \psi(x, 0) = \zeta^*(x)u(x).$$

Thus,

$$J_1(u) = \int_\Omega (\phi(x, u) - \phi(x, 0))dx \le \int_\Omega |\xi^*||u|dx \le c_\phi \int_\Omega s_0^{q(x)-1}|u|^{q(x)}dx$$

$$\le c_\phi \int_\Omega |u|^{q(x)}dx \le c_\phi \int_\Omega \|u\|_\infty^{q(x)}dx,$$

and

$$J_2(u) = \int_{\partial\Omega} (\psi(x, u) - \psi(x, 0))d\sigma \le \int_{\partial\Omega} |\zeta^*||u|d\sigma \le c_\psi \int_{\partial\Omega} s_1^{r(x)-1}|u|^{r(x)}d\sigma$$

$$\le c_\psi \int_{\partial\Omega} |u|^{r(x)}d\sigma \le c_\psi \int_{\partial\Omega} \|u\|_\infty^{r(x)}d\sigma.$$

Hence for $u \in X$ with $\|u\| > 1$ and $\|u\|_\infty > 1$ we have

$$\mathcal{E}_\lambda(u) = \int_\Omega \frac{1}{p(x)}\left[|\nabla u|^{p(x)} + |u|^{p(x)}\right]dx - \int_\Omega \phi(x, u)dx - \lambda \int_{\partial\Omega} \psi(x, u)d\sigma$$

$$\ge \frac{1}{p^+}\|u\|^{p^-} - c_\phi \mathrm{meas}(\Omega)\|u\|_\infty^{q^+} - \lambda c_\psi \mathrm{meas}(\Omega)\|u\|_\infty^{r^+}$$

$$\ge \frac{1}{p^+}\|u\|^{p^-} - c_\phi \mathrm{meas}(\Omega)c_\infty^{q^+}\|u\|^{q^+} - \lambda c_\psi \mathrm{meas}(\Omega)c_\infty^{r^+}\|u\|^{r^+}.$$

We conclude that $\mathcal{E}_\lambda(u) \to \infty$ as $\|u\| \to \infty$ since $r^+ < p^-$ and $q^+ < p^-$.

STEP 4.   *There exists $\bar{u}_0 \in X$ such that $\mathcal{E}_\lambda(\bar{u}_0) < 0$.*

Choosing $\bar{u}_0(x) := \eta$ for all $x \in \overline{\Omega}$ and taking into account (H^3) we conclude that

$$\mathcal{E}_\lambda(\bar{u}_0) = \int_\Omega \frac{1}{p(x)}\eta^{p(x)}dx - \int_\Omega \phi(x, \eta)dx - \lambda \int_{\partial\Omega} \psi(x, \eta)d\sigma < 0.$$

STEP 5.   *There exists $R_0 > 0$ s.t. $J_1(u) \le L(u)$ and $J_2(u) \le 0 \; \forall u \in B(0; R_0) \setminus \{0\}$.*

Let us define $R_0 < \min\left\{\frac{\delta_1}{c_\infty}; \frac{\delta_2}{c_\infty}\right\}$ where c_∞ is given in (6.26) and δ_1, δ_2 are given in (H_1) and (H_2), respectively. For an arbitrarily fixed $u \in B(0; R_0)$, taking into account the way we defined the operators T and S, we have

$$|u(x)| \le \|u\|_\infty \le c_\infty \|u\| \le c_\infty R_0 < \delta_1, \quad \forall x \in \Omega$$

and

$$|u(x)| \le \|u\|_\infty \le c_\infty \|u\| \le c_\infty R_0 < \delta_2, \quad \forall x \in \partial\Omega.$$

Hypotheses (H_1) and (H_2) ensure that $\phi(x, u(x)) \leq 0$ and $\psi(x, u(x)) \leq 0$ for all $u \in B(0; R_0)$, therefore $J_1(u) \leq 0 < L(u)$ and $J_2(u) \leq 0$ for all $u \in B(0; R_0) \setminus \{0\}$.
STEP 6. *There exists $\rho \in \mathbb{R}$ such that*

$$\sup_{\lambda>0} \inf_{u\in X} \lambda\,[L(u) - J_1(u) + \rho] - J_2(u) < \inf_{u\in X} \sup_{\lambda>0} \lambda\,[L(u) - J_1(u) + \rho] - J_2(u).$$

Using the same arguments as Ricceri [34] (see the proof of Theorem 6.2) we conclude that it suffices to find $\rho \in \mathbb{R}$ and $\bar{u}_1, \bar{u}_2 \in X$ such that

$$L(\bar{u}_1) - J_1(\bar{u}_1) < \rho < L(\bar{u}_2) - J_1(\bar{u}_2) \tag{6.28}$$

and

$$\frac{\sup_{u\in A} J_2(u) - J_2(\bar{u}_1)}{\rho - L(\bar{u}_1) + J_1(\bar{u}_1)} < \frac{\sup_{u\in A} J_2(u) - J_2(\bar{u}_2)}{\rho - L(\bar{u}_2) + J_1(\bar{u}_2)}, \tag{6.29}$$

where $A := (L - J_1)^{-1}((-\infty, \rho])$.
Let us define $\bar{u}_1 \equiv \eta$ and choose $\bar{u}_2$ such that

$$\psi(x, \bar{u}_2(x)) > \sup_{|t|\leq \mu} \psi(x, t).$$

We point out the fact that a $\bar{u}_2$ satisfying the above relation exists due to (H_5). Next we define

$$\rho := \min\left\{ \frac{1}{p^+}\left(\frac{\mu}{c_\infty}\right)^{p^+} - \|m\|_{L^1(\Omega)}; \ \frac{1}{p^+}\left(\frac{\mu}{c_\infty}\right)^{p^-} - \|m\|_{L^1(\Omega)} \right\}$$

and observe that $\rho > 0$.
We shall prove next that for any $u \in A$ we have $\|u\|_\infty \leq \mu$. In order to do this, let us fix $u \in A$. Keeping in mind (H_4) and the way we defined ρ we distinguish the following cases:

CASE 1. $\|u\| \leq 1$.
  Then $\|u\|^{p^+} \leq I(u)$ and we obtain the following estimates:

$$\frac{1}{p^+}\|u\|^{p^+} \leq \frac{1}{p^+} I(u) \leq \int_\Omega \frac{1}{p(x)}\left[|\nabla u|^{p(x)} + |u|^{p(x)}\right] dx \leq \rho + \int_\Omega \phi(x, u)dx$$

$$\leq \rho + \int_\Omega m(x)dx \leq \frac{1}{p^+}\left(\frac{\mu}{c_\infty}\right)^{p^+}.$$

We conclude from above that $\|u\| \leq \frac{\mu}{c_\infty}$ therefore we must have $\|u\|_\infty \leq \mu$.

CASE 2. $\|u\| > 1$.

In this case we have $\|u\|^{p^-} \leq I(u)$ and we obtain the following estimates:

$$\frac{1}{p^+}\|u\|^{p^-} \leq \frac{1}{p^+}I(u) \leq \int_\Omega \frac{1}{p(x)}\left[|\nabla u|^{p(x)} + |u|^{p(x)}\right]dx \leq \rho + \int_\Omega \phi(x, u)dx$$

$$\leq \rho + \int_\Omega m(x)dx \leq \frac{1}{p^+}\left(\frac{\mu}{c_\infty}\right)^{p^-}.$$

The above computations enable us to conclude that $\|u\| \leq \frac{\mu}{c_\infty}$ therefore we must have $\|u\|_\infty \leq \mu$.

We only have to check that (6.28) and (6.29) hold for $\bar{u}_1$ and $\bar{u}_2$ chosen as above. From above we conclude that $\bar{u}_2 \notin A$ and thus

$$\sup_{u\in A} J_2(u) \leq \sup_{\|u\|_\infty \leq \mu} J_2(u) \leq J_2(\bar{u}_1), \quad \sup_{u\in A} J_2(u) \leq \sup_{\|u\|_\infty \leq \mu} J_2(u) \leq J_2(\bar{u}_2),$$

and

$$L(\bar{u}_1) - J_1(\bar{u}_1) \leq 0 < \rho < L(\bar{u}_2) - J_1(\bar{u}_2).$$

STEP 7.   *Any critical point of the functional $\mathcal{E}_\lambda$ is a solution of problem (P_λ).*

It is easy to check that $u \in W^{1,p(\cdot)}(\Omega)$ is a solution of problem (P_λ), if and only if there exist $\xi(x) \in \partial_C^2 \phi(x, u(x))$ and $\zeta(x) \in \partial_C^2 \psi(x, u(x))$ such that for all $v \in W^{1,p(\cdot)}(\Omega)$

$$0 = \int_\Omega \left(|\nabla u|^{p(x)-2}\nabla u \cdot \nabla v + |u|^{p(x)-2}uv\right)dx + \int_\Omega \xi(-v)dx + \int_{\partial\Omega} \zeta(-\lambda v)d\sigma.$$

Moreover,

$$J_1^0(y_1; y_2) \leq \int_\Omega \phi^0(x, y_1(x); y_2(x))dx, \quad \forall y_1, y_2 \in Y,$$

and

$$J_2^0(w_1; w_2) \leq \int_{\partial\Omega} \psi^0(x, w_1(x); w_2(x))dx, \quad \forall w_1, w_2 \in Z.$$

Let $u \in X$ be a critical point of $\mathcal{E}_\lambda$ and $v \in X$ be fixed. Taking into account the properties of the generalized directional derivative we obtain

$$0 \leq \mathcal{E}_\lambda^0(u; v) = (L - J_1 - \lambda J_2)^0(u; v) \leq L^0(u; v) + (-J_1)^0(u; v) + \lambda(-J_2)^0(u; v)$$

$$\leq \langle L'(u), v \rangle + J_1^0(u; -v) + J_2^0(u; -\lambda v) \leq \int_\Omega |\nabla u|^{p(x)-2}\nabla u \cdot \nabla v + |u|^{p(x)-2}uvdx$$

$$+ \int_\Omega \phi^0(x, u(x); -v(x))dx + \int_{\partial\Omega} \psi^0(x, u(x); -\lambda v(x))d\sigma.$$

On the other hand, Proposition 2.4 ensures that for almost every $x \in \Omega$ there exists $\xi(x) \in \partial_C^2\phi(x, u(x))$ such that, for all $t \in \mathbb{R}$, we have

$$\phi^0(x, u(x); t) = \xi(x)t = \max\left\{ zt : z \in \partial_C^2\phi(x, u(x)) \right\}.$$

In a similar way we deduce that for almost every $x \in \partial\Omega$ there exist $\zeta(x) \in \partial\psi(x, u(x))$ such that

$$\psi^0(x, u(x); t) = \zeta(x)t = \max\left\{ \tilde{z} : \tilde{z} \in \partial_C^2\psi(x, u(x)) \right\}.$$

Combining the above relations we conclude that any critical point u of $\mathcal{E}_\lambda$ satisfies

$$0 \leq \int_\Omega |\nabla u|^{p(x)-2}\nabla u \cdot \nabla v + |u|^{p(x)-2}uvdx + \int_\Omega \xi(-v)dx + \int_{\partial\Omega} \zeta(-\lambda v)d\sigma.$$

Replacing v with $-v$ in the above relation we get

$$0 = \int_\Omega |\nabla u|^{p(x)-2}\nabla u \cdot \nabla v + |u|^{p(x)-2}uvdx + \int_\Omega \xi(-v)dx + \int_{\partial\Omega} \zeta(-\lambda v)d\sigma,$$

which shows that u is a solution of (P_λ) □

6.5   Dirichlet Differential Inclusions Driven by the Φ−Laplacian

6.5.1   Variational Setting and Existence Results

Throughout this section Ω is a bounded domain of $\mathbb{R}^N$, $N \geq 3$, with Lipschitz boundary $\partial\Omega$. Consider the problem

$$(\mathcal{P}) : \quad \begin{cases} -\Delta_\Phi u \in \partial_C^2 f(x, u), & \text{in } \Omega, \\ u = 0, & \text{on } \partial\Omega, \end{cases}$$

where $\Phi : \mathbb{R} \to [0, \infty)$ is the N-function given by $\Phi(t) := \int_0^t a(|s|)s\,ds$ and $\Delta_\Phi u := \operatorname{div}(a(|\nabla u|)\nabla u)$ is the Φ-Laplace operator. The function $a : (0, \infty) \to (0, \infty)$ is prescribed, $f : \Omega \times \mathbb{R} \to \mathbb{R}$ is a locally Lipschitz functional w.r.t. the second variable and $\partial_C^2 f(x, t)$ denotes the Clarke subdifferential of the mapping $t \mapsto f(x, t)$.

Following Clément, de Pagter, Sweers and de Thélin [7] we say that a function $\varphi : \mathbb{R} \to \mathbb{R}$ is *admissible* if it is continuous, odd, strictly increasing and onto. In this particular case, φ has an inverse and the complementary N-function of Φ is given by

$$\Phi^*(s) = \int_0^s \varphi^{-1}(\tau)d\tau.$$

In addition, if we assume that

$$1 < \varphi^- \leq \varphi^+ < \infty,$$

where

$$\varphi^- := \inf_{t>0} \frac{t\varphi(t)}{\Phi(t)} \text{ and } \varphi^+ := \sup_{t>0} \frac{t\varphi(t)}{\Phi(t)},$$

then both Φ and Φ^* satisfy the Δ_2-condition (see Clément et al. [7, Lemma C.6]), hence $E^\Phi(\Omega) = L^\Phi(\Omega)$ and $L^\Phi(\Omega), L^{\Phi^*}(\Omega)$ are reflexive Banach spaces and each is the dual of the other. Moreover, if $1 < \varphi^- \leq \varphi^+ < \infty$, then the following relations between the Luxemburg norm $|\cdot|_\Phi$ and the integral $\int_\Omega \Phi(|\cdot|)dx$ can be established (see Clément et al. [7, Lemma C.7])

$$|u|_\Phi^{\varphi^+} \leq \int_\Omega \Phi(|u|)dx \leq |u|_\Phi^{\varphi^-}, \forall u \in L^\Phi(\Omega), |u|_\Phi < 1, \tag{6.30}$$

$$|u|_\Phi^{\varphi^-} \leq \int_\Omega \Phi(|u|)dx \geq |u|_\Phi^{\varphi^-}, \forall u \in L^\Phi(\Omega), |u|_\Phi > 1. \tag{6.31}$$

In this section we establish the existence of weak solutions for problem $(\mathcal{P})$ provided $t \mapsto a(|t|)t$ defines an admissible function. In this case the appropriate function space for problem $(\mathcal{P})$ is $W_0^1 L^\Phi(\Omega)$ with Φ being the N-function generated by $\varphi : \mathbb{R} \to \mathbb{R}$, defined as follows

$$\varphi(t) := \begin{cases} 0, & \text{if } t = 0, \\ a(|t|)t, & \text{otherwise.} \end{cases} \tag{6.32}$$

Using the definition of the Clarke subdifferential, one can define weak solutions of problem $(\mathcal{P})$ in terms of hemivariational inequalities as follows: *a function* $u \in W_0^1 L^\Phi(\Omega)$ *is a weak solution of problem* $(\mathcal{P})$ *if*

$$\int_\Omega a(|\nabla u|)\nabla u \cdot \nabla v \mathrm{d}x \leq \int_\Omega f^0(x, u(x); v(x))\mathrm{d}x, \forall v \in W_0^1 L^\Phi(\Omega). \qquad (6.33)$$

We formulate below the basic assumptions that will be used in this section.

(H_1) $a : (0, \infty) \to (0, \infty)$ is s.t. the function φ defined in (6.32) is admissible and

$$1 < \varphi^- \leq \varphi^+ < \infty.$$

(H_2) $f : \Omega \times \mathbb{R} \to \mathbb{R}$ is a Carathéodory function s.t.
    (i) $f(x, 0) = 0$ for a.e. $x \in \Omega$;
    (ii) $t \mapsto f(x, t)$ is locally Lipschitz for a.e. $x \in \Omega$;
    (iii) there exist an admissible function $\psi : \mathbb{R} \to \mathbb{R}$ s.t. $1 < \psi^- \leq \psi^+ < \infty$ and

$$|\zeta| \leq \psi(|t|),$$

    for a.e. $x \in \Omega$, all $t \in \mathbb{R}$ and all $\zeta \in \partial_C^2 f(x, t)$.

(H_3) If

$$\int_1^\infty \frac{\Phi^{-1}(s)}{s^{\frac{N+1}{N}}} \mathrm{d}s = \infty,$$

then we assume that $\Psi(t) := \int_0^t \psi(s)\mathrm{d}s$ grows essentially more slowly than Φ_*.

Let us consider the functionals $I : W_0^1 L^\Phi(\Omega) \to \mathbb{R}$ and $F : L^\Psi(\Omega) \to \mathbb{R}$ defined by

$$I(u) := \int_\Omega \Phi(|\nabla u|)\mathrm{d}x,$$

and

$$F(w) := \int_\Omega f(x, w(x))\mathrm{d}x.$$

The *energy functional* corresponding to problem $(\mathcal{P})$, $E : W_0^1 L^\Phi(\Omega) \to \mathbb{R}$, is given by

$$E(u) := I(u) - F(u). \qquad (6.34)$$

The following lemma guarantees the fact that, in order to solve problem $(\mathcal{P})$, it suffices to seek for critical points of the energy functional associated with our problem.

Lemma 6.1 *Assume* $(H_1) - (H_3)$ *hold. Then the functional* $E : W_0^1 L^\Phi(\Omega) \to \mathbb{R}$ *defined in (6.34) has the following properties:*

(i) *E is locally Lipschitz;*
(ii) *E is weakly lower semicontinuous;*
(iii) *each critical point of E is a weak solution of problem* $(\mathcal{P})$.

Proof

(i) According to García-Huidobro et al. [20, Lemma 3.4] the functional I belongs to $C^1 \left(W_0^1 L^\Phi(\Omega), \mathbb{R} \right)$ and

$$\langle I'(u), v \rangle = \int_\Omega a(|\nabla u|) \nabla u \cdot \nabla v dx, \tag{6.35}$$

hence I is locally Lipschitz (see, e.g., Clarke [6, Section 2.2, p. 32]).

Since the embedding $W_0^1 L^\Phi(\Omega) \hookrightarrow L^\Psi(\Omega)$ is compact there exists $C_\Psi > 0$ such that

$$|u|_\Psi \leq C_\Psi \|u\|, \forall u \in W_0^1 L^\Phi(\Omega). \tag{6.36}$$

Let us fix now $u_0 \in W_0^1 L^\Phi(\Omega)$ and prove that there exists $r > 0$ sufficiently small such that F is Lipschitz continuous on $B_{W_0^1 L^\Phi(\Omega)}(u_0, r) := \{v \in W_0^1 L^\Phi(\Omega) : \|v - u_0\| < r\}$. Theorem 2.7 ensures the existence of an $r_0 > 0$ such that F is Lipschitz continuous on $B_{L^\Psi(\Omega)}(u_0, r_0)$, hence there exists a positive constant L such that

$$|F(w_1) - F(w_2)| \leq L|w_1 - w_2|_\Psi, \forall w_1, w_2 \in B_{L^\Psi(\Omega)}(u_0, r_0). \tag{6.37}$$

From (6.36) and (6.37) we get

$$|F(u_1) - F(u_2)| \leq LC_\Psi \|u_1 - u_2\|, \forall u_1, u_2 \in B_{W_0^1 L^\Phi(\Omega)}(u_0, r_0/C_\Psi). \tag{6.38}$$

(ii) Let us consider $\{u_n\} \subset W_0^1 L^\Phi(\Omega)$ such that $u_n \rightharpoonup u$ in $W_0^1 L^\Phi(\Omega)$. It is known that I is weakly lower semicontinuous (see García-Huidobro et al. [20, Lemma 3.2]). On the other hand, $u_n \to u$ in $L^\Psi(\Omega)$ and by Fatou's lemma

$$\limsup_{n\to\infty} F(u_n) = \limsup_{n\to\infty} \int_\Omega f(x, u_n(x))dx \leq \int_\Omega \limsup_{n\to\infty} f(x, u_n(x))dx$$

$$= \int_\Omega f(x, u(x))dx = F(u),$$

which shows that $F|_{W_0^1 L^\Phi(\Omega)}$ is weakly upper semicontinuous.

(iii) Let $u \in W_0^1 L^\Phi(\Omega)$ be a critical point of E. Basic subdifferential calculus ensures that

$$0 \in \partial_C E(u) \subseteq I'(u) - \partial_C \left(F|_{W_0^1 L^\Phi(\Omega)} \right)(u),$$

and $\partial_C \left(F|_{W_0^1 L^\Phi(\Omega)} \right)(u) = \partial_C F(u)$ in the sense that any element of $\partial_C \left(F|_{W_0^1 L^\Phi(\Omega)} \right)(u)$ admits a unique extension to an element of $\partial_C F(u)$. Hence there exists $\xi \in \partial_C F(u)$ such that

$$I'(u) = \xi, \text{ in } \left(W_0^1 L^\Phi(\Omega) \right)^*. \tag{6.39}$$

On the other hand, Theorem 2.7 ensures the existence of a $\zeta \in L^{\Psi^*}(\Omega)$ which satisfies

$$\begin{cases} \zeta(x) \in \partial_C^2 f(x, u(x)), & \text{for a.e. } x \in \Omega, \\ \langle \xi, w \rangle = \int_\Omega \zeta(x)w(x)dx, \ \forall w \in L^\Psi(\Omega). \end{cases} \tag{6.40}$$

It follows from (6.35), (6.39), and (6.40) that

$$\int_\Omega a(|\nabla u|)\nabla u \cdot \nabla v dx = \int_\Omega \zeta(x)v(x)dx \leq \int_\Omega f^0(x, u(x); v(x))dx, \ \forall v \in W_0^1 L^\Phi(\Omega).$$

$\square$

Theorem 6.5 ([10]) *Suppose* $(H_1) - (H_3)$ *hold and assume in addition that* $\psi^+ < \varphi^-$. *Then problem* $(\mathcal{P})$ *has at least one weak solution.*

Proof Let $u \in W_0^1 L^\Phi(\Omega)$ be such that $\|u\| > 1$. Then, from (6.31), we have

$$I(u) = \int_\Omega \Phi(|\nabla u|)dx \geq ||\nabla u||_\Phi^{\varphi^-} = ||u||^{\varphi^-}. \tag{6.41}$$

On the other hand, condition (H_2)-(iii) ensures that

$$|f(x,t)| \leq \Psi(2|t|), \quad \text{for a.e } x \in \Omega \text{ and all } t \in \mathbb{R}. \tag{6.42}$$

Indeed, Lebourg's mean value theorem and the Δ_2-condition ensure that for some $r \in \{\mu t : \mu \in (0,1)\}$ there exists $\zeta \in \partial_C^2 f(x,r)$ such that

$$|f(x,t)| = |f(x,t) - f(x,0)| = |\zeta t| \leq \psi(|r|)|t| \leq |t|\psi(|t|) \leq \int_{|t|}^{2|t|} \psi(s)ds \leq \Psi(2|t|).$$

Thus,

$$|F(u)| \leq \int_{\Omega} \Psi(|2u|)dx \leq k_1 \left(|u|_{\Psi}^{\psi^-} + |u|_{\Psi}^{\psi^+} \right) \leq k_2 \|u\|^{\psi^+},$$

for some suitable constant $k_2 > 0$. Therefore

$$E(u) = I(u) - F(u) \geq \|u\|^{\varphi^-} - k_2\|u\|^{\psi^+} \to \infty \text{ as } \|u\| \to \infty.$$

The fact that E is weakly lower semicontinuous and coercive ensures that there exists $u_0 \in W_0^1 L^{\Phi}(\Omega)$ (see Theorem 1.7) such that

$$E(u_0) = \inf_{v \in W_0^1 L^{\Phi}(\Omega)} E(v),$$

which means that u_0 is a critical point of E.   □

In the proof of the above theorem it is shown that E possesses a global minimizer, but it may happen that

$$\inf_{v \in W_0^1 L^{\Phi}(\Omega)} E(v) = 0 = E(0),$$

hence our problem might possess only the trivial solution. In order to avoid this it suffices to impose conditions that ensure the existence of at least one point $u \in W_0^1 L^{\Phi}(\Omega) \setminus \{0\}$ such that $E(u) \leq 0$. An example is given below.

(H_4) There exist $\theta \in \left(1, \varphi^-\right)$ and an open subset of positive measure $\omega \subset \Omega$ s.t.

$$\liminf_{t \to 0} \frac{\inf_{x \in \omega} f(x,t)}{|t|^{\theta}} > 0.$$

Lemma 6.2 *Suppose* $(H_1) - (H_4)$ *hold. Then there exist* $u_* \in W_0^1 L^\Phi(\Omega) \setminus \{0\}$ *and* $T_0 \in (0, 1)$ *such that*

$$E(tu_*) < 0, \forall t \in (0, T_0). \tag{6.43}$$

Proof Let ω_0 be such that $\omega_0 \subset\subset \omega$ and meas$(\omega_0) > 0$. Then there exists $u_* \in C_0^\infty(\omega)$ such that $\omega_0 \subset \text{supp}(u_*)$, $u_*(x) = 1$ on $\bar{\omega}_0$ and $0 \leq u_*(x) \leq 1$ on $\omega \setminus \bar{\omega}_0$. Obviously $u_* \in W_0^1 L^\Phi(\Omega) \setminus \{0\}$, hence $\|u_*\| > 0$. On the other hand, (H_4) ensures that for sufficiently small $\varepsilon > 0$ there exists $\delta > 0$ such that

$$f(x, t) \geq \varepsilon |t|^\theta, \forall (x, t) \in \omega \times [-\delta, \delta],$$

thus, for any $0 < t < \min\left\{1, \delta, \frac{1}{\|u_*\|}\right\}$ we have

$$E(tu_*) = I(tu_*) - F(tu_*) = \int_\Omega \Phi(t|\nabla u_*|) dx - \int_\omega f(x, tu_*(x)) dx \leq t^{\varphi^-} \|u_*\|^{\varphi^-}$$

$$- \int_{\bar{\omega}_0} \varepsilon |t|^\theta dx = t^\theta \left(t^{\varphi^- - \theta} \|u_*\|^{\varphi^-} - \varepsilon \text{meas}(\omega_0)\right),$$

which shows that (6.43) holds with $T_0 := \min\left\{1, \delta, \frac{1}{\|u_*\|}, \left(\frac{\varepsilon \text{meas}(\omega_0)}{\|u_*\|^{\varphi^-}}\right)^{\frac{1}{\varphi^- - \theta}}\right\}$. $\square$

Corollary 6.1 *Assume* $(H_1) - (H_4)$ *hold. If* $\psi^+ < \varphi^-$, *then problem* $(\mathcal{P})$ *has at least one nontrivial weak solution.*

In order to find critical points which are not necessarily global minimizers of E, instead of (H_1) we shall use the following more restrictive assumption:

(H_1') $a : (0, \infty) \to (0, \infty)$ is a non-decreasing function s.t. φ is admissible and

$$1 < \varphi^- \leq \varphi^+ < \infty.$$

The reasoning behind this is given by the following theorem.

Theorem 6.6 ([10]) *Assume* (H_1') *holds. Then the following assertions hold:*

(i) *The space* $\left(W_0^1 L^\Phi(\Omega), \|\cdot\|\right)$ *is uniformly convex;*
(ii) $I' : W_0^1 L^\Phi(\Omega) \to \left(W_0^1 L^\Phi(\Omega)\right)^*$ *satisfies the* $(S)_+$-*condition.*

Proof

(i) Let $\varepsilon \in (0, 2]$ be fixed and assume $u, v \in W_0^1 L^\Phi(\Omega)$ are such that $\|u\| = \|v\| = 1$ and $\|u - v\| \geq \varepsilon$. Keeping in mind the way $|\cdot|_\Phi$ was defined and relations (6.30)–(6.31) one can easily check that for any $w \in L^\Phi(\Omega)$

$$|w|_\Phi < 1 (> 1, = 1) \text{ if and only if } \int_\Omega \Phi(|w|)dx < 1 (> 1, = 1).$$

Thus

$$\int_\Omega \Phi(|\nabla u|)dx = \int_\Omega \Phi(|\nabla v|)dx = 1,$$

and

$$\int_\Omega \Phi\left(\frac{|\nabla u - \nabla v|}{2}\right) dx \geq \min\left\{ \left(\frac{\varepsilon}{2}\right)^{\varphi^-}, \left(\frac{\varepsilon}{2}\right)^{\varphi^+} \right\}.$$

On the other hand, the fact that a is non-decreasing implies that

$$0 \leq \frac{1}{2}\left(a(\sqrt{t}) - a(\sqrt{s})\right) = \Phi(\sqrt{\cdot})'(t) - \Phi(\sqrt{\cdot})'(s), \forall t \geq s > 0.$$

Thus the mapping $t \mapsto \Phi(\sqrt{t})$ is convex on $[0, \infty)$ and according to Lamperti [25, Theorem 2.1]

$$\Phi(|\zeta + \eta|) + \Phi(|\zeta - \eta|) \geq 2\Phi(|\zeta|) + 2\Phi(|\eta|), \forall \zeta, \eta \in \mathbb{R}^N. \tag{6.44}$$

Taking $\zeta := (\nabla u + \nabla v)/2, \eta := (\nabla u - \nabla v)/2$ and integrating over Ω we get

$$\int_\Omega \Phi\left(\left|\frac{\nabla u + \nabla v}{2}\right|\right) dx \leq \int_\Omega \frac{\Phi(|\nabla u|) + \Phi(|\nabla v|)}{2} - \Phi\left(\left|\frac{\nabla u - \nabla v}{2}\right|\right) dx, \tag{6.45}$$

that is,

$$1 > 1 - \gamma \geq \int_\Omega \Phi\left(\left|\frac{\nabla u + \nabla v}{2}\right|\right) dx \geq \left\|\frac{u + v}{2}\right\|^{\varphi^+},$$

with $\gamma := \min\{(\varepsilon/2)^{\varphi^-}, (\varepsilon/2)^{\varphi^+}\}$. Then $\delta := 1 - (1 - \gamma)^{\frac{1}{\varphi^+}}$.

(ii) Arguing by contradiction, assume there exist $\varepsilon_0 > 0$, $\{u_n\} \subset W_0^1 L^\Phi(\Omega)$ and $u \in W_0^1 L^\Phi(\Omega)$ such that $u_n \rightharpoonup u$ as $n \to \infty$, $\|u_n - u\| \geq \varepsilon_0$ for all $n \geq 1$ and

$$\limsup_{n \to \infty} \langle I'(u_n), u_n - u \rangle \leq 0.$$

Then

$$0 < \min\left\{\left(\frac{\varepsilon_0}{2}\right)^{\varphi^-}, \left(\frac{\varepsilon_0}{2}\right)^{\varphi^+}\right\} \le \int_\Omega \Phi\left(\left|\frac{\nabla u_n - \nabla u}{2}\right|\right) dx, \forall n \ge 1.$$

The convexity of I implies that $I(u) - I(u_n) \ge \langle I'(u_n), u - u_n\rangle$, for all $n \ge 1$, therefore (see (6.45)) the following estimates hold

$$I\left(\frac{u_n + u}{2}\right) = \int_\Omega \Phi\left(\left|\frac{\nabla u_n + \nabla u}{2}\right|\right) dx \le \int_\Omega \frac{\Phi(|\nabla u_n|) + \Phi(|\nabla u|)}{2} dx$$

$$-\int_\Omega \Phi\left(\left|\frac{\nabla u_n - \nabla u}{2}\right|\right) dx = \frac{I(u_n) + I(u)}{2} - \min\left\{\left(\frac{\varepsilon_0}{2}\right)^{\varphi^-}, \left(\frac{\varepsilon_0}{2}\right)^{\varphi^+}\right\}$$

$$\le I(u) + \frac{1}{2}\langle I'(u_n), u_n - u\rangle - \min\left\{\left(\frac{\varepsilon_0}{2}\right)^{\varphi^-}, \left(\frac{\varepsilon_0}{2}\right)^{\varphi^+}\right\}.$$

Keeping in mind that I is weakly lower semicontinuos and taking the superior limit we get

$$I(u) \le I(u) - \min\left\{\left(\frac{\varepsilon_0}{2}\right)^{\varphi^-}, \left(\frac{\varepsilon_0}{2}\right)^{\varphi^+}\right\},$$

which clearly is a contradiction.   □

A key ingredient in applying the Mountain Pass Theorem is to prove that E satisfies the (PS)-condition. The above theorem is useful in this regard as we have the following result concerning bounded (PS)-sequences for E.

Lemma 6.3 *Assume (H_1'), (H_2), and (H_3) and let $\{u_n\} \subset X$ be a bounded (PS)-sequence for E. Then $\{u_n\}$ possesses a (strongly) convergent subsequence.*

Proof The space $W_0^1 L^\Phi(\Omega)$ is uniformly convex, hence reflexive, thus there exist a subsequence $\{u_{n_k}\}$ of $\{u_n\}$ and $u \in W_0^1 L^\Phi(\Omega)$ such that

$$u_{n_k} \rightharpoonup u, \text{ in } W_0^1 L^\Phi(\Omega), \text{ and } u_{n_k} \to u, \text{ in } L^\Psi(\Omega).$$

Since $\lambda_E(u_{n_k}) \to 0$, one gets the following estimates

$$0 \le E^0(u_{n_k}; u - u_{n_k}) + \varepsilon_{n_k}\|u - u_{n_k}\| = (I - F)^0(u_{n_k}; u - u_{n_k}) + \varepsilon_{n_k}\|u - u_{n_k}\|$$

$$\le \langle I'(u_{n_k}), u - u_{n_k}\rangle + (-F)^0(u_{n_k}; u - u_{n_k}) + \varepsilon_{n_k}\|u - u_{n_k}\|$$

$$\le \langle I'(u_{n_k}), u - u_{n_k}\rangle + F^0(u_{n_k}; u_{n_k} - u) + \varepsilon_{n_k}\|u - u_{n_k}\|.$$

Using the fact that $F^0(\cdot, \cdot)$ is upper semicontinuous and $\varepsilon_{n_k} \to 0$ we get

$$\limsup_{k \to \infty} \langle I'(u_{n_k}), u_{n_k} - u \rangle \leq \limsup_{k \to \infty} F^0(u_{n_k}; u_{n_k} - u) + \limsup_{k \to \infty} \varepsilon_{n_k} \|u - u_{n_k}\| \leq 0.$$

The (S_+)-property of I' allows us to conclude that

$$u_{n_k} \to u, \text{ in } W_0^1 L^\Phi(\Omega),$$

completing the proof. $\qquad\square$

The previous lemma shows that, in order to prove E satisfies the (PS)-condition, we only need to impose conditions which ensure the boundedness of all (PS)-sequences $\{u_n\} \subset X$ for which $\{E(u_n)\}$ is bounded. Obviously this is the case if E is coercive, or equivalently bounded below as a locally Lipschitz functional which satisfies the (PS)-condition is bounded below if and only if it is coercive (see, e.g., Motreanu and Motreanu [30, Corollary 2]). However, this case is not of interest here as the existence of a global minimizer of E can be proved under weaker conditions via Theorem 6.5. Therefore, in the remainder of this section we discuss only the case when E is unbounded below. Let us consider the following nonsmooth counterpart of the Ambrosetti and Rabinowitz condition (see [2])

(H_5) There exist $\sigma > \varphi^+$ and $\mu > 0$ such that

$$\sigma f(x, t) \leq t\zeta, \text{ for a.e.} x \in \Omega,$$

whenever $|t| \geq \mu$ and $\zeta \in \partial_C^2 f(x, t)$.

Theorem 6.7 ([10]) *Assume (H_1'), (H_2), (H_3) and (H_5) hold. If E is unbounded below and $\varphi^+ < \psi^-$, then problem $(\mathcal{P})$ possesses a nontrivial weak solution.*

Proof We carry out the proof in several steps as follows.

STEP 1. *E satisfies the (PS)-condition.*

Let $\{u_n\} \subset W_0^1 L^\Phi(\Omega)$ be such that $\{E(u_n)\}$ is bounded and $\lambda_E(u_n) \to 0$ as $n \to \infty$. According to Lemma 6.3 it suffices to prove that $\{u_n\}$ is bounded. It is readily seen that there exists a sequence $\{\xi_n\} \subset \left(W_0^1 L^\Phi(\Omega)\right)^*$ such that $\xi_n \in \partial_C F(u_n)$ and

$$I'(u_n) - \xi_n \to 0, \text{ as } n \to \infty.$$

On the other hand, Theorem 2.7 ensures that there exists $\zeta_n \in L^{\Psi^*}(\Omega)$ such that $\zeta_n(x) \in \partial_C^2 f(x, u_n(x))$ for a.e. $x \in \Omega$ and

$$\langle \xi_n, v \rangle = \int_\Omega \zeta_n(x) v(x) dx, \forall v \in W_0^1 L^\Phi(\Omega).$$

Thus, there exists $N \in \mathbb{N}$ such that

$$\left| \langle J_\Phi'(u_n), u_n \rangle - \int_\Omega \zeta_n(x) u_n(x) \mathrm{d}x \right| \le \frac{1}{n}, \forall n \ge N. \tag{6.46}$$

Fix $n \ge N$ and define $\Omega_n := \{x \in \Omega : |u_n| \ge \mu\}$ and $\Omega_n^c := \Omega \setminus \Omega_n$. If $x \in \Omega_n^c$, then $|u_n(x)| < \mu$ and by (6.42) we have

$$\int_{\Omega_n^c} f(x, u_n(x)) \mathrm{d}x \le \int_{\Omega_n^c} \Psi(2\mu) \mathrm{d}x \le \Psi(2\mu) \mathrm{meas}(\Omega) =: c_1.$$

If $x \in \Omega_n$, then $|u_n(x)| \ge \mu$ and

$$\int_{\Omega_n} f(x, u_n) \mathrm{d}x \le \frac{1}{\sigma} \int_{\Omega_n} \zeta_n(x) u_n(x) \mathrm{d}x = \frac{1}{\sigma} \int_\Omega \zeta_n(x) u_n(x) \mathrm{d}x - \frac{1}{\sigma} \int_{\Omega_n^c} \zeta_n(x) u_n(x) \mathrm{d}x.$$

Hypothesis (H_2) implies that

$$\left| \int_{\Omega_n^c} \zeta_n(x) u_n(x) \mathrm{d}x \right| \le \int_{\Omega_n^c} \psi(|u_n(x)|)|u_n(x)| \mathrm{d}x \le \int_{\Omega_n^c} \psi(\mu)\mu \mathrm{d}x \le \mu\psi(\mu) \mathrm{meas}(\Omega) =: c_2.$$

But, $\{E(u_n)\}$ is bounded, hence there exists $M > 0$ such that

$$M \ge E(u_n) = I(u_n) - \int_\Omega f(x, u_n) \mathrm{d}x = I(u_n) - \int_{\Omega_n} f(x, u_n) \mathrm{d}x - \int_{\Omega_n^c} f(x, u_n) \mathrm{d}x$$

$$\ge I(u_n) - c_1 - \frac{c_2}{\sigma} - \frac{1}{\sigma} \int_\Omega \zeta_n(x) u_n(x) \mathrm{d}x.$$

Combining this with (6.46) we get

$$I(u_n) - \frac{1}{\sigma} \langle I'(u_n), u_n \rangle \le M + c_1 + \frac{c_2}{\sigma} + \frac{1}{n\sigma}, \forall n \ge N. \tag{6.47}$$

On the other hand, the definition of φ^+ shows that

$$t\varphi(t) \le \varphi^+ \Phi(t), \forall t \ge 0,$$

therefore,

$$\frac{1}{\sigma} \langle I'(u_n), u_n \rangle = \frac{1}{\sigma} \int_\Omega \varphi(|\nabla u_n|)|\nabla u_n| \mathrm{d}x \le \frac{\varphi^+}{\sigma} I(u_n). \tag{6.48}$$

Now, (6.30), (6.31), (6.47), and (6.48) and the fact that $\sigma > \varphi^+$ show that there exists a positive constant c_3 such that

$$\min\left\{\|u_n\|^{\varphi^-}, \|u_n\|^{\varphi^+}\right\} \le I(u_n) \le c_3, \forall n \ge N, \tag{6.49}$$

which shows that $\{u_n\}$ is indeed bounded in $W_0^1 L^\Phi(\Omega)$.

STEP 2.   *There exists $r > 0$ such that*

$$E(u) > 0, \forall u \in S_r := \{u \in W_0^1 L^\Phi(\Omega) : \|u\| = r\}.$$

Let $0 < r < \min\left\{\frac{1}{2}, \frac{1}{2C_\Psi}\right\}$. Then

$$\max\{\|u\|, |2u|_\Psi\} < 1, \forall u \in S_r.$$

Thus, for all $u \in S_r$ estimate (6.42) ensures that

$$E(u) = I(u) - \int_\Omega f(x, u(x))dx > \|u\|^{\varphi^+} - \int_\Omega \Psi(2|u|)dx > \|u\|^{\varphi^+} - |u|_\Psi^{\psi^-}$$

$$\ge \|u\|^{\varphi^+} - C_\Psi^{\psi^-} \|u\|^{\psi^-} = r^{\varphi^+}\left(1 - C_\Psi^{\psi^-} r^{\psi^- - \varphi^+}\right).$$

Obviously $E(u) > 0$ whenever $\|u\| = r$ and $0 < r < \min\left\{1, \frac{1}{2C_\Psi}, C_\Psi^{-\frac{\psi^-}{\psi^- - \varphi^+}}\right\}$.

STEP 3.   *The functional E maps bounded sets into bounded sets.*

Let $W \subset W_0^1 L^\Phi(\Omega)$ and $M > 1$ be such that

$$\|u\| \le M, \forall u \in W.$$

Then (6.30), (6.31), and (6.42) show that for all $u \in W$ we have

$$|E(u)| \le \int_\Omega \Phi(|\nabla u|)dx + \int_\Omega |f(x, u(x)|dx \le \max\left\{\|u\|^{\varphi^-}, \|u\|^{\varphi^+}\right\} + \int_\Omega \Psi(2|u|)dx$$

$$\le M^{\varphi^+} + \max\left\{2^{\psi^-}|u|_\Psi^{\psi^-}, 2^{\psi^+}|u|_\Psi^{\psi^+}\right\} \le M^{\varphi^+} + 2^{\psi^+}M^{\psi^+}\max\left\{C_\Psi^{\psi^-}, C_\Psi^{\psi^+}\right\}.$$

Since E is unbounded below, it follows that there exists $\{v_n\} \subset W_0^1 L^\Phi(\Omega)$ such that

$$E(v_n) \to -\infty, \text{ as } n \to \infty.$$

STEP 3 ensures that $\{v_n\}$ is unbounded, thus there exists $n_0 \ge 1$ such that

$$\|v_{n_0}\| > r \text{ and } E(v_{n_0}) \le 0,$$

with $r > 0$ being given by STEP 1. Applying Corollary 5.4 with $e := v_{n_0}$ we conclude that E possesses a nontrivial critical point. □

We point out the fact that, in the previous theorem, the requirement "E unbounded below" can be dropped if we use the following stronger version of (H_5):

(H_5') There exist $\sigma > \varphi^+$ and $\mu > 0$ such that

$$\sigma f(x, t) \leq t\zeta, \text{ for a.e. } x \in \Omega,$$

whenever $|t| \geq \mu$ and $\zeta \in \partial_C^2 f(x, t)$ and $f(x, t) > 0$, if $t \geq \mu$ or $t \leq -\mu$.

Corollary 6.2 *Assume (H_1'), (H_2), (H_3), and (H_5') hold. If $\varphi^+ < \psi^-$, then problem $(\mathcal{P})$ has at least one nontrivial weak solution.*

Proof We need to prove that (H_5') implies that E is unbounded below. Assume $f(x, t) > 0$ for $t \geq \mu$. We claim that there exists $\alpha \in L^1(\Omega)$, $\alpha > 0$ such that

$$f(x, t) \geq \alpha(x)t^\sigma, \text{ for a.e. } x \in \Omega, \text{ and all } t \geq \mu. \tag{6.50}$$

With this end in mind, let us consider $g : \Omega \times [\mu, \infty) \to \mathbb{R}$ defined by

$$g(x, t) := \frac{f(x, t)}{t^\sigma}.$$

Then, according to Clarke [6, Proposition 2.3.14], the functional g is locally Lipschitz with respect to the second variable and

$$\partial_C^2 g(x, t) \subseteq \frac{t\partial_C^2 f(x, t) - \sigma f(x, t)}{t^{\sigma+1}}. \tag{6.51}$$

Thus, for any $t > \mu$, Lebourg's mean value theorem ensures that there exist $s \in (\mu, t)$ and $\xi \in \partial_C^2 g(x, s)$ such that

$$g(x, t) - g(x, \mu) = \xi(t - \mu) \geq 0,$$

which shows that (6.50) holds with $\alpha(x) := \mu^{-\sigma} f(x, \mu)$, whenever $t \geq \mu$.

Let $\omega_0 \subset\subset \Omega$ be such that $\text{meas}(\omega_0) > 0$. Then there exists $u^* \in C_0^\infty(\Omega)$ such that $u^*(x) = 1$ on $\bar{\omega}_0$ and $0 \leq u^*(x) \leq 1$ on $\Omega \setminus \bar{\omega}_0$. Obviously $u^* \in W_0^1 L^\Phi(\Omega) \setminus \{0\}$ and for any $t > \max\left\{1, \mu, \frac{1}{\|u^*\|}\right\}$ we have

$$\omega_t := \left\{x \in \Omega : tu^*(x) \geq \mu\right\} \supset \bar{\omega}_0,$$

and

$$I(tu^*) \leq t^{\varphi^+} \|u^*\|^{\varphi^+}.$$

On the other hand,

$$F(tu^*) = \int_\Omega f(x, tu^*(x))dx = \int_{\omega_t} f(x, tu^*(x))dx + \int_{\Omega \setminus \omega_t} f(x, tu^*(x))dx$$

$$\geq \int_{\omega_0} f(x, tu^*)dx - \int_{\Omega \setminus \omega_t} \Psi(2\mu)dx \geq t^\sigma \int_{\omega_0} \alpha(x)dx - \Psi(2\mu)\mathrm{meas}(\Omega),$$

which shows that $E(tu^*) \to -\infty$ as $t \to \infty$. A similar argument can be employed if $f(x, t) > 0$ for $t \leq -\mu$. □

If the nonlinearity f satisfies (H_5), but does not satisfy (H_5'), then we can use the following assumption

(H_6) There exist $\Theta > \varphi^+$ and an open subset of positive measure $\omega \subset \Omega$ such that either

$$\liminf_{t \to \infty} \frac{\inf_{x \in \omega} f(x, t)}{t^\Theta} > 0, \tag{6.52}$$

or,

$$\liminf_{t \to -\infty} \frac{\inf_{x \in \omega} f(x, t)}{|t|^\Theta} > 0. \tag{6.53}$$

Corollary 6.3 *Assume (H_1'), (H_2), (H_3), (H_5), and (H_6) hold. If $\varphi^+ < \psi^-$, then problem $(\mathcal{P})$ has at least one nontrivial weak solution.*

Proof Condition (H_6) implies that for any sufficiently small $\varepsilon > 0$ there exists $\delta > 0$ such that

$$f(x, t) \geq \varepsilon t^\Theta, \forall (x, t) \in \omega \times [\delta, \infty),$$

if (6.52) holds and

$$f(x, t) \geq \varepsilon |t|^\Theta, \forall (x, t) \in \omega \times (-\infty, -\delta],$$

if (6.53) is satisfied.

Reasoning as in the proof of Corollary 6.3, one can easily prove that there exists $u^* \in W_0^1 L^\Phi(\Omega) \setminus \{0\}$ such that either $E(tu^*) \to -\infty$ as $t \to \infty$, or $E(tu^*) \to -\infty$ as $t \to -\infty$. □

6.5.2   Dropping the Ambrosetti-Rabinowitz Type Condition

Although in general (PS)-sequences do not lead to critical points we have seen in the previous section (see Lemma 6.3) that, under some reasonable assumptions, any bounded (PS)-sequence possesses a subsequence converging to a critical point of our energy functional. However, the Ambrosetti-Rabinowitz type condition (H_5), which ensures the boundedness of every (PS)-sequence, is quite restrictive and many nonlinearities fail to fulfil it. Consequently, it is natural to ask ourselves if bounded (PS)-sequences can be obtained without this condition, or even if the energy functional does not satisfy the (PS)-condition at all.

Let us consider the following eigenvalue problem obtained by perturbing $(\mathcal{P})$ with the duality mapping

$$(\mathcal{P}_\lambda) : \quad \begin{cases} -\Delta_\Phi u \in \lambda J_a(u) + \partial_C^2 f(x, u), & \text{in } \Omega, \\ u = 0, & \text{on } \partial\Omega, \end{cases}$$

where J_a is the duality mapping on $W_0^1 L^\Phi(\Omega)$ corresponding to the normalization function $[0, \infty) \ni t \mapsto a(t)t$.

Note that, if (H_1) holds, then the norm $\| \cdot \|$ is Fréchet-differentiable on $W_0^1 L^\Phi(\Omega) \setminus \{0\}$ (see, e.g., Dincă and Matei [11, Theorem 3.6]) and for $u \neq 0$

$$\langle \| \cdot \|'(u), v \rangle = \|u\| \frac{\int_\Omega a\left(\frac{|\nabla u|}{\|u\|}\right) \nabla u \cdot \nabla v \, dx}{\int_\Omega a\left(\frac{|\nabla u|}{\|u\|}\right) |\nabla u|^2 dx}, \quad \forall v \in W_0^1 L^\Phi(\Omega). \tag{6.54}$$

Consequently,

$$\langle J_a(u), v \rangle = \langle \Phi(\| \cdot \|)'(u), v \rangle = \begin{cases} 0, & \text{if } u = 0, \\ a(\|u\|)\|u\|^2 \dfrac{\int_\Omega a\left(\frac{|\nabla u|}{\|u\|}\right) \nabla u \cdot \nabla v \, dx}{\int_\Omega a\left(\frac{|\nabla u|}{\|u\|}\right) |\nabla u|^2 dx}, & \text{otherwise.} \end{cases}$$

Definition 6.5 A number $\lambda \in \mathbb{R}$ is called *eigenvalue* of problem $(\mathcal{P}_\lambda)$ if there exists $u_\lambda \in W_0^1 L^\Phi(\Omega) \setminus \{0\}$ such that

$$\int_\Omega a(|\nabla u_\lambda|)\nabla u_\lambda \cdot \nabla v dx - \lambda \langle J_a(u_\lambda), v\rangle \leq \int_\Omega f^0(x, u_\lambda(x); v(x))dx, \forall v \in W_0^1 L^\Phi(\Omega).$$

The function u_λ is called an eigenfunction corresponding to λ.

Reasoning as in Lemma 6.1 one can easily check that in order to find eigenvalues of problem $(\mathcal{P})$ it suffices to seek for nontrivial critical points of the locally Lipschitz functional $\mathcal{E}_\lambda : W_0^1 L^\Phi(\Omega) \to \mathbb{R}$

$$\mathcal{E}_\lambda(u) := E(u) - \lambda \Phi(\|u\|),$$

or equivalently to solve the following differential inclusion

$$\lambda J_a(u) \in \partial_C E(u), \tag{6.55}$$

with E being the energy functional corresponding to problem $(\mathcal{P})$. Obviously, any eigenfunction corresponding to $\lambda_0 := 0$ is a nontrivial solution of problem $(\mathcal{P})$. We prove that either problem $(\mathcal{P})$ possesses multiple nontrivial weak solutions or problem $(\mathcal{P}_\lambda)$ possesses a rich family of negative eigenvalues. Note that, under (H_1), the space $W_0^1 L^\Phi(\Omega)$ is reflexive, which combined with (6.54) ensures that $(W_0^1 L^\Phi(\Omega))^*$ is strictly convex, hence the results from Sect. 5.4 are indeed applicable here. In order to establish the main result of this section we assume, among others, that $E(u_0) \leq 0$ for some $u_0 \in W_0^1 L^\Phi(\Omega) \setminus \{0\}$. A simple condition ensuring this is

(H_7) There exists $u_0 \in W_0^1 L^\Phi(\Omega) \setminus \{0\}$ such that

$$\max\left\{\|u_0\|^{\varphi^+}, \|u_0\|^{\varphi^-}\right\} \leq \int_\Omega f(x, u_0(x))dx.$$

Theorem 6.8 *Assume (H_1'), (H_2), (H_3) and (H_7) hold. If $\varphi^+ < \psi^-$, then the following alternative takes place:*

(A_1) *problem $(\mathcal{P})$ possesses at least two nontrivial weak solutions;*
   or,
(A_2) *for each $R \in (\|u_0\|, \infty)$ there exists an eigenpair (λ, u_λ) of problem $(\mathcal{P}_\lambda)$ satisfying $\lambda < 0$ and $\|u_\lambda\| = R$.*

Proof Let $R > \|u_0\|$ be fixed. Employing the same arguments as in the proof of Theorem 6.7, we conclude that E maps bounded sets into bounded sets and there exists $\rho \in (0, \|u_0\|)$ such that

$$E(u) \geq \rho^{\varphi^+} \left(1 - C_{\Psi}^{\psi^-} \rho^{\psi^- - \varphi^+}\right) =: \alpha > 0, \quad \forall u \in S_\rho. \tag{6.56}$$

Let us define $A := \{0, u_0\}$, $B := S_\rho$, with $\rho > 0$ given by (6.56), $b_0 := \inf_B E \geq \alpha > 0$, $m_R := \inf_{B_R} E$ and $c_R := \inf_{\Gamma \in \Phi} \sup_{\substack{t \in [0,1] \\ u \in A}} E(\Gamma(t, u))$. Then the set A links B w.r.t Φ in the sense of Definition E.5 (see Example E.1 and Remark E.1 in Appendix E) and

$$-\infty < m_R \leq 0 < b_0 \leq c_R < \infty. \tag{6.57}$$

We prove next that E satisfies the assumptions of Theorem 5.13.

STEP 1. *There exists $\Lambda_R > 0$ such that $|\langle \zeta, u \rangle| \leq \Lambda_R$, for all $u \in S_R$ and all $\zeta \in \partial_C E(u)$.*

Let $u \in S_R$ and $\zeta \in \partial_C E(u)$ be fixed. According to Theorem 2.7 there exists $\xi \in L^{\Psi^*}(\Omega)$ such that $\xi(x) \in \partial_C^2 f(x, u(x))$ for a.e. $x \in \Omega$ and

$$|\langle \zeta, u \rangle| = \left| \langle I'(u), u \rangle - \int_\Omega \xi u \, dx \right| \leq \int_\Omega a(|\nabla u|)|\nabla u|^2 dx + \int_\Omega |\xi||u| dx$$

$$\leq \int_\Omega \varphi(|\nabla u|)|\nabla u| dx + \int_\Omega \psi(|u|)|u| dx \leq \int_\Omega \Phi(2|\nabla u|) dx + \int_\Omega \Psi(2|u|) dx$$

$$\leq \max\left\{ (2\|u\|)^{\varphi^+}, (2\|u\|)^{\varphi^-} \right\} + \max\left\{ (2|u|_\Psi)^{\psi^+}, (2|u|_\Psi)^{\psi^-} \right\}$$

$$\leq \max\left\{ (2R)^{\varphi^+}, (2R)^{\varphi^-} \right\} + \max\left\{ (2C_\Psi R)^{\psi^+}, (2C_\Psi R)^{\psi^-} \right\} =: \Lambda_R.$$

STEP 2. *The functional E satisfies $(SPS)_c$ in $\overline{B}_R$ for all $c \in \mathbb{R}$.*

Let $\{u_n\} \subset \overline{B}_R$ and $c \in \mathbb{R}$ be such that

- $E(u_n) \to c$ as $n \to \infty$;
- there exist $\zeta_n \in \partial_C E(u_n)$ and $\nu \leq 0$ s.t. $\|\pi_{u_n}(\zeta_n)\| \to 0$ and $\langle \zeta_n, u_n \rangle \to \nu$.

The boundedness of $\{u_n\}$ and the fact that $W_0^1 L^\Phi(\Omega)$ is reflexive ensure there exist $u \in W_0^1 L^\Phi(\Omega)$ and subsequence of $\{u_n\}$, again denoted $\{u_n\}$, such that

$$u_n \rightharpoonup u \text{ in } W_0^1 L^\Phi(\Omega) \text{ and } u_n \to u \text{ in } L^\Psi(\Omega).$$

Without loss of generality we may assume that $\|u_n\| \to r$. If $r = 0$, then $u_n \to 0$ and the proof of the Claim is complete. If $r > 0$, then

$$0 = \lim_{n\to\infty} \langle \pi_{u_n}(\zeta_n), u_n - u \rangle$$

$$= \lim_{n\to\infty} \langle \zeta_n, u_n - u \rangle - \lim_{n\to\infty} \frac{\langle \zeta_n, u_n \rangle}{a(\|u_n\|)\|u_n\|^2} \langle J_a(u_n), u_n - u \rangle.$$

Keeping in mind the definition of J_a and the fact that the duality mapping is demicontinuous on reflexive Banach spaces, we get

$$\lim_{n\to\infty} \langle \zeta_n, u_n - u \rangle = \nu \left(1 - \frac{a(\|u\|)\|u\|^2}{a(r)r^2} \right). \tag{6.58}$$

On the other hand, for each $n \in \mathbb{N}$ there exists $\xi_n \in \partial_C F(u_n)$ such that

$$\zeta_n = I'(u_n) - \xi_n. \tag{6.59}$$

Since $L^\Psi(\Omega)$ is reflexive, $u_n \to u$ in $L^\Psi(\Omega)$ and $\xi_n \in \partial_C F(u_n)$, it follows (see Costea et al. [9, Proposition 2]) that there exist $\xi \in \partial_C F(u)$ and a subsequence of $\{\xi_n\}$ (for simplicity we do not relabel) such that $\xi_n \rightharpoonup \xi$ in $L^{\Psi^*}(\Omega)$, hence

$$\lim_{n\to\infty} \langle \xi_n, u_n - u \rangle = 0. \tag{6.60}$$

Combining (6.58)–(6.60) we get

$$\lim_{n\to\infty} \langle I'(u_n), u_n - u \rangle = \nu \left(1 - \frac{\varphi(\|u\|)\|u\|}{\varphi(r)r} \right) \leq 0,$$

as $\nu \leq 0$, φ is strictly incresing and $\|u\| \leq \liminf_{n\to\infty} \|u_n\| = r$. Therefore $u_n \to u$ in $W_0^1 L^\Phi(\Omega)$, due to the (S_+) property of I'.

The above steps show that E satisfies the conditions of Theorem 5.13. Consequently, there exist $u_1, u_2 \in \overline{B}_R$ and $\lambda_1, \lambda_2 \leq 0$ such that

$$E(u_1) = m_R, \quad E(u_2) = c_R \text{ and } \lambda_k J_a(u_k) \in \partial_C^2 E(u_k), \quad k = 1, 2.$$

Relation (6.57) implies that $u_1 \neq u_2$ and $0 \notin \{u_1, u_2\}$. If $\lambda_1 = \lambda_2 = 0$, then (A_1) is obtained. Otherwise, at least one eigenvalue is negative which forces the corresponding eigenfunction to belong to S_R.   □

6.6 Differential Inclusions Involving Oscillatory Terms

Let $F, G : \mathbb{R}_+ \to \mathbb{R}$ be locally Lipschitz functions and as usual, let us denote by $\partial_C F$ and $\partial_C G$ their generalized gradients in the sense of Clarke. Hereafter, $\mathbb{R}_+ = [0, \infty)$. Let $p > 0, \lambda \geq 0$ and $\Omega \subset \mathbb{R}^N$ be a bounded open domain, and consider the elliptic differential inclusion problem

$$
\begin{cases}
-\Delta u(x) \in \partial_C F(u(x)) + \lambda \partial_C G(u(x)) & \text{in } \Omega; \\
u \geq 0 & \text{in } \Omega; \\
u = 0, & \text{on } \partial\Omega.
\end{cases}
\tag{$\mathcal{D}_\lambda$}
$$

In the sequel, we provide a quite complete picture about the competition concerning the terms $s \mapsto \partial_C F(s)$ and $s \mapsto \partial_C G(s)$, respectively. In fact, we distinguish the cases when $\partial_C F$ oscillates near the *origin* or at *infinity*; we follow the results of Kristály, Mezei and Szilák [24]. Before stating such competition phenomena, we provide a general localization result.

6.6.1 Localization: A Generic Result

We consider the following differential inclusion problem

$$
\begin{cases}
-\Delta u(x) + ku(x) \in \partial_C A(u(x)), & u(x) \geq 0 \quad x \in \Omega, \\
u(x) = 0 & x \in \partial\Omega,
\end{cases}
\tag{D_A^k}
$$

where $k > 0$ and

(H_A^1): $A : [0, \infty) \to \mathbb{R}$ is a locally Lipschitz function with $A(0) = 0$, and there is $M_A > 0$ such that

$$
\max\{|\partial_C A(s)|\} := \max\{|\xi| : \xi \in \partial_C A(s)\} \leq M_A
$$

for every $s \geq 0$;

(H_A^2): there are $0 < \delta < \eta$ such that $\max\{\xi : \xi \in \partial_C A(s)\} \leq 0$ for every $s \in [\delta, \eta]$.

For simplicity, we extend the function A by $A(s) = 0$ for $s \leq 0$; the extended function is locally Lipschitz on the whole $\mathbb{R}$. The natural energy functional $\mathcal{T} : H_0^1(\Omega) \to \mathbb{R}$ associated with the differential inclusion problem (D_A^k) is defined by

$$
\mathcal{T}(u) := \frac{1}{2}\|u\|_{H_0^1}^2 + \frac{k}{2}\int_\Omega u^2 dx - \int_\Omega A(u(x))dx.
$$

The energy functional $\mathcal{T}$ is well defined and locally Lipschitz on $H_0^1(\Omega)$, while its critical points in the sense of Chang are precisely the weak solutions of the differential inclusion problem

$$\begin{cases} -\Delta u(x) + ku(x) \in \partial_C A(u(x)), & x \in \Omega, \\ u(x) = 0 & x \in \partial\Omega; \end{cases} \qquad (\text{D}_A^{k,0})$$

note that at this stage we have no information on the sign of u. Indeed, if $0 \in \partial_C \mathcal{T}(u)$, then for every $v \in H_0^1(\Omega)$ we have

$$\int_\Omega \nabla u(x) \nabla v(x) dx - k \int_\Omega u(x) v(x) dx - \int_\Omega \xi_x(x) v(x) dx = 0,$$

where $\xi_x \in \partial_C A(u(x))$ a.e. $x \in \Omega$, see e.g. Motreanu and Panagiotopoulos [31]. By using the divergence theorem for the first term at the left hand side (and exploring the Dirichlet boundary condition), we obtain that

$$\int_\Omega \nabla u(x) \nabla v(x) dx = - \int_\Omega \text{div}(\nabla u(x)) v(x) dx = - \int_\Omega \Delta u(x) v(x) dx.$$

Accordingly, we have that

$$-\int_\Omega \Delta u(x) v(x) dx + k \int_\Omega u(x) v(x) = \int_\Omega \xi_x v(x) dx$$

for every test function $v \in H_0^1(\Omega)$ which means that $-\Delta u(x) + ku(x) \in \partial_C A(u(x))$ in the weak sense in Ω, as claimed before.

Let us consider the number $\eta \in \mathbb{R}$ from (H_A^2) and the set

$$W^\eta = \{u \in H_0^1(\Omega) : \|u\|_{L^\infty} \leq \eta\}.$$

Our localization result reads as follows (see Kristály and Moroşanu [22, Theorem 2.1] for its smooth form):

Theorem 6.9 *Let $k > 0$ and assume that hypotheses (H_A^1) and (H_A^2) hold. Then*

(i) *the energy functional $\mathcal{T}$ is bounded from below on W^η and its infimum is attained at some $\tilde{u} \in W^\eta$;*

(ii) *$\tilde{u}(x) \in [0, \delta]$ for a.e. $x \in \Omega$;*

(iii) *$\tilde{u}$ is a weak solution of the differential inclusion (D_A^k).*

Proof

(i) Due to (H^1_A), it is clear that the energy functional $\mathcal{T}$ is bounded from below on $H^1_0(\Omega)$. Moreover, due to the compactness of the embedding $H^1_0(\Omega) \subset L^q(\Omega)$, $q \in [2, 2^*)$, it turns out that $\mathcal{T}$ is sequentially weak lower semi-continuous on $H^1_0(\Omega)$. In addition, the set W^η is weakly closed, being convex and closed in $H^1_0(\Omega)$. Thus, there is $\tilde{u} \in W^\eta$ which is a minimum point of $\mathcal{T}$ on the set W^η.

(ii) We introduce the set $L = \{x \in \Omega : \tilde{u}(x) \notin [0, \delta]\}$ and suppose indirectly that $m(L) > 0$. Define the function $\gamma : \mathbb{R} \to \mathbb{R}$ by $\gamma(s) = \min(s_+, \delta)$, where $s_+ = \max(s, 0)$. Now, set $w = \gamma \circ \tilde{u}$. It is clear that γ is a Lipschitz function and $\gamma(0) = 0$. Accordingly, based on the superposition theorem of Marcus and Mizel [27], one has that $w \in H^1_0(\Omega)$. Moreover, $0 \le w(x) \le \delta$ for a.e. Ω. Consequently, $w \in W^\eta$.

Let us introduce the sets

$$L_1 = \{x \in L : \tilde{u}(x) < 0\} \text{ and } L_2 = \{x \in L : \tilde{u}(x) > \delta\}.$$

In particular, $L = L_1 \cup L_2$, and by definition, it follows that $w(x) = \tilde{u}(x)$ for all $x \in \Omega \setminus L$, $w(x) = 0$ for all $x \in L_1$, and $w(x) = \delta$ for all $x \in L_2$. In addition, one has

$$\mathcal{T}(w) - \mathcal{T}(\tilde{u}) = \frac{1}{2}\left[\|w\|^2_{H^1_0} - \|\tilde{u}\|^2_{H^1_0}\right] + \frac{k}{2}\int_\Omega \left[w^2 - \tilde{u}^2\right] - \int_\Omega [A(w(x)) - A(\tilde{u}(x))]$$

$$= -\frac{1}{2}\int_L |\nabla \tilde{u}|^2 + \frac{k}{2}\int_L [w^2 - \tilde{u}^2] - \int_L [A(w(x)) - A(\tilde{u}(x))].$$

On account of $k > 0$, we have

$$k\int_L [w^2 - \tilde{u}^2] = -k\int_{L_1} \tilde{u}^2 + k\int_{L_2} [\delta^2 - \tilde{u}^2] \le 0.$$

Since $A(s) = 0$ for all $s \le 0$, we have

$$\int_{L_1} [A(w(x)) - A(\tilde{u}(x))] = 0.$$

By means of the Lebourg's mean value theorem, for a.e. $x \in L_2$, there exists $\theta(x) \in [\delta, \tilde{u}(x)] \subseteq [\delta, \eta]$ such that

$$A(w(x)) - A(\tilde{u}(x)) = A(\delta) - A(\tilde{u}(x)) = a(\theta(x))(\delta - \tilde{u}(x)),$$

where $a(\theta(x)) \in \partial_C A(\theta(x))$. Due to (H_A^2), it turns out that

$$\int_{L_2} [A(w(x)) - A(\tilde{u}(x))] \geq 0.$$

Therefore, we obtain that $\mathcal{T}(w) - \mathcal{T}(\tilde{u}) \leq 0$. On the other hand, since $w \in W^\eta$, then $\mathcal{T}(w) \geq \mathcal{T}(\tilde{u}) = \inf_{W^\eta} \mathcal{T}$, thus every term in the difference $\mathcal{T}(w) - \mathcal{T}(\tilde{u})$ should be zero; in particular,

$$\int_{L_1} \tilde{u}^2 = \int_{L_2} [\tilde{u}^2 - \delta^2] = 0.$$

The latter relation implies in particular that $m(L) = 0$, which is a contradiction, completing the proof of (ii).

(iii) Since $\tilde{u}(x) \in [0, \delta]$ for a.e. $x \in \Omega$, an arbitrarily small perturbation $\tilde{u} + \epsilon v$ of $\tilde{u}$ with $0 < \epsilon \ll 1$ and $v \in C_0^\infty(\Omega)$ still implies that $\mathcal{T}(\tilde{u} + \varepsilon v) \geq \mathcal{T}(\tilde{u})$; accordingly, $\tilde{u}$ is a minimum point for $\mathcal{T}$ in the strong topology of $H_0^1(\Omega)$, thus $0 \in \partial_C \mathcal{T}(\tilde{u})$. Consequently, it follows that $\tilde{u}$ is a weak solution of the differential inclusion (D_A^k). $\qquad\square$

In the sequel, we need a truncation function of $H_0^1(\Omega)$. To construct this function, let $B(x_0, r) \subset \Omega$ be the N-dimensional ball with radius $r > 0$ and center $x_0 \in \Omega$. For $s > 0$, define

$$w_s(x) = \begin{cases} 0, & \text{if } x \in \Omega \setminus B(x_0, r); \\ s, & \text{if } x \in B(x_0, r/2); \\ \frac{2s}{r}(r - |x - x_0|), & \text{if } x \in B(x_0, r) \setminus B(x_0, r/2). \end{cases} \tag{6.61}$$

Note that that $w_s \in H_0^1(\Omega)$, $\|w_s\|_{L^\infty} = s$ and

$$\|w_s\|_{H_0^1}^2 = \int_\Omega |\nabla w_s|^2 = 4r^{N-2}(1 - 2^{-N})\omega_N n s^2 \equiv C(r, Nn)s^2 > 0; \tag{6.62}$$

hereafter ω_N stands for the volume of $B(0, 1) \subset \mathbb{R}^N$.

6.6.2   Oscillation Near the Origin

We assume:

(F_0^0) $F(0) = 0$;

(F_1^0) $-\infty < \liminf_{s \to 0^+} \frac{F(s)}{s^2}$; $\limsup_{s \to 0^+} \frac{F(s)}{s^2} = +\infty$;

(F_2^0) $l_0 := \liminf_{s \to 0+} \frac{\max\{\xi : \xi \in \partial_C F(s)\}}{s} < 0.$

(G_0^0) $G(0) = 0;$

(G_1^0) There exist $p > 0$ and $\underline{c}, \overline{c} \in \mathbb{R}$ such that

$$\underline{c} = \liminf_{s \to 0+} \frac{\min\{\xi : \xi \in \partial_C G(s)\}}{s^p} \leq \limsup_{s \to 0+} \frac{\max\{\xi : \xi \in \partial_C G(s)\}}{s^p} = \overline{c}.$$

Remark 6.3 Hypotheses (F_1^0) and (F_2^0) imply a strong oscillatory behavior of $\partial_C F$ near the origin. Moreover, it turns out that $0 \in \partial_C F(0)$; indeed, if we assume the contrary, by the upper semicontinuity of $\partial_C F$ we also have that $0 \notin \partial_C F(s)$ for every small $s > 0$. Thus, by (F_2^0) we have that $\partial_C F(s) \subset (-\infty, 0]$ for these values of $s > 0$. By using (F_0^0) and Lebourg's mean value theorem, it follows that $F(s) = F(s) - F(0) = \xi s \leq 0$ for some $\xi \in \partial_C F(\theta s) \subset (-\infty, 0]$ with $\theta \in (0, 1)$. The latter inequality contradicts the second assumption from (F_1^0). Similarly, one obtains that $0 \in \partial_C G(0)$ by exploring (G_0^0) and (G_1^0), respectively.

In conclusion, since $0 \in \partial_C F(0)$ and $0 \in \partial_C G(0)$, it turns out that $0 \in H_0^1(\Omega)$ is a solution of the differential inclusion $(\mathcal{D}_\lambda)$. Clearly, we are interested in nonzero solutions of $(\mathcal{D}_\lambda)$.

Example 6.2 Let us consider $F_0(s) = \int_0^s f_0(t), s \geq 0$, where $f_0(t) = \sqrt{t}(\frac{1}{2} + \sin t^{-1})$, $t > 0$ and $f_0(0) = 0$, or some of its jumping variants. One can prove that $\partial_C F_0 = f_0$ verifies the assumptions $(F_0^0) - (F_2^0)$. For a fixed $p > 0$, let $G_0(s) = \ln(1 + s^{p+2}) \max\{0, \cos s^{-1}\}, s > 0$ and $G_0(0) = 0$. It is clear that G_0 is not of class C^1 and verifies (G_1^0) with $\underline{c} = -1$ and $\overline{c} = 1$, respectively; see Fig. 6.1 representing both f_0 and G_0 (for $p = 2$).

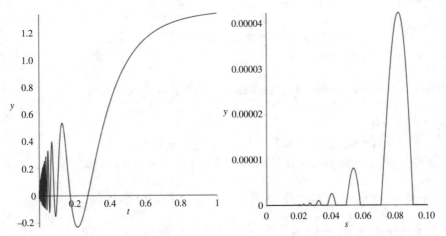

Fig. 6.1 Graphs of f_0 and G_0 around the origin, respectively

First, we are going to show that when $p \geq 1$ then the 'leading' term is the oscillatory function $\partial_C F$; roughly speaking, one can say that the effect of $s \mapsto \partial_C G(s)$ is negligible in this competition. More precisely, we prove the following result.

Theorem 6.10 ([24]) (Case $p \geq 1$) *Assume that $p \geq 1$ and the locally Lipschitz functions $F, G : \mathbb{R}_+ \to \mathbb{R}$ satisfy $(F_0^0) - (F_2^0)$ and $(G_0^0) - (G_1^0)$. If*

(i) either $p = 1$ and $\lambda \bar{c} < -l_0$ (with $\lambda \geq 0$),
(ii) or $p > 1$ and $\lambda \geq 0$ is arbitrary,

then the differential inclusion problem $(\mathcal{D}_\lambda)$ admits a sequence $\{u_i\}_i \subset H_0^1(\Omega)$ of distinct weak solutions such that

$$\lim_{i \to \infty} \|u_i\|_{H_0^1} = \lim_{i \to \infty} \|u_i\|_{L^\infty} = 0. \tag{6.63}$$

In the case when $p < 1$, the perturbation term $\partial_C G$ may compete with the oscillatory function $\partial_C F$; namely, we have:

Theorem 6.11 ([24]) (Case $0 < p < 1$) *Assume $0 < p < 1$ and that the locally Lipschitz functions $F, G : \mathbb{R}_+ \to \mathbb{R}$ satisfy $(F_0^0) - (F_2^0)$ and $(G_0^0) - (G_1^0)$. Then, for every $k \in \mathbb{N}$, there exists $\lambda_k > 0$ such that the differential inclusion $(\mathcal{D}_\lambda)$ has at least k distinct weak solutions $\{u_{1,\lambda}, \ldots, u_{k,\lambda}\} \subset H_0^1(\Omega)$ whenever $\lambda \in [0, \lambda_k]$. Moreover,*

$$\|u_{i,\lambda}\|_{H_0^1} < i^{-1} \quad and \quad \|u_{i,\lambda}\|_{L^\infty} < i^{-1} \quad for\ any\ i = \overline{1, k};\ \lambda \in [0, \lambda_k]. \tag{6.64}$$

Before giving the proof of Theorems 6.10 and 6.11, we study the differential inclusion problem

$$\begin{cases} -\Delta u(x) + k u(x) \in \partial_C A(u(x)), & u(x) \geq 0 \quad x \in \Omega, \\ u(x) = 0 & x \in \partial\Omega, \end{cases} \tag{D_A^k}$$

where $k > 0$ and the locally Lipschitz function $A : \mathbb{R}_+ \to \mathbb{R}$ verifies

(H_0^0): $A(0) = 0$;
(H_1^0): $-\infty < \liminf_{s \to 0^+} \frac{A(s)}{s^2}$ and $\limsup_{s \to 0^+} \frac{A(s)}{s^2} = +\infty$;
(H_2^0): there are two sequences $\{\delta_i\}, \{\eta_i\}$ with $0 < \eta_{i+1} < \delta_i < \eta_i$, $\lim_{i \to \infty} \eta_i = 0$, and

$$\max\{\partial_C A(s)\} := \max\{\xi : \xi \in \partial_C A(s)\} \leq 0$$

for every $s \in [\delta_i, \eta_i]$, $i \in \mathbb{N}$.

Theorem 6.12 *Let $k > 0$ and assume hypotheses* (H_0^0), (H_1^0) *and* (H_2^0) *hold. Then there exists a sequence* $\{u_i^0\}_i \subset H_0^1(\Omega)$ *of distinct weak solutions of the differential inclusion problem* (D_A^k) *such that*

$$\lim_{i \to \infty} \|u_i^0\|_{H_0^1} = \lim_{i \to \infty} \|u_i^0\|_{L^\infty} = 0. \tag{6.65}$$

Proof We may assume that $\{\delta_i\}_i$, $\{\eta_i\}_i \subset (0, 1)$. For any fixed number $i \in \mathbb{N}$, we define the locally Lipschitz function $A_i : \mathbb{R} \to \mathbb{R}$ by

$$A_i(s) = A(\tau_{\eta_i}(s)), \tag{6.66}$$

where $A(s) = 0$ for $s \leq 0$ and $\tau_\eta : \mathbb{R} \to \mathbb{R}$ denotes the truncation function $\tau_\eta(s) = \min(\eta, s)$, $\eta > 0$. For further use, we introduce the energy functional $\mathcal{T}_i : H_0^1(\Omega) \to \mathbb{R}$ associated with problem $(D_{A_i}^k)$.

We notice that for $s \geq 0$, the chain rule gives

$$\partial_C A_i(s) = \begin{cases} \partial_C A(s) & \text{if } s < \eta_i, \\ \overline{\text{co}}\{0, \partial_C A(\eta_i)\} & \text{if } s = \eta_i, \\ \{0\} & \text{if } s > \eta_i. \end{cases}$$

It turns out that on the compact set $[0, \eta_i]$, the upper semicontinuous set-valued map $s \mapsto \partial_C A_i(s)$ attains its supremum; therefore, there exists $M_{A_i} > 0$ such that

$$\max |\partial_C A_i(s)| := \max\{|\xi| : \xi \in \partial_C A_i(s)\} \leq M_{A_i}$$

for every $s \geq 0$, i.e., $(H_{A_i}^1)$ holds. The same is true for $(H_{A_i}^2)$ by using (H_2^0) on $[\delta_i, \eta_i]$, $i \in \mathbb{N}$.

Accordingly, the assumptions of Theorem 6.9 are verified for every $i \in \mathbb{N}$ with $[\delta_i, \eta_i]$, thus there exists $u_i^0 \in W^{\eta_i}$ such that

$$u_i^0 \text{ is the minimum point of the functional } \mathcal{T}_i \text{ on } W^{\eta_i}, \tag{6.67}$$

$$u_i^0(x) \in [0, \delta_i] \text{ for a.e. } x \in \Omega, \tag{6.68}$$

$$u_i^0 \text{ is a solution of } (D_{A_i}^k). \tag{6.69}$$

On account of relations (6.66), (6.68), and (6.69), u_i^0 is a weak solution also for the differential inclusion problem (D_A^k).

We are going to prove that there are infinitely many distinct elements in the sequence $\{u_i^0\}_i$. To conclude it, we first prove that

$$\mathcal{T}_i(u_i^0) < 0 \quad \text{for all } i \in \mathbb{N}; \text{ and} \tag{6.70}$$

$$\lim_{i \to \infty} \mathcal{T}_i(u_i^0) = 0. \tag{6.71}$$

The left part of (H_1^0) implies the existence of some $l_0 > 0$ and $\zeta \in (0, \eta_1)$ such that

$$A(s) \geq -l_0 s^2 \quad \text{for all } s \in (0, \zeta). \tag{6.72}$$

One can choose $L_0 > 0$ such that

$$\frac{1}{2} C(r, N) + \left(\frac{k}{2} + l_0\right) m(\Omega) < L_0 (r/2)^n \omega_n, \tag{6.73}$$

where $r > 0$ and $C(r, N) > 0$ come from (6.62). Based on the right part of (H_1^0), one can find a sequence $\{\tilde{s}_i\}_i \subset (0, \zeta)$ such that $\tilde{s}_i \leq \delta_i$ and

$$A(\tilde{s}_i) > L_0 \tilde{s}_i^2 \quad \text{for all } i \in \mathbb{N}. \tag{6.74}$$

Let $i \in \mathbb{N}$ be a fixed number and let $w_{\tilde{s}_i} \in H_0^1(\Omega)$ be the function from (6.61) corresponding to the value $\tilde{s}_i > 0$. Then $w_{\tilde{s}_i} \in W^{\eta_i}$, and due to (6.72), (6.74), and (6.62) one has

$$\mathcal{T}_i(w_{\tilde{s}_i}) = \frac{1}{2}\|w_{\tilde{s}_i}\|_{H_0^1}^2 + \frac{k}{2}\int_\Omega w_{\tilde{s}_i}^2 - \int_\Omega A_i(w_{\tilde{s}_i}(x))dx = \frac{1}{2}C(r, N)\tilde{s}_i^2 + \frac{k}{2}\int_\Omega w_{\tilde{s}_i}^2$$

$$- \int_{B(x_0, r/2)} A(\tilde{s}_i)dx - \int_{B(x_0, r) \setminus B(x_0, r/2)} A(w_{\tilde{s}_i}(x))dx$$

$$\leq \left[\frac{1}{2}C(r, N) + \frac{k}{2}m(\Omega) - L_0(r/2)^n \omega_n + l_0 m(\Omega)\right]\tilde{s}_i^2.$$

Accordingly, with (6.67) and (6.73), we conclude that

$$\mathcal{T}_i(u_i^0) = \min_{W^{\eta_i}} \mathcal{T}_i \leq \mathcal{T}_i(w_{\tilde{s}_i}) < 0 \tag{6.75}$$

which completes the proof of (6.70).

Now, we prove (6.71). For every $i \in \mathbb{N}$, by using the Lebourg's mean value theorem, relations (6.66) and (6.68) and (H_0^0), we have

$$\mathcal{T}_i(u_i^0) \geq - \int_\Omega A_i(u_i^0(x))dx = - \int_\Omega A_1(u_i^0(x))dx \geq -M_{A_1}m(\Omega)\delta_i.$$

Since $\lim_{i \to \infty} \delta_i = 0$, the latter estimate and (6.75) provides relation (6.71).

Based on (6.66) and (6.68), we have that $\mathcal{T}_i(u_i^0) = \mathcal{T}_1(u_i^0)$ for all $i \in \mathbb{N}$. This relation with (6.70) and (6.71) means that the sequence $\{u_i^0\}_i$ contains infinitely many distinct elements.

We now prove (6.65). One can prove the former limit by (6.68), i.e. $\|u_i^0\|_{L^\infty} \leq \delta_i$ for all $i \in \mathbb{N}$, combined with $\lim_{i \to \infty} \delta_i = 0$. For the latter limit, we use $k > 0$, (6.75), (6.66) and (6.68) to get for all $i \in \mathbb{N}$ that

$$\frac{1}{2}\|u_i^0\|_{H_0^1}^2 \leq \frac{1}{2}\|u_i^0\|_{H_0^1}^2 + \frac{k}{2}\int_\Omega (u_i^0)^2 < \int_\Omega A_i(u_i^0(x)) = \int_\Omega A_1(u_i^0(x)) \leq M_{A_1}m(\Omega)\delta_i,$$

which completes the proof.   □

Proof of Theorem 6.10 We split the proof into two parts.

(i) *Case $p = 1$.* Let $\lambda \geq 0$ with $\lambda\bar{c} < -l_0$ and fix $\tilde{\lambda}_0 \in \mathbb{R}$ such that $\lambda\bar{c} < \tilde{\lambda}_0 < -l_0$. With these choices we define

$$k := \tilde{\lambda}_0 - \lambda\bar{c} > 0 \text{ and } A(s) := F(s) + \frac{\tilde{\lambda}_0}{2}s^2 + \lambda\left(G(s) - \frac{\bar{c}}{2}s^2\right) \text{ for every } s \in [0, \infty).$$

$$(6.76)$$

It is clear that $A(0) = 0$, i.e., (H_0^0) is verified. Since $p = 1$, by (G_1^0) one has

$$\underline{c} = \liminf_{s \to 0^+} \frac{\min\{\partial_C G(s)\}}{s} \leq \limsup_{s \to 0^+} \frac{\max\{\partial_C G(s)\}}{s} = \bar{c}.$$

In particular, for sufficiently small $\epsilon > 0$ there exists $\gamma = \gamma(\epsilon) > 0$ such that

$$\max\{\partial_C G(s)\} - \bar{c}s < \epsilon s, \ \forall s \in [0, \gamma],$$

and

$$\min\{\partial_C G(s)\} - \underline{c}s > -\epsilon s, \ \forall s \in [0, \gamma].$$

For $s \in [0, \gamma]$, Lebourg's mean value theorem and $G(0) = 0$ implies that there exists $\xi_s \in \partial_C G(\theta_s s)$ for some $\theta_s \in [0, 1]$ such that $G(s) - G(0) = \xi_s s$. Accordingly, for every $s \in [0, \gamma]$ we have that

$$(\underline{c} - \epsilon)s^2 \leq G(s) \leq (\overline{c} + \epsilon)s^2. \tag{6.77}$$

$$\liminf_{s \to 0^+} \frac{A(s)}{s^2} \geq \liminf_{s \to 0^+} \frac{F(s)}{s^2} + \frac{\tilde{\lambda}_0 - \lambda\overline{c}}{2} + \lambda \liminf_{s \to 0^+} \frac{G(s)}{s^2}$$

$$\geq \liminf_{s \to 0^+} \frac{F(s)}{s^2} + \frac{\tilde{\lambda}_0 - \lambda\overline{c}}{2} + \lambda\underline{c} > -\infty$$

and

$$\limsup_{s \to 0^+} \frac{A(s)}{s^2} \geq \limsup_{s \to 0^+} \frac{F(s)}{s^2} + \frac{\tilde{\lambda}_0 - \lambda\overline{c}}{2} + \lambda \liminf_{s \to 0^+} \frac{G(s)}{s^2} = +\infty,$$

i.e., (H_1^0) is verified.

Since

$$\partial_C A(s) \subseteq \partial_C F(s) + \tilde{\lambda}_0 s + \lambda(\partial_C G(s) - \overline{c}s), \tag{6.78}$$

and $\lambda \geq 0$, we have that

$$\max\{\partial_C A(s)\} \leq \max\{\partial_C F(s) + \tilde{\lambda}_0 s\} + \lambda \max\{\partial_C G(s) - \overline{c}s\}. \tag{6.79}$$

Since

$$\limsup_{s \to 0^+} \frac{\max\{\partial_C G(s)\}}{s} = \overline{c},$$

cf. (G_1^0), and

$$\liminf_{s \to 0^+} \frac{\max\{\partial_C F(s)\}}{s} = l_0 < 0,$$

cf. (F_2^0), it turns out by (6.79) that

$$\liminf_{s \to 0^+} \frac{\max\{\partial_C A(s)\}}{s} \leq \liminf_{s \to 0^+} \frac{\max\{\partial_C F(s)\}}{s} + \tilde{\lambda}_0 - \lambda\overline{c} + \lambda \limsup_{s \to 0^+} \frac{\max\{\partial_C G(s)\}}{s}$$

$$\leq l_0 + \tilde{\lambda}_0 < 0.$$

Therefore, one has a sequence $\{s_i\}_i \subset (0, 1)$ converging to 0 such that $\frac{\max\{\partial_C A(s_i)\}}{s_i} < 0$ i.e., $\max\{\partial_C A(s_i)\} < 0$ for all $i \in \mathbb{N}$. By using the upper semicontinuity of $s \mapsto \partial_C A(s)$, we may choose two numbers $\delta_i, \eta_i \in (0, 1)$ with $\delta_i < s_i < \eta_i$ such that $\partial_C A(s) \subset \partial_C A(s_i) + [-\epsilon_i, \epsilon_i]$ for every $s \in [\delta_i, \eta_i]$, where $\epsilon_i := -\max\{\partial_C A(s_i)\}/2 > 0$. In particular, $\max\{\partial_C A(s)\} \leq 0$ for all $s \in [\delta_i, \eta_i]$. Thus, one may fix two sequences $\{\delta_i\}_i, \{\eta_i\}_i \subset (0, 1)$ such that $0 < \eta_{i+1} < \delta_i < s_i < \eta_i$, $\lim_{i \to \infty} \eta_i = 0$, and $\max\{\partial_C A(s)\} \leq 0$ for all $s \in [\delta_i, \eta_i]$ and $i \in \mathbb{N}$. Accordingly, (H_2^0) is verified as well. Let us apply Theorem 6.12 with the choice (6.76), i.e., there exists a sequence $\{u_i\}_i \subset H_0^1(\Omega)$ of different elements such that

$$
\begin{cases}
-\Delta u_i(x) + (\tilde{\lambda}_0 - \lambda \bar{c})u_i(x) \in \partial_C F(u_i(x)) + \tilde{\lambda}_0 u_i(x) \\
\qquad\qquad\qquad\qquad +\lambda(\partial_C G(u_i(x)) - \bar{c}u_i(x)) & x \in \Omega, \\
u_i(x) \geq 0 & x \in \Omega, \\
u_i(x) = 0 & x \in \partial\Omega,
\end{cases}
$$

where we used the inclusion (6.78). In particular, u_i solves problem $(\mathcal{D}_\lambda)$, $i \in \mathbb{N}$, which completes the proof of (i).

(ii) *Case* $p > 1$. Let $\lambda \geq 0$ be arbitrary fixed and choose a number $\lambda_0 \in (0, -l_0)$. Let

$$
k := \lambda_0 > 0 \quad \text{and} \quad A(s) := F(s) + \lambda G(s) + \lambda_0 \frac{s^2}{2} \text{ for every } s \in [0, \infty). \tag{6.80}
$$

Since $F(0) = G(0) = 0$, hypothesis (H_0^0) clearly holds. By (G_1^0) one has

$$
\underline{c} = \liminf_{s \to 0^+} \frac{\min\{\partial_C G(s)\}}{s^p} \leq \limsup_{s \to 0^+} \frac{\max\{\partial_C G(s)\}}{s^p} = \bar{c}.
$$

In particular, since $p > 1$, then

$$
\lim_{s \to 0^+} \frac{\min\{\partial_C G(s)\}}{s} = \lim_{s \to 0^+} \frac{\max\{\partial_C G(s)\}}{s} = 0 \tag{6.81}
$$

and for sufficiently small $\epsilon > 0$ there exists $\gamma = \gamma(\epsilon) > 0$ such that

$$
\max\{\partial_C G(s)\} - \bar{c}s^p < \epsilon s^p, \ \forall s \in [0, \gamma]
$$

and

$$
\min\{\partial_C G(s)\} - \underline{c}s^p > -\epsilon s^p, \ \forall s \in [0, \gamma].
$$

For a fixed $s \in [0, \gamma]$, by Lebourg's mean value theorem and $G(0) = 0$ we conclude again that $G(s) - G(0) = \xi_s s$. Accordingly, for sufficiently small $\epsilon > 0$ there exists

$\gamma = \gamma(\epsilon) > 0$ such that $(\underline{c} - \epsilon)s^{p+1} \le G(s) \le (\overline{c} + \epsilon)s^{p+1}$ for every $s \in [0, \gamma]$. Thus, since $p > 1$,

$$\lim_{s \to 0^+} \frac{G(s)}{s^2} = \lim_{s \to 0^+} \frac{G(s)}{s^{p+1}} s^{p-1} = 0.$$

Therefore, by using (6.80) and (F_1^0), we conclude that

$$\liminf_{s \to 0^+} \frac{A(s)}{s^2} = \liminf_{s \to 0^+} \frac{F(s)}{s^2} + \lambda \lim_{s \to 0^+} \frac{G(s)}{s^2} + \frac{\lambda_0}{2} > -\infty,$$

and

$$\limsup_{s \to 0^+} \frac{A(s)}{s^2} = \infty,$$

i.e., (H_0^1) holds. Since

$$\partial_C A(s) \subseteq \partial_C F(s) + \lambda \partial_C G(s) + \lambda_0 s,$$

and $\lambda \ge 0$, we have that

$$\max\{\partial_C A(s)\} \le \max\{\partial_C F(s)\} + \max\{\lambda \partial_C G(s) + \lambda_0 s\}.$$

Since

$$\limsup_{s \to 0^+} \frac{\max\{\partial_C G(s)\}}{s^p} = \overline{c},$$

cf. (G_1^0), and

$$\liminf_{s \to 0^+} \frac{\max\{\partial_C F(s)\}}{s} = l_0,$$

cf. (F_2^0), by relation (6.81) it turns out that

$$\liminf_{s \to 0^+} \frac{\max\{\partial_C A(s)\}}{s} = \liminf_{s \to 0^+} \frac{\max\{\partial_C F(s)\}}{s} + \lambda \lim_{s \to 0^+} \frac{\max\{\partial_C G(s)\}}{s} + \lambda_0$$

$$= l_0 + \lambda_0 < 0,$$

and the upper semicontinuity of $\partial_C A$ implies the existence of two sequences $\{\delta_i\}_i$ and $\{\eta_i\}_i \subset (0, 1)$ such that $0 < \eta_{i+1} < \delta_i < s_i < \eta_i$, $\lim_{i \to \infty} \eta_i = 0$, and $\max\{\partial_C A(s)\} \le 0$ for all $s \in [\delta_i, \eta_i]$ and $i \in \mathbb{N}$. Therefore, hypothesis (H_2^0) holds.

Now, we can apply Theorem 6.12, i.e., there is a sequence $\{u_i\}_i \subset H_0^1(\Omega)$ of different elements such that

$$
\begin{cases}
-\Delta u_i(x) + \lambda_0 u_i(x) \in \partial_C A(u_i(x)) \\
\qquad\qquad \subseteq \partial_C F(u_i(x)) + \lambda \partial_C G(u_i(x)) + \lambda_0 u_i(x) & x \in \Omega, \\
u_i(x) \geq 0 & x \in \Omega, \\
u_i(x) = 0 & x \in \partial\Omega,
\end{cases}
$$

which means that u_i solves problem $(\mathcal{D}_\lambda)$, $i \in \mathbb{N}$. This completes the proof of Theorem 6.10. $\qquad\qquad\qquad\qquad\qquad\qquad\qquad\qquad\qquad\qquad\qquad\qquad\square$

Proof of Theorem 6.11 The proof is done in two steps:

(i) Let $\lambda_0 \in (0, -l_0)$, $\lambda \geq 0$ and define

$$
k := \lambda_0 > 0 \text{ and } A^\lambda(s) := F(s) + \lambda G(s) + \lambda_0 \frac{s^2}{2} \text{ for every } s \in [0, \infty). \qquad (6.82)
$$

One can observe that $\partial_C A^\lambda(s) \subseteq \partial_C F(s) + \lambda_0 s + \lambda \partial_C G(s)$ for every $s \geq 0$. On account of (F_2^0), there is a sequence $\{s_i\}_i \subset (0, 1)$ converging to 0 such that

$$
\max\{\partial_C A^{\lambda=0}(s_i)\} \leq \max\{\partial_C F(s_i)\} + \lambda_0 s_i < 0.
$$

Thus, due to the upper semicontinuity of $(s, \lambda) \mapsto \partial_C A^\lambda(s)$, we can choose three sequences $\{\delta_i\}_i, \{\eta_i\}_i, \{\lambda_i\}_i \subset (0, 1)$ such that $0 < \eta_{i+1} < \delta_i < s_i < \eta_i$, $\lim_{i \to \infty} \eta_i = 0$, and

$$
\max\{\partial_C A^\lambda(s)\} \leq 0 \text{ for all } \lambda \in [0, \lambda_i], s \in [\delta_i, \eta_i], i \in \mathbb{N}.
$$

Without any loss of generality, we may choose

$$
\delta_i \leq \min\{i^{-1}, 2^{-1} i^{-2}[1 + m(\Omega)(\max_{s \in [0,1]} |\partial_C F(s)| + \max_{s \in [0,1]} |\partial_C G(s)|)]^{-1}\}. \qquad (6.83)
$$

For every $i \in \mathbb{N}$ and $\lambda \in [0, \lambda_i]$, let $A_i^\lambda : [0, \infty) \to \mathbb{R}$ be defined as

$$
A_i^\lambda(s) = A^\lambda(\tau_{\eta_i}(s)), \qquad (6.84)
$$

and the energy functional $\mathcal{T}_{i,\lambda} : H_0^1(\Omega) \rightarrow \mathbb{R}$ associated with the differential inclusion problem$(D_{A_i^\lambda}^k)$ is given by

$$\mathcal{T}_{i,\lambda}(u) = \frac{1}{2}\|u\|_{H_0^1}^2 + \frac{k}{2}\int_\Omega u^2 dx - \int_\Omega A_i^\lambda(u(x))dx.$$

One can easily check that for every $i \in \mathbb{N}$ and $\lambda \in [0, \lambda_i]$, the function A_i^λ verifies the hypotheses of Theorem 6.9. Accordingly, for every $i \in \mathbb{N}$ and $\lambda \in [0, \lambda_i]$:

$$\mathcal{T}_{i,\lambda} \text{ attains its infinum on } W^{\eta_i} \text{ at some } u_{i,\lambda}^0 \in W^{\eta_i} \qquad (6.85)$$

$$u_{i,\lambda}^0(x) \in [0, \delta_i] \text{for a.e.} x \in \Omega; \qquad (6.86)$$

$$u_{i,\lambda}^0 \text{ is a weak solution of} (D_{A_i^\lambda}^k). \qquad (6.87)$$

By the choice of the function A^λ and $k > 0$, $u_{i,\lambda}^0$ is also a solution to the differential inclusion problem $(D_{A^\lambda}^k)$, so $(\mathcal{D}_\lambda)$.

(ii) It is clear that for $\lambda = 0$, the set-valued map $\partial_C A_i^\lambda = \partial_C A_i^0$ verifies the hypotheses of Theorem 6.12. In particular, $\mathcal{T}_i := \mathcal{T}_{i,0}$ is the energy functional associated with problem $(D_{A_i^0}^k)$. Consequently, the elements $u_i^0 := u_{i,0}^0$ verify not only (6.85)–(6.87) but also

$$\mathcal{T}_i(u_i^0) = \min_{W^{\eta_i}} \mathcal{T}_i \leq \mathcal{T}_i(w_{\tilde{s}_i}) < 0 \text{for all } i \in \mathbb{N}. \qquad (6.88)$$

Similarly to Kristály and Moroşanu [22], let $\{\theta_i\}_i$ be a sequence with negative terms such that $\lim_{i\to\infty} \theta_i = 0$. Due to (6.88) we may assume that

$$\theta_i < \mathcal{T}_i(u_i^0) \leq \mathcal{T}_i(w_{\tilde{s}_i}) < \theta_{i+1}. \qquad (6.89)$$

Let us choose

$$\lambda_i' = \frac{\theta_{i+1} - \mathcal{T}_i(w_{\tilde{s}_i})}{m(\Omega)\max_{s\in[0,1]}|G(s)| + 1} \text{and} \lambda_i'' = \frac{\mathcal{T}_i(u_i^0) - \theta_i}{m(\Omega)\max_{s\in[0,1]}|G(s)| + 1}, i \in \mathbb{N}, \qquad (6.90)$$

and for a fixed $k \in \mathbb{N}$, set

$$\lambda_k^0 = \min(1, \lambda_1, \ldots, \lambda_k, \lambda_1', \ldots, \lambda_k', \lambda_1'', \ldots, \lambda_k'') > 0. \qquad (6.91)$$

Having in our mind these choices, for every $i \in \{1, \ldots, k\}$ and $\lambda \in [0, \lambda_k^0]$ one has

$$\mathcal{T}_{i,\lambda}(u_{i,\lambda}^0) \leq \mathcal{T}_{i,\lambda}(w_{\tilde{s}_i}) = \frac{1}{2}\|w_{\tilde{s}_i}\|_{H_0^1}^2 - \int_\Omega F(w_{\tilde{s}_i}(x))dx - \lambda \int_\Omega G(w_{\tilde{s}_i}(x))dx$$

$$= \mathcal{T}_i(w_{\tilde{s}_i}) - \lambda \int_\Omega G(w_{\tilde{s}_i}(x))dx$$

$$< \theta_{i+1}, \tag{6.92}$$

and due to $u_{i,\lambda}^0 \in W^{\eta i}$ and to the fact that u_i^0 is the minimum point of $\mathcal{T}_i$ on the set $W^{\eta i}$, by (6.89) we also have

$$\mathcal{T}_{i,\lambda}(u_{i,\lambda}^0) = \mathcal{T}_i(u_{i,\lambda}^0) - \lambda \int_\Omega G(u_{i,\lambda}^0(x))dx \geq \mathcal{T}_i(u_i^0) - \lambda \int_\Omega G(u_{i,\lambda}^0(x))dx > \theta_i. \tag{6.93}$$

Therefore, by (6.92) and (6.93), for every $i \in \{1, \ldots, k\}$ and $\lambda \in [0, \lambda_k^0]$, one has

$$\theta_i < \mathcal{T}_{i,\lambda}(u_{i,\lambda}^0) < \theta_{i+1},$$

thus

$$\mathcal{T}_{1,\lambda}(u_{1,\lambda}^0) < \ldots < \mathcal{T}_{k,\lambda}(u_{k,\lambda}^0) < 0.$$

We notice that $u_i^0 \in W^{\eta_1}$ for every $i \in \{1, \ldots, k\}$, so $\mathcal{T}_{i,\lambda}(u_{i,\lambda}^0) = \mathcal{T}_{1,\lambda}(u_{i,\lambda}^0)$ because of (6.84). Therefore, we conclude that for every $\lambda \in [0, \lambda_k^0]$,

$$\mathcal{T}_{1,\lambda}(u_{1,\lambda}^0) < \ldots < \mathcal{T}_{1,\lambda}(u_{k,\lambda}^0) < 0 = \mathcal{T}_{1,\lambda}(0).$$

Based on these inequalities, it turns out that the elements $u_{1,\lambda}^0, \ldots, u_{k,\lambda}^0$ are distinct and non-trivial whenever $\lambda \in [0, \lambda_k^0]$.

Now, we are going to prove the estimate (6.64). We have for every $i \in \{1, \ldots, k\}$ and $\lambda \in [0, \lambda_k^0]$:

$$\mathcal{T}_{1,\lambda}(u_{i,\lambda}^0) = \mathcal{T}_{i,\lambda}(u_{i,\lambda}^0) < \theta_{i+1} < 0.$$

By Lebourg's mean value theorem and (6.83), we have for every $i \in \{1, \ldots, k\}$ and $\lambda \in [0, \lambda_k^0]$ that

$$\frac{1}{2}\|u_{i,\lambda}^0\|_{H_0^1}^2 < \int_\Omega F(u_{i,\lambda}^0(x))dx + \lambda \int_\Omega G(u_{i,\lambda}^0(x))dx$$

$$\leq m(\Omega)\delta_i[\max_{s\in[0,1]} |\partial_C F(s)| + \max_{s\in[0,1]} |\partial_C G(s)|]$$

$$\leq \frac{1}{2i^2}.$$

This completes the proof of Theorem 6.11.   □

6.6.3   Oscillation at Infinity

Let assume:

(F_0^∞) $F(0) = 0$;

(F_1^∞) $-\infty < \liminf_{s\to\infty} \frac{F(s)}{s^2}$; $\limsup_{s\to\infty} \frac{F(s)}{s^2} = +\infty$;

(F_2^∞) $l_\infty := \liminf_{s\to\infty} \frac{\max\{\xi:\xi\in\partial_C F(s)\}}{s} < 0$.

(G_0^∞) $G(0) = 0$;

(G_1^∞) There exist $p > 0$ and $\underline{c}, \overline{c} \in \mathbb{R}$ such that

$$\underline{c} = \liminf_{s\to\infty} \frac{\min\{\xi : \xi \in \partial_C G(s)\}}{s^p} \leq \limsup_{s\to\infty} \frac{\max\{\xi : \xi \in \partial_C G(s)\}}{s^p} = \overline{c}.$$

Remark 6.4 Hypotheses (F_1^∞) and (F_2^∞) imply a strong oscillatory behavior of the set-valued map $\partial_C F$ at infinity.

Example 6.3 We consider $F_\infty(s) = \int_0^s f_\infty(t), s \geq 0$, where $f_\infty(t) = \sqrt{t}(\frac{1}{2} + \sin t)$, $t \geq 0$, or some of its jumping variants; one has that F_∞ verifies the assumptions $(F_0^\infty) - (F_2^\infty)$. For a fixed $p > 0$, let $G_\infty(s) = s^p \max\{0, \sin s\}, s \geq 0$; it is clear that G_∞ is a typically locally Lipschitz function on $[0, \infty)$ (not being of class C^1) and verifies (G_1^∞) with $\underline{c} = -1$ and $\overline{c} = 1$; see Fig. 6.2 representing both f_∞ and G_∞ (for $p = 2$), respectively.

In the sequel, we investigate the competition at infinity concerning the terms $s \mapsto \partial_C F(s)$ and $s \mapsto \partial_C G(s)$, respectively. First, we show that when $p \leq 1$ then the 'leading' term is the oscillatory function F, i.e., the effect of $s \mapsto \partial_C G(s)$ is negligible. More precisely, we prove the following result:

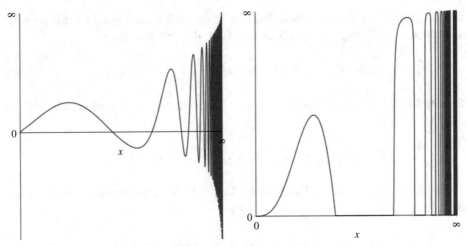

Fig. 6.2 Graphs of f_∞ and G_∞ at infinity, respectively

Theorem 6.13 ([24]) (Case $p \leq 1$) *Assume that $p \leq 1$ and the locally Lipschitz functions $F, G : \mathbb{R}_+ \to \mathbb{R}$ satisfy $(F_0^\infty) - (F_2^\infty)$ and $(G_0^\infty) - (G_1^\infty)$. If*

(i) *either $p = 1$ and $\lambda \bar{c} \leq -l_0$ (with $\lambda \geq 0$),*
(ii) *or $p < 1$ and $\lambda \geq 0$ is arbitrary,*

then the differential inclusion $(\mathcal{D}_\lambda)$ admits a sequence $\{u_i\}_i \subset H_0^1(\Omega)$ of distinct weak solutions such that

$$\lim_{i \to \infty} \|u_i^\infty\|_{L^\infty} = \infty. \tag{6.94}$$

Remark 6.5 Let 2^* be the usual critical Sobolev exponent. In addition to (6.94), we also have $\lim_{i \to \infty} \|u_i^\infty\|_{H_0^1} = \infty$ whenever

$$\sup_{s \in [0, \infty)} \frac{\max\{|\xi| : \xi \in \partial_C F(s)\}}{1 + s^{2^*-1}} < \infty. \tag{6.95}$$

In the case when $p > 1$, it turns out that the perturbation term $\partial_C G$ may compete with the oscillatory function $\partial_C F$; more precisely, we have:

Theorem 6.14 ([24]) (Case $p > 1$) *Assume that $p > 1$ and the locally Lipschitz functions $F, G : \mathbb{R}_+ \to \mathbb{R}$ satisfy $(F_0^\infty) - (F_2^\infty)$ and $(G_0^\infty) - (G_1^\infty)$. Then, for every $k \in \mathbb{N}$,* *there exists* $\lambda_k^\infty > 0$ *such that the differential inclusion* $(\mathcal{D}_\lambda)$ *has at least* k *distinct weak solutions* $\{u_{1,\lambda}, \ldots, u_{k,\lambda}\} \subset H_0^1(\Omega)$ *whenever* $\lambda \in [0, \lambda_k^\infty]$. *Moreover,*

$$\|u_{i,\lambda}\|_{L^\infty} > i - 1 \quad for\ any\ i = \overline{1,k};\ \lambda \in [0, \lambda_k^\infty]. \tag{6.96}$$

Remark 6.6 If (6.95) holds and $p \leq 2^* - 1$ in Theorem 6.95, then we have in addition that

$$\|u_{i,\lambda}^\infty\|_{H_0^1} > i - 1 \quad \text{for any } i = \overline{1,k};\ \lambda \in [0, \lambda_k^\infty].$$

Before giving the proof of Theorems 6.13 and 6.14, we consider again the differential inclusion problem

$$\begin{cases} -\Delta u(x) + ku(x) \in \partial_C A(u(x)), \quad u(x) \geq 0 \quad x \in \Omega, \\ u(x) = 0 \qquad\qquad\qquad\qquad\qquad\qquad x \in \partial\Omega, \end{cases} \tag{D_A^k}$$

where $k > 0$ and the locally Lipschitz function $A : \mathbb{R}_+ \to \mathbb{R}$ verifies

(H_0^∞): $A(0) = 0$;
(H_1^∞): $-\infty < \liminf_{s \to \infty} \frac{A(s)}{s^2}$ and $\limsup_{s \to \infty} \frac{A(s)}{s^2} = +\infty$;
(H_2^∞): there are two sequences $\{\delta_i\}, \{\eta_i\}$ with $0 < \delta_i < \eta_i < \delta_{i+1}$, $\lim_{i \to \infty} \delta_i = \infty$, and

$$\max\{\partial_C A(s)\} := \max\{\xi : \xi \in \partial_C A(s)\} \leq 0$$

for every $s \in [\delta_i, \eta_i]$, $i \in \mathbb{N}$.

The counterpart of Theorem 6.12 reads as follows.

Theorem 6.15 *Let* $k > 0$ *and assume the hypotheses* (H_0^∞), (H_1^∞), *and* (H_2^∞) *hold. Then the differential inclusion problem* (D_A^k) *admits a sequence* $\{u_i^\infty\}_i \subset H_0^1(\Omega)$ *of distinct weak solutions such that*

$$\lim_{i \to \infty} \|u_i^\infty\|_{L^\infty} = \infty. \tag{6.97}$$

Proof The proof is similar to the one performed in Theorem 6.12; we shall show the differences only. We associate the energy functional $\mathcal{T}_i : H_0^1(\Omega) \to \mathbb{R}$ with problem $(D_{A_i}^k)$, where $A_i : \mathbb{R} \to \mathbb{R}$ is given by

$$A_i(s) = A(\tau_{\eta_i}(s)), \tag{6.98}$$

with $A(s) = 0$ for $s \leq 0$. One can show that there exists $M_{A_i} > 0$ such that

$$\max |\partial_C A_i(s)| := \max\{|\xi| : \xi \in \partial_C A_i(s)\} \leq M_{A_i}$$

for all $s \geq 0$, i.e, hypothesis $(H^1_{A_i})$ holds. Moreover, $(H^2_{A_i})$ follows by (H^∞_2). Thus Theorem 6.12 can be applied for all $i \in \mathbb{N}$, i.e., we have an element $u^\infty_i \in W^{\eta_i}$ such that

$$u^\infty_i \text{ is the minimum point of the functional } \mathcal{T}_i \text{ on } W^{\eta_i}, \tag{6.99}$$

$$u^\infty_i(x) \in [0, \delta_i] \text{ for a.e. } x \in \Omega, \tag{6.100}$$

$$u^\infty_i \text{ is a weak solution of } (D^k_{A_i}). \tag{6.101}$$

By (6.98), u^∞_i turns to be a weak solution also for differential inclusion problem (D^k_A).

We shall prove that there are infinitely many distinct elements in the sequence $\{u^\infty_i\}_i$ by showing that

$$\lim_{i \to \infty} \mathcal{T}_i(u^\infty_i) = -\infty. \tag{6.102}$$

By the left part of (H^∞_1) we can find $l^A_\infty > 0$ and $\zeta > 0$ such that

$$A(s) \geq -l^A_\infty \text{ for all } s > \zeta. \tag{6.103}$$

Let us choose $L^A_\infty > 0$ large enough such that

$$\frac{1}{2}C(r, n) + \left(\frac{k}{2} + l^A_\infty\right) m(\Omega) < L^A_\infty (r/2)^n \omega_n. \tag{6.104}$$

On account of the right part of (H^∞_1), one can fix a sequence $\{\tilde{s}_i\}_i \subset (0, \infty)$ such that $\lim_{i \to \infty} \tilde{s}_i = \infty$ and

$$A(\tilde{s}_i) > L^A_\infty \tilde{s}_i^2 \text{ for every } i \in \mathbb{N}. \tag{6.105}$$

We know from (H^∞_2) that $\lim_{i \to \infty} \delta_i = \infty$, therefore one has a subsequence $\{\delta_{m_i}\}_i$ of $\{\delta_i\}_i$ such that $\tilde{s}_i \leq \delta_{m_i}$ for all $i \in \mathbb{N}$. Let $i \in \mathbb{N}$, and recall $w_{s_i} \in H^1_0(\Omega)$ from (6.61) with $s_i := \tilde{s}_i > 0$. Then $w_{\tilde{s}_i} \in W^{\eta_{m_i}}$ and according to (6.62), (6.103), and (6.105) we have

$$\mathcal{T}_{m_i}(w_{\tilde{s}_i}) = \frac{1}{2}\|w_{\tilde{s}_i}\|^2_{H^1_0} + \frac{k}{2}\int_\Omega w^2_{\tilde{s}_i} - \int_\Omega A_{m_i}(w_{\tilde{s}_i}(x))dx = \frac{1}{2}C(r, n)\tilde{s}_i^2 + \frac{k}{2}\int_\Omega w^2_{\tilde{s}_i}$$

$$- \int_{B(x_0, r/2)} A(\tilde{s}_i)dx - \int_{(B(x_0, r)\backslash B(x_0, r/2))\cap\{w_{\tilde{s}_i} > \zeta\}} A(w_{\tilde{s}_i}(x))dx$$

$$
-\int_{(B(x_0,r)\setminus B(x_0,r/2))\cap\{w_{\tilde{s}_i}\leq\zeta\}} A(w_{\tilde{s}_i}(x))dx
$$

$$
\leq \left[\frac{1}{2}C(r,n)+\frac{k}{2}m(\Omega)-L_\infty^A(r/2)^n\omega_n+l_\infty^A m(\Omega)\right]\tilde{s}_i^2+\tilde{M}_A m(\Omega)\zeta,
$$

where $\tilde{M}_A=\max\{|A(s)| : s\in[0,\zeta]\}$ does not depend on $i\in\mathbb{N}$. This estimate combined by (6.104) and $\lim_{i\to\infty}\tilde{s}_i=\infty$ yields that

$$
\lim_{i\to\infty}\mathcal{T}_{m_i}(w_{\tilde{s}_i})=-\infty. \tag{6.106}
$$

By Eq. (6.99), one has

$$
\mathcal{T}_{m_i}(u_{m_i}^\infty)=\min_{W^{\eta m_i}}\mathcal{T}_{m_i}\leq\mathcal{T}_{m_i}(w_{\tilde{s}_i}). \tag{6.107}
$$

It follows by (6.106) that $\lim_{i\to\infty}\mathcal{T}_{m_i}(u_{m_i}^\infty)=-\infty$.

We notice that the sequence $\{\mathcal{T}_i(u_i^\infty)\}_i$ is non-increasing. Indeed, let $i<k$; due to (6.98) one has that

$$
\mathcal{T}_i(u_i^\infty)=\min_{W^{\eta_i}}\mathcal{T}_i=\min_{W^{\eta_i}}\mathcal{T}_k\geq\min_{W^{\eta_k}}\mathcal{T}_k=\mathcal{T}_k(u_k^\infty), \tag{6.108}
$$

which completes the proof of (6.102). The proof of (6.97) follows easily. □

Proof of Theorem 6.13 We split the proof into two parts.

(i) *Case $p=1$.* Let $\lambda\geq 0$ with $\lambda\bar{c}<-l_\infty$ and fix $\tilde{\lambda}_\infty\in\mathbb{R}$ such that $\lambda\bar{c}<\tilde{\lambda}_\infty<-l_\infty$. With these choices, we define

$$
k:=\tilde{\lambda}_\infty-\lambda\bar{c}>0\ \text{ and }\ A(s):=F(s)+\frac{\tilde{\lambda}_\infty}{2}s^2+\lambda\left(G(s)-\frac{\bar{c}}{2}s^2\right)\ \text{ for every } s\in[0,\infty). \tag{6.109}
$$

It is clear that $A(0)=0$, i.e., (H_0^∞) is verified. A similar argument for the p-order perturbation $\partial_C G$ as before shows that

$$
\liminf_{s\to\infty}\frac{A(s)}{s^2}\geq\liminf_{s\to\infty}\frac{F(s)}{s^2}+\frac{\tilde{\lambda}_\infty-\lambda\bar{c}}{2}+\lambda\liminf_{s\to\infty}\frac{G(s)}{s^2}
$$

$$
\geq\liminf_{s\to\infty}\frac{F(s)}{s^2}+\frac{\tilde{\lambda}_\infty-\lambda\bar{c}}{2}+\lambda\underline{c}>-\infty,
$$

and

$$\limsup_{s \to \infty} \frac{A(s)}{s^2} \geq \limsup_{s \to \infty} \frac{F(s)}{s^2} + \frac{\tilde{\lambda}_\infty - \lambda\bar{c}}{2} + \lambda \liminf_{s \to \infty} \frac{G(s)}{s^2} = +\infty,$$

i.e., (H_1^∞) is verified.

Since

$$\partial_C A(s) \subseteq \partial_C F(s) + \tilde{\lambda}_\infty s + \lambda(\partial_C G(s) - \bar{c}s), \quad s \geq 0, \tag{6.110}$$

it turns out that

$$\liminf_{s \to \infty} \frac{\max\{\partial_C A(s)\}}{s} \leq \liminf_{s \to \infty} \frac{\max\{\partial_C F(s)\}}{s} + \tilde{\lambda}_\infty - \lambda\bar{c} + \lambda \limsup_{s \to \infty} \frac{\max\{\partial_C G(s)\}}{s}$$

$$= l_\infty + \tilde{\lambda}_\infty < 0.$$

By using the upper semicontinuity of $s \mapsto \partial_C A(s)$, one may fix two sequences $\{\delta_i\}_i, \{\eta_i\}_i \subset (0, \infty)$ such that $0 < \delta_i < s_i < \eta_i < \delta_{i+1}$, $\lim_{i \to \infty} \delta_i = \infty$, and $\max\{\partial_C A(s)\} \leq 0$ for all $s \in [\delta_i, \eta_i]$ and $i \in \mathbb{N}$. Thus, (H_2^∞) is verified as well. By applying the inclusion (6.110) and Theorem 6.12 with the choice (6.109), there exists a sequence $\{u_i\}_i \subset H_0^1(\Omega)$ of different elements such that

$$\begin{cases} -\Delta u_i(x) + (\tilde{\lambda}_\infty - \lambda\bar{c})u_i(x) \in \partial_C F(u_i(x)) + \tilde{\lambda}_\infty u_i(x) \\ \qquad\qquad\qquad\qquad + \lambda(\partial_C G(u_i(x)) - \bar{c}u_i(x)) & x \in \Omega, \\ u_i(x) \geq 0 & x \in \Omega, \\ u_i(x) = 0 & x \in \partial\Omega, \end{cases}$$

i.e., u_i solves problem $(\mathcal{D}_\lambda)$, $i \in \mathbb{N}$.

(ii) *Case $p < 1$.* Let $\lambda \geq 0$ be arbitrary fixed and choose a number $\lambda_\infty \in (0, -l_\infty)$. Let

$$k := \lambda_\infty > 0 \text{ and } A(s) := F(s) + \lambda G(s) + \lambda_\infty \frac{s^2}{2} \text{ for every } s \in [0, \infty). \tag{6.111}$$

Since $F(0) = G(0) = 0$, hypothesis (H_0^∞) clearly holds. Moreover, by (G_1^∞), for sufficiently small $\epsilon > 0$ there exists $s_0 > 0$, such that $(\underline{c} - \epsilon)s^{p+1} \leq G(s) \leq (\bar{c} + \epsilon)s^{p+1}$ for every $s > s_0$. Thus, since $p < 1$,

$$\lim_{s \to \infty} \frac{G(s)}{s^2} = \lim_{s \to \infty} \frac{G(s)}{s^{p+1}} s^{p-1} = 0.$$

Accordingly, by using (6.111) we obtain that hypothesis (H_1^∞) holds. A similar argument as above implies that

$$\liminf_{s\to\infty} \frac{\max\{\partial_C A(s)\}}{s} \le l_0 + \lambda_\infty < 0,$$

and the upper semicontinuity of $\partial_C A$ implies the existence of two sequences $\{\delta_i\}_i$ and $\{\eta_i\}_i \subset (0,1)$ such that $0 < \delta_i < s_i < \eta_i < \delta_{i+1}$, $\lim_{i\to\infty} \delta_i = \infty$, and $\max\{\partial_C A(s)\} \le 0$ for all $s \in [\delta_i, \eta_i]$ and $i \in \mathbb{N}$. Therefore, hypothesis (H_2^∞) holds. Now, we can apply Theorem 6.12, i.e., there is a sequence $\{u_i\}_i \subset H_0^1(\Omega)$ of different elements such that

$$\begin{cases} -\Delta u_i(x) + \lambda_\infty u_i(x) \in \partial_C A(u_i(x)) \\ \qquad\qquad\qquad \subseteq \partial_C F(u_i(x)) + \lambda\partial_C G(u_i(x)) + \lambda_\infty u_i(x) \text{ in } \Omega, \\ u_i(x) \ge 0 \qquad\qquad x \in \Omega, \\ u_i(x) = 0 \qquad\qquad x \in \partial\Omega, \end{cases}$$

which means that u_i solves problem $(\mathcal{D}_\lambda)$, $i \in \mathbb{N}$, which completes the proof. □

Proof of Theorem 6.14 The proof is done in two steps:

(*i*) Let $\lambda_\infty \in (0, -l_\infty)$, $\lambda \ge 0$ and define

$$k := \lambda_\infty > 0 \text{ and } A^\lambda(s) := F(s) + \lambda G(s) + \lambda_\infty \frac{s^2}{2} \text{ for every } s \in [0, \infty). \qquad (6.112)$$

One has clearly that $\partial_C A^\lambda(s) \subseteq \partial_C F(s) + \lambda_\infty s + \lambda\partial_C G(s)$ for every $s \in \mathbb{R}$. On account of (F_2^∞), there is a sequence $\{s_i\}_i \subset (0, \infty)$ converging to ∞ such that

$$\max\{\partial_C A^{\lambda=0}(s_i)\} \le \max\{\partial_C F(s_i)\} + \lambda_\infty s_i < 0.$$

By the upper semicontinuity of $(s, \lambda) \mapsto \partial_C A^\lambda(s)$, we can choose the sequences $\{\delta_i\}_i, \{\eta_i\}_i, \{\lambda_i\}_i \subset (0, \infty)$ such that $0 < \delta_i < s_i < \eta_i < \delta_{i+1}$, $\lim_{i\to\infty} \delta_i = \infty$, and

$$\max\{\partial_C A^\lambda(s)\} \le 0$$

for all $\lambda \in [0, \lambda_i]$, $s \in [\delta_i, \eta_i]$ and $i \in \mathbb{N}$.

For every $i \in \mathbb{N}$ and $\lambda \in [0, \lambda_i]$, let $A_i^\lambda : [0, \infty) \to \mathbb{R}$ be defined by

$$A_i^\lambda(s) = A^\lambda(\tau_{\eta_i}(s)), \qquad (6.113)$$

and accordingly, the energy functional $\mathcal{T}_{i,\lambda} : H_0^1(\Omega) \to \mathbb{R}$ associated with the differential inclusion problem $(\mathrm{D}_{A_i^\lambda}^k)$ is

$$\mathcal{T}_{i,\lambda}(u) = \frac{1}{2}\|u\|_{H_0^1}^2 + \frac{k}{2}\int_\Omega u^2 dx - \int_\Omega A_i^\lambda(u(x)) dx.$$

Then for every $i \in \mathbb{N}$ and $\lambda \in [0, \lambda_i]$, the function A_i^λ clearly verifies the hypotheses of Theorem 6.9. Accordingly, for every $i \in \mathbb{N}$ and $\lambda \in [0, \lambda_i]$ there exists

$$\mathcal{T}_{i,\lambda} \text{ attains its infimum at some} \tilde{u}_{i,\lambda}^\infty \in W^{\eta_i} \tag{6.114}$$

$$\tilde{u}_{i,\lambda}^\infty \in [0, \delta_i] \text{ for a.e. } x \in \Omega; \tag{6.115}$$

$$\tilde{u}_{i,\lambda}^\infty(x) \text{ is a weak solution of } (\mathrm{D}_{A_i^\lambda}^k). \tag{6.116}$$

Due to (6.113), $\tilde{u}_{i,\lambda}^\infty$ is not only a solution to $(\mathrm{D}_{A_i^\lambda}^k)$ but also to the differential inclusion problem $(\mathrm{D}_{A^\lambda}^k)$, so $(\mathcal{D}_\lambda)$.

(ii) For $\lambda = 0$, the function $\partial_C A_i^\lambda = \partial_C A_i^0$ verifies the hypotheses of Theorem 6.12. Moreover, $\mathcal{T}_i := \mathcal{T}_{i,0}$ is the energy functional associated with problem $(\mathrm{D}_{A_i^0}^k)$. Consequently, the elements $u_i^\infty := u_{i,0}^\infty$ verify not only (6.114)–(6.116) but also

$$\mathcal{T}_{m_i}(u_{m_i}^\infty) = \min_{W^{\eta m_i}} (\mathcal{T}_{m_i}) \le \mathcal{T}_{m_i}(w_{\tilde{s}_i}) \text{ for all } i \in \mathbb{N}, \tag{6.117}$$

where the subsequence $\{u_{m_i}^\infty\}_i$ of $\{u_i^\infty\}_i$ and $w_{\tilde{s}_i} \in W^{\eta_i}$ appear in the proof of Theorem 6.15.

Similarly to [22], let $\{\theta_i\}_i$ be a sequence with negative terms such that $\lim_{i\to\infty} \theta_i = -\infty$. On account of (6.117) we may assume that

$$\theta_{i+1} < \mathcal{T}_{m_i}(u_{m_i}^\infty) \le \mathcal{T}_{m_i}(w_{\tilde{s}_i}) < \theta_i. \tag{6.118}$$

Let

$$\lambda_i' = \frac{\theta_i - \mathcal{T}_{m_i}(w_{\tilde{s}_i})}{m(\Omega) \max_{s\in[0,1]} |G(s)| + 1} \text{ and } \lambda_i'' = \frac{\mathcal{T}_{m_i}(u_{m_i}^\infty) - \theta_{i+1}}{m(\Omega) \max_{s\in[0,1]} |G(s)| + 1}, i \in \mathbb{N}, \tag{6.119}$$

and for a fixed $k \in \mathbb{N}$, we set

$$\lambda_k^\infty = \min(1, \lambda_1, \dots, \lambda_k, \lambda_1', \dots, \lambda_k', \lambda_1'', \dots, \lambda_k'') > 0. \tag{6.120}$$

Then, for every $i \in \{1, \ldots, k\}$ and $\lambda \in [0, \lambda_k^\infty]$, due to (6.118) we have that

$$\mathcal{T}_{m_i, \lambda}(\tilde{u}_{m_i, \lambda}^\infty) \leq \mathcal{T}_{m_i, \lambda}(w_{\tilde{s}_i}) = \frac{1}{2}\|w_{\tilde{s}_i}\|_{H_0^1}^2 - \int_\Omega F(w_{\tilde{s}_i}(x))dx - \lambda \int_\Omega G(w_{\tilde{s}_i}(x))dx$$

$$= \mathcal{T}_{m_i}(w_{\tilde{s}_i}) - \lambda \int_\Omega G(w_{\tilde{s}_i}(x))dx$$

$$< \theta_i. \tag{6.121}$$

Similarly, since $\tilde{u}_{m_i, \lambda}^\infty \in W^{\eta m_i}$ and $u_{m_i}^\infty$ is the minimum point of $\mathcal{T}_i$ on the set $W^{\eta m_i}$, on account of (6.118) we have

$$\mathcal{T}_{m_i, \lambda}(\tilde{u}_{m_i, \lambda}^\infty) = \mathcal{T}_{m_i}(\tilde{u}_{m_i, \lambda}^\infty) - \lambda \int_\Omega G(\tilde{u}_{m_i, \lambda}^\infty)dx \geq \mathcal{T}_{m_i}(u_{m_i}^\infty) - \lambda \int_\Omega G(\tilde{u}_{m_i, \lambda}^\infty)dx > \theta_{i+1}. \tag{6.122}$$

Therefore, for every $i \in \{1, \ldots, k\}$ and $\lambda \in [0, \lambda_k^\infty]$,

$$\theta_{i+1} < \mathcal{T}_{m_i, \lambda}(\tilde{u}_{m_i, \lambda}^\infty) < \theta_i < 0, \tag{6.123}$$

thus

$$\mathcal{T}_{m_k, \lambda}(\tilde{u}_{m_k, \lambda}^\infty) < \ldots < \mathcal{T}_{m_1, \lambda}(\tilde{u}_{m_1, \lambda}^\infty) < 0. \tag{6.124}$$

Because of (6.113), we notice that $\tilde{u}_{m_i, \lambda}^\infty \in W^{\eta m_k}$ for every $i \in \{1, \ldots, k\}$, thus $\mathcal{T}_{m_i, \lambda}(\tilde{u}_{m_i, \lambda}^\infty) = \mathcal{T}_{m_k, \lambda}(\tilde{u}_{i, \lambda}^\infty)$. Therefore, for every $\lambda \in [0, \lambda_k^\infty]$,

$$\mathcal{T}_{m_k, \lambda}(\tilde{u}_{m_k, \lambda}^\infty) < \ldots < \mathcal{T}_{m_k, \lambda}(\tilde{u}_{m_1, \lambda}^\infty) < 0 = \mathcal{T}_{m_k, \lambda}(0),$$

i.e, the elements $\tilde{u}_{m_1, \lambda}^\infty, \ldots, \tilde{u}_{m_k, \lambda}^\infty$ are distinct and non-trivial whenever $\lambda \in [0, \lambda_k^\infty]$. The estimate (6.96) follows in a similar manner as in [22]. $\qquad\square$

References

1. R.A. Adams, *Sobolev Spaces* (Academic Press, New York, 1975)
2. A. Ambrosetti, P. Rabinowitz, Dual variational methods in critical point theory and applications. J. Funct. Anal. **14**, 349–381 (1973)
3. I. Babuška, J. Osborn, Eigenvalue problems, in *Handbook of Numerical Analysis*, vol. 2 (North-Holland, Amsterdam, 1991), pp. 641–787
4. H. Brezis, *Functional Analysis, Sobolev Spaces and Partial Differential Equations* (Springer, Berlin, 2011)
5. K.-C. Chang, Variational methods for non-differentiable functionals and their applications to partial differential equations. J. Math. Anal. Appl. **80**, 102–129 (1981)

6. F.H. Clarke, Optimization and nonsmooth analysis, in *Classics in Applied Mathematics, Society for Industrial and Applied Mathematics* (1990)

7. P. Clément, B. de Pagter, G. Sweers, F. de Thélin, Existence of solutions to a semilinear elliptic system through Orlicz-Sobolev spaces. Mediterr. J. Math. **3**, 241–267 (2004)

8. N. Costea, C. Varga, Multiple critical points for non-differentiable parametrized functionals and applications to differential inclusions. J. Global Optim. **56**, 399–416 (2013)

9. N. Costea, M. Csirik, C. Varga, Linking-type results in nonsmooth critical point theory and applications. Set-Valued Var. Anal. **25**, 333–356 (2017)

10. N. Costea, G. Moroşanu, C. Varga, Weak solvability for Dirichlet partial differential inclusions in Orlicz-Sobolev spaces. Adv. Differential Equations **23**, 523–554 (2018)

11. G. Dincă, P. Matei, Variational and topological methods for operator equations involving duality mappings on Orlicz-Sobolev spaces. Electron. J. Differential Equations **2007**, 1–47 (2007)

12. D. Edmunds, J. Rákosník, Density of smooth functions in $W^{k,p(x)}(\Omega)$. Proc. R. Soc. Lond. Ser. A. **437**, 229–236 (1992)

13. D. Edmunds, J. Rákosník, Sobolev embedding with variable exponent. Studia Math. **143**, 267–293 (2000)

14. D. Edmunds, J. Lang, A. Nekvinda, On $L^{p(x)}$ norms. Proc. R. Soc. Lond. Ser. A. **455**, 219–225 (1999)

15. K. Fan, Some properties of convex sets related to fixed point theorems. Math. Ann. **266**, 519–537 (1984)

16. X. Fan, Q. Zhang, Existence of solutions for $p(x)$−Laplacian Dirichlet problem. Nonlinear Anal. **52**, 1843–1853 (2003)

17. X.L. Fan, Y.Z. Zhao, Linking and multiplicity results for the p-Laplacian on unbounded cylinder. J. Math. Anal. Appl. **260**, 479–489 (2001)

18. X. Fan, J. Shen, D. Zhao, Sobolev embedding theorems for spaces $W^{k,p(x)}(\Omega)$. J. Math. Anal. Appl. **262**, 749–760 (2001)

19. J. Fernández-Bonder, J. Rossi, Existence results for the p−Laplacian with nonlinear boundary conditions. J. Math. Anal. Appl. **263**, 195–223 (2001)

20. M. García-Huidobro, V.K. Le, R. Manásevich, K. Schmitt, On principal eigenvalues for quasilinear elliptic operators: an Orlicz-Sobolev space setting. NoDEA Nonlinear Differential Equations Appl. **6**, 207–225 (1999)

21. O. Kováčik, J. Rákosník, On spaces $l^{p(x)}$ and $w^{1,p(x)}$. Czechoslovak Math. J. **41**, 592–618 (1991)

22. A. Kristály, G. Moroşanu, New competition phenomena in Dirichlet problems. J. Math. Pures Appl. **94**(9), 555–570 (2010)

23. A. Kristály, W. Marzantowicz, C. Varga, A non-smooth three critical points theorem with applications in differential inclusions. J. Global Optim. **46**, 49–62 (2010)

24. A. Kristály, I.I. Mezei, K. Szilák, Differential inclusions involving oscillatory terms. Nonlinear Anal. **197**, 111834 (2020)

25. J.W. Lamperti, On the isometries of certain function-spaces. Pacific J. Math. **8**, 459–466 (1958)

26. J. Lions, *Quelques Méthodes de résolution des Problèmes Aux Limites Non Linéaires* (Collection études Mathématiques, Dunod, 1969)

27. M. Marcus, V.J. Mizel, Every superposition operator mapping one Sobolev space into another is continuous. J. Funct. Anal. **33**, 217–229 (1979)

28. S. Martinez, J. Rossi, Isolation and simplicity for the first eigenvalue of the p−Laplacian with a nonlinear boundary condition. Abstr. Appl. Anal. **7**, 287–293 (2002)

29. M. Mihăilescu, V. Rădulescu, A multiplicity result for a nonlinear degenerate problem arising in the theory of electrorheological fluids. Proc. R. Soc. London Ser. A **462**, 2625–2641 (2006)

30. D. Motreanu, V.V. Motreanu, Coerciveness property for a class of non-smooth functionals. Z. Anal. Anwend **19**, 1087–1093 (2000)

31. D. Motreanu, P. Panagiotopoulos, *Minimax Theorems and Qualitative Properties of the Solutions of Hemivariational Inequalities*. Nonconvex Optimization and Its Applications (Kluwer Academic Publishers, Dordrecht, 1999)

32. D. Motreanu, C. Varga, Some critical point results for locally Lipschitz functionals. Commun. Appl. Nonlinear Anal. **4**, 17–33 (1997)

33. J. Musielak, Orlicz spaces and modular spaces, in *Lecture Notes in Mathematics*, vol. 1034 (Springer, Berlin, 1983)

34. B. Ricceri, Multiplicity of global minima for parametrized functions, Atti Accad. Naz. Lincei Cl. Sci. Fis. Mat. Natur. Rend. Lincei (9) Mat. Appl. **21**, 47–57 (2010)

Hemivariational Inequalities and Differential Inclusions on Unbounded Domains

7

7.1 Hemivariational Inequalities Involving the Duality Mapping

Let $\Omega \subset \mathbb{R}^N$ ($N \geq 2$) be an unbounded domain with smooth boundary $\partial\Omega$ and $p \in (1, N)$ be a real number. Throughout this section X denotes a separable, uniformly convex Banach space with strictly convex topological dual; moreover, we assume that

(X) X is compactly embedded in $L^r(\Omega)$ for some $r \in [p, p^*)$,

$p^* := Np/(N - p)$ being the Sobolev critical exponent. We denote by $\| \cdot \|_r$ the norm in $L^r(\Omega)$ and by c_r the embedding constant. Also let J_ϕ be the duality mapping corresponding to the normalization function $\phi(t) := t^{p-1}$.

Condition (X) is equivalent to the assumption that X is a linear subspace of $L^r(\Omega)$, endowed with a norm $\| \cdot \|$ such that the identity is a compact operator from $(X, \| \cdot \|)$ into $(L^r(\Omega), \| \cdot \|_r)$.

Let $f : \mathbb{R} \to \mathbb{R}$ be a locally Lipschitz function satisfying

(f) $f(0) = 0$ and there exist $k > 0$, $q \in (0, p - 1)$ such that

$$|\xi| \leq k|s|^q, \quad \forall s \in \mathbb{R}, \forall \xi \in \partial_C f(s).$$

Let $b : \Omega \to \mathbb{R}$ be a nonnegative, nonzero function such that

(b) $b \in L^1(\Omega) \cap L^\infty(\Omega) \cap L^\nu(\Omega)$, where $\nu := \frac{r}{r-(q+1)}$.

We shall prove next that under suitable assumptions, there exist u_0 and $\lambda > 0$ such that the inequality problem

$(P_{u_0,\lambda})$ Find $u \in X$ such that

$$\langle J_\phi(u - u_0), v \rangle + \lambda \int_\Omega b(x) f^0(u(x); -v(x)) dx \geq 0, \quad \forall v \in X,$$

possesses at least three solutions.

Let us define the functional $F : X \to \mathbb{R}$ by

$$F(u) := \int_\Omega b(x) f(u(x)) dx$$

for all $u \in X$. The next Lemma summarizes the properties of F:

Lemma 7.1 *The functional F is well-defined, locally Lipschitz, sequentially weakly continuous and satisfies*

$$F^0(u; v) \leq \int_\Omega b(x) f^0(u(x); v(x)) dx, \quad \forall u, v \in X.$$

Proof We begin by giving an estimate of the integral which defines F: from Lebourg's mean value theorem it follows that for all $s \in \mathbb{R}$ there exist $t \in \mathbb{R}$, with $0 < |t| < |s|$, and $\xi \in \partial_C f(t)$ such that $f(s) = \xi s$, so, by (f),

$$|f(s)| \leq k|s|^{q+1}. \tag{7.1}$$

Thus, for all $u \in X$ we get by applying Hölder's inequality that

$$\left| \int_\Omega b(x) f(u(x)) dx \right| \leq k \int_\Omega b(x) |u(x)|^{q+1} dx \leq k \|b\|_\nu \|u\|_r^{q+1} \leq K \|u\|^{q+1},$$

where $K = c_r^{q+1} k \|b\|_\nu$. Hence, F is well-defined.

By means of (7.1) it is can also proved that F is Lipschitz on bounded sets. Let us choose $M > 0$ and $u, v \in X$ with $\|u\|, \|v\| < M$: then we have for all $x \in \Omega$

$$|f(u(x)) - f(v(x))| \leq k \left(|u(x)|^q + |v(x)|^q \right) |u(x) - v(x)|,$$

hence, again by Hölder's inequality,

$$|F(u) - F(v)| \leq k \|b\|_\nu \left(\int_\Omega \left(|u(x)|^q + |v(x)|^q \right)^{\frac{r}{q}} dx \right)^{\frac{q}{r}} \|u - v\|_r$$

$$\leq 2k \|b\|_\nu \left(\|u\|_r^r + \|v\|_r^r \right)^{\frac{q}{r}} \|u - v\|_r$$

$$\leq 2^{\frac{r+q}{r}} K M^q \|u - v\|.$$

We prove now that F is sequentially weakly continuous: let $\{u_n\}$ be a sequence in X, weakly convergent to some $\bar{u} \in X$. Due to condition (X), there is a subsequence, still denoted by $\{u_n\}$, such that $\|u_n - \bar{u}\|_r \to 0$; then, by well-known results, we may assume that $u_n \to \bar{u}$ a.e. in Ω and there exists a positive function $g \in L^r(\Omega)$ such that $|u_n(x)| \leq g(x)$ for all $n \in \mathbb{N}$ and almost all $x \in \Omega$. By Lebesgue's dominated convergence theorem, $\{F(u_n)\}$ tends to $F(\bar{u})$.

By Proposition 3.3 of [8], the inequality in the thesis follows, and the proof is concluded.   □

Given $u_0 \in X$ and $\lambda > 0$, the energy functional $E : X \to \mathbb{R}$ is defined by

$$E(u) := \frac{\|u - u_0\|^p}{p} - \lambda F(u).$$

We observe that Theorem C.1 and Proposition C.1 ensure that the the convex functional $u \to \|u - u_0\|^p/p$ is Gâteaux differentiable with derivative $J_\phi(u - u_0)$, so it is locally Lipschitz. Hence, E is locally Lipschitz too.

Lemma 7.2 *Let $u_0 \in$ and $\lambda > 0$ be fixed. If u is a critical point of E, then u is a solution of $(P_{u_0,\lambda})$.*

Proof It follows at once that

$$E^0(u, v) \leq \langle J_\phi(u - u_0), v \rangle + \lambda(-F)^0(u; v) = \langle J_\phi(u - u_0), v \rangle + \lambda F^0(u; -v)$$

$$\leq \langle J_\phi(u - u_0), v \rangle + \lambda \int_\Omega b(x) f^0(u(x); -v(x)) \mathrm{d}x.$$

But, u is a critical point of E, therefore

$$E^0(u; v) \geq 0, \quad \forall v \in X,$$

and this shows that u solves $(P_{u_0,\lambda})$.   □

First we prove the following multiplicity alternative concerning $(P_{u_0,\lambda})$.

Theorem 7.1 ([13]) *Assume (X), (f) and (b) are fulfilled. Then, for every $\sigma \in (\inf_X F, \sup_X F)$ and every $u_0 \in F^{-1}((-\infty, \sigma))$ one of the following conditions is true:*

(B_1) *there exists $\lambda > 0$ such that the problem $(P_{u_0,\lambda})$ has at least three solutions;*
(B_2) *there exists $v \in F^{-1}(\sigma)$ such that, for all $u \in F^{-1}([\sigma, +\infty))$, $u \neq v$,*

$$\|u - u_0\| > \|v - u_0\|.$$

Proof Fix σ and u_0 as in the thesis, and assume that (B_1) does not hold: we shall prove that (B_2) is true.

Setting $\Lambda := [0, +\infty)$ and endowing X with the weak topology, we define the function $g : X \times \Lambda \to \mathbb{R}$ by

$$g(u, \lambda) := \frac{\|u - u_0\|^p}{p} + \lambda(\sigma - F(u)),$$

which satisfies all the hypotheses of Theorem D.12. Indeed, conditions (c_1), (c_3) are trivial.

In examining condition (c_2), let $\lambda \geq 0$ be fixed: we first observe that, by Lemma 7.1, the functional $g(\cdot, \lambda)$ is sequentially weakly lower semicontinuous (l.s.c.).

Moreover, $g(\cdot, \lambda)$ is coercive: indeed, for all $u \in X$ we have

$$g(u, \lambda) \geq \|u\|^p \left(\frac{\|u - u_0\|^p}{p \, \|u\|^p} - \lambda \, k \, c_r^{q+1} \|b\|_\nu \|u\|^{(q+1)-p} \right) + \lambda \sigma,$$

and the latter goes to $+\infty$ as $\|u\| \to +\infty$. As a consequence of the Eberlein-Smulyan theorem, the outcome is that $g(\cdot, \lambda)$ is weakly l.s.c.

We need to check that every local minimum of $g(\cdot, \lambda)$ is a global minimum. Arguing by contradiction, suppose that $g(\cdot, \lambda)$ admits a local, non-global minimum; besides, being coercive, it has a global minimum too, that is, it has two strong local minima.

We now prove that $g(\cdot, \lambda)$ satisfies the $(PS)-$condition: let $\{u_n\}$ be a Palais-Smale sequence such that $\{g(u_n, \lambda)\}$ is bounded. The coercivity of $g(\cdot, \lambda)$ ensures that $\{u_n\}$ is bounded, hence we can find a subsequence, which we still denote $\{u_n\}$, weakly convergent to a point $\bar{u} \in X$. By condition (X) we can choose $\{u_n\}$ to be convergent to $\bar{u}$ with respect to the norm of $L^r(\Omega)$.

Fix $\varepsilon > 0$. For $n \in \mathbb{N}$ large enough we have

$$\lambda_g(u_n, \lambda)\|u_n - \bar{u}\| < \frac{\varepsilon}{2},$$

so, from Lemma 7.1 it follows

$$0 \leq g^0(u_n, \lambda; \bar{u} - u_n) + \frac{\varepsilon}{2} \leq \langle J_\phi(u_n - u_0), \bar{u} - u_n \rangle$$

$$+\lambda \int_\Omega b(x) f^0(u_n(x); u_n(x) - \bar{u}(x))\mathrm{d}x + \frac{\varepsilon}{2}$$

$(g^0(\cdot, \lambda; \cdot)$ denotes the generalized directional derivative of the locally Lipschitz functional $g(\cdot, \lambda)$). Moreover, for n large enough

$$\left| \int_\Omega b(x)\ f^0(u_n(x); u_n(x) - \bar{u}(x)) \mathrm{d}x \right| \le k \int_\Omega b(x) |u_n(x)|^q |u_n(x) - \bar{u}(x)| \mathrm{d}x$$

$$\le k\ c_r^q \|b\|_\nu \|u_n\|^q \|u_n - \bar{u}\|_r < \frac{\varepsilon}{2\lambda}.$$

Hence

$$\langle J_\phi(u_n - u_0), u_n - \bar{u}\rangle < \varepsilon$$

for $n \in \mathbb{N}$ large enough. On the other hand, $\langle J_\phi(\bar{u} - u_0), u_n - \bar{u}\rangle$ tends to zero as n goes to infinity. From the previous computations, it follows that

$$\limsup_n \langle J_\phi(u_n - u_0) - J_\phi(\bar{u} - u_0), u_n - \bar{u}\rangle \le 0. \tag{7.2}$$

Using the properties of the duality mapping and keeping in mind that $\phi(t) = t^{p-1}$ we get

$$\langle J_\phi(u_n - u_0) - J_\phi(\bar{u} - u_0), u_n - \bar{u}\rangle \ge$$

$$\left(\|u_n - u_0\|^{p-1} - \|\bar{u} - u_0\|^{p-1} \right) (\|u_n - u_0\| - \|\bar{u} - u_0\|) \ge 0.$$

From the previous inequality and (7.2), we deduce that $\{\|u_n - u_0\|\}$ tends to $\|\bar{u} - u_0\|$ and this, together with the weak convergence, implies that $\{u_n\}$ tends to $\bar{u}$ in X, that is, the Palais-Smale condition is fulfilled.

Then, we can apply Corollary 5.4, deducing that $g(\cdot, \lambda)$ (or equivalently the energy functional E) admits a third critical point: by Lemma 7.2, the inequality $(P_{u_0,\lambda})$ should have at least three solutions in X, against our assumption. Thus, condition (c_2) is fulfilled.

Now Theorem D.12 assures that

$$\sup_{\lambda \in \Lambda} \inf_{u \in X} g(u, \lambda) = \inf_{u \in X} \sup_{\lambda \in \Lambda} g(u, \lambda) =: \alpha. \tag{7.3}$$

Notice that the function $\lambda \mapsto \inf_{u \in X} g(u, \lambda)$ is upper semicontinuous in Λ, and tends to $-\infty$ as $\lambda \to +\infty$ (since $\sigma < \sup_X F$): hence, it attains its supremum in $\lambda^* \in \Lambda$, that is

$$\alpha = \inf_{u \in X} \left(\frac{\|u - u_0\|^p}{p} + \lambda^*(\sigma - F(u)) \right). \tag{7.4}$$

The infimum in the right hand side of (7.3) is easily determined as

$$\alpha = \inf_{u \in F^{-1}([\sigma, +\infty))} \frac{\|u - u_0\|^p}{p} = \frac{\|v - u_0\|^p}{p}$$

for some $v \in F^{-1}([\sigma, +\infty))$.

It is easily seen that $v \in F^{-1}(\sigma)$. Hence

$$\alpha = \inf_{u \in F^{-1}(\sigma)} \frac{\|u - u_0\|^p}{p} \qquad \text{(in particular } \alpha > 0\text{)}. \tag{7.5}$$

By (7.4) and (7.5) it follows that

$$\inf_{u \in X} \left(\frac{\|u - u_0\|^p}{p} - \lambda^* F(u) \right) = \inf_{u \in F^{-1}(\sigma)} \left(\frac{\|u - u_0\|^p}{p} - \lambda^* F(u) \right). \tag{7.6}$$

We deduce that $\lambda^* > 0$: if $\lambda^* = 0$, indeed, (7.6) would become $\alpha = 0$, against (7.5).

Now we can prove (B_2). Arguing by contradiction, let $w \in F^{-1}([\sigma, +\infty)) \setminus \{v\}$ be such that $\|w - u_0\| = \|v - u_0\|$. As above, we have that $w \in F^{-1}(\sigma)$, and so both w and v are global minima of the functional I (for $\lambda = \lambda^*$) over $F^{-1}(\sigma)$, hence, by (7.6), over X. Thus, applying Corollary 5.4, we obtain that I has at least three critical points, against the assumption that (B_1) does not hold (recall that λ^* is positive). This concludes the proof.   □

In the next Corollary, the alternative of Theorem 7.1 is resolved, under a very general assumption on the functional F ensuring option (B_2) cannot occur and so we are led to a multiplicity result for the hemivariational inequality $(P_{u_0, \lambda})$ (for suitable data u_0, λ).

Corollary 7.1 *Let Ω, p, X, f, b be as in Theorem 7.1 and let S be a convex, dense subset of X. Moreover, let $F^{-1}([\sigma, +\infty))$ be not convex for some $\sigma \in (\inf_X F, \sup_X F)$. Then, there exist $u_0 \in F^{-1}((-\infty, \sigma)) \cap S$ and $\lambda > 0$ such that problem $(P_{u_0, \lambda})$ admits at least three solutions.*

Proof Since F is sequentially weakly continuous (see Lemma 7.1), the set $M := F^{-1}([\sigma, +\infty))$ is sequentially weakly closed.

According to an well known result in approximation theory (see, e.g., [9,23]), for some $u_0 \in S$, there exist two distinct points v_1, $v_2 \in M$ satisfying

$$\|v_1 - u_0\| = \|v_2 - u_0\| = \text{dist}(u_0, M).$$

Clearly $u_0 \notin M$, that is, $F(u_0) < \sigma$. In the framework of Theorem 7.1, condition (B_2) is false, so (B_1) must be true: there exists $\lambda > 0$ such that $(P_{u_0\lambda})$ has at least three solutions. $\qquad\square$

7.2 Hemivariational Inequalities in $\mathbb{R}^N$

In this section we investigate the existence multiplicity of solutions for an abstract hemivariational inequality, formulated in $\mathbb{R}^N$. Specific forms of this inequality will be also discussed at the end of the section.

Let $(X, || \cdot ||)$ be a real, separable, reflexive Banach space, $(X^*, || \cdot ||_*)$ its dual and we suppose that the inclusion $X \hookrightarrow L^l(\mathbb{R}^N)$ is continuous with the embedding constant $C(l)$, where $l \in [p, p^*]$ ($p \geq 2$, $p^* = \frac{Np}{N-p}$), $N > p$. Let us denote by $|| \cdot ||_l$ the norm of $L^l(\mathbb{R}^N)$. Let $A : X \to X^*$ be a potential operator with the potential $a : X \to \mathbb{R}$, i.e., a is Gateaux differentiable and

$$\lim_{t \to 0} \frac{a(u + tv) - a(u)}{t} = \langle A(u), v \rangle,$$

for every $u, v \in X$. Here $\langle \cdot, \cdot \rangle$ denotes the duality pairing between X^* and X. For a potential we always assume that $a(0) = 0$. We suppose that $A : X \to X^*$ satisfies the following properties:

(A_1) A is *hemicontinuous*, i.e., A is continuous on line segments in X and X^* equipped with the weak (star) topology;

(A_2) A is *homogeneous of degree $p - 1$*, i.e., for every $u \in X$ and $t > 0$ we have $A(tu) = t^{p-1} A(u)$. Consequently, for a hemicontinuous homogeneous operator of degree $p - 1$, we have $a(u) = \frac{1}{p} \langle A(u), u \rangle$;

(A_3) $A : X \to X^*$ is a *strongly monotone* operator, i.e., there exists a continuous function $\kappa : [0, \infty) \to [0, \infty)$ which is strictly positive on $(0, \infty)$, $\kappa(0) = 0$, and $\lim_{t \to \infty} \kappa(t) = \infty$ and such that

$$\langle A(v) - A(u), v - u \rangle \geq \kappa(||v - u||) ||v - u||, \quad \forall u, v \in X.$$

Now, we formulate the hemivariational inequality problem, which will be studied in this section.

(P) Find $u \in X$ such that

$$\langle Au, v \rangle + \int_{\mathbb{R}^N} F^0(x, u(x); -v(x)) dx \geq 0, \quad \forall v \in X.$$

7.2.1   Existence and Multiplicity Results

To study the existence of solutions of the problem (P) we introduce the functional Ψ : $X \rightarrow \mathbb{R}$ defined by $\Psi(u) := a(u) - \Phi(u)$, where $a(u) := \frac{1}{p}\langle A(u), u\rangle$ and $\Phi(u) :=$ $\int_{\mathbb{R}^N} F(x, u(x))dx$. We prove that the critical points of the functional Ψ are solutions of the problem (P).

Proposition 7.1 *If* $0 \in \partial_C \Psi(u)$, *then* u *solves the problem* (P).

Proof Because $0 \in \partial_C \Psi(u)$, we have $\Psi^0(u; v) \geq 0$ for every $v \in X$. Using Proposition 2.16 and a property of Clarke's derivative we obtain

$$0 \leq \Psi^0(u; v) = \langle A(u), v\rangle + (-\Phi)^0(u; v) = \langle A(u), v\rangle + \Phi^0(u; -v)$$

$$\leq \langle A(u), v\rangle + \int_{\mathbb{R}^N} F^0(x, u(x), -v(x))dx,$$

for every $v \in X$. $\square$

In order to study the existence of the critical points of the function Ψ it is necessary to impose some further conditions on the function F.

(F_1) $F : \mathbb{R}^N \times \mathbb{R} \rightarrow \mathbb{R}$ is defined by

$$F(x, t) := \int_0^t f(x, s)ds$$

and

$$|f(x, s)| \leq c(|s|^{p-1} + |s|^{r-1});$$

(F_1') The embedding $X \hookrightarrow L^r(\mathbb{R}^N)$ is compact for each $r \in [p, p^*)$;
(F_2) There exist $\alpha > p$, $\lambda \in \left[0, \frac{\kappa(1)(\alpha-p)}{C^p(p)}\right)$ and a continuous function $g : \mathbb{R} \rightarrow \mathbb{R}_+$, such that for a.e. $x \in \mathbb{R}^N$ and for all $s \in \mathbb{R}$ we have

$$\alpha F(x, s) + F^0(x, s; -s) \leq g(s), \qquad (7.7)$$

where $\lim_{|s|\to\infty} \frac{g(s)}{|s|^p} = \lambda$.

(F_2') There exists $\alpha \in \left(\max\left\{ p, p^* \frac{r-p}{p^*-p} \right\}, p^* \right)$ and a constant $C > 0$ such that for a.e. $x \in \mathbb{R}^N$ and for all $s \in \mathbb{R}$ we have

$$-C|s|^\alpha \geq F(x, s) + \frac{1}{p} F^0(x, s; -s). \qquad (7.8)$$

Before imposing further assumptions on F, let us we recall that

$$f_-(x, s) := \lim_{\delta \to 0^+} \operatorname{essinf}\{ f(x, t) : |t - s| < \delta \},$$

$$f_+(x, s) := \lim_{\delta \to 0^+} \operatorname{esssup}\{ f(x, t) : |t - s| < \delta \},$$

for every $s \in \mathbb{R}$ and for a.e. $x \in \mathbb{R}^N$. It is clear that the function $f_-(x, \cdot)$ is lower semicontinuous and $f_+(x, \cdot)$ is upper semicontinuous.

(F_3) The functions f_-, f_+ are N-measurable, i.e., for every measurable function $u : \mathbb{R}^N \to \mathbb{R}$ the functions $x \mapsto f_-(x, u(x)), x \mapsto f_+(x, u(x))$ are measurable.

(F_4) For every $\varepsilon > 0$, there exists $c(\varepsilon) > 0$ such that for a.e. $x \in \mathbb{R}^N$ and for every $s \in \mathbb{R}$ we have

$$|f(x, s)| \leq \varepsilon|s|^{p-1} + c(\varepsilon)|s|^{r-1}.$$

(F_5) There exist $\alpha \in (p, p^*)$ satisfying condition (F_2) and $c^* > 0$ such that for a.e. $x \in \mathbb{R}^N$ and all $s \in \mathbb{R}$ we have

$$F(x, s) \geq c^*(|s|^\alpha - |s|^p).$$

Remark 7.1 We observe that if we impose

(F_4') $\lim_{\varepsilon \to 0^+} \operatorname{esssup} \left\{ \frac{|f(x,s)|}{|s|^p} : (x, s) \in \mathbb{R}^N \times (-\varepsilon, \varepsilon) \right\} = 0,$

then this condition together with (F_1) implies (F_4).

Proposition 7.2 *Let $\{u_n\} \subset X$ be a sequence such that $\Psi(u_n) \to c$ and $\lambda_\Psi(u_n) \to 0$ for some $c \in \mathbb{R}$. If the conditions (F_1) and (F_2) are fulfilled, then the sequence $\{u_n\}$ is bounded in X.*

Proof Let $\{u_n\} \subset X$ be a sequence with the required properties. From the condition $\Psi(u_n) \to c$ we get in particular $c + 1 \geq \Psi(u_n)$ for sufficiently large $n \in \mathbb{N}$. Since $\lambda_\Psi(u_n) \to 0$ then $\|u_n\| \geq \|u_n\|\lambda_\Psi(u_n)$ for every sufficiently large $n \in \mathbb{N}$. From the definition of $\lambda_\Psi(u_n)$ it follows the existence of an element $\zeta_{u_n} \in \partial_C \Psi(u_n)$ such that $\lambda_\Psi(u_n) = ||\zeta_{u_n}||_*$. For every $v \in X$, we have $|\langle \zeta_{u_n}, v \rangle| \leq ||\zeta_{u_n}||_*||v||$, therefore $||\zeta_{u_n}||_*||v|| \geq -\langle \zeta_{u_n}, v \rangle$. If we take $v = u_n$, then $||\zeta_{u_n}||_*||u_n|| \geq -\langle \zeta_{u_n}, u_n \rangle$.

Using the property $\Psi^0(u; v) = \max\{\langle \zeta, v \rangle : \zeta \in \partial_C \Psi(u)\}$ for every $v \in X$, we have $-\langle \zeta, v \rangle \geq -\Psi^0(u; v)$ for all $\zeta \in \partial_C \Psi(u)$ and $v \in X$. If we take $u = v = u_n$ and $\zeta := \zeta_{u_n}$, we get $-\langle \zeta_{u_n}^*, u_n \rangle \geq -\Psi^0(u_n; u_n)$. Therefore for every $\alpha > 0$, we have

$$\frac{1}{\alpha}||u_n|| \geq \frac{1}{\alpha}||\zeta_{u_n}||_*||u_n|| \geq -\frac{1}{\alpha}\Psi^0(u_n; u_n).$$

If we add the above inequality with $c + 1 \geq \Psi(u_n)$, we obtain

$$c + 1 + \frac{1}{\alpha}||u_n|| \geq \Psi(u_n) - \frac{1}{\alpha}\Psi^0(u_n; u_n).$$

Using the above inequality, the relation $\Psi^0(u; v) = \langle A(u), v \rangle + \Phi^0(u; -v)$ and Proposition 2.16, one has

$$c + 1 + \frac{1}{\alpha}||u_n|| \geq \Psi(u_n) - \frac{1}{\alpha}\Psi^0(u_n; u_n)$$

$$= \frac{1}{p}\langle A(u_n), u_n \rangle - \Phi(u_n) - \frac{1}{\alpha}\left(\langle A(u_n), u_n \rangle + \Phi^0(u_n; -u_n) \right)$$

$$\geq \left(\frac{1}{p} - \frac{1}{\alpha} \right) \langle A(u_n), u_n \rangle - \int_{\mathbb{R}^N} \left[F(x, u_n(x)) + \frac{1}{\alpha}F_y^0(x, u_n(x); -u_n(x)) \right] dx$$

$$\geq \left(\frac{1}{p} - \frac{1}{\alpha} \right) \langle A(u_n), u_n \rangle - \frac{1}{\alpha}\int_{\mathbb{R}^N} g(u_n(x))dx.$$

Fix $0 < \varepsilon < \frac{\kappa(1)(\alpha - p)}{C^p(p)} - \lambda$. The relation $\lim_{|u| \to \infty} \frac{g(u)}{|u|^p} = \lambda$ assures the existence of a constant M, such that

$$\int_{\mathbb{R}^N} g(u_n(x))dx \leq M + (\lambda + \varepsilon)\int_{\mathbb{R}^N} |u_n(x)|^p dx.$$

If we use again that the inclusion $X \hookrightarrow L^p(\mathbb{R}^N)$ is continuous, and the facts that $a(u) = \frac{1}{p}\langle A(u), u \rangle$ and $a(u) = ||u||^p \left\langle A\left(\frac{u}{||u||} \right), \frac{u}{||u||} \right\rangle \geq \kappa(1)||u||^p$ we get

$$c + 1 + ||u_n|| \geq \left(\frac{1}{p} - \frac{1}{\alpha} \right) \langle A(u_n), u_n \rangle - \frac{(\lambda + \varepsilon)C^p(p)}{\alpha}||u_n||^p - \frac{M}{\alpha}$$

$$\geq \frac{\kappa(1)(\alpha - p) - (\lambda + \varepsilon)C^p(p)}{\alpha}||u_n||^p - \frac{M}{\alpha}.$$

From the above inequality it follows that the sequence $\{u_n\}$ is bounded. $\qquad\square$

Proposition 7.3 *Let $\{u_n\} \subset X$ be a sequence such that*

$$\Psi(u_n) \to c \text{ and } (1 + \|u_n\|)\lambda_\Psi(u_n) \to 0$$

for some $c \in \mathbb{R}$. If the conditions (F_1), (F_2') and (F_4) are fulfilled, then the sequence $\{u_n\}$ is bounded in X.

Proof Let $\{u_n\} \subset X$ be a sequence with the above properties. As in Proposition 7.2, there exists $\zeta_{u_n} \in \partial\Psi(u_n)$ such that

$$\frac{1}{p}\|\zeta_{u_n}\|_*\|u_n\| \geq -\Psi^0\left(u_n; \frac{1}{p}u_n\right).$$

From this inequality, Proposition 2.16, condition (F_2') and the property $\Psi^0(u; v) = \langle Au, v\rangle + \Phi^0(u; -v)$ we obtain

$$c + 1 \geq \Psi(u_n) - \frac{1}{p}\Psi^0(u_n; u_n) \geq a(u_n) - \Phi(u_n) - \frac{1}{p}\Big[\langle Au_n, u_n\rangle + \Phi^0(u_n; -u_n)\Big]$$

$$- \int_{\mathbb{R}^N}\left[F(x, u_n(x)) + \frac{1}{p}F^0(x, u_n(x); -u_n(x))\right]dx$$

$$\geq C\|u_n\|_\alpha^\alpha.$$

Therefore the sequence $\{u_n\}$ is bounded in $L^\alpha(\mathbb{R}^N)$. From the condition (F_4) follows that, for every $\varepsilon > 0$, there exists $c(\varepsilon) > 0$, such that for a.e. $x \in \mathbb{R}^N$

$$F(x, u(x)) \leq \frac{\varepsilon}{p}|u(x)|^p + \frac{c(\varepsilon)}{r}|u(x)|^r.$$

After integration, we obtain

$$\Phi(u) \leq \frac{\varepsilon}{p}\|u\|_p^p + \frac{c(\varepsilon)}{r}\|u\|_r^r.$$

Using the above inequality, the expression of Ψ and the inequality $\|u\|_p \leq C(p)\|u\|$, we obtain

$$\frac{\kappa(1) - \varepsilon C^p(p)}{p}\|u\|^p \leq \Psi(u) + \frac{c(\varepsilon)}{r}\|u\|_r^r \leq c + 1 + \|u\|_r^r.$$

First of all, we fix an $\varepsilon \in \left(0, \frac{\kappa(1)}{C^p(p)}\right)$. Now, we study the behaviour of the sequence $\{||u_n||_r\}$. We have the following situations:

(i) If $r = \alpha$, then obviously the sequence $\{||u_n||_r\}$ is bounded and so is $\{u_n\}$;

(ii) If $r \in (\alpha, p^*)$ and $\alpha > p^* \frac{r-p}{p^*-p}$, then we have

$$||u||_r^r \leq ||u||_\alpha^{(1-s)\alpha} \cdot ||u||_{p^*}^{sp^*},$$

where $r := (1-s)\alpha + sp^*, s \in (0, 1)$.

Using the inequality $||u||_{p^*}^{sp^*} \leq C^{sp^*}(p)||u||^{sp^*}$ we obtain

$$\frac{\kappa(1) - \varepsilon C^p(p)}{p}||u||^p \leq c + 1 + \frac{c(\varepsilon)}{r}||u||_\alpha^{(1-s)\alpha}||u||^{sp^*}. \tag{7.9}$$

Since $sp^* < p$, we obtain that the sequence $\{u_n\}$ is bounded in X. $\square$

In the next result we give conditions, when the function Ψ satisfies the $(PS)_c$ and $(C)_c$ conditions.

Theorem 7.2 ([8])

(i) *If conditions* (F_1), (F_1') *and* $(F_2) - (F_4)$ *hold, the function* Ψ *satisfies the* $(PS)_c$ *condition for every* $c \in \mathbb{R}$;

(ii) *If conditions* (F_1), (F_1'), (F_2'), (F_3), (F_4) *hold, the function* Ψ *satisfies the* $(C)_c$ *condition for every* $c > 0$.

Proof Let $\{u_n\} \subset X$ be a sequence from Propositions 7.2, 7.3, respectively. It follows that it is a bounded sequence in X. Since X is a reflexive Banach space, there exists an element $u \in X$ such that $u_n \to u$ weakly in X. Because the inclusion $X \hookrightarrow L^r(\mathbb{R}^N)$ is compact, we have that $u_n \to u$ strongly in $L^r(\mathbb{R}^N)$.

In the following we provide useful estimate for the sequences $I_n^1 := \Psi^0(u_n; u - u_n)$ and $I_n^2 := \Psi^0(u; u_n - u)$.

We know that $\Psi^0(u; v) = \max\{\langle \zeta, v \rangle : \zeta \in \partial_C \Psi(u)\}$, $\forall v \in X$. Therefore, for every $\zeta_u \in \partial_C \Psi(u)$ we have $\Psi^0(u; u_n - u) \geq \langle \zeta_u, u_n - u \rangle$. From the above relation and from the fact that $u_n \to u$ weakly in X, we get

$$\liminf_{n \to \infty} I_n^2 \geq 0.$$

Now, we estimate the expression $I_n^1 = \Psi^0(u_n; u - u_n)$. Since $I_n^1 \geq -\|\zeta_{u_n}\|_* \|u_n - u\|$, and using $\|\zeta_{u_n}\|_* = \lambda_\Psi(u_n) \to 0$ it follows that

$$\liminf_{n \to \infty} I_n^1 \geq 0.$$

Finally, we estimate the expression $I_n := \Phi^0(u_n; u_n - u) + \Phi^0(u; u - u_n)$. For the simplicity in calculus we introduce the notations $h_1(s) := |s|^{p-1}$ and $h_2(s) := |s|^{r-1}$. For this we observe that if we use the continuity of the functions h_1 and h_2, the condition (F_4) implies that for every $\varepsilon > 0$, there exists a $c(\varepsilon) > 0$ such that

$$\max\{|f_-(x, s)|, |f_+(x, s)|\} \leq \varepsilon h_1(s) + c(\varepsilon)h_2(s), \tag{7.10}$$

for a.e. $x \in \mathbb{R}^N$ and for all $s \in \mathbb{R}$. Using this relation and Proposition 2.16, we have

$$
\begin{aligned}
I_n =& \Phi^0(u_n; u_n - u) + \Phi^0(u; u - u_n) \\
\leq& \int_{\mathbb{R}^N} \left[F^0(x, u_n(x); u_n(x) - u(x)) + F^0(x, u(x); u(x) - u_n(x)) \right] dx \\
\leq& \int_{\mathbb{R}^N} [|f_-(x, u_n(x))| + |f_+(x, u(x))|] \, |u(x) - u_n(x)| dx \leq \\
\leq& 2\varepsilon \int_{\mathbb{R}^N} [h_1(u_n(x)) + h_1(u(x))] \, |u(x) - u_n(x)| dx \\
& + 2c_\varepsilon \int_{\mathbb{R}^N} [(h_2(u_n(x)) + h_2(u(x))] \, |u(x) - u_n(x)| dx.
\end{aligned}
$$

If we use the Hölder inequality and the fact that the inclusion $X \hookrightarrow L^p(\mathbb{R}^N)$ is continuous, we obtain:

$$
\begin{aligned}
I_n \leq & 2\varepsilon C(p)\|u_n - u\|(\|h_1(u)\|_{p'} + \|h_1(u_n)\|_{p'}) \\
& + 2c(\varepsilon)\|u_n - u\|_r (\|h_2(u)\|_{r'} + \|h_2(u_n)\|_{r'}),
\end{aligned}
$$

where $\frac{1}{p} + \frac{1}{p'} = 1$ and $\frac{1}{r} + \frac{1}{r'} = 1$.

Using the fact that the inclusion $X \hookrightarrow L^r(\mathbb{R}^N)$ is compact, we get that $\|u_n - u\|_r \to 0$ as $n \to \infty$. For $\varepsilon \to 0^+$ and $n \to \infty$ we obtain that

$$\limsup_{n \to \infty} I_n \leq 0.$$

One clearly has $\langle A(u), v \rangle = \Phi^0(u; v) - \Psi^0(u; -v)$. If in the above inequality we replace u and v by $u = u_n, v = u_n - u$ and then $u = u, v = u - u_n$ we get

$$\langle A(u_n), u_n - u \rangle = \Phi^0(u_n; u_n - u) - \Psi^0(u_n; u - u_n);$$

$$\langle A(u), u - u_n \rangle = \Phi^0(u; u - u_n) - \Psi^0(u; u_n - u).$$

Adding these relations, we have the following inequality:

$$||u_n - u||\kappa(||u_n - u||) \leq \langle A(u_n - u), u_n - u \rangle$$

$$= \left[\Phi^0(u_n; u_n - u) + \Phi^0(u; u - u_n) \right]$$

$$- \Psi^0(u_n; u - u_n) - \Psi^0(u; u_n - u)$$

$$= I_n - I_n^1 - I_n^2.$$

Using the above relation and the estimates for I_n, I_n^1 and I_n^2, we easily have that $||u_n - u|| \to 0$, thanks to the properties of the function κ.   □

Remark 7.2 It is worth to noticing that the above results remain true if we replace the Banach space X with every closed subspace Y of X, and we restrict the functional Ψ to Y.

Theorem 7.3 ([8])

(i) If $(F_1), (F_1')$ and $(F_2) - (F_5)$ hold, then (P) has at least a nontrivial solution;
(ii) If $(F_1), (F_1'), (F_2'), (F_3)$ and (F_4) hold, then (P) has at least a nontrivial solution.

Proof

(i) Using (i) of Theorem 7.2 , from the conditions $(F_1) - (F_4)$ follows that the functional $\Psi(u) := \frac{1}{p}\langle A(u), u \rangle - \Phi(u)$ satisfies the $(PS)_c$ condition for every $c \in \mathbb{R}$. For the sake of simplicity, we introduce the notations $S_\rho(0) := \{u \in X : ||u|| = \rho\}$ and $B_\rho(0) := \{u \in X : ||u|| \leq \rho\}$. From Theorem 5.2 we only need to verify the following geometric hypotheses (the mountain pass geometry of the energy functional):

$$\exists \beta, \rho > 0 \text{ such that } \Psi(u) \geq \beta \text{ on } S_\rho(0); \tag{7.11}$$

$$\Psi(0) = 0 \text{ and } \exists v \in X \setminus B_\rho(0) \text{ such that } \Psi(v) \leq 0. \tag{7.12}$$

For the proof of relation (7.11), we use the relation (F_4), i.e., $|f(x, s)| \leq \varepsilon|s|^{p-1} + c(\varepsilon)|s|^{r-1}$. Integrating this inequality and using that the inclusions

$X \hookrightarrow L^p(\mathbb{R}^N)$, $X \hookrightarrow L^r(\mathbb{R}^N)$ are continuous, we get that

$$\Psi(u) \geq \frac{\kappa(1) - \varepsilon C(p)}{p} \langle A(u), u \rangle - \frac{1}{r} c(\varepsilon) C(r) ||u||_r^r$$

$$\geq \frac{\kappa(1) - \varepsilon C(p)}{p} ||u||^p - \frac{1}{r} c(\varepsilon) C(r) ||u||^r.$$

The right-hand side member of the inequality is a function $\chi : \mathbb{R}_+ \to \mathbb{R}$ of the form $\chi(t) := At^p - Bt^r$, where $A := \frac{\kappa(1) - \varepsilon C(p)}{p}$, $B := \frac{1}{r} c(\varepsilon) C(r)$. The function χ attains its global maximum in the point $t_M := (\frac{pA}{rB})^{\frac{1}{r-p}}$. If we take $\rho := t_M$ and $\beta \in (0, \chi(t_M)]$, it is easy to see that the condition (7.11) is fulfilled.

From the condition (F_5) we have

$$\Psi(u) \leq \frac{1}{p} \langle A(u), u \rangle + c^* ||u||_p^p - c^* ||u||_\alpha^\alpha.$$

If we fix an element $v \in X \setminus \{0\}$ and in place of u we put tv, then we have

$$\Psi(tv) \leq \left(\frac{1}{p} \langle A(v), v \rangle + c^* ||v||_p^p \right) t^p - c^* t^\alpha ||v||_\alpha^\alpha.$$

From this we see that if t is large enough, $tv \notin B_\rho(0)$ and $\Psi(tv) < 0$. So, the condition (7.12) is satisfied and Theorem 5.2 assures the existence of a nontrivial critical point of Ψ.

(ii) Now if we use (ii) of Theorem 7.2, from the condition (F_1), (F_2') and (F_3), (F_4) we get that the function Ψ satisfies the condition $(C)_c$ for every $c > 0$. We use again the Theorem 5.2, which ensures the existence of a nontrivial critical point for the function Ψ. It is sufficient to prove only the relation (7.12), because (7.11) can be proved in the same way as above.

To prove the relation (7.12) we fix an element $u \in X$ and we define the function $h : (0, +\infty) \to \mathbb{R}$ by

$$h(t) := \frac{1}{t} F(x, t^{\frac{1}{p}} u) - C \frac{p}{\alpha - p} t^{\frac{\alpha}{p} - 1} |u|^\alpha.$$

The function h is locally Lipschitz. We fix a number $t > 1$, and from the Lebourg's mean value theorem follows the existence of an element $\tau \in (1, t)$ such that

$$h(t) - h(1) \in \partial_C^t h(\tau)(t - 1),$$

where ∂_C^t denotes the generalized gradient of Clarke with respect to $t \in \mathbb{R}$. From the Chain Rules we have

$$\partial_C^t t F(x, t^{\frac{1}{p}} u) \subset \frac{1}{p} \partial_C F(x, t^{\frac{1}{p}} u) t^{\frac{1}{p}-1} u.$$

We also have

$$\partial_C^t h(t) \subset -\frac{1}{t^2} F(x, t^{\frac{1}{p}} u) + \frac{1}{t} \partial_C F(x, t^{\frac{1}{p}} u) t^{\frac{1}{p}-1} u - C t^{\frac{\alpha}{p}-2} |u|^\alpha.$$

Therefore, we have

$$h(t) - h(1) \subset \partial_C^t h(\tau)(t-1)$$

$$\subset -\frac{1}{t^2} \left[F(x, t^{\frac{1}{p}} u) - t^{\frac{1}{p}} u \partial_C F(x, t^{\frac{1}{p}} u) + C|t^{\frac{1}{p}} u|^\alpha \right](t-1).$$

Using the relation (F_2'), we obtain that $h(t) \geq h(1)$, therefore

$$\frac{1}{t} F(x, t^{\frac{1}{p}} u) - C \frac{p}{\alpha - p} t^{\frac{\alpha}{p}-1} |u|^\alpha \geq F(x, u) - C \frac{p}{\alpha - p} |u|^\alpha.$$

From this we get

$$F(x, t^{\frac{1}{p}} u) \geq t F(x, u) + C \frac{p}{\alpha - p} \left[t^{\frac{\alpha}{p}} - t \right] |u|^\alpha, \tag{7.13}$$

for every $t > 1$ and $u \in \mathbb{R}$.

Let us fix an element $u_0 \in X \setminus \{0\}$; then for every $t > 1$, we have

$$\Psi(t^{\frac{1}{p}} u_0) = \frac{1}{p} \langle A(t^{\frac{1}{p}} u_0), t^{\frac{1}{p}} u_0 \rangle - \int_{\mathbb{R}^N} F(x, t^{\frac{1}{p}} u_0(x)) dx \frac{t}{p} \langle A u_0, u_0 \rangle$$

$$- t \int_{\mathbb{R}^N} F(x, u_0(x)) dx - C \frac{p}{\alpha - p} \left[t^{\frac{\alpha}{p}} - t \right] \|u_0\|_\alpha^\alpha.$$

If t is sufficiently large, then for $v_0 = t^{\frac{1}{p}} u_0$ we have $\Psi(v_0) \leq 0$. This completes the proof. $\square$

Now we will treat a special case, i.e., when $X = H$ is a Hilbert space with the inner product $\langle \cdot, \cdot \rangle$. The norm of H induced by $\langle \cdot, \cdot \rangle$ is denoted by $\| \cdot \|$. In this case $p = 2$ and the problem (P) takes the form

Find $u \in H$ such that

$$\langle u, v \rangle + \int_{\mathbb{R}^N} F^0(x, u(x); -v(x)) dx \geq 0, \quad \forall v \in H. \tag{P'}$$

Finally, we impose the following condition

(F_7) $f(x, -s) = -f(x, s)$, for a.e. $x \in \mathbb{R}^N$ and all $s \in \mathbb{R}$.

Theorem 7.4

(i) If the conditions (F_1), (F_1'), $(F_2) - (F_5)$ and (F_7) hold, then the problem (P') has infinitely many distinct solutions.

(ii) If the conditions (F_1), (F_1'), (F_2'), (F_3), (F_4), (F_5) and (F_7) hold, then the problem (P') has infinitely many distinct solutions.

Proof We prove that the function Ψ verifies the conditions from Theorem 5.6. Using Theorem 7.2, the conditions $(F_1) - (F_1')$, $(F_2) - (F_4)$, we obtain that the function Ψ satisfies the $(PS)_c$ for every $c \in \mathbb{R}$ and from $(F_1) - (F_1')$, $(F_3) - (F_4)$ we obtain that Ψ satisfies the $(C)_c$ condition for every $c > 0$.

From the assumption (F_7) we easily have that the function Ψ is even. To prove the assertion of this theorem we verify that the conditions of Theorem 5.6 hold.

Let us choose an orthonormal basis $\{e_j\}_{j \in \mathbb{N}}$ of H and define the set

$$H_k := \mathrm{span}\{e_1, \ldots, e_k\}.$$

As above we have $\Psi(v) \leq (c^* C(\alpha) + \frac{1}{2})\|v\|_H^2 - c^*\|v\|_\alpha^\alpha$. Thus, we have $\Psi(0) = 0$. Using the fact that the inclusion $H \hookrightarrow L^\alpha(\mathbb{R}^N)$ is continuous we have that $\| \cdot \|_\alpha|_{H_k}$ is continuous. Because on a finite dimensional space the continuous norms are equivalent and since $\alpha > 2$, there exists an $R_k > 0$ large enough such that for every $u \in H$ with $\|u\|_H \geq R_k$, we have $\Psi(u) \leq \Psi(0) = 0$. Therefore the condition (f_1') from Theorem 5.6 is verified.

Now, we verify the condition (f_2') from Theorem 5.6. For every $u \in H_k^\perp$ and $k \in \mathbb{N}^*$ we consider the real numbers $\beta_k := \sup_{u \in H_k^\perp \setminus \{0\}} \frac{\|u\|_p}{\|u\|_H}$. As in [3, Lemma 3.3] we get $\beta_k \to 0$, if $k \to \infty$. As in the proof of relation (7.11) we have

$$\Psi(u) \geq \left(\frac{1 - \varepsilon C(2)}{2}\right) \|u\|_H^2 - \frac{1}{p} c_\varepsilon \|u\|_p^p.$$

From the definition of number β_k we have $\|u\|_p \leq \beta_k \|u\|_H$ and combining this with the above relation we get

$$\Psi(u) \geq \left(\frac{1 - \varepsilon C(2)}{2}\right) \|u\|_H^2 - \frac{1}{p} c_\varepsilon \beta_k^p \|u\|_H^p.$$

If we choose $0 < \varepsilon < \frac{1}{C(2)} \frac{p-2}{p}$ and $r_k \in \left((c_\varepsilon \beta_k^p)^{\frac{1}{2-p}} \right]$, then we have

$$\Psi(u) \geq \left(\frac{1 - \varepsilon C(2)}{2} - \frac{1}{p} \right) r_k^2,$$

for every $u \in H_k^\perp$ with $\|u\|_H = r_k$. Due to the choice of ε and since $\beta_k \to 0$, the assumptions of Theorem 5.6 are verified. Therefore there exists a sequence of unbounded critical values of Ψ, which completes the proof. $\qquad\square$

In the sequel let G be the compact topological group $O(N)$ or a subgroup of $O(N)$. We suppose that G acts continuously and linear isometric on the Banach space X. We denote by

$$X^G := \{ u \in H : \ gu = u \text{ for all } g \in G \}$$

the fixed point set of the action G on X. It is well known that X^G is a closed subspace of X. We suppose that the potential $a : X \to \mathbb{R}$ of the operator $A : X \to X^*$ is G-invariant and the next condition for the function $f : \mathbb{R}^N \times \mathbb{R} \to \mathbb{R}$ holds: (F_6) for a.e. $x \in \mathbb{R}^N$ and for every $g \in G, s \in \mathbb{R}$ we have $f(gx, s) = f(x, s)$.

In several applications the (F_1') is replaced by the following condition: (F_1'') the embeddings $X^G \hookrightarrow L^r(\mathbb{R}^N)$ are compact ($p < r < p^*$).

Now, if we use the principle of symmetric criticality for locally Lipschitz functions, i.e., $(PSCL)$ from the above theorem we obtain the following corollary, which is very useful in the applications.

Corollary 7.2 *We suppose that the potential $a : X \to \mathbb{R}$ is G-invariant and the condition (F_6) is satisfied. Then the following assertions hold.*

(a) *If the conditions (F_1), (F_1'') and $(F_2) - (F_5)$ are fulfilled, then the problem (P) has a nontrivial solution;*

(b) *If the conditions (F_1), (F_1''), (F_2'), (F_3) and (F_4) are fulfilled, then the problem (P) has a nontrivial solution.*

Now we return to the case, when $X = H$ is a Hilbert space with the inner product $\langle \cdot, \cdot \rangle$. We suppose that G acts continuously and linear isometric on the Hilbert space X. Applying again $(PSCL)$ we obtain the next useful result.

Corollary 7.3 *We suppose that the condition (F_6) is satisfied. Then the following assertions hold.*

(a) *If the conditions $(f_1), (F_1''), (F_2) - (F_5)$ and (F_7) hold, then the problem (P') has infinitely many distinct solutions;*

(b) *If the conditions $(F_1), (F_1''), (F_2'), (F_3), (F_4), (F_5)$ and (F_7) hold, then the problem (P') has infinitely many distinct solutions.*

Further, we give an example of a discontinuous function F for which the problem (P) has a nontrivial solution.

Example 7.1 Let $\{a_n\} \subset \mathbb{R}$ be a sequence with $a_0 = 0, a_n > 0, n \in \mathbb{N}$ such that the series $\sum_{n=0}^{\infty} a_n$ is convergent and $\sum_{n=0}^{\infty} a_n > 1$. We introduce the following notations

$$A_n := \sum_{k=0}^{n} a_k, \ A := \sum_{k=0}^{\infty} a_k.$$

With these notations we have $A > 1$ and $A_n = A_{n-1} + a_n$ for every $n \in \mathbb{N}^*$. Let $f : \mathbb{R} \to \mathbb{R}$ defined by

$$f(s) := s|s|^{p-2} \left(|s|^{r-p} + A_n \right),$$

for all $s \in (-n-1, -n] \cup [n, n+1), n \in \mathbb{N}$ and $r, s \in \mathbb{R}$ with $r > p > 2$. The function f defined above satisfies the properties $(F_1), (F_2'), (F_3)$ and (F_4). The discontinuity set of f is

$$\mathcal{D}_f = \mathbb{Z} \setminus \{0\}.$$

It is easy to see that the function f satisfies the conditions (F_1) and (F_4'), therefore (F_4). Let $F : \mathbb{R} \to \mathbb{R}$ be the function defined by

$$F(t) := \int_0^t f(s)ds, \ \text{with } u \in [n, n+1),$$

when $n \geq 1$. Because $F(u) = F(-u)$, it is sufficient to consider the case $u > 0$. We have

$$F(u) = \sum_{k=0}^{n-1} \int_k^{k+1} f(s)ds + \int_n^u f(s)ds.$$

Therefore for every

$$F(u) = \frac{1}{r}u^r + \frac{1}{p}A_n u^p - \frac{1}{p}\sum_{k=0}^{n} a_k k^p, \ \text{for every } u \in [n, n+1].$$

It is easy to see that $F^0(u; -u) = -uf(u)$ for every $u \in (n, n + 1]$, i.e.,

$$F^0(u, -u) = -u^r - A_n u^p.$$

Thus,

$$F(u) + \frac{1}{p}F^0(u, -u) = -\left(\frac{1}{p} - \frac{1}{r}\right)u^r - \frac{1}{p}\sum_{k=0}^{n} a_k k^p \le -\left(\frac{1}{p} - \frac{1}{r}\right)u^r.$$

If we choose $C := \frac{1}{p} - \frac{1}{r}, \alpha = r > 2$, the condition (F_2') is fulfilled.

7.2.2 Applications

(1) Let $f : \mathbb{R}^N \times \mathbb{R} \to \mathbb{R}$ be a measurable function and $b : \mathbb{R}^N \times \mathbb{R} \to \mathbb{R}$ be a continuous function. For b we shall first assume the following.
   (b_1) $b_0 := \inf_{x \in \mathbb{R}^N} b(x) > 0$;
   (b_2) For every $M > 0$, $\text{meas}(\{x \in \mathbb{R}^N : b(x) \le M\}) < \infty$.

We consider the Hilbert space

$$H := \left\{u \in W^{1,2}(\mathbb{R}^N) : \int_{\mathbb{R}^N}(|\nabla u|^2 + b(x)u^2)dx < \infty\right\},$$

with the inner product

$$\langle u, v\rangle_H := \int_{\mathbb{R}^N}(\nabla u \cdot \nabla v + b(x)uv)dx.$$

In the paper of Bartsch and Wang [2] is proved that the inclusion $H \hookrightarrow L^s(\mathbb{R}^N)$ is compact for $p \in [2, \frac{2N}{N-2})$. Now we formulate the problem.
   Find a positive $u \in H$ such that

$$\int_{\mathbb{R}^N}(\nabla u \cdot \nabla v + b(x)uv)dx + \int_{\mathbb{R}^N} F^0(x, u(x); -v(x))dx \ge 0, \quad \forall v \in H. \tag{P_1}$$

The following corollary extends the results of Gazzola and Rădulescu [14] and Bartsch and Wang [2].

Corollary 7.4 *The following assertions are true*

(i) *If the conditions $(F_1) - (F_5)$ and $(b_1) - (b_2)$ hold, then the problem (P_1) has a positive solution;*

(ii) *If the conditions (F_1), (F_2'), (F_3), (F_4) and $(b_1) - (b_2)$ hold then the problem (P_1) has a positive solution;*

(iii) *If the conditions $(F_1) - (F_5)$, $(b_1) - (b_2)$ and (F_7) hold, then the problem (P_1) has infinitely many distinct positive solutions;*

(iv) *If the conditions (F_1), (F_2'), $(F_3) - (F_5)$, $(b_1) - (b_2)$ and (F_7) hold then, the problem (P_1) has infinitely many positive solutions.*

Proof We replace the function f by $f^+ : \mathbb{R}^N \times \mathbb{R} \to \mathbb{R}$ defined by

$$f^+(x, u) = \begin{cases} f(x, u), & \text{if } u \geq 0, \\ 0, & \text{if } u < 0, \end{cases} \tag{7.14}$$

and we apply Theorems 7.3 and 7.4. $\qquad\square$

(2) Now, we consider $Au := -\Delta u + |x|^2 u$ for $u \in D(A)$, where

$$D(A) := \left\{ u \in L^2(\mathbb{R}^N) : \ Au \in L^2(\mathbb{R}^N) \right\}.$$

Here $|\cdot|$ denotes the Euclidian norm of $\mathbb{R}^N$. In this case the Hilbert space H is defined by

$$H := \left\{ u \in L^2(\mathbb{R}^N) : \ \int_{\mathbb{R}^N} (|\nabla u|^2 + |x|^2 u^2) dx < \infty \right\},$$

with the inner product

$$\langle u, v \rangle := \int_{\mathbb{R}^N} (\nabla u \nabla v + |x|^2 uv) dx.$$

The inclusion $H \hookrightarrow L^s(\mathbb{R}^N)$ is compact for $s \in [2, \frac{2N}{N-2})$ (see, e.g., Kavian [16, Exercise 20, pp. 278]). Therefore the condition (F_1') is satisfied.

We formulate the problem.

Find a positive $u \in H$ such that

$$\int_{\mathbb{R}^N} (\nabla u \nabla v + |x|^2 uv) dx + \int_{\mathbb{R}^N} F^0(x, u(x); -v(x)) dx \geq 0, \quad \forall v \in H. \tag{P_2}$$

Corollary 7.5 *The following assertions are true*

(i) *If the conditions $(F_1) - (F_5)$ hold, then the problem (P_2) has a positive solution;*

(ii) *If the conditions $(F_1), (F_2'), (F_3), (F_4)$ hold, then the problem (P_2) has a positive solution;*

(iii) *If the conditions $(F_1) - (F_5)$ and (F_7) hold, then the problem (P_2) has infinitely many distinct positive solutions;*

(iv) *If the conditions $(F_1), (F_2'), (F_3) - (F_5)$ and (F_7) hold then, the problem (P_2) has infinitely many positive solutions.*

(3) For this application we consider the Hilbert space H given by

$$H := H^1(\mathbb{R}^N) = \left\{ u \in L^2(\mathbb{R}^N) : \nabla u \in L^2(\mathbb{R}^N) \right\}$$

with the inner product

$$\langle u, v \rangle_H := \int_{\mathbb{R}^N} (\nabla u \nabla v + uv) dx.$$

Let use consider $G := O(N)$, $N \geq 3$. The group G acts linearly and orthogonal on $\mathbb{R}^N$. The action of G on H is defined by $gu(x) := u(g^{-1}x)$ for all $g \in G$ and $x \in \mathbb{R}^N$. The fixed point set of this action is

$$H^G := \{ u \in H^1(\mathbb{R}^N) : gu = u \}.$$

According to a result of Lions [20] the inclusion $H^G \hookrightarrow L^s(\mathbb{R}^N)$ is compact for $s \in \left(2, \frac{2N}{N-2} \right)$. Thus, condition (F_1'') is satisfied.

The proposed problem read as follows.

Find $u \in H$ such that

$$\int_{\mathbb{R}^N} (\nabla u \nabla v + uv) dx + \int_{\mathbb{R}^N} F^0(x, u(x); -v(x)) dx \geq 0, \quad \forall v \in H. \qquad (P_3)$$

Corollary 7.6 ([18])

(i) *If the conditions $(F_1) - (F_7)$ hold, then the problem (P_3) has infinitely many distinct solutions;*

(ii) *If the conditions $(F_1), (F_2')$ and $(F_3) - (F_7)$ hold, then the problem (P_3) has infinitely many distinct solutions.*

Remark 7.3 By the construction, the above solutions are radially symmetric. In [18] we actually guaranteed also the existence of infinitely many radially non-symmetric solutions of (P_3) in the case when $N = 4$ or $N \geq 6$.

In the final part of this section, we present an example provided by Kristály [18].

Example 7.2 We denote by $\lfloor u \rfloor$ the nearest integer to $u \in \mathbb{R}$, if $u + \frac{1}{2} \notin \mathbb{Z}$; otherwise we put $\lfloor u \rfloor = u$. Let $f : \mathbb{R} \to \mathbb{R}$ be defined by

$$f(s) := \lfloor |s| s \rfloor.$$

Then the conclusion of Corollary 7.6 holds for $N \in \{3, 4, 5\}$.

Proof We will verify the hypotheses from the first item for $p := 2$, $r = \alpha = 3$. To have $r < 2^*$, we need $N \in \{3, 4, 5\}$. It is easy to show that f is an odd function, i.e. (F_7) holds. It is easy to verify that (F_1) holds too, while (F_3) and (F_6) become trivial facts. Moreover, (F_4) is also obvious, since $f(s) = 0$ for every $s \in \left[-\frac{1}{\sqrt{2}}, \frac{1}{\sqrt{2}}\right]$, see Remark 7.1. Thus, it remains to verify (F_2) and (F_5). To this end, we recall two nice inequalities for every $n \in \mathbb{N}$, i.e.,

$$2n\sqrt{\frac{2n+1}{2}} - 3 \cdot \frac{1 + \sqrt{3} + \cdots + \sqrt{2n-1}}{\sqrt{2}} - \frac{2n+1}{8} \leq 0, ; \qquad (I_{\leq}^n)$$

and

$$\frac{4n+1}{2}\sqrt{\frac{2n-1}{2}} - 3 \cdot \frac{1 + \sqrt{3} + \cdots + \sqrt{2n-1}}{\sqrt{2}} + \frac{2n-1}{2} \geq 0.. \qquad (I_{\geq}^n)$$

Let $F(s) := \int_0^s f(t)dt$. Since F is even, it is enough to verify both relations only for non-negative numbers. One has

$$F(s) = \begin{cases} 0, & s \in \left[0, \frac{1}{\sqrt{2}}\right], \\ F_n(s), & s \in I_n, \end{cases} \qquad (7.15)$$

where $I_n = \left(\sqrt{\frac{2n-1}{2}}, \sqrt{\frac{2n+1}{2}}\right]$, $n \in \mathbb{N}$ and $F_n(s) = ns - \frac{1+\sqrt{3}+\cdots+\sqrt{2n-1}}{\sqrt{2}}$, $s \in I_n$.

Let us fix $s \geq 0$. If $s \in \left[0, \frac{1}{\sqrt{2}}\right]$, then the two inequalities are trivial. Let $\kappa := \mathrm{id}_{\mathbb{R}_+}$, and $g(s) := \frac{s^2}{4}$. Now, we are in the position to prove the main part of (f_2). We may assume that there exists a unique $n \in \mathbb{N}$ such that $s \in I_n$. If $s \in \mathrm{int}I_n$ then $F^0(s; -s) = -ns$ and due to (7.15), we need

$$3\left(ns - \frac{1 + \sqrt{3} + \cdots + \sqrt{2n-1}}{\sqrt{2}}\right) - ns - \frac{s^2}{4} \leq 0,$$

which follows precisely by $(I_{\leq}^n)$. If $s_n = \sqrt{\frac{2n+1}{2}}$, then $F^0(s_n; -s_n) = -n\sqrt{\frac{2n+1}{2}}$. In this case, (F_2) reduces exactly to $(I_{\leq}^n)$.

Since the function $x \mapsto \frac{1}{3}(x^3 - x^2) - nx$ is decreasing in I_n, $n \in \mathbb{N}$, to show (F_5), it is enough to verify that

$$\frac{1}{3}\left(\left(\frac{2n-1}{2}\right)^{\frac{3}{2}} - \frac{2n-1}{2}\right) \leq n\sqrt{\frac{2n-1}{2}} - \frac{1 + \sqrt{3} + \cdots + \sqrt{2n-1}}{\sqrt{2}},$$

which is exactly $(I_{\geq}^n)$. This completes the proof. $\qquad\square$

7.3　Hemivariational Inequalities in $\Omega = \omega \times \mathbb{R}^l, l \geq 2$

Let ω be a bounded open set in $\mathbb{R}^m$ with smooth boundary and let $\Omega := \omega \times \mathbb{R}^l$ be a *strip-like domain* ; $m \geq 1, l := N - m \geq 2$. Let $F : \Omega \times \mathbb{R} \to \mathbb{R}$ be a Carathéodory function which is locally Lipschitz in the second variable such that

(F_1)　$F(x, 0) = 0$, and there exist $c_1 > 0$ and $p \in (2, 2^*)$ such that

$$|\xi| \leq c_1(|s| + |s|^{p-1}), \quad \forall s \in \mathbb{R}, \forall \xi \in \partial_C^2 F(x, s) \text{ and a.e. } x \in \Omega.$$

Here, and hereafter, we denote by $2^* := 2N/(N-2)$ the Sobolev critical exponent.

In this section we study the following *eigenvalue problem for hemivariational inequalities*.

$(EPHI_\mu)$ Find $u \in H_0^1(\Omega)$ such that

$$\int_\Omega \nabla u \cdot \nabla v \mathrm{d}x + \mu \int_\Omega F^0(x, u(x); -v(x)) \mathrm{d}x \geq 0, \quad \forall v \in H_0^1(\Omega).$$

The expression $F^0(x, s; t)$ stands for the generalized directional derivative of $F(x, \cdot)$ at the point $s \in \mathbb{R}$ in the direction $t \in \mathbb{R}$.

The motivation to study this type of problem comes from mathematical physics. Moreover, if we particularize the form of F (see Remark 7.5), then $(EPHI_\mu)$ reduces to the following eigenvalue problem

$$-\Delta u = \mu f(x, u) \text{ in } \Omega, \quad u \in H_0^1(\Omega), \qquad (EP_\mu)$$

which is a simplified form of certain stationary waves in the non-linear Klein-Gordon or Schrödinger equations (see for instance Amick [1]). Under some restrictive conditions on the nonlinear term f, (EP_μ) has been firstly studied by Esteban [10]. Further investigations, closely related to [10] can be found in the papers of Burton [5], Fan and Zhao [12], Schindler [22].

Although related problems to (EP_μ) have been extensively studied on *bounded domains*, in *unbounded domains* the problem is more delicate, due to the lack of compactness in the Sobolev embeddings. In order to solve (EP_μ), Esteban [10] used a minimization procedure via axially symmetric functions. In the case of strip-like domains, the space of axially symmetric functions has been the main tool in the investigations, due to its 'good behavior' concerning the compact embeddings (do not forget that $N \geq m+2$, see [20]); this is the reason why many authors used this space in their works (see for instance [10, 12]). On the other hand, no attention has been paid in the literature to the existence of axially *non*-symmetric solutions, even in the classical case (EP_μ). Thus, the study of existence of axially non-symmetric solutions for $(EPHI_\mu)$ constitutes one of the main tasks of this section. A non-smooth version of the fountain theorem of Bartsch provides not only infinitely many axially symmetric solutions but also axially non-symmetric solutions, when $N := m + 4$ or $N \geq m + 6$.

Throughout this section, $H_0^1(\Omega)$ denotes the usual Sobolev space endowed with the inner product

$$\langle u, v \rangle_0 := \int_\Omega \nabla u \cdot \nabla v dx$$

and norm $\| \cdot \|_0 = \sqrt{\langle \cdot, \cdot \rangle_0}$, while the norm of $L^\alpha(\Omega)$ will be denoted by $\| \cdot \|_\alpha$. Since Ω has the cone property, we have the continuous embedding $H_0^1(\Omega) \hookrightarrow L^\alpha(\Omega), \alpha \in [2, 2^*]$, that is, there exists $k_\alpha > 0$ such that $\|u\|_\alpha \leq k_\alpha \|u\|_0$ for all $u \in H_0^1(\Omega)$.

We say that a function $h : \Omega \to \mathbb{R}$ is *axially symmetric*, if $h(x, y) = h(x, gy)$ for all $x \in \omega$, $y \in \mathbb{R}^{N-m}$ and $g \in O(N - m)$. In particular, we denote by $H_{0,s}^1(\Omega)$ the closed subspace of axially symmetric functions of $H_0^1(\Omega)$. $u \in H_0^1(\Omega)$ is called *axially non-symmetric*, if it is not axially symmetric.

We require the following assumptions on nonlinearity F.

(F_2) $\lim_{s \to 0} \frac{\max\{|\xi|: \xi \in \partial_C^2 F(x,s)\}}{s} = 0$ uniformly for a.e. $x \in \Omega$;

(F_3) There exist $\nu \geq 1$ and $\gamma \in L^\infty(\Omega)$ with $\text{essinf}_{x \in \Omega} \gamma(x) = \gamma_0 > 0$ such that

$$2F(x, s) + F^0(x, s; -s) \leq -\gamma(x)|s|^\nu,$$

for all $s \in \mathbb{R}$ and a.e. $x \in \Omega$.

The following theorem can be considered an extension of Bartsch and Willem's result (see [3]) to the case of strip-like domains.

Theorem 7.5 ([17]) *Let $F : \Omega \times \mathbb{R} \to \mathbb{R}$ be a function which satisfies (F_1), (F_2), and (F_3) for some $v > \max\{2, N(p-2)/2\}$. If F is axially symmetric in the first variable and even in the second variable then (EPHI_μ) has infinitely many axially symmetric solutions for every $\mu > 0$. In addition, if $N = m + 4$ or $N \geq m + 6$, (EPHI_μ) has infinitely many axially non-symmetric solutions.*

Remark 7.4 The inequality from (F_3) is a non-smooth version of one introduced by Costa and Magalhães [7]. Let us suppose for a moment that F is autonomous. Note that (F_3) is implied in many cases by the following condition (of Ambrosetti-Rabinowitz type):

$$vF(s) + F^0(s; -s) \leq 0 \text{ for all } s \in \mathbb{R}, \tag{7.16}$$

where $v > 2$. Indeed, from (7.16) and Lebourg's mean value theorem, applied to the locally Lipschitz function $g : (0, +\infty) \to \mathbb{R}$, $g(t) := t^{-v} F(tu)$ (with arbitrary fixed $u \in \mathbb{R}$) we obtain that $t^{-v} F(tu) \geq s^{-v} F(su)$ for all $t \geq s > 0$. If we assume in addition that $\liminf_{s \to 0} \frac{F(s)}{|s|^v} \geq a_0 > 0$, from the above relation (substituting $t = 1$) we have for $u \neq 0$ that $F(u) \geq \liminf_{s \to 0^+} \frac{F(su)}{|su|^v} |u|^v \geq a_0 |u|^v$. So, $2F(u) + F^0(u; -u) \leq (2 - v)F(u) \leq -\gamma_0 |u|^v$, where $\gamma_0 = a_0(v - 2) > 0$.

Remark 7.5 Let $f : \Omega \times \mathbb{R} \to \mathbb{R}$ be a measurable (not necessarily continuous) function and suppose that there exists $c > 0$ such that $|f(x, s)| \leq c(|s| + |s|^{p-1})$ for all $s \in \mathbb{R}$ and a.e. $x \in \Omega$. Define $F : \Omega \times \mathbb{R} \to \mathbb{R}$ by $F(x, s) := \int_0^s f(x, t)dt$. Then F is a Carathéodory function which is locally Lipschitz in the second variable which satisfies the growth condition from (F_1). Indeed, since $f(x, \cdot) \in L_{loc}^\infty(\mathbb{R})$, by [21, Proposition 1.7] we have $\partial_C^2 F(x, s) = [f_-(x, s), f_+(x, s)]$ for all $s \in \mathbb{R}$ and a.e. $x \in \Omega$, where $f_-(x, s) = \lim_{\delta \to 0^+} \text{essinf}_{|t-s|<\delta} f(x, t)$ and $f_+(x, s) = \lim_{\delta \to 0^+} \text{esssup}_{|t-s|<\delta} f(x, t)$.

Moreover, if f is continuous in the second variable, then $\partial_C^2 F(x, s) = \{f(x, s)\}$ for all $s \in \mathbb{R}$ and a.e. $x \in \Omega$. Therefore, the inequality from (EPHI_μ) takes the form

$$\int_\Omega \nabla u \cdot \nabla v dx - \mu \int_\Omega f(x, u(x))v(x)dx = 0, \text{ for all } v \in H_0^1(\Omega),$$

that is, $u \in H_0^1(\Omega)$ is a weak solution of (EP_μ) in the usual sense.

Remark 7.6 In view of Remark 7.5, under appropriate hypotheses on f, corresponding to $(F_1) - (F_3)$, it is possible to state the smooth counterpart of Theorem 7.5.

The remainder of this section is dedicated to the proof of Theorem 7.5. We have the following auxiliary results.

Lemma 7.3 *If $F : \Omega \times \mathbb{R} \to \mathbb{R}$ satisfies (F_1) and (F_2), for every $\varepsilon > 0$ there exists $c(\varepsilon) > 0$ such that*

(i) $|\xi| \leq \varepsilon|s| + c(\varepsilon)|s|^{p-1}$ *for all $s \in \mathbb{R}$, $\xi \in \partial_C^2 F(x, s)$ and a.e. $x \in \Omega$;*
(ii) $|F(x, s)| \leq \varepsilon s^2 + c(\varepsilon)|s|^p$ *for all $s \in \mathbb{R}$ and a.e. $x \in \Omega$.*

Proof

(i) Let $\varepsilon > 0$ be fixed. Condition (F_2) implies that there exists $\delta := \delta(\varepsilon) > 0$ such that $|\xi| \leq \varepsilon|s|$ for $|s| < \delta$, $\xi \in \partial_C^2 F(x, s)$ and a.e. $x \in \Omega$. If $|s| \geq \delta$, then (F_1) implies that $|\xi| \leq c_1(|s|^{2-p} + 1)|s|^{p-1} \leq c(\delta)|s|^{p-1}$ for all $\xi \in \partial_C^2 F(x, s)$ and a.e. $x \in \Omega$. Combining the above relations we get the required inequality.

(ii) We use Lebourg's mean value theorem, obtaining $|F(x, s)| = |F(x, s) - F(x, 0)| = |\xi_{\theta s} s|$ for some $\xi_{\theta s} \in \partial_C^2 F(x, \theta s)$ where $\theta \in (0, 1)$. Now, we apply *(i)* to complete the proof. $\qquad\square$

Define $\mathcal{F}, \Psi(\cdot, \mu) : H_0^1(\Omega) \to \mathbb{R}$ by

$$\mathcal{F}(u) := \int_\Omega F(x, u(x)) dx$$

and

$$\Psi(u, \mu) := \frac{1}{2}\|u\|_0^2 - \mu \mathcal{F}(u)$$

for $\mu \geq 0$. The following result plays a crucial role in the study of (EPHI_μ).

Lemma 7.4 *Let $F : \Omega \times \mathbb{R} \to \mathbb{R}$ be a locally Lipschitz function satisfying (F_1). Then $\mathcal{F}$ is well-defined and locally Lipschitz. Moreover, let E be a closed subspace of $H_0^1(\Omega)$ and $\mathcal{F}_E$ the restriction of $\mathcal{F}$ to E. Then*

$$\mathcal{F}_E^0(u; v) \leq \int_\Omega F^0(x, u(x); v(x)) dx, \quad \forall u, v \in E. \tag{7.17}$$

Proof The proof is similar to that of Proposition 2.16, but for the sake of completeness we will give it. Let us fix $s_1, s_2 \in \mathbb{R}$ arbitrary. By Lebourg's mean value theorem, there exist $\theta \in (0, 1)$ and $\xi_\theta \in \partial_C^2 F(x, \theta s_1 + (1 - \theta)s_2)$ such that $F(x, s_1) - F(x, s_2) = \xi_\theta(s_1 - s_2)$. By (F_1) we conclude that

$$|F(x, s_1) - F(x, s_2)| \leq d|s_1 - s_2| \cdot \left[|s_1| + |s_2| + |s_1|^{p-1} + |s_2|^{p-1}\right] \tag{7.18}$$

for all s_1, $s_2 \in \mathbb{R}$ and a.e. $x \in \Omega$, where $d = d(c_1, p) > 0$. In particular, if $u \in H_0^1(\Omega)$, we obtain that

$$|\mathcal{F}(u)| \leq \int_\Omega |F(x, u(x))|dx \leq d(\|u\|_2^2 + \|u\|_p^p) \leq d(k_2^2\|u\|_0^2 + k_p^p\|u\|_0^p) < +\infty,$$

that is, the function $\mathcal{F}$ is well-defined. Moreover, by (7.18), there exists $d_0 > 0$ such that for every $u, v \in H_0^1(\Omega)$

$$|\mathcal{F}(u) - \mathcal{F}(v)| \leq d_0\|u - v\|_0 \left[\|u\|_0 + \|v\|_0 + \|u\|_0^{p-1} + \|v\|_0^{p-1}\right].$$

Therefore, $\mathcal{F}$ is a locally Lipschitz function on $H_0^1(\Omega)$.

Let us fix u and w in E. Since $F(x, \cdot)$ is continuous, $F^0(x, u(x); v(x))$ can be expressed as the upper limit of $\frac{F(x, y+tv(x)) - F(x,y)}{t}$, where $t \to 0^+$ takes rational values and $y \to u(x)$ takes values in a countable dense subset of $\mathbb{R}$. Therefore, the map $x \mapsto F^0(x, u(x); v(x))$ is measurable as the "countable limsup" of measurable functions of x. According to (F_1), the map $x \mapsto F^0(x, u(x); v(x))$ belongs to $L^1(\Omega)$.

Since E is separable, there are functions $u_n \in E$ and numbers $t_n \to 0^+$ such that $u_n \to u$ in E and

$$\mathcal{F}_E^0(u; v) = \lim_{n \to +\infty} \frac{\mathcal{F}_E(u_n + t_n v) - \mathcal{F}_E(u_n)}{t_n},$$

and without loss of generality, we may assume that there exist $h_2 \in L^2(\Omega, \mathbb{R}_+)$ and $h_p \in L^p(\Omega, \mathbb{R}_+)$ such that $|u_n(x)| \leq \min\{h_2(x), h_p(x)\}$ and $u_n(x) \to u(x)$ a.e. in Ω, as $n \to +\infty$.

We define $g_n : \Omega \to \mathbb{R} \cup \{+\infty\}$ by

$$g_n(x) := -\frac{F(x, u_n(x) + t_n v(x)) - F(x, u_n(x))}{t_n} + d|v(x)|[|u_n(x) + t_n v(x)| +$$

$$+ |u_n(x)| + |u_n(x) + t_n v(x)|^{p-1} + |u_n(x)|^{p-1}].$$

The maps g_n are measurable and non-negative, see (7.18). From Fatou's lemma we have

$$I = \limsup_{n \to +\infty} \int_\Omega [-g_n(x)]dx \leq \int_\Omega \limsup_{n \to +\infty} [-g_n(x)]dx = J.$$

Let $B_n := A_n + g_n$, where

$$A_n(x) := \frac{F(x, u_n(x) + t_n v(x)) - F(x, u_n(x))}{t_n}.$$

By the Lebesgue dominated convergence theorem, we have

$$\lim_{n \to +\infty} \int_\Omega B_n dx = 2d \int_\Omega |v(x)|(|u(x)| + |u(x)|^{p-1}) dx.$$

Therefore, we have

$$I = \limsup_{n \to +\infty} \frac{\mathcal{F}_E(u_n + t_n v) - \mathcal{F}_E(u_n)}{t_n} - \lim_{n \to +\infty} \int_\Omega B_n dx$$

$$= \mathcal{F}_E^0(u; v) - 2d \int_\Omega |v(x)| \left(|u(x)| + |u(x)|^{p-1} \right) dx.$$

Now, we estimate J. We have $J \leq J_A - J_B$, where

$$J_A := \int_\Omega \limsup_{n \to +\infty} A_n(x) dx \quad \text{and} \quad J_B := \int_\Omega \liminf_{n \to +\infty} B_n(x) dx.$$

Since $u_n(x) \to u(x)$ a.e. in Ω and $t_n \to 0^+$, we have

$$J_B = 2d \int_\Omega |v(x)| \left(|u(x)| + |u(x)|^{p-1} \right) dx.$$

On the other hand,

$$J_A = \int_\Omega \limsup_{n \to +\infty} \frac{F(x, u_n(x) + t_n v(x)) - F(x, u_n(x))}{t_n} dx$$

$$\leq \int_\Omega \limsup_{y \to u(x), t \to 0^+} \frac{F(x, y + t v(x)) - F(x, y)}{t} dx$$

$$= \int_\Omega F^0(x, u(x); v(x)) dx.$$

From the above estimations we obtain (7.17), which concludes the proof. $\quad\square$

We suppose that assumptions of Theorem 7.5 are fulfilled. Let us denote by $\mathcal{F}_E$, $\Psi_E(\cdot, \mu)$, $\langle \cdot, \cdot \rangle_E$ and $\| \cdot \|_E$ the restrictions of $\mathcal{F}$, $\Psi(\cdot, \mu)$, $\langle \cdot, \cdot \rangle_0$ and $\| \cdot \|_0$, respectively, to a closed subspace E of $H_0^1(\Omega)$, $(\mu \geq 0)$.

Lemma 7.5 *If the embedding $E \hookrightarrow L^p(\Omega)$ is compact then $\Psi_E(\cdot, \mu)$ satisfies $(C)_c$ for all $\mu, c > 0$.*

Proof Let us fix a $\mu > 0$ and a sequence $\{u_n\}$ from E such that $\Psi_E(u_n, \mu) \to c > 0$ and

$$(1 + \|u_n\|_E)\lambda_{\Psi_E(\cdot, \mu)}(u_n) \to 0 \tag{7.19}$$

as $n \to +\infty$. We shall prove firstly that $\{u_n\}$ is bounded in E. Let $\zeta_n \in \partial_C \Psi_E(\cdot, \mu)(u_n)$ such that $\|\zeta_n\|_E = \lambda_{\Psi_E(\cdot, \mu)}(u_n)$; it is clear that $\|\zeta_n\|_E \to 0$ as $n \to +\infty$. Moreover, $\Psi_E(\cdot, \mu)^0(u_n; u_n) \geq \langle \zeta_n, u_n \rangle_E \geq -\|z_n\|_E \|u_n\|_E \geq -(1 + \|u_n\|_E)\lambda_{\Psi_E(\cdot, \mu)}(u_n)$. Therefore, by Lemma 7.4, for n large enough

$$
\begin{aligned}
2c + 1 &\geq 2\Psi_E(u_n, \mu) - \Psi_E(\cdot, \mu)^0(u_n; u_n) \\
&= -2\mu \mathcal{F}_E(u_n) - \mu(-\mathcal{F}_E)^0(u_n; u_n) \\
&\geq -\mu \int_\Omega [2F(x, u_n(x)) + F^0(x, u_n(x); -u_n(x))]dx \\
&\geq \mu \gamma_0 \|u_n\|_\nu^\nu.
\end{aligned}
$$

Thus,

$$\{u_n\} \text{ is bounded in } L^\nu(\Omega). \tag{7.20}$$

After integration in Lemma 7.3 $ii)$, we obtain that for all $\varepsilon > 0$ there exists $c(\varepsilon) > 0$ such that $\mathcal{F}_E(u_n) \leq \varepsilon \|u_n\|_E^2 + c(\varepsilon)\|u_n\|_p^p$ (note that $\|u\|_2^2 \leq k_2^2 \|u\|_0^2$). For n large, one has

$$c + 1 \geq \Psi_E(u_n, \mu) = \frac{1}{2}\|u_n\|_E^2 - \mu \mathcal{F}_E(u_n) \geq \left(\frac{1}{2} - \varepsilon\mu\right)\|u_n\|_E^2 - \mu c(\varepsilon)\|u_n\|_p^p.$$

Choosing $\varepsilon > 0$ small enough, we will find $c_2, c_3 > 0$ such that

$$c_2\|u_n\|_E^2 \leq c + 1 + c_3\|u_n\|_p^p. \tag{7.21}$$

Since $\nu \leq p$ (compare Lemma 7.3 $ii)$ and (7.22) below), we distinguish two cases.

(I) If $\nu = p$ it is clear from (7.21) and (7.20) that $\{u_n\}$ is bounded in E.
(II) If $\nu < p$, we have the interpolation inequality

$$\|u_n\|_p \leq \|u_n\|_\nu^{1-\delta}\|u_n\|_{2^*}^\delta \leq k_{2^*}^\delta \|u_n\|_\nu^{1-\delta}\|u_n\|_E^\delta$$

(since $u_n \in E \hookrightarrow L^\nu(\Omega) \cap L^{2^*}(\Omega)$), where $1/p = (1 - \delta)/\nu + \delta/2^*$. Since $\nu > N(p - 2)/2$, we have $\delta p < 2$. According again to (7.20) and (7.21), we conclude that $\{u_n\}$ should be bounded in E.

Since $E \hookrightarrow L^p(\Omega)$ is compact, up to a subsequence, $u_n \rightharpoonup u$ in E and $u_n \to u$ in $L^p(\Omega)$. Moreover, we have

$$\Psi_E(\cdot, \mu)^0(u_n; u - u_n) = \langle u_n, u - u_n \rangle_E + \mu(-\mathcal{F}_E)^0(u_n; u - u_n),$$

$$\Psi_E(\cdot, \mu)^0(u; u_n - u) = \langle u, u_n - u \rangle_E + \mu(-\mathcal{F}_E)^0(u; u_n - u).$$

Adding these two relations, we obtain

$$\|u_n - u\|_E^2 = \mu[\mathcal{F}_E^0(u_n; -u + u_n) + \mathcal{F}_E^0(u; -u_n + u)]$$
$$- \Psi_E(\cdot, \mu)^0(u_n; u - u_n) - \Psi_E(\cdot, \mu)^0(u; u_n - u).$$

On the other hand, there exists $\zeta_n \in \partial_C \Psi_E(\cdot, \mu)(u_n)$ such that $\|\zeta_n\|_E = \lambda_{\Psi_E(\cdot, \mu)}(u_n)$. Here, we used the Riesz representation theorem. By (7.19), one has $\|\zeta_n\|_E \to 0$ as $n \to +\infty$. Since $u_n \rightharpoonup u$ in E, fixing an element $\zeta \in \partial_C \Psi_E(\cdot, \mu)(u)$, we have $\langle \zeta, u_n - u \rangle_E \to 0$. Therefore, by the inequality (7.17) and Lemma 7.3 i), we have

$$\|u_n - u\|_E^2 \leq \mu \int_\Omega [F^0(x, u_n(x); -u(x) + u_n(x)) + F^0(x, u(x); -u_n(x) + u(x))]dx$$

$$- \langle \zeta_n, u - u_n \rangle_E - \langle \zeta, u_n - u \rangle_E$$

$$= \mu \int_\Omega \max \left\{ \xi_n(x)(-u(x) + u_n(x)) : \xi_n(x) \in \partial_C^2 F(x, u_n(x)) \right\} dx$$

$$+ \mu \int_\Omega \max \left\{ \xi(x)(-u_n(x) + u(x)) : \xi(x) \in \partial_C^2 F(x, u(x)) \right\} dx$$

$$- \langle \zeta_n, u - u_n \rangle_E - \langle \zeta, u_n - u \rangle_E$$

$$\leq \mu \int_\Omega \varepsilon (|u_n(x)| + |u(x)|) |u_n(x) - u(x)| dx$$

$$+ \mu \int_\Omega c(\varepsilon) \left(|u_n(x)|^{p-1} + |u(x)|^{p-1} \right) |u_n(x) - u(x)| dx$$

$$+ \|\zeta_n\|_E \|u - u_n\|_E - \langle \zeta, u_n - u \rangle_E$$

$$\leq 2\varepsilon\mu k_2^2(\|u_n\|_E^2 + \|u\|_E^2) + \mu c(\varepsilon) \left(\|u_n\|_p^{p-1} + \|u\|_p^{p-1} \right) \|u_n - u\|_p$$

$$+ \|\zeta_n\|_E \|u - u_n\|_E - \langle \zeta, u_n - u \rangle_E.$$

Due to the arbitrariness of $\varepsilon > 0$, we have that $\|u_n - u\|_E^2 \to 0$ as $n \to +\infty$. Thus, $\{u_n\}$ converges strongly to u in E. This concludes the proof. $\quad\square$

Proof of Theorem 7.5 For the first part, we verify the conditions of Theorem 5.6, choosing $E := H_{0,s}^1(\Omega)$ and $f := \Psi_E(\cdot, \mu)$, where $\Psi_E(\cdot, \mu)$ denotes the restriction of

$\Psi(\cdot, \mu)$ to E, $\mu > 0$ being arbitrary fixed. By assumption, F is even in the second variable, so $\Psi_E(\cdot, \mu)$ is also even, and by Lemma 7.4 it is a locally Lipschitz function.

Clearly, $\Psi_E(0, \mu) = 0$, while condition $(C)_c$ is verified, due to Lemma 7.5. Indeed, $H^1_{0,s}(\Omega)$ is compactly embedded into $L^p(\Omega)$.

In order to prove (f'_2) of Theorem 5.6, we consider $g : \Omega \times (0, +\infty) \to \mathbb{R}$ defined by

$$g(x, t) := t^{-2} F(x, t) - \frac{\gamma(x)}{\nu - 2} t^{\nu - 2}.$$

Let us fix $x \in \Omega$. Clearly, $g(x, \cdot)$ is a locally Lipschitz function and by the Chain Rule we have

$$\partial g(x, t) \subseteq -2t^{-3} F(x, t) + t^{-2} \partial F(x, t) - \gamma(x) t^{\nu - 3}, \quad t > 0.$$

Let $t > s > 0$. By Lebourg's mean value theorem, there exist $\tau := \tau(x) \in (s, t)$ and $w_\tau := w_\tau(x) \in \partial g(x, \tau)$ such that $g(x, t) - g(x, s) = w_\tau(t - s)$. Therefore, there exists $\xi_\tau := \xi_\tau(x) \in \partial_C^2 F(x, \tau)$ such that $w_\tau = -2\tau^{-3} F(x, \tau) + \tau^{-2} \xi_\tau - \gamma(x)\tau^{\nu - 3}$ and

$$g(x, t) - g(x, s) \geq -\tau^{-3} [2F(x, \tau) + F^0(x, \tau; -\tau) + \gamma(x)\tau^\nu](t - s).$$

By (F_3) one has $g(x, t) \geq g(x, s)$. On the other hand, by Lemma 7.3 we have that $F(x, s) = o(s^2)$ as $s \to 0$ for a.e. $x \in \Omega$. If $s \to 0^+$ in the above inequality, we have that $F(x, t) \geq \gamma(x)t^\nu/(\nu - 2)$ for all $t > 0$ and a.e. $x \in \Omega$. Since $F(x, 0) = 0$ and $F(x, \cdot)$ is even, we obtain that

$$F(x, t) \geq \frac{\gamma(x)}{\nu - 2} |t|^\nu, \quad \forall t \in \mathbb{R} \text{ and a.e.} x \in \Omega. \tag{7.22}$$

Now, let $\{e_i\}$ be a fixed orthonormal basis of E and $E_k = \{e_1, \ldots, e_k\}$, $k \geq 1$. Denoting by $\| \cdot \|_E$ the restriction of $\| \cdot \|_0$ to E, from (7.22) one has

$$\Psi_E(u, \mu) \leq \frac{1}{2} \|u\|_E^2 - \frac{\mu\gamma_0}{\nu - 2} \|u\|_\nu^\nu, \quad \forall u \in E.$$

Let us fix $k \geq 1$ arbitrary. Since $\nu > 2$ and on the finite dimensional space E_k all norms are equivalent (in particular $\| \cdot \|_0$ and $\| \cdot \|_\nu$), choosing a large $R_k > 0$, we have $\Psi_E(u, \mu) \leq \Psi_E(0, \mu) = 0$ if $\|u\|_E \geq R_k$, $u \in E_k$. This proves (f'_2).

Again, from Lemma 7.3 $ii)$ we have that for all $\varepsilon > 0$ there exists $c(\varepsilon) > 0$ such that $\mathcal{F}_E(u) \leq \varepsilon \|u\|_E^2 + c(\varepsilon)\|u\|_p^2$ for all $u \in E$. Let $\beta_k = \sup\{\|u\|_p/\|u\|_E : u \in E_k^\perp, u \neq 0\}$. As in [3, Lemma 3.3], it can be proved that $\beta_k \to 0$ as $k \to +\infty$. For $u \in E_k^\perp$, one has

$$\Psi_E(u, \mu) \geq \left(\frac{1}{2} - \varepsilon\mu\right) \|u\|_E^2 - \mu c(\varepsilon)\|u\|_p^p \geq \left(\frac{1}{2} - \varepsilon\mu\right) \|u\|_E^2 - \mu c(\varepsilon)\beta_k^p \|u\|_E^p.$$

Choosing $\varepsilon < (p-2)(2p\mu)^{-1}$ and $\rho_k := (p\mu c(\varepsilon)\beta_k^p)^{\frac{1}{2-p}}$, we have

$$\Psi_E(u, \mu) \geq \left(\frac{1}{2} - \frac{1}{p} - \varepsilon\mu\right)\rho_k^2$$

for every $u \in E_k^\perp$ with $\|u\|_E = \rho_k$. Since $\beta_k \to 0$, then $\rho_k \to +\infty$ as $k \to +\infty$. Thus (f_1') of Theorem 5.6 is concluded.

Hence, $\Psi_E(\cdot, \mu)$ has infinitely many critical points on $E = H_{0,s}^1(\Omega)$. Using the principle (PSCL), these points are actually critical point for the original functions $\Psi(\cdot, \mu)$. Now, using Lemma 7.4, the above points will be precisely solutions for (EPHI$_\mu$).

Now, we deal with the second part. The following construction is inspired by [3]. Let $N := m + 4$ or $N \geq m + 6$. In both cases we find at least a number $k \in [2, \frac{N-m}{2}] \cap \mathbb{N} \setminus \{\frac{N-m-1}{2}\}$. For a such $k \in \mathbb{N}$, we have $\Omega := \omega \times \mathbb{R}^k \times \mathbb{R}^k \times \mathbb{R}^{N-2k-m}$. Let $H := id^m \times O(k) \times O(k) \times O(N - 2k - m)$ and define

$$G_\tau := \langle H \cup \{\tau\}\rangle,$$

where $\tau(x_1, x_2, x_3, x_4) = (x_1, x_3, x_2, x_4)$, for every $x_1 \in \omega$, $x_2, x_3 \in \mathbb{R}^k$, $x_4 \in \mathbb{R}^{N-2k-m}$. G_τ will be a subgroup of $O(N)$ and its elements can be written uniquely as h or $h\tau$ with $h \in H$. The action of G_τ on $H_0^1(\Omega)$ is defined by

$$gu(x_1, x_2, x_3, x_4) = \pi(g)u(x_1, g_2x_2, g_3x_3, g_4x_4) \tag{7.23}$$

for all $g = id^m \times g_2 \times g_3 \times g_4 \in G_\tau$, $(x_1, x_2, x_3, x_4) \in \omega \times \mathbb{R}^k \times \mathbb{R}^k \times \mathbb{R}^{N-2k-m}$, where $\pi : G_\tau \to \{\pm 1\}$ is the canonical epimorphism, that is, $\pi(h) = 1$ and $\pi(h\tau) = -1$ for all $h \in H$. The group G_τ acts linear isometrically on $H_0^1(\Omega)$, and $\Psi(\cdot, \mu)$ is G_τ-invariant, since F is axially symmetric in the first variable and even in the second variable. Let

$$H_{0,ns}^1(\Omega) := \{u \in H_0^1(\Omega) : gu = u, \forall g \in G_\tau\}.$$

Clearly, $H_{0,ns}^1(\Omega)$ is a closed subspace of $H_0^1(\Omega)$ and

$$H_{0,ns}^1(\Omega) \subset H_0^1(\Omega)^H \stackrel{df.}{:=} \{u \in H_0^1(\Omega) : hu = u, \forall h \in H\}.$$

On the other hand, $H_0^1(\Omega)^H \hookrightarrow L^p(\Omega)$ is compact (see [20, Théorème III.2.]), hence $H_{0,ns}^1(\Omega) \hookrightarrow L^p(\Omega)$ is also compact.

Now, repeating the proof of the first part for $E = H_{0,ns}^1(\Omega)$ instead of $H_{0,s}^1(\Omega)$, we obtain infinitely many solutions for (EPHI$_\mu$), which belong to $H_{0,ns}^1(\Omega)$. But we observe that 0 is the only axially symmetric function of $H_{0,ns}^1(\Omega)$. Indeed, let $u \in H_{0,ns}^1(\Omega) \cap H_{0,s}^1(\Omega)$. Since $gu = u$ for all $g \in G_\tau$, choosing in particular $\tau \in G_\tau$ and using (7.23), we have that $u(x_1, x_2, x_3, x_4) = -u(x_1, x_3, x_2, x_4)$ for all $(x_1, x_2, x_3, x_4) \in \omega \times \mathbb{R}^k \times$

$\mathbb{R}^k \times \mathbb{R}^{N-2k-m}$. Since u is axially symmetric and $|(x_2, x_3, x_4)| = |(x_3, x_2, x_4)|$, ($|\cdot|$ being the norm on $\mathbb{R}^{N-m}$), it follows that u must be 0. Therefore, the above solutions are axially non-symmetric functions. This concludes the proof.   □

Remark 7.7 The reader can observe that we considered only $N \geq m + 2$. In fact, in this case $H^1_{0,s}(\Omega)$ can be embedded compactly into $L^p(\Omega)$, $p \in (2, 2^*)$ which was crucial in the verification of the Cerami condition. When $N := m + 1$ the above embedding is no longer compact. In the latter case it is recommended to construct the closed convex cone (see [11, Theorem 2]), defined by

$$\mathcal{K} := \left\{ \begin{array}{l} u \in H^1_0(\omega \times \mathbb{R}) : u \geq 0, \ u(x, y) \text{ is nonincreasing in } y \text{ for } x \in \omega, \ y \geq 0 \\ \text{and } u(x, y) \text{ is nondecreasing in } y \text{ for } x \in \omega, \ y \leq 0 \end{array} \right\},$$

because the Sobolev embedding from $H^1_0(\omega \times \mathbb{R})$ into $L^p(\omega \times \mathbb{R})$ transforms the bounded closed sets of $\mathcal{K}$ into relatively compact sets of $L^p(\omega \times \mathbb{R})$, $p \in (2, 2^*)$ (note that $2^* = +\infty$, if $m = 1$). Since $\mathcal{K}$ is not a subspace of $H^1_0(\omega \times \mathbb{R})$, the above described machinery does not work. However, we will treat a closely related form of the above problem in the next section.

In the final part of this section, we provide two examples, which highlight the applicability of the main result.

Example 7.3 Let $p \in (2, 2^*)$. Then, for all $\mu > 0$, the problem

$$-\Delta u = \mu |u|^{p-2} u \text{ in } \Omega, \ u \in H^1_0(\Omega),$$

has infinitely many axially symmetric solutions. Moreover, if $N := m + 4$ or $N \geq m + 6$, the problem has infinitely many axially non-symmetric solutions.

Indeed, consider the (continuously differentiable) function $F(x, s) = F(s) := |s|^p$, which verifies obviously the assumptions of Theorem 7.5 (choose $\nu = p$).

Example 7.4 We denote by $\lfloor u \rfloor$ the nearest integer to $u \in \mathbb{R}$, if $u + \frac{1}{2} \notin \mathbb{Z}$; otherwise we put $\lfloor u \rfloor = u$. Let $N \in \{3, 4, 5\}$ and let $F : \Omega \times \mathbb{R} \to \mathbb{R}$ be defined by

$$F(x, s) = F(s) := \int_0^s \lfloor t|t| \rfloor dt + |s|^3.$$

It is clear that F is a locally Lipschitz, even function. Due to the first part of Remark 7.5, F verifies (F_1) with the choice $p := 3$ while (F_2) follows from the fact that $\lfloor t|t| \rfloor = 0$ if $t \in (-2^{-1/2}, 2^{-1/2})$. Since F is even (in particular, $F^0(s; -s) = F^0(-s; s)$ for all $s \in \mathbb{R}$), it is enough to very (F_3) for $s \geq 0$. We have that $F(s) = s^3$ if $s \in [0, a_1]$, and $F(s) = s^3 + ns - \sum_{k=1}^n \sqrt{2k-1}/\sqrt{2}$ if $s \in (a_n, a_{n+1}]$, where $a_n = (2n-1)^{1/2}2^{-1/2}$,

$n \in \mathbb{N} \setminus \{0\}$. Moreover, $F^0(s; -s) = -3s^3 - ns$ when $s \in (a_n, a_{n+1})$ while $F^0(a_n; -a_n) = -3a_n^3 - (n-1)a_n$, since $\partial_C F(a_n) = [3a_n^2 + n - 1, 3a_n^2 + n]$, $n \in \mathbb{N} \setminus \{0\}$. Choosing $\gamma(x) = \gamma_0 = 1/3$ and $\nu = 3$, from the above expressions the required inequality yields. Therefore $(EPHI_\mu)$ has infinitely many axially symmetric solutions for every $\mu > 0$. Moreover, if $\Omega := \omega \times \mathbb{R}^4$, where ω is an open bounded interval in $\mathbb{R}$, then $(EPHI_\mu)$ has infinitely many axially non-symmetric solutions for every $\mu > 0$.

7.4   Variational Inequalities in $\Omega = \omega \times \mathbb{R}$

In this section we will continue our study on the strip-like domains, but contrary to the previous section, we consider domains of the form $\Omega := \omega \times \mathbb{R}$, where $\omega \subset \mathbb{R}^m (m \geq 1)$ is a bounded open subset. This section is based on the paper of Kristály, Varga and Varga [19].

As we pointed out in the previous section, Lions [20, Théorème III. 2] (see also [11, Theorem 2]) observed that defining the closed convex cone

$$\mathcal{K} := \left\{ u \in H_0^1(\omega \times \mathbb{R}) : \begin{array}{l} u \text{ is nonnegative,} \\ y \mapsto u(x, y) \text{ is nonincreasing for } x \in \omega, \ y \geq 0, \\ y \mapsto u(x, y) \text{ is nondecreasing for } x \in \omega, \ y \leq 0, \end{array} \right\} \qquad \text{(K)}$$

the bounded subsets of $\mathcal{K}$ are relatively compact in $L^p(\omega \times \mathbb{R})$ whenever $p \in (2, 2^*)$. Note that $2^* = \infty$, if $m = 1$. Burton [5] was the first who exploited in its entirety the above "compactness"; namely, by means of a version of the Mountain Pass theorem (due to Hofer [15] for an order-preserving operator on Hilbert spaces), he was able to establish the existence of a nontrivial solution for an elliptic equation on domains of the type $\omega \times \mathbb{R}$. The main ingredient in his proof was the symmetric decreasing rearrangement of the suitable functions, proving that the cone $\mathcal{K}$ remains invariant under a carefully chosen nonlinear operator, which is an indispensable hypothesis in the Hofer's result.

The main goal of this section is to give a new approach to treat elliptic (eigenvalue) problems in cylinders of the type $\Omega := \omega \times \mathbb{R}$. The genesis of our method relies on the Szulkin type functionals. Indeed, since the indicator function of a closed convex subset of a vector space (so, in particular $\mathcal{K}$ in $H_0^1(\omega \times \mathbb{R})$) is convex, lower semicontinuous and proper, this approach arises in a natural manner as it was already forecasted in [17]. We point out that in [19] we considered a much general problem; instead of a Szulkin type functional we considered the Motreanu-Panagiotopoulos type function (see [21, Chapter 3]). In order to formulate our problem, we shall consider a continuous function $f : (\omega \times \mathbb{R}) \times \mathbb{R} \to \mathbb{R}$ such that

(F_1) $f(x, 0) = 0$, and there exist $c_1 > 0$ and $p \in (2, 2^*)$ such that

$$|f(x, s)| \leq c_1(|s| + |s|^{p-1}), \quad \forall(x, s) \in (\omega \times \mathbb{R}) \times \mathbb{R}.$$

Let $a \in L^1(\omega \times \mathbb{R}) \cap L^\infty(\omega \times \mathbb{R})$ with $a \geq 0$, $a \not\equiv 0$, and $q \in (1, 2)$. For $\lambda > 0$, we consider the following *variational inequality problem*:

(P_λ) Find $u \in \mathcal{K}$ such that

$$\int_{\omega \times \mathbb{R}} \nabla u(x) \cdot \nabla(v(x) - u(x)) dx + \int_{\omega \times \mathbb{R}} f(x, u(x))(-v(x) + u(x)) dx$$

$$\geq \lambda \int_{\omega \times \mathbb{R}} a(x)|u(x)|^{q-2} u(x)(v(x) - u(x)) dx, \ \forall v \in \mathcal{K}.$$

For the sake of simplicity, we introduce $\Omega := \omega \times \mathbb{R}$. Define $F : \Omega \times \mathbb{R} \to \mathbb{R}$ by $F(x, s) := \int_0^s f(x, t) dt$ and beside of (F_1) we assume:

(F_2) $\lim_{s \to 0} \frac{f(x,s)}{s} = 0$, uniformly for every $x \in \Omega$;

(F_3) There exists $\nu > 2$ such that

$$\nu F(x, s) - s f(x, s) \leq 0, \ \forall (x, s) \in \Omega \times \mathbb{R};$$

(F_4) There exists $R > 0$ such that

$$\inf\{F(x, s) : (x, |s|) \in \omega \times [R, \infty)\} > 0.$$

Remark 7.8 It is readily seen that if the conditions (F_1) and (F_2) hold, then for every $\varepsilon > 0$ there exists $c(\varepsilon) > 0$ such that

(i) $|f(x, s)| \leq \varepsilon|s| + c(\varepsilon)|s|^{p-1}, \ \forall (x, s) \in \Omega \times \mathbb{R}$;
(ii) $|F(x, s)| \leq \varepsilon s^2 + c(\varepsilon)|s|^p, \ \forall (x, s) \in \Omega \times \mathbb{R}$.

Lemma 7.6 *If the functions $f, F : \Omega \times \mathbb{R} \to \mathbb{R}$ satisfies (F_1), (F_3) and (F_4) then there exist $c_2, c_3 > 0$ such that*

$$F(x, s) \geq c_2|s|^\nu - c_3 s^2, \ \forall (x, s) \in \Omega \times \mathbb{R}.$$

Proof First, for arbitrary fixed $(x, u) \in \Omega \times \mathbb{R}$ we consider the function $g : (0, +\infty) \to \mathbb{R}$ defined by

$$g(t) := t^{-\nu} F(x, tu).$$

Clearly, g is a function of class C^1 and we have

$$g'(t) = -\nu t^{-\nu-1} F(x, tu) + t^{-\nu} u f(x, tu), \ t > 0.$$

For $t > 1$, by mean value theorem, there exist $\tau := \tau(x, u) \in (1, t)$ such that $g(t) - g(1) = g'(\tau)(t - 1)$. Therefore, $g'(\tau) = -\nu \tau^{-\nu-1} F(x, \tau u) + \tau^{-\nu} u f(x, \tau u)$, thus

$$g(t) - g(1) = -\tau^{-\nu-1}[\nu F(x, \tau u) - \tau u f(x, \tau u)](t - 1).$$

By (F_3) one has $g(t) \geq g(1)$, i.e., $F(x, tu) \geq t^\nu F(x, u)$, for every $t \geq 1$. Define $c_R := \inf\{F(x, s) : (x, |s|) \in \omega \times [R, \infty)\}$, which is a strictly positive number, due to (F_4). Combining the above facts we derive

$$F(x, s) \geq \frac{c_R}{R^\nu}|s|^\nu, \quad \forall (x, s) \in \Omega \times \mathbb{R} \text{ with } |s| \geq R. \tag{7.24}$$

On the other hand, by (F_1) we have $|F(x, s)| \leq c_1(s^2 + |s|^p)$ for every $(x, s) \in \Omega \times \mathbb{R}$. In particular, we have

$$-F(x, s) \leq c_1(s^2 + |s|^p) \leq c_1(1 + R^{p-2} + R^{\nu-2})s^2 - c_1|s|^\nu$$

for every $(x, s) \in \Omega \times \mathbb{R}$ with $|s| \leq R$. Combining the above inequality with (7.24), the desired inequality yields if one chooses $c_2 := \min\{c_1, c_R/R^\nu\}$ and $c_3 := c_1(1 + R^{p-2} + R^{\nu-2})$. □

Remark 7.9 In particular, Lemma 7.6 ensures that $2 < \nu < p$.

To investigate the existence of solutions of (P_λ) we shall construct a functional $\mathcal{J}_\lambda : H_0^1(\Omega) \to \mathbb{R}$ associated to (P_λ) which is defined by

$$\mathcal{J}_\lambda(u) := \frac{1}{2}\int_\Omega |\nabla u|^2 dx - \int_\Omega F(x, u(x))dx - \frac{\lambda}{q}\int_\Omega a(x)|u|^q dx + \psi_{\mathcal{K}}(u),$$

where $\psi_{\mathcal{K}}$ is the indicator function of the set $\mathcal{K}$.

If we consider the function $\mathcal{F} : H_0^1(\Omega) \to \mathbb{R}$ defined by

$$\mathcal{F}(u) := \int_\Omega F(x, u(x))dx,$$

then $\mathcal{F}$ is of class C^1 and

$$\langle \mathcal{F}'(u), v \rangle_{H_0^1(\Omega)} = \int_\Omega f(x, u(x))v(x)dx, \quad \forall u, v \in H_0^1(\Omega).$$

By standard arguments we have that the functionals $A_1, A_2 : H_0^1(\Omega) \to \mathbb{R}$, defined by $A_1(u) := \|u\|_0^2$ and $A_2(u) := \int_\Omega a(x)|u|^q dx$ are of class C^1 with derivatives

$$\langle A_1'(u), v \rangle_{H_0^1(\Omega)} = 2\langle u, v \rangle_0$$

and

$$\langle A_2'(u), v \rangle_{H_0^1(\Omega)} = q \int_\Omega a(x)|u|^{q-2}uv\,dx.$$

Therefore the function

$$\mathcal{H}_\lambda(u) := \frac{1}{2}\|u\|_0^2 - \frac{\lambda}{q}\int_\Omega a(x)|u|^q dx - \mathcal{F}(u)$$

on $H_0^1(\Omega)$ is of class C^1. On the other hand, the indicator function of the set $\mathcal{K}$, i.e.,

$$\psi_\mathcal{K}(u) := \begin{cases} 0, & \text{if } u \in \mathcal{K}, \\ +\infty, & \text{if } u \notin \mathcal{K}, \end{cases}$$

is convex, proper, and lower semicontinuous. In conclusion, $\mathcal{J}_\lambda = \mathcal{H}_\lambda + \psi_\mathcal{K}$ is a Szulkin type functional.

Proposition 7.4 *Fix $\lambda > 0$ arbitrary. Every critical point $u \in H_0^1(\Omega)$ of $\mathcal{J}_\lambda$ (in the sense of Szulkin) is a solution of (P_λ).*

Proof Since $u \in H_0^1(\Omega)$ is a critical point of $\mathcal{J}_\lambda = \mathcal{H}_\lambda + \psi_\mathcal{K}$, one has

$$\langle \mathcal{H}_\lambda'(u), v - u \rangle_{H_0^1(\Omega)} + \psi_\mathcal{K}(v) - \psi_\mathcal{K}(u) \geq 0, \quad \forall v \in H_0^1(\Omega).$$

We have immediately that u belongs to $\mathcal{K}$. Otherwise, we would have $\psi_\mathcal{K}(u) = +\infty$ which led us to a contradiction, letting for instance $v = 0 \in \mathcal{K}$ in the above inequality. Now, we fix $v \in \mathcal{K}$ arbitrary and we obtain the desired inequality. $\qquad\square$

Remark 7.10 It is easy to see that $0 \in \mathcal{K}$ is a trivial solution of (P_λ) for every $\lambda \in \mathbb{R}$.

Proposition 7.5 *If the conditions $(F_1) - (F_3)$ hold, then $\mathcal{J}_\lambda$ satisfies the (PS)−condition (in the sense of Szulkin) for every $\lambda > 0$.*

Proof Let $\lambda > 0$ and $c \in \mathbb{R}$ be some fixed numbers and let $\{u_n\}$ be a sequence from $H_0^1(\Omega)$ such that

$$\mathcal{J}_\lambda(u_n) = \mathcal{H}_\lambda(u_n) + \psi_{\mathcal{K}}(u_n) \to c; \tag{7.25}$$

$$\langle \mathcal{H}_\lambda'(u_n), v - u_n \rangle_{H_0^1(\Omega)} + \psi_{\mathcal{K}}(v) - \psi_{\mathcal{K}}(u_n) \geq -\varepsilon_n \|v - u_n\|_0, \, \forall v \in H_0^1(\Omega), \tag{7.26}$$

for a sequence $\{\varepsilon_n\}$ in $[0, \infty)$ with $\varepsilon_n \to 0$. By (7.25) one concludes that the sequence $\{u_n\}$ lies entirely in $\mathcal{K}$. Setting $v := 2u_n$ in (7.26), we obtain

$$\langle \mathcal{H}_\lambda'(u_n), u_n \rangle_{H_0^1(\Omega)} \geq -\varepsilon_n \|u_n\|_0.$$

From the above inequality we derive

$$\|u_n\|_0^2 - \lambda \int_\Omega a(x)|u_n|^q \mathrm{d}x - \int_\Omega f(x, u_n(x))u_n(x)\mathrm{d}x \geq -\varepsilon_n \|u_n\|_0. \tag{7.27}$$

By (7.25) one has for large $n \in \mathbb{N}$ that

$$c + 1 \geq \frac{1}{2}\|u_n\|_0^2 - \frac{\lambda}{q} \int_\Omega a(x)|u_n|^q \mathrm{d}x - \int_\Omega F(x, u_n(x))\mathrm{d}x \tag{7.28}$$

Multiplying (7.27) by v^{-1} and adding this one to (7.28), by Hölder's inequality we have for large $n \in \mathbb{N}$

$$c + 1 + \frac{1}{v}\|u_n\|_0 \geq (\frac{1}{2} - \frac{1}{v})\|u_n\|_0^2 - \lambda(\frac{1}{q} - \frac{1}{v}) \int_\Omega a(x)|u_n|^q$$

$$- \frac{1}{v} \int_\Omega [vF(x, u_n(x)) - u_n(x)f(x, u_n(x))]\mathrm{d}x$$

$$\overset{(F_3)}{\geq} (\frac{1}{2} - \frac{1}{v})\|u_n\|_0^2 - \lambda(\frac{1}{q} - \frac{1}{v})\|a\|_{v/(v-q)}\|u_n\|_v^q$$

$$\geq (\frac{1}{2} - \frac{1}{v})\|u_n\|_0^2 - \lambda(\frac{1}{q} - \frac{1}{v})\|a\|_{v/(v-q)}k_v^q\|u_n\|_0^q.$$

In the above inequalities we used the Remark 7.9 and the hypothesis $a \in L^1(\Omega) \cap L^\infty(\Omega)$ thus, in particular, $a \in L^{v/(v-q)}(\Omega)$. Since $q < 2 < v$, from the above estimate we derive that the sequence $\{u_n\}$ is bounded in $\mathcal{K}$. Therefore, $\{u_n\}$ is relatively compact in $L^p(\Omega)$, $p \in (2, 2^*)$. Up to a subsequence, we can suppose that

$$u_n \to u \text{ weakly in } H_0^1(\Omega); \tag{7.29}$$

$$u_n \to u \text{ strongly in } L^\mu(\Omega), \, \mu \in (2, 2^*). \tag{7.30}$$

Since $\mathcal{K}$ is (weakly) closed then $u \in \mathcal{K}$. Setting $v := u$ in (7.26), we have

$$\langle u_n, u - u_n \rangle_0 + \int_\Omega f(x, u_n(x))(u_n(x) - u(x))dx$$

$$-\lambda \int_\Omega a(x)|u_n|^{q-2}u_n(u - u_n) \geq -\varepsilon_n \|u - u_n\|_0.$$

Therefore, in view of Remark 7.8 i) we derive

$$\|u - u_n\|_0^2 \leq \langle u, u - u_n \rangle_0 + \int_\Omega f(x, u_n(x))(u_n(x) - u(x))dx$$

$$-\lambda \int_\Omega a(x)|u_n|^{q-2}u_n(u - u_n) + \varepsilon_n \|u - u_n\|_0$$

$$\leq \langle u, u - u_n \rangle_0 + \lambda \|a\|_{v/(v-q)} \|u_n\|_v^{q-1} \|u - u_n\|_v + \varepsilon_n \|u - u_n\|_0$$

$$+ \varepsilon \|u_n\|_0 \|u_n - u\|_0 + c(\varepsilon) \|u_n\|_p^{p-1} \|u_n - u\|_p,$$

where $\varepsilon > 0$ is arbitrary small. Taking into account relations (7.29) and (7.30), the facts that $v, p \in (2, 2^*)$, the arbitrariness of $\varepsilon > 0$ and $\varepsilon_n \to 0^+$, one has that $\{u_n\}$ converges strongly to u in $H_0^1(\Omega)$. $\square$

Proposition 7.6 *If the conditions $(F_1) - (F_4)$ are verified, then there exists a $\lambda_0 > 0$ such that for every $\lambda \in (0, \lambda_0)$ the function $\mathcal{J}_\lambda$ satisfies the Mountain Pass Geometry, i.e., the following assertions are true:*

(i) there exist constants $\alpha_\lambda > 0$ and $\rho_\lambda > 0$ such that $\mathcal{J}_\lambda(u) \geq \alpha_\lambda$, for all $\|u\|_0 = \rho_\lambda$;
(ii) there exists $e_\lambda \in H_0^1(\Omega)$ with $\|e_\lambda\|_0 > \rho_\lambda$ and $\mathcal{J}_\lambda(e_\lambda) \leq 0$.

Proof

(i) Due to Remark 7.8 ii), for every $\varepsilon > 0$ there exists $c(\varepsilon) > 0$ such that $\mathcal{F}(u) \leq \varepsilon \|u\|_0^2 + c(\varepsilon)\|u\|_p^p$ for every $u \in H_0^1(\Omega)$. It suffices to restrict our attention to elements u which belong to $\mathcal{K}$; otherwise $\mathcal{J}_\lambda(u)$ will be $+\infty$, i.e., (i) holds trivially. Fix $\varepsilon_0 \in (0, \frac{1}{2})$. One has

$$\mathcal{J}_\lambda(u) \geq \left(\frac{1}{2} - \varepsilon_0\right) \|u\|_0^2 - k_p^p c(\varepsilon_0) \|u\|_0^p - \frac{\lambda k_p^q}{q} \|a\|_{p/(p-q)} \|u\|_0^q \quad (7.31)$$

$$= \left(A - B\|u\|_0^{p-2} - \lambda C \|u\|_0^{q-2}\right) \|u\|_0^2,$$

where $A := (\frac{1}{2} - \varepsilon_0) > 0$, $B := k_p^p c(\varepsilon_0) > 0$ and $C := k_p^q \|a\|_{p/(p-q)}/q > 0$.

For every $\lambda > 0$, let us define a function $g_\lambda : (0, \infty) \to \mathbb{R}$ by

$$g_\lambda(t) = A - Bt^{p-2} - \lambda Ct^{q-2}.$$

Clearly, $g'_\lambda(t_\lambda) = 0$ if and only if $t_\lambda = (\lambda \frac{2-q}{p-2} \frac{C}{B})^{\frac{1}{p-q}}$. Moreover, $g_\lambda(t_\lambda) = A - D\lambda^{\frac{p-2}{p-q}}$, where $D := D(p, q, B, C) > 0$. Choosing $\lambda_0 > 0$ such that $g_{\lambda_0}(t_{\lambda_0}) > 0$, one clearly has for every $\lambda \in (0, \lambda_0)$ that $g_\lambda(t_\lambda) > 0$. Therefore, for every $\lambda \in (0, \lambda_0)$, setting $\rho_\lambda := t_\lambda$ and $\alpha_\lambda := g_\lambda(t_\lambda)t_\lambda^2$, the assertion from (i) holds true.

(ii) By Lemma 7.6 we have $\mathcal{F}(u) \geq c_2 \|u\|_\nu^\nu - c_3 \|u\|_2^2$ for every $u \in H_0^1(\Omega)$. Let us fix $u \in \mathcal{K}$. Then we have

$$\mathcal{J}_\lambda(u) \leq (\frac{1}{2} + c_3 k_2^2) \|u\|_0^2 - c_2 \|u\|_\nu^\nu + \frac{\lambda}{q} \|a\|_{\nu/(\nu-q)} k_\nu^q \|u\|_0^q. \tag{7.32}$$

Fix arbitrary $u_0 \in \mathcal{K} \setminus \{0\}$. Letting $u := su_0$ ($s > 0$) in (7.32), we have that $\mathcal{J}_\lambda(su_0) \to -\infty$ as $s \to +\infty$, since $\nu > 2 > q$. Thus, for every $\lambda \in (0, \lambda_0)$, it is possible to set $s := s_\lambda$ so large that for $e_\lambda := s_\lambda u_0$, we have $\|e_\lambda\|_0 > \rho_\lambda$ and $\mathcal{J}_\lambda(e_\lambda) \leq 0$. □

The main result of this section can be read as follows.

Theorem 7.6 ([19]) *Let $f : \Omega \times \mathbb{R} \to \mathbb{R}$ be a function which satisfies $(F_1) - (F_4)$. Then there exists $\lambda_0 > 0$ such that (P_λ) has at least two nontrivial, distinct solutions $u_\lambda^1, u_\lambda^2 \in \mathcal{K}$ whenever $\lambda \in (0, \lambda_0)$.*

Proof In the first step we prove the existence of the first nontrivial solution of (P_λ). By Proposition 7.5, the functional $\mathcal{J}_\lambda$ satisfies (PS) and clearly $\mathcal{J}_\lambda(0) = 0$ for every $\lambda > 0$. Let us fix $\lambda \in (0, \lambda_0)$, λ_0 being from Proposition 7.6. It follows that there are constants $\alpha_\lambda, \rho_\lambda > 0$ and $e_\lambda \in H_0^1(\Omega)$ such that $\mathcal{J}_\lambda$ fulfills the properties (i) and (ii) from Theorem 5.14. Therefore, the number

$$c_\lambda^1 := \inf_{\gamma \in \Gamma} \sup_{t \in [0,1]} \mathcal{J}_\lambda(\gamma(t)),$$

where $\Gamma := \{\gamma \in C([0, 1], H_0^1(\Omega)) : \gamma(0) = 0, \gamma(1) = e_\lambda\}$, is a critical value of $\mathcal{J}_\lambda$ with $c_\lambda^1 \geq \alpha_\lambda > 0$. It is clear that the critical point $u_\lambda^1 \in H_0^1(\Omega)$ which corresponds to c_λ^1 cannot be trivial since $\mathcal{J}_\lambda(u_\lambda^1) = c_\lambda^1 > 0 = \mathcal{J}_\lambda(0)$. It remains to apply Proposition 7.4 which concludes that u_λ^1 is actually an element of $\mathcal{K}$ and it is a solution of (P_λ).

In the next step we prove the existence of the second solution of the problem (P_λ). For this let us fix $\lambda \in (0, \lambda_0)$ arbitrary, λ_0 being from the first step. By Proposition 7.6, there exists $\rho_\lambda > 0$ such that

$$\inf_{\|u\|_0 = \rho_\lambda} \mathcal{J}_\lambda(u) > 0. \tag{7.33}$$

On the other hand, since $a \geq 0$, $a \not\equiv 0$, there exists $u_0 \in \mathcal{K}$ such that $\int_\Omega a(x)|u_0(x)|^q dx > 0$. Thus, for $t > 0$ small one has

$$\mathcal{J}_\lambda(tu_0) \leq t^2 \left(\frac{1}{2} + c_3 k_2^2\right) \|u_0\|_0^2 - c_2 t^\nu \|u_0\|_\nu^\nu - \frac{\lambda}{q} t^q \int_\Omega a(x)|u_0(x)|^q dx < 0.$$

For $r > 0$, let us denote by $\bar{B}_r := \{u \in H_0^1(\Omega) : \|u\|_0 \leq r\}$ and $S_r := \partial \bar{B}_r$. With these notations, relation (7.33) and the above inequality can be summarized as

$$c_\lambda^2 := \inf_{u \in B_{\rho_\lambda}} \mathcal{J}_\lambda(u) < 0 < \inf_{u \in S_{\rho_\lambda}} \mathcal{J}_\lambda(u). \tag{7.34}$$

We point out that c_λ^2 is finite, due to (7.31). Moreover, we will show that c_λ^2 is another critical point of $\mathcal{J}_\lambda$. To this end, let $n \in \mathbb{N} \setminus \{0\}$ such that

$$\frac{1}{n} < \inf_{u \in S_{\rho_\lambda}} \mathcal{J}_\lambda(u) - \inf_{u \in B_{\rho_\lambda}} \mathcal{J}_\lambda(u). \tag{7.35}$$

By Ekeland's variational principle, applied to the lower semicontinuous functional $\mathcal{J}_{\lambda|B_{\rho_\lambda}}$, which is bounded below (see (7.34)), there is $u_{\lambda,n} \in B_{\rho_\lambda}$ such that

$$\mathcal{J}_\lambda(u_{\lambda,n}) \leq \inf_{u \in B_{\rho_\lambda}} \mathcal{J}_\lambda(u) + \frac{1}{n}; \tag{7.36}$$

$$\mathcal{J}_\lambda(w) \geq \mathcal{J}_\lambda(u_{\lambda,n}) - \frac{1}{n}\|w - u_{\lambda,n}\|_0, \ \forall w \in B_{\rho_\lambda}. \tag{7.37}$$

By (7.35) and (7.36) we have that $\mathcal{J}_\lambda(u_{\lambda,n}) < \inf_{u \in S_{\rho_\lambda}} \mathcal{J}_\lambda(u)$; therefore $\|u_{\lambda,n}\|_0 < \rho_\lambda$.

Fix an element $v \in H_0^1(\Omega)$. It is possible to choose $t > 0$ so small such that $w := u_{\lambda,n} + t(v - u_{\lambda,n}) \in B_{\rho_\lambda}$. Putting this element into (7.37), using the convexity of $\psi_\mathcal{K}$ and dividing by $t > 0$, one concludes

$$\frac{\mathcal{H}_\lambda(u_{\lambda,n} + t(v - u_{\lambda,n})) - \mathcal{H}_\lambda(u_{\lambda,n})}{t} + \psi_\mathcal{K}(v) - \psi_\mathcal{K}(u_{\lambda,n}) \geq -\frac{1}{n}\|v - u_{\lambda,n}\|_0.$$

Letting $t \to 0^+$, we derive

$$\langle \mathcal{H}_\lambda'(u_{\lambda,n}), v - u_{\lambda,n}\rangle_{H_0^1(\Omega)} + \psi_\mathcal{K}(v) - \psi_\mathcal{K}(u_{\lambda,n}) \geq -\frac{1}{n}\|v - u_{\lambda,n}\|_0. \tag{7.38}$$

By (7.34) and (7.36) we obtain that

$$\mathcal{J}_\lambda(u_{\lambda,n}) = \mathcal{H}_\lambda(u_{\lambda,n}) + \psi_{\mathcal{K}}(u_{\lambda,n}) \to c_\lambda^2 \tag{7.39}$$

as $n \to \infty$. Since v was arbitrary fixed in (7.38), the sequence $\{u_{\lambda,n}\}$ fulfills (7.25) and (7.26), respectively. Hence, it is possible to prove in a similar manner as in Proposition 7.5 that $\{u_{\lambda,n}\}$ contains a convergent subsequence; denote it again by $\{u_{\lambda,n}\}$ and its limit point by u_λ^2. It is clear that u_λ^2 belongs to B_{ρ_λ}. By the lower semicontinuity of $\psi_{\mathcal{K}}$ we have $\psi_{\mathcal{K}}(u_\lambda^2) \leq \liminf_{n\to\infty} \psi_{\mathcal{K}}(u_{\lambda,n})$. Combining this inequality with $\lim_{n\to\infty} \langle \mathcal{H}_\lambda'(u_{\lambda,n}), v - u_{\lambda,n}\rangle_{H_0^1(\Omega)} = \langle \mathcal{H}_\lambda'(u_\lambda^2), v - u_\lambda^2\rangle$ and (7.38) we have

$$\langle \mathcal{H}_\lambda'(u_\lambda^2), v - u_\lambda^2\rangle_{H_0^1(\Omega)} + \psi_{\mathcal{K}}(v) - \psi_{\mathcal{K}}(u_\lambda^2) \geq 0, \ \forall v \in H_0^1(\Omega),$$

i.e. u_λ^2 is a critical point of $\mathcal{J}_\lambda$. Moreover,

$$c_\lambda^2 \overset{(7.34)}{=} \inf_{u \in B_{\rho_\lambda}} \mathcal{J}_\lambda(u) \leq \mathcal{J}_\lambda(u_\lambda^2) \leq \liminf_{n\to\infty} \mathcal{J}_\lambda(u_{\lambda,n}) \overset{(7.39)}{=} c_\lambda^2,$$

i.e. $\mathcal{J}_\lambda(u_\lambda^2) = c_\lambda^2$. Since $c_\lambda^2 < 0$, it follows that u_λ^2 is not trivial. We apply again Proposition 7.4, concluding that u_λ^2 is a solution of (P_λ) which differs from u_λ^1. This completes the proof of Theorem 7.6. □

In the next we give a simple example which satisfies the conditions $(F_1) - (F_4)$ from Theorem 7.6.

Example 7.5 $F(x, s) = F(s) := -s^3/3$ if $s \leq 0$, and $F(x, s) = F(s) := s^3 \ln(2 + s)$ if $s \geq 0$. One can choose arbitrary $p \in (2, 2^*)$ and $v \in (2, 3]$ in (F_1) and (F_3), respectively.

7.5 Differential Inclusions in $\mathbb{R}^N$

In this section we are going to study the differential inclusion problem

$$\begin{cases} -\triangle_p u + |u|^{p-2}u \in \alpha(x)\partial_C F(u(x)), \ x \in \mathbb{R}^N, \\ u \in W^{1,p}(\mathbb{R}^N), \end{cases} \tag{DI}$$

where $2 \leq N < p < +\infty$, $\alpha \in L^1(\mathbb{R}^N) \cap L^\infty(\mathbb{R}^N)$ is radially symmetric, and $\partial_C F$ stands for the generalized gradient of a locally Lipschitz function $F : \mathbb{R} \to \mathbb{R}$. This class of inclusions have been first studied in the paper of Kristály [18].

By a solution of (DI) it will be understood an element $u \in W^{1,p}(\mathbb{R}^N)$ for which there corresponds a mapping $\mathbb{R}^N \ni x \mapsto \zeta_x$ with $\zeta_x \in \partial_C F(u(x))$ for almost every $x \in \mathbb{R}^N$

having the property that for every $v \in W^{1,p}(\mathbb{R}^N)$, the function $x \mapsto \alpha(x)\zeta_x v(x)$ belongs to $L^1(\mathbb{R}^N)$ and

$$\int_{\mathbb{R}^N}(|\nabla u|^{p-2}\nabla u \nabla v + |u|^{p-2}uv)dx = \int_{\mathbb{R}^N} \alpha(x)\zeta_x v(x)dx. \tag{7.40}$$

Under suitable oscillatory assumptions on the potential F at zero or at infinity, we show the existence of infinitely many, radially symmetric solutions of (DI).

For $l = 0$ or $l = +\infty$, set

$$F_l := \limsup_{|\rho| \to l} \frac{F(\rho)}{|\rho|^p}. \tag{7.41}$$

Problem (DI) will be studied in the following four cases:

- $0 < F_l < +\infty$, whenever $l = 0$ or $l = +\infty$ and
- $F_l = +\infty$, whenever $l = 0$ or $l = +\infty$.

We assume that:

(H) • $F : \mathbb{R} \to \mathbb{R}$ is locally Lipschitz, $F(0) = 0$, and $F(s) \geq 0$, $\forall s \in \mathbb{R}$;
    • $\alpha \in L^1(\mathbb{R}^N) \cap L^\infty(\mathbb{R}^N)$ is radially symmetric, and $\alpha(x) \geq 0$, $\forall x \in \mathbb{R}^N$.

Let $\mathcal{F} : L^\infty(\mathbb{R}^N) \to \mathbb{R}$ be a function defined by

$$\mathcal{F}(u) = \int_{\mathbb{R}^N} \alpha(x)F(u(x))dx.$$

Since F is continuous and $\alpha \in L^1(\mathbb{R}^N)$, we easily seen that $\mathcal{F}$ is well-defined. Moreover, if we fix a $u \in L^\infty(\mathbb{R}^N)$ arbitrarily, there exists $k_u \in L^1(\mathbb{R}^N)$ such that for every $x \in \mathbb{R}^N$ and $v_i \in \mathbb{R}$ with $|v_i - u(x)| < 1$, ($i \in \{1, 2\}$) one has

$$|\alpha(x)F(v_1) - \alpha(x)F(v_2)| \leq k_u(x)|v_1 - v_2|.$$

Indeed, if we fix some small open intervals I_j ($j \in J$), such that $F|_{I_j}$ is Lipschitz function (with Lipschitz constant $L_j > 0$) and $[-\|u\|_{L^\infty}-1, \|u\|_{L^\infty}+1] \subset \bigcup_{j\in J} I_j$, then we choose $k_u = \alpha \max_{j\in J} L_j$. (Here, without losing the generality, we supposed that card$J < +\infty$.) Thus, we are in the position to apply Theorem 2.7.3 from Clarke [6]; namely, $\mathcal{F}$ is a locally Lipschitz function on $L^\infty(\mathbb{R}^N)$ and for every closed subspace E of $L^\infty(\mathbb{R}^N)$ we have

$$\partial_C(\mathcal{F}|_E)(u) \subseteq \int_{\mathbb{R}^N} \alpha(x)\partial_C F(u(x))dx, \quad \text{for every } u \in E, \tag{7.42}$$

where $\mathcal{F}|_E$ stands for the restriction of $\mathcal{F}$ to E. The interpretation of (7.42) is as follows (see also Clarke [6]): for every $\zeta \in \partial_C(\mathcal{F}|_E)(u)$ there corresponds a mapping $\mathbb{R}^N \ni x \mapsto \zeta_x$ such that $\zeta_x \in \partial_C F(u(x))$ for almost every $x \in \mathbb{R}^N$ having the property that for every $v \in E$ the function $x \mapsto \alpha(x)\zeta_x v(x)$ belongs to $L^1(\mathbb{R}^N)$ and

$$\langle \zeta, v \rangle_E = \int_{\mathbb{R}^N} \alpha(x)\zeta_x v(x)dx.$$

Now, let $\mathcal{E} : W^{1,p}(\mathbb{R}^N) \to \mathbb{R}$ be the energy functional associated to our problem (DI), i.e., for every $u \in W^{1,p}(\mathbb{R}^N)$ set

$$\mathcal{E}(u) = \frac{1}{p}\|u\|_{W^{1,p}}^p - \mathcal{F}(u).$$

It is clear that $\mathcal{E}$ is locally Lipschitz on $W^{1,p}(\mathbb{R}^N)$ and we have

Proposition 7.7 *Any critical point* $u \in W^{1,p}(\mathbb{R}^N)$ *of* $\mathcal{E}$ *is a solution of* (DI).

Proof Combining $0 \in \partial_C\mathcal{E}(u) = -\Delta_p u + |u|^{p-2}u - \partial_C(\mathcal{F}|_{W^{1,p}(\mathbb{R}^N)})(u)$ with the interpretation of (7.42), the desired requirement yields, see (7.40). $\square$

We denote by

$$W_{\text{rad}}^{1,p}(\mathbb{R}^N) = \{u \in W^{1,p}(\mathbb{R}^N) : gu = u \text{ for all } g \in O(N)\},$$

the subspace of radially symmetric functions of $W^{1,p}(\mathbb{R}^N)$.

Proposition 7.8 ([18]) *The embedding* $W_{\text{rad}}^{1,p}(\mathbb{R}^N) \hookrightarrow L^\infty(\mathbb{R}^N)$ *is compact whenever* $2 \le N < p < \infty$.

Proof Let $\{u_n\}$ be a bounded sequence in $W_{\text{rad}}^{1,p}(\mathbb{R}^N)$. Up to a subsequence, $u_n \rightharpoonup u$ in $W^{1,p}(\mathbb{R}^N)$ for some $u \in W_{\text{rad}}^{1,p}(\mathbb{R}^N)$. Let $\rho > 0$ be an arbitrarily fixed number. Due to the radially symmetric properties of u and u_n, we have

$$\|u_n - u\|_{W^{1,p}(B_N(g_1 y, \rho))} = \|u_n - u\|_{W^{1,p}(B_N(g_2 y, \rho))} \tag{7.43}$$

for every $g_1, g_2 \in O(N)$ and $y \in \mathbb{R}^N$. For a fixed $y \in \mathbb{R}^N$, we can define

$$m(y, \rho) = \sup\{n \in \mathbb{N} : \exists g_i \in O(N), \ i \in \{1, \ldots, n\} \text{ such that}$$

$$B_N(g_i y, \rho) \cap B_N(g_j y, \rho) = \emptyset, \ \forall i \ne j\}.$$

By virtue of (7.43), for every $y \in \mathbb{R}^N$ and $n \in \mathbb{N}$, we have

$$\|u_n - u\|_{W^{1,p}(B_N(y,\rho))} \leq \frac{\|u_n - u\|_{W^{1,p}}}{m(y,\rho)} \leq \frac{\sup_{n \in \mathbb{N}} \|u_n\|_{W_{1,p}} + \|u\|_{W^{1,p}}}{m(y,\rho)}.$$

The right hand side does not depend on n, and $m(y,\rho) \to +\infty$ whenever $|y| \to +\infty$ (ρ is kept fixed, and $N \geq 2$). Thus, for every $\varepsilon > 0$ there exists $R_\varepsilon > 0$ such that for every $y \in \mathbb{R}^N$ with $|y| \geq R_\varepsilon$ one has

$$\|u_n - u\|_{W^{1,p}(B_N(y,\rho))} < \varepsilon(2S_\rho)^{-1} \quad \text{for every } n \in \mathbb{N}, \tag{7.44}$$

where $S_\rho > 0$ is the embedding constant of $W^{1,p}(B_N(0,\rho)) \hookrightarrow C^0(B_N[0,\rho])$. Moreover, we observe that the embedding constant for $W^{1,p}(B_N(y,\rho)) \hookrightarrow C^0(B_N[y,\rho])$ can be chosen S_ρ as well, *independent* of the position of the point $y \in \mathbb{R}^N$. This fact can be concluded either by a simple translation of the functions $u \in W^{1,p}(B_N(y,\rho))$ into $B_N(0,\rho)$, i.e. $\tilde{u}(\cdot) = u(\cdot - y) \in W^{1,p}(B_N(0,\rho))$ (thus $\|u\|_{W^{1,p}(B_N(y,\rho))} = \|\tilde{u}\|_{W^{1,p}(B_N(0,\rho))}$ and $\|u\|_{C^0(B_N[y,\rho])} = \|\tilde{u}\|_{C^0(B_N[0,\rho])}$); or, by the invariance with respect to rigid motions of the cone property of the balls $B_N(y,\rho)$ when ρ is kept fixed. Thus, in view of (7.44), one has that

$$\sup_{|y| \geq R_\varepsilon} \|u_n - u\|_{C^0(B_N[y,\rho])} \leq \varepsilon/2 \quad \text{for every } n \in \mathbb{N}. \tag{7.45}$$

On the other hand, since $u_n \rightharpoonup u$ in $W_{\mathrm{rad}}^{1,p}(\mathbb{R}^N)$, then in particular, by Rellich theorem it follows that $u_n \to u$ in $C^0(B_N[0, R_\varepsilon])$, i.e., there exists $n_\varepsilon \in \mathbb{N}$ such that

$$\|u_n - u\|_{C^0(B_N[0,R_\varepsilon])} < \varepsilon \quad \text{for every } n \geq n_\varepsilon. \tag{7.46}$$

Combining (7.45) with (7.46), one concludes that $\|u_n - u\|_{L^\infty} < \varepsilon$ for every $n \geq n_\varepsilon$, i.e., $u_n \to u$ in $L^\infty(\mathbb{R}^N)$. This ends the proof.   □

Remark 7.11 We can give an alternate proof of Proposition 7.8 as follows. Lions [Lemme II.1] [20] provided a Strauss-type estimation for radially symmetric functions of $W^{1,p}(\mathbb{R}^N)$; namely, for every $u \in W_{\mathrm{rad}}^{1,p}(\mathbb{R}^N)$ we have

$$|u(x)| \leq p^{1/p}(\mathrm{Area}S^{N-1})^{-1/p}\|u\|_{W^{1,p}}|x|^{(1-N)/p}, \quad x \neq 0, \tag{7.47}$$

where S^{N-1} is the N-dimensional unit sphere. Now, let $\{u_n\}$ be a sequence in $W_{\mathrm{rad}}^{1,p}(\mathbb{R}^N)$ which converges weakly to some $u \in W_{\mathrm{rad}}^{1,p}(\mathbb{R}^N)$. By applying inequality (7.47) for $u_n - u$, and taking into account that $\|u_n - u\|_{W^{1,p}}$ is bounded and $N \geq 2$, for every $\varepsilon > 0$ there exists $R_\varepsilon > 0$ such that

$$\|u_n - u\|_{L^\infty(|x|\geq R_\varepsilon)} \leq C|R_\varepsilon|^{(1-N)/p} < \varepsilon, \quad \forall n \in \mathbb{N},$$

where $C > 0$ does not depend on n. The rest is similar as above.

Since α is radially symmetric, then $\mathcal{E}$ is $O(N)$-invariant, i.e. $\mathcal{E}(gu) = \mathcal{E}(u)$ for every $g \in O(N)$ and $u \in W^{1,p}(\mathbb{R}^N)$, we are in the position to apply the Principle of Symmetric Criticality for locally Lipschitz functions, see Theorem 3.3. Therefore, we have

Proposition 7.9 *Any critical point of $\mathcal{E}_r = \mathcal{E}|_{W^{1,p}_{rad}(\mathbb{R}^N)}$ will be also a critical point of $\mathcal{E}$.*

Remark 7.12 In view of Propositions 7.7 and 7.9, it is enough to find critical points of $\mathcal{E}_r$ in order to guarantee solutions for (DI). This fact will be carried out by means of Theorem 5.17, by setting

$$X := W^{1,p}_{rad}(\mathbb{R}^N), \ \ \tilde{X} := L^\infty(\mathbb{R}^N), \ \ \Phi := -\mathcal{F}, \ \text{ and } \Psi := \| \cdot \|_r^p, \tag{7.48}$$

where the notation $\| \cdot \|_r$ stands for the restriction of $\| \cdot \|_{W^{1,p}}$ into $W^{1,p}_{rad}(\mathbb{R}^N)$. A few assumptions are already verified. Indeed, the embedding $X \hookrightarrow \tilde{X}$ is compact (cf. Theorem 7.8), $\Phi = -\mathcal{F}$ is locally Lipschitz, while $\Psi = \| \cdot \|_r^p$ is of class C^1 (thus, locally Lipschitz as well), coercive and weakly sequentially lower semicontinuous (see Brezis [4]). Moreover, $\mathcal{E}_r \equiv \Phi|_{W^{1,p}_{rad}(\mathbb{R}^N)} + \frac{1}{p}\Psi$. According to (7.48), the function φ (defined in (5.51)) becomes

$$\varphi(\rho) = \inf_{\|u\|_r^p < \rho} \frac{\sup_{\|v\|_r^p \leq \rho} \mathcal{F}(v) - \mathcal{F}(u)}{\rho - \|u\|_r^p}, \quad \rho > 0. \tag{7.49}$$

The investigation of the numbers γ and δ (defined in (5.52)), as well as the cases (A) and (B) from Theorem 5.17 constitute our objective. The first result reads as follows.

Theorem 7.7 (([18], $0 < F_l < +\infty$)) *Let $l = 0$ or $l = +\infty$, and let $2 \leq N < p < +\infty$. Let $F : \mathbb{R} \to \mathbb{R}$ and $\alpha : \mathbb{R}^N \to \mathbb{R}$ be two functions which satisfy the hypotheses (H) and $0 < F_l < +\infty$. Assume that $\|\alpha\|_{L^\infty} F_l > 2^N p^{-1}$ and there exists a number $\beta_l \in]2^N(pF_l)^{-1}, \|\alpha\|_{L^\infty}[$ such that*

$$\frac{2}{(2^{-N} p\beta_l F_l - 1)^{1/p}} < \sup\{r : \text{meas}(B_N(0, r) \setminus \alpha^{-1}(]\beta_l, +\infty[)) = 0\}. \tag{7.50}$$

Assume further that there are sequences $\{a_k\}$ and $\{b_k\}$ in $]0, +\infty[$ with $a_k < b_k$, $\lim_{k\to+\infty} b_k = l$, $\lim_{k\to+\infty} \frac{b_k}{a_k} = +\infty$ such that

$$\sup\{\operatorname{sign}(s)\xi : \xi \in \partial_C F(s), |s| \in]a_k, b_k[\} \leq 0. \tag{7.51}$$

Then problem (DI) possesses a sequence $\{u_n\}$ of solutions which are radially symmetric and

$$\lim_{n\to+\infty} \|u_n\|_{W^{1,p}} = l.$$

In addition, if $F(s) = 0$ for every $s \in]-\infty, 0[$, then the elements u_n are non-negative.

Proof Since $\lim_{k\to+\infty} b_k = +\infty$, instead of the sequence $\{b_k\}$, we may consider a non-decreasing subsequence of it, denoted again by $\{b_k\}$. Fix an $s \in \mathbb{R}$ such that $|s| \in]a_k, b_k]$. By using Lebourg's mean value theorem (see Theorem 2.1), there exists $\theta \in]0, 1[$ and $\xi_\theta \in \partial_C F(\theta s + (1-\theta)\operatorname{sign}(s)a_k)$ such that

$$F(s) - F(\operatorname{sign}(s)a_k) = \xi_\theta(s - \operatorname{sign}(s)a_k) = \operatorname{sign}(s)\xi_\theta(|s| - a_k)$$

$$= \operatorname{sign}(\theta s + (1-\theta)\operatorname{sign}(s)a_k)\xi_\theta(|s| - a_k).$$

Due to (7.51), we obtain that $F(s) \leq F(\operatorname{sign}(s)a_k)$ for every $s \in \mathbb{R}$ complying with $|s| \in]a_k, b_k]$. In particular, we are led to $\max_{[-a_k,a_k]} F = \max_{[-b_k,b_k]} F$ for every $k \in \mathbb{N}$. Therefore, one can fix a $\overline{\rho}_k \in [-a_k, a_k]$ such that

$$F(\overline{\rho}_k) = \max_{[-a_k,a_k]} F = \max_{[-b_k,b_k]} F. \tag{7.52}$$

Moreover, since $\{b_k\}$ is non-decreasing, the sequence $\{|\overline{\rho}_k|\}$ can be chosen non-decreasingly as well. In view of (7.50) we can choose a number μ such that

$$\frac{2}{(2^{-N}p\beta_\infty F_\infty - 1)^{1/p}} < \mu < \tag{7.53}$$

$$< \sup\{r : \operatorname{meas}(B_N(0, r) \setminus \alpha^{-1}(]\beta_\infty, +\infty[)) = 0\}.$$

In particular, one has

$$\alpha(x) > \beta_\infty, \quad \text{for a.e. } x \in B_N(0, \mu). \tag{7.54}$$

For every $k \in \mathbb{N}$ we define

$$u_k(x) = \begin{cases} 0, & \text{if } x \in \mathbb{R}^N \setminus B_N(0, \mu); \\ \overline{\rho}_k, & \text{if } x \in B_N(0, \frac{\mu}{2}); \\ \frac{2\overline{\rho}_k}{\mu}(\mu - |x|), & \text{if } x \in B_N(0, \mu) \setminus B_N(0, \frac{\mu}{2}). \end{cases} \tag{7.55}$$

It is easy to see that u_k belongs to $W^{1,p}(\mathbb{R}^N)$ and it is radially symmetric. Thus, $u_k \in W^{1,p}_{\text{rad}}(\mathbb{R}^N)$. Let $\rho_k = (\frac{b_k}{c_\infty})^p$, where c_∞ is the embedding constant of $W^{1,p}(\mathbb{R}^N) \hookrightarrow L^\infty(\mathbb{R}^N)$. $\qquad \square$

CLAIM 1   There exists a $k_0 \in \mathbb{N}$ such that $\|u_k\|_r^p < \rho_k$ for every $k > k_0$.
   Since $\lim_{k \to +\infty} \frac{b_k}{a_k} = +\infty$, there exists a $k_0 \in \mathbb{N}$ such that

$$\frac{b_k}{a_k} > c_\infty (\mu^N \omega_N K(p, N, \mu))^{1/p}, \quad \text{for every } k > k_0, \tag{7.56}$$

where ω_N denotes the volume of the N-dimensional unit ball and

$$K(p, N, \mu) := \frac{2^p}{\mu^p}\left(1 - \frac{1}{2^N}\right) + 1. \tag{7.57}$$

Thus, for every $k > k_0$ one has

$$\|u_k\|_r^p = \int_{\mathbb{R}^N} |\nabla u_k|^p dx + \int_{\mathbb{R}^N} |u_k|^p dx$$

$$\leq \left(\frac{2|\overline{\rho}_k|}{\mu}\right)^p (\text{vol} B_N(0, \mu) - \text{vol} B_N(0, \frac{\mu}{2})) + |\overline{\rho}_k|^p \text{vol} B_N(0, \mu)$$

$$= |\overline{\rho}_k|^p \mu^N \omega_N K(p, N, \mu) \leq a_k^p \mu^N \omega_N K(p, N, \mu)$$

$$< (\frac{b_k}{c_\infty})^p = \rho_k,$$

which proves Claim 1.
   Now, let φ from (7.49) and $\gamma = \liminf_{\rho \to +\infty} \varphi(\rho)$ defined in (5.52).

CLAIM 2 $\gamma = 0$. By definition, $\gamma \geq 0$. Suppose that $\gamma > 0$. Since $\lim_{k \to +\infty} \frac{\rho_k}{|\overline{\rho}_k|^p} = +\infty$, there is a number $k_1 \in \mathbb{N}$ such that for every $k > k_1$ we have

$$\frac{\rho_k}{|\overline{\rho}_k|^p} > \frac{2}{\gamma}(F_\infty + 1)(\|\alpha\|_{L^1} - \beta_\infty \overline{\mu}^N \omega_N) + \mu^N \omega_N K(p, N, \mu), \tag{7.58}$$

where $\overline{\mu}$ is an arbitrary fixed number complying with

$$0 < \overline{\mu} < \min\left\{\left(\frac{\|\alpha\|_{L^1}}{\beta_\infty \omega_N}\right)^{1/N}, \frac{\mu}{2}\right\}. \tag{7.59}$$

Moreover, since $|\overline{\rho}_k| \to +\infty$ as $k \to +\infty$ (otherwise we would have $F_\infty = 0$), by the definition of F_∞, see (7.41), there exists a $k_2 \in \mathbb{N}$ such that

$$\frac{F(\overline{\rho}_k)}{|\overline{\rho}_k|^p} < F_\infty + 1, \quad \text{for every } k > k_2. \tag{7.60}$$

Now, let $v \in W_{\text{rad}}^{1,p}(\mathbb{R}^N)$ arbitrarily fixed with $\|v\|_r^p \leq \rho_k$. Due to the continuous embedding $W^{1,p}(\mathbb{R}^N) \hookrightarrow L^\infty(\mathbb{R}^N)$, we have $\|v\|_{L^\infty}^p \leq c_\infty^p \rho_k = b_k^p$. Therefore, one has

$$\sup_{x \in \mathbb{R}^N} |v(x)| \leq b_k.$$

In view of (7.52), we obtain

$$F(v(x)) \leq \max_{[-b_k, b_k]} F = F(\overline{\rho}_k), \quad \text{for every } x \in \mathbb{R}^N. \tag{7.61}$$

Hence, for every $k > \max\{k_0, k_1, k_2\}$, one has

$$\sup_{\|v\|_r^p \leq \rho_k} \mathcal{F}(v) - \mathcal{F}(u_k) = \sup_{\|v\|_r^p \leq \rho_k} \int_{\mathbb{R}^N} \alpha(x) F(v(x)) dx - \int_{\mathbb{R}^N} \alpha(x) F(u_k(x)) dx$$

$$\leq F(\overline{\rho}_k) \|\alpha\|_{L^1} - \int_{B_N(0,\overline{\mu})} \alpha(x) F(u_k(x)) dx$$

$$\leq F(\overline{\rho}_k)(\|\alpha\|_{L^1} - \beta_\infty \overline{\mu}^N \omega_N)$$

$$\leq (F_\infty + 1)|\overline{\rho}_k|^p(\|\alpha\|_{L^1} - \beta_\infty \overline{\mu}^N \omega_N)$$

$$\leq \frac{\gamma}{2}(\rho_k - |\overline{\rho}_k|^p \mu^N \omega_N K(p, N, \mu))$$

$$\leq \frac{\gamma}{2}(\rho_k - \|u_k\|_r^p).$$

Since $\|u_k\|_r^p < \rho_k$ (cf. Claim 1), and $\rho_k \to +\infty$ as $k \to +\infty$, we obtain

$$\gamma = \liminf_{\rho \to +\infty} \varphi(\rho) \leq \liminf_{k \to +\infty} \varphi(\rho_k) \leq \liminf_{k \to +\infty} \frac{\sup_{\|v\|_r^p \leq \rho_k} \mathcal{F}(v) - \mathcal{F}(u_k)}{\rho_k - \|u_k\|_r^p} \leq \frac{\gamma}{2},$$

contradiction. This proves Claim 2.

CLAIM 3 $\mathcal{E}_r$ is not bounded from below on $W_{\text{rad}}^{1,p}(\mathbb{R}^N)$.

By (7.53), we find a number ε_∞ such that

$$0 < \varepsilon_\infty < F_\infty - \frac{2^N}{p\beta_\infty}\left(\left(\frac{2}{\mu}\right)^p + 1\right). \tag{7.62}$$

In particular, for every $k \in \mathbb{N}$, $\sup_{|\rho|\geq k} \frac{F(\rho)}{|\rho|^p} > F_\infty - \varepsilon_\infty$. Therefore, we can fix $\tilde{\rho}_k$ with $|\tilde{\rho}_k| \geq k$ such that

$$\frac{F(\tilde{\rho}_k)}{|\tilde{\rho}_k|^p} > F_\infty - \varepsilon_\infty. \tag{7.63}$$

Now, define $w_k \in W^{1,p}_{\mathrm{rad}}(\mathbb{R}^N)$ in the same way as u_k, see (7.55), replacing $\overline{\rho}_k$ by $\tilde{\rho}_k$. We obtain

$$\mathcal{E}_r(w_k) = \frac{1}{p}\|w_k\|^p_r - \mathcal{F}(w_k)$$

$$\leq \frac{1}{p}|\tilde{\rho}_k|^p \mu^N \omega_N K(p, N, \mu) - \int_{B_N(0,\frac{\mu}{2})} \alpha(x) F(w_k(x))dx$$

$$\leq \frac{1}{p}|\tilde{\rho}_k|^p \mu^N \omega_N K(p, N, \mu) - (F_\infty - \varepsilon_\infty)|\tilde{\rho}_k|^p \beta_\infty \omega_N \left(\frac{\mu}{2}\right)^N$$

$$= |\tilde{\rho}_k|^p \mu^N \omega_N \left(\frac{1}{p}K(p, N, \mu) - \frac{1}{2^N}(F_\infty - \varepsilon_\infty)\beta_\infty\right)$$

$$< -\frac{1}{p}|\tilde{\rho}_k|^p \omega_N \left(\frac{2}{\mu}\right)^{p-N}.$$

Since $|\tilde{\rho}_k| \to +\infty$ as $k \to +\infty$, we obtain $\lim_{k\to+\infty} \mathcal{E}_r(w_k) = -\infty$, which ends the proof of Claim 3.

The Case $0 < F_\infty < +\infty$ It is enough to apply Remark 7.12. Indeed, since $\gamma = 0$ (cf. Claim 2) and the function $\mathcal{E}_r \equiv -\mathcal{F}|_{W^{1,p}_{\mathrm{rad}}(\mathbb{R}^N)} + \frac{1}{p}\|\cdot\|^p_r$ is not bounded below (cf. Claim 3), the alternative (A1) from Theorem 5.17, applied to $\lambda = \frac{1}{p}$, is excluded. Thus, there exists a sequence $\{u_n\} \subset W^{1,p}_{\mathrm{rad}}(\mathbb{R}^N)$ of critical points of $\mathcal{E}_r$ with $\lim_{n\to+\infty} \|u_n\|_r = +\infty$.

Now, let us suppose that $F(s) = 0$ for every $s \in]-\infty, 0[$, and let u be a solution of (DI). Denote $S = \{x \in \mathbb{R}^N : u(x) < 0\}$, and assume that $S \neq \emptyset$; it is clear that S is open. Define $u_S : \mathbb{R}^N \to \mathbb{R}$ by $u_S = \min\{u, 0\}$. Applying (7.40) for $v := u_S \in W^{1,p}(\mathbb{R}^N)$ and taking into account that $\zeta_x \in \partial_C F(u(x)) = \{0\}$ for every $x \in S$, one has

$$0 = \int_{\mathbb{R}^N}(|\nabla u|^{p-2}\nabla u \nabla u_S + |u|^{p-2}uu_S)dx = \int_S(|\nabla u|^p + |u|^p)dx = \|u\|^p_{W^{1,p}(S)},$$

which contradicts the choice of the set S. This ends the proof in this case.

Remark 7.13 A closer inspection of the proof allows us to replace hypothesis (7.50) by a weaker, but a more technical condition. More specifically, it is enough to require that $p\|\alpha\|_{L^\infty} F_l > 1$, and instead of (7.50), put

$$\sup_M \left\{ N_{\beta_l} - \frac{1}{(1-\sigma)(p\beta_l F_l \sigma^N - 1)^{1/p}} \right\} > 0, \qquad (7.64)$$

where

$$M = \{(\sigma, \beta_l) : \sigma \in](p\|\alpha\|_{L^\infty} F_l)^{-1/N}, 1[, \ \beta_l \in](p F_l \sigma^N)^{-1}, \|\alpha\|_{L^\infty}[\}$$

and

$$N_{\beta_l} = \sup\{r : \operatorname{meas}(B_N(0, r) \setminus \alpha^{-1}(]\beta_l, +\infty[)) = 0\}.$$

Now, in the construction of the functions w_k we replace the radius $\frac{\mu}{2}$ of the ball by $\sigma\mu$, where σ is chosen according to (7.64).

The Case $0 < F_0 < +\infty$ The proof works similarly as in the case $0 < F_\infty < +\infty$ and we will show only the differences. The sequence $\{\rho_k\}$ defined as above, converges now to 0, while the same holds for $\{\overline{\rho}_k\}$. Instead of Claim 2, we can prove that $\delta = \liminf_{\rho \to 0^+} \varphi(\rho) = 0$. Since 0 is the unique global minimum of $\Psi = \|\cdot\|_r^p$, it would be enough to show that 0 is not a local minimum of $\mathcal{E}_r \equiv -\mathcal{F}|_{W_{\mathrm{rad}}^{1,p}(\mathbb{R}^N)} + \frac{1}{p}\|\cdot\|_r^p$, in order to exclude alternative (B1) from Theorem 5.17. To this end, we fix $\tilde{\rho}_k$ with $|\tilde{\rho}_k| \leq \frac{1}{k}$ such that

$$\frac{F(\tilde{\rho}_k)}{|\tilde{\rho}_k|^p} > F_0 - \varepsilon_0,$$

where ε_0 is fixed in a similar manner as in (7.62), replacing β_∞, F_∞ by β_0, F_0, respectively. If we take w_k as in case $0 < F_\infty < +\infty$, then it is clear that $\{w_k\}$ strongly converges now to 0 in $W_{\mathrm{rad}}^{1,p}(\mathbb{R}^N)$, while $\mathcal{E}_r(w_k) < -\frac{1}{p}|\tilde{\rho}_k|^p \omega_N (2/\mu)^{p-N} < 0 = \mathcal{E}_r(0)$. Thus, 0 is not a local minimum of $\mathcal{E}_r$. So, there exists a sequence $\{u_n\} \subset W_{\mathrm{rad}}^{1,p}(\mathbb{R}^N)$ of critical points of $\mathcal{E}_r$ such that $\lim_{n \to +\infty} \|u_n\|_r = 0 = \inf_{W_{\mathrm{rad}}^{1,p}(\mathbb{R}^N)} \Psi$. This concludes completely the proof of Theorem 7.7.

In the next result we trait the case when the function F has oscillation at infinity. We have the following result.

Theorem 7.8 ([18], $F_l = +\infty$) *Let $l = 0$ or $l = +\infty$, and let $2 \leq N < p < +\infty$. Let $F : \mathbb{R} \to \mathbb{R}$ and $\alpha : \mathbb{R}^N \to \mathbb{R}$ be two functions which satisfy (H) and $F_l = +\infty$. Assume* *that* $\|\alpha\|_{L^\infty} > 0$, *and there exist* $\mu > 0$ *and* $\beta_l \in]0, \|\alpha\|_{L^\infty}[$ *such that*

$$\text{meas}(B_N(0, \mu) \setminus \alpha^{-1}(]\beta_l, +\infty[)) = 0, \tag{7.65}$$

and there are sequences $\{a_k\}$ *and* $\{b_k\}$ *in* $]0, +\infty[$ *with* $a_k < b_k$, $\lim_{k \to +\infty} b_k = l$, $\lim_{k \to +\infty} \frac{b_k}{a_k} = +\infty$ *such that*

$$\sup\{\text{sign}(s)\xi : \xi \in \partial_C F(s), |s| \in]a_k, b_k[\} \leq 0,$$

and

$$\limsup_{k \to +\infty} \frac{\max_{[-a_k, a_k]} F}{b_k^p} < (pc_\infty^p \|\alpha\|_{L^1})^{-1}, \tag{7.66}$$

where c_∞ *is the embedding constant of* $W^{1,p}(\mathbb{R}^N) \hookrightarrow L^\infty(\mathbb{R}^N)$. *Then the conclusions of Theorem 7.7 hold.*

Proof *The case* $F_\infty = +\infty$. *Due to* (7.65),

$$\alpha(x) > \beta_\infty, \quad \text{for a.e. } x \in B_N(0, \mu). \tag{7.67}$$

Let $\overline{\rho}_k$ and ρ_k as in the proof of Theorem 7.7, as well as u_k, defined this time by means of $\mu > 0$ from (7.67).

CLAIM 1' There exists a $k_0 \in \mathbb{N}$ such that $\|u_k\|_r^p < \rho_k$, for every $k > k_0$.
  The proof is similarly as in the proof of Theorem 7.7.

CLAIM 2' $\gamma < \frac{1}{p}$.
  Note that $F(\overline{\rho}_k) = \max_{[-a_k, a_k]} F$, cf. (7.52). Since $|\overline{\rho}_k| \leq a_k$, then $\lim_{k \to +\infty} \frac{|\overline{\rho}_k|}{b_k} = 0$. Combining this fact with (7.66), and choosing $\varepsilon > 0$ sufficiently small, one has

$$\limsup_{k \to +\infty} \frac{F(\overline{\rho}_k) + |\overline{\rho}_k|^p \mu^N \omega_N p^{-1} \|\alpha\|_{L^1}^{-1} K(p, N, \mu)}{b_k^p} < ((p + \varepsilon)c_\infty^p \|\alpha\|_{L^1})^{-1},$$

where $K(p, N, \mu)$ is from (7.57). According to the above inequality, there exists $k_3 \in \mathbb{N}$ such that for every $k > k_3$ we readily have

$$F(\overline{\rho}_k)\|\alpha\|_{L^1} \leq (p + \varepsilon)^{-1} c_\infty^{-p} b_k^p - p^{-1}|\overline{\rho}_k|^p \mu^N \omega_N K(p, N, \mu)$$

$$\leq \frac{1}{p + \varepsilon}\left(\rho_k - \frac{p + \varepsilon}{p}\|u_k\|_r^p\right) < \frac{1}{p + \varepsilon}\left(\rho_k - \|u_k\|_r^p\right).$$

Thus, for every $k > k_3$, one has

$$\sup_{\|v\|_r^p \le \rho_k} \mathcal{F}(v) - \mathcal{F}(u_k) < F(\overline{\rho}_k)\|\alpha\|_{L^1} < \frac{1}{p+\varepsilon}\left(\rho_k - \|u_k\|_r^p\right).$$

Hence $\gamma \le \frac{1}{p+\varepsilon} < \frac{1}{p}$, which concludes the proof of Claim 2'.

CLAIM 3' $\mathcal{E}_r$ is not bounded below on $W_{\mathrm{rad}}^{1,p}(\mathbb{R}^N)$.

Since $F_\infty = +\infty$, for an arbitrarily large number $M > 0$, we can fix $\tilde{\rho}_k$ with $|\tilde{\rho}_k| \ge k$ such that

$$\frac{F(\tilde{\rho}_k)}{|\tilde{\rho}_k|^p} > M. \tag{7.68}$$

Define $w_k \in W_{\mathrm{rad}}^{1,p}(\mathbb{R}^N)$ as in (7.55), putting $\tilde{\rho}_k$ instead of $\overline{\rho}_k$. We obtain

$$\mathcal{E}_r(w_k) = \frac{1}{p}\|w_k\|_r^p - \mathcal{F}(w_k)$$

$$\le \frac{1}{p}\mu^N \omega_N |\tilde{\rho}_k|^p K(p, N, \mu) - \int_{B_N(0, \frac{\mu}{2})} \alpha(x) F(w_k(x))dx$$

$$\le |\tilde{\rho}_k|^p \mu^N \omega_N \left(\frac{1}{p}K(p, N, \mu) - \frac{1}{2^N}M\beta_\infty\right).$$

Since $|\tilde{\rho}_k| \to +\infty$ as $k \to +\infty$, and M is large enough we obtain that $\lim_{k \to +\infty} \mathcal{E}_r(w_k) = -\infty$. The proof of Claim 3' is concluded.

Proof Concluded Since $\gamma < \frac{1}{p}$ (cf. Claim 2'), we can apply Theorem 5.17 (A) for $\lambda = \frac{1}{p}$. The rest is the same as in Theorem 7.7.   □

The Case $F_0 = +\infty$ We follow the line of the proof for $F_\infty = +\infty$. The sequences $\{\rho_k\}$, $\{\overline{\rho}_k\}$ are defined as above; they converge to 0. Let $\mu > 0$ be as in (7.67), replacing β_∞ by β_0. Instead of Claim 2', we may prove that $\delta = \liminf_{\rho \to 0^+} \varphi(\rho) < \frac{1}{p}$. Now, we are in the position to apply Theorem 5.17 (B) with $\lambda = \frac{1}{p}$. Since $F_0 = +\infty$, for an arbitrarily large number $M > 0$, we may choose $\tilde{\rho}_k$ with $|\tilde{\rho}_k| \le \frac{1}{k}$ such that

$$\frac{F(\tilde{\rho}_k)}{|\tilde{\rho}_k|^p} > M.$$

Define $w_k \in W_{\mathrm{rad}}^{1,p}(\mathbb{R}^N)$ by means of $\tilde{\rho}_k$ as above. It is clear that $\{w_k\}$ strongly converges to 0 in $W_{\mathrm{rad}}^{1,p}(\mathbb{R}^N)$ while

$$\mathcal{E}_r(w_k) \le |\tilde{\rho}_k|^p \mu^N \omega_N \left(\frac{1}{p} K(p, N, \mu) - \frac{1}{2^N} M\beta_0 \right) < 0 = \mathcal{E}_r(0).$$

Consequently, in spite of the fact that 0 is the unique global minimum of $\Psi = \| \cdot \|_r^p$, it is not a local minimum of $\mathcal{E}_r$; thus, (B1) can be excluded. The rest is the same as in the proof of Theorem 7.7. This completes the proof of Theorem 7.8.

In the sequel we give some examples where the results apply; we suppose that $2 \le N < p < +\infty$.

Example 7.6 Let $F : \mathbb{R} \to \mathbb{R}$ be defined by

$$F(s) = \frac{2^{N+p+3}}{p} |s|^p \max\{0, \sin \ln(\ln(|s| + 1) + 1)\},$$

and $\alpha : \mathbb{R}^N \to \mathbb{R}$ by

$$\alpha(x) = \frac{1}{(1 + |x|^N)^2}. \tag{7.69}$$

Then (DI) has an unbounded sequence of radially symmetric solutions.

Proof The functions F and α clearly fulfill (H). Moreover, $F_\infty = \frac{2^{N+p+3}}{p}$. Since $\|\alpha\|_{L^\infty} = 1$, we may fix $\beta_\infty = 1/4$ which verifies (7.50). For every $k \in \mathbb{N}$ let

$$a_k = e^{e^{(2k-1)\pi} - 1} - 1 \quad \text{and} \quad b_k = e^{e^{2k\pi} - 1} - 1.$$

If $a_k \le |s| \le b_k$, then $(2k - 1)\pi \le \ln(\ln(|s| + 1) + 1) \le 2k\pi$, thus $F(s) = 0$ for every $s \in \mathbb{R}$ complying with $a_k \le |s| \le b_k$. So, $\partial_C F(s) = \{0\}$ for every $|s| \in]a_k, b_k[$ and (7.51) is verified. Thus, all the assumptions of Theorem 7.7 are satisfied. $\square$

Example 7.7 Fix $\sigma \in \mathbb{R}$. Let $F : \mathbb{R} \to \mathbb{R}$ be defined by

$$F(s) = \begin{cases} \frac{8^{N+1}}{p} s^{p-\sigma} \max\{0, \sin \ln \ln \frac{1}{s}\}, & s \in]0, e^{-1}[; \\ 0, & s \notin]0, e^{-1}[, \end{cases}$$

and let $\alpha : \mathbb{R}^N \to \mathbb{R}$ be as in (7.69). Then, for every $\sigma \in [0, \min\{p - 1, p(1 - e^{-\pi})\}[$, (DI) admits a sequence of non-negative, radially symmetric solutions which strongly converges to 0 in $W^{1,p}(\mathbb{R}^N)$.

Proof Since $\sigma < p - 1$, (H) is verified. We distinguish two cases: $\sigma = 0$, and $\sigma \in$ $]0, \min\{p - 1, p(1 - e^{-\pi})\}[$.

Case 1. $\sigma = 0$ We have $F_0 = \frac{8^{N+1}}{p}$. If we choose $\beta_0 = (1 + 2^N)^{-2}$, this clearly verifies (7.50). For every $k \in \mathbb{N}$ set

$$a_k = e^{-e^{2k\pi}} \quad \text{and} \quad b_k = e^{-e^{(2k-1)\pi}}. \tag{7.70}$$

For every $s \in [a_k, b_k]$, one has $(2k - 1)\pi \le \ln\ln\frac{1}{s} \le 2k\pi$; thus $F(s) = 0$. So, $\partial_C F(s) = \{0\}$ for every $s \in]a_k, b_k[$ and (7.51) is verified. Now, we apply Theorem 7.7.

Case 2. $\sigma \in]0, \min\{p - 1, p(1 - e^{-\pi})\}[$ We have $F_0 = +\infty$. In order to verify (7.65), we fix for instance $\beta_0 = (1 + 2^N)^{-2}$ and $\mu = 2$. Take $\{a_k\}$ and $\{b_k\}$ in the same way as in (7.70). The inequality in (7.66) becomes obvious since

$$\limsup_{k \to +\infty} \frac{\max_{[-a_k, a_k]} F}{b_k^p} \le \frac{8^{N+1}}{p} \limsup_{k \to +\infty} \frac{a_k^{p-\sigma}}{b_k^p} =$$

$$= \frac{8^{N+1}}{p} \lim_{k \to +\infty} e^{[p - e^\pi (p-\sigma)]e^{(2k-1)\pi}} = 0.$$

Therefore, we may apply Theorem 7.8. $\qquad\qquad\square$

Example 7.8 Let $\{a_k\}$ and $\{b_k\}$ be two sequences such that $a_1 = 1$, $b_1 = 2$ and $a_k = k^k$, $b_k = k^{k+1}$ for every $k \ge 2$. Define, for every $s \in \mathbb{R}$ the function

$$f(s) = \begin{cases} \frac{b_{k+1}^p - b_k^p}{a_{k+1} - b_k}, & \text{if } s \in [b_k, a_{k+1}[; \\ 0, & \text{otherwise.} \end{cases}$$

Then the problem

$$\begin{cases} -\Delta_p u + |u|^{p-2} u \in \frac{\sigma}{(1+|x|^N)^2} [\underline{f}(u(x)), \overline{f}(u(x))], & x \in \mathbb{R}^N, \\ u \in W^{1,p}(\mathbb{R}^N), \end{cases}$$

possesses an unbounded sequence of non-negative, radially symmetric solutions whenever $0 < \sigma < \frac{N}{p} \left(\frac{p-N}{2p}\right)^p (\text{Area} S^{N-1})^{-1}$.

Proof Let $F(s) = \int_0^s f(t)dt$. Since the function f is locally (essentially) bounded, F is locally Lipschitz. A more explicit expression of F is

$$F(s) = \begin{cases} b_k^p - b_1^p + \frac{b_{k+1}^p - b_k^p}{a_{k+1} - b_k}(s - b_k), & \text{if } s \in [b_k, a_{k+1}[; \\ b_k^p - b_1^p, & \text{if } s \in [a_k, b_k[; \\ 0, & \text{otherwise.} \end{cases}$$

An easy calculation shows, as we expect, that $\partial_C F(s) = [\underline{f}(s), \overline{f}(s)]$ for every $s \in \mathbb{R}$. Taking $\alpha(x) = \frac{\sigma}{(1+|x|^N)^2}$, (H) is verified, and $\|\alpha\|_{L^1} = \frac{\sigma}{N}\mathrm{Area}S^{N-1}$. Moreover,

$$F_\infty = \limsup_{|s| \to +\infty} \frac{F(s)}{|s|^p} \geq \lim_{k \to +\infty} \frac{F(a_k)}{a_k^p} = \lim_{k \to +\infty} \frac{b_k^p - b_1^p}{a_k^p} = +\infty.$$

Choosing $\mu = 1$ and $\beta_\infty = \sigma/4$, (7.65) is verified, while (7.51) becomes trivial. Since $\max_{[-a_k, a_k]} F = F(a_k) = b_k^p - b_1^p$, relation (7.66) reduces to $pc_\infty^p\|\alpha\|_{L^1} < 1$. It remains to apply Theorem 7.8. □

References

1. C.J. Amick, Semilinear elliptic eigenvalue problems on an infinite strip with an application to stratified fluids. Ann. Sc. Norm. Super. Pisa Cl. Sci. **11**, 441–499 (1984)
2. T. Bartsch, Z.-Q. Wang, Existence and multiplicity results for some superlinear elliptic problems in $\mathbb{R}^N$. Commun. Partial Differ. Equ. **20**, 1725–1741 (1995)
3. T. Bartsch, M. Willem, Infinitely many non-radial solutions of an Euclidian scalar field equation. J. Funct. Anal. **117**, 447–460 (1993)
4. H. Brezis, *Functional Analysis, Sobolev Spaces and Partial Differential Equations* (Springer, Berlin, 2011)
5. G.R. Burton, Semilinear elliptic equations on unbounded domains. Math. Z. **190**, 519–525 (1985)
6. F.H. Clarke, in *Optimization and Nonsmooth Analysis*. Classics in Applied Mathematics (Society for Industrial and Applied Mathematics, 1990)
7. D.G. Costa, C.A. Magalhaes, A unified approach to a class of strongly indefinite functionals. J. Differ. Equ. **125**, 521–547 (1996)
8. Z. Dályai, C. Varga, An existence result for hemivariational inequalities. Electron. J. Differ. Equ. **37**, 1–17 (2004)
9. N.V. Efimov, S.B. Stechkin, Approximate compactness and Chebyshev sets. Sov. Math. Dokl. **2**, 1226–1228 (1961)
10. M.J. Esteban, Nonlinear elliptic problems in strip-like domains: symmetry of positive vortex rings. Nonlinear Anal. **7**, 365–379 (1983)
11. M.J. Esteban, P.-L. Lions, A compactness lemma. Nonlinear Anal. **7**, 381–385 (1983)
12. X.L. Fan, Y.Z. Zhao, Linking and multiplicity results for the p-Laplacian on unbounded cylinder. J. Math. Anal. Appl. **260**, 479–489 (2001)
13. F. Faraci, A. Iannizzotto, H. Lisei, C. Varga, A multiplicity result for hemivariational inequalities. J. Math. Anal. Appl. **330**, 683–698 (2007)

14. F. Gazolla, V. Rădulescu, A nonsmooth critical point approach to some nonlinear elliptic equations in $\mathbb{R}^N$. Differ. Integr. Equ. **13**, 47–60 (2000)
15. H. Hofer, Variational and topological methods in partially ordered Hilbert spaces. Math. Ann. **261**, 493–514 (1982)
16. O. Kavian, *Introduction à la Théorie des Point Critique et Applications aux Proble'emes Elliptique* (Springer, Berlin, 1995)
17. A. Kristály, Multiplicity results for an eigenvalue problem for Hemivariational inequalities in strip-like domains. Set-Valued Anal. **13**, 85–103 (2005)
18. A. Kristály, Infinitely many solutions for a differential inclusion problem in $\mathbb{R}^n$. J. Differ. Equ. **220**, 511–530 (2006)
19. A. Kristály, C. Varga, V. Varga, An eigenvalue problem for hemivariational inequalities with combined nonlinearities on an infinite strip. Nonlinear Anal. **63**, 260–272 (2005)
20. P.-L. Lions, Symétrie et compacité dans les espaces de Sobolev. J. Funct. Anal. **49**, 312–334 (1982)
21. D. Motreanu, P.D. Panagiotopoulos, in *Minimax Theorems and Qualitative Properties of the Solutions of Hemivariational Inequalities and Applications*, vol. 29 of Nonconvex Optimization and its Applications (Kluwer Academic Publishers, Boston, Dordrecht, London, 1999)
22. J.T. Schwartz, Generalizing the Lusternik-Schnirelmann theory of critical points. Commun. Pure Appl. Math. **17**, 307–315 (1964)
23. I.G. Tsar'kov, Nonunique solvability of certain differential equations and their connection with geometric approximation theory. Math. Notes **75**, 259–271 (2004)

Fixed Point Approach

8

8.1   A Set-Valued Approach to Hemivariational Inequalities

In this part we use fixed point and KKM theories in order to guarantee solutions for various differential inclusions and nonlinear hemivariational inequalities. The results are mainly based on the papers by Kristály and Varga [10], and Costea and Varga [5]. Let X be a Banach space, X^* its dual, and let $T : X \to L^p(\Omega, \mathbb{R}^k)$ be a linear continuous operator, where $1 \leq p < \infty$, $k \in \mathbb{N}^*$, Ω being a bounded open set in $\mathbb{R}^N$. Let K be a subset of X and let $\mathcal{A} : K \rightsquigarrow X^*$ a set-valued map with nonempty values. We denote by $\sigma(\mathcal{A}(u), \cdot)$ the support function of $\mathcal{A}(u)$, that is

$$\sigma(\mathcal{A}(u), h) := \sup_{u^* \in \mathcal{A}(u)} \langle u^*, v \rangle, \quad \forall v \in X.$$

Definition 8.1 Let X be a Banach space, and let K be a nonempty subset of X. A set-valued map $\mathcal{A} : K \rightsquigarrow X^*$ with bounded values is said to be *upper demicontinuous* at $u_0 \in K$ (u.d.c. at $u_0 \in K$) if, for any $v \in X$, the real valued function $K \ni u \mapsto \sigma(\mathcal{A}(u), v)$ is upper semicontinuous at u_0.
$\mathcal{A}$ is upper demicontinuous on K (u.d.c. on K) if it is u.d.c. at every $u \in K$.

Remark 8.1 If $\mathcal{A}(u) := \{A(u)\}$ for all $u \in K$, that is $\mathcal{A}$ is a single-valued map, then $\mathcal{A}$ is u.d.c. at $u_0 \in K$ if and only if the map $A : K \to X^*$ is *w*-demicontinuous* at $u_0 \in K$, i.e., for each sequence $\{u_n\} \in K$ converging to u_0 in the strong topology, the image sequence $\{A(u_n)\}$ converges to $A(u_0)$ in the weak*-topology of X^*. It is easy to verify that, for all $u \in K$, the function $v \in X \mapsto \sigma(\mathcal{A}(u), v)$ is lower semicontinuous, subadditive and positive homogeneous.

Using Banach-Steinhaus theorem, we can state the following the following useful result.

Proposition 8.1 *Let K be a nonempty subset of a Banach space X and assume $\mathcal{A} : K \rightsquigarrow X^*$ is an upper demicontinuous set-valued map with bounded values. Then the function $u \mapsto \sigma(\mathcal{A}(u), v - u)$ is upper semicontinuous for all $v \in K$.*

In the following we consider the following set-valued maps $\mathcal{A} : K \rightsquigarrow X^*$, $G : K \times X \rightsquigarrow \mathbb{R}$ and $F : \Omega \times \mathbb{R}^k \times \mathbb{R}^k \rightsquigarrow \mathbb{R}$ with nonempty values such that the following conditions hold:

(H_1) $x \in \Omega \rightsquigarrow F(x, Tu(x), Tv(x) - Tu(x))$ is a measurable set-valued map for all $u, v \in K$;

(H_2) There exist $h_1 \in L^{p/p-1}(\Omega, \mathbb{R}_+)$ and $h_2 \in L^\infty(\Omega, \mathbb{R}_+)$ such that

$$\mathrm{dist}(0, F(x, y, z)) \leq (h_1(x) + h_2(x)|y|^{p-1})|z|, \quad \text{for a.e. } x \in \Omega,$$

for every $y, z \in \mathbb{R}^k$;

(H_3) $X \ni w \rightsquigarrow G(u, w)$ and $\mathbb{R}^k \ni z \rightsquigarrow F(x, y, z)$ are convex for all $u \in K$, $x \in \Omega$, $y \in \mathbb{R}^k$;

(H_4) $G(u, 0) \subseteq \mathbb{R}_+$ and $F(x, y, 0) \subseteq \mathbb{R}_+$ for all $u \in K$, $x \in \Omega$, $y \in \mathbb{R}^k$;

(H_5) $K \times X \ni (u, w) \rightsquigarrow G(u, w)$ is lower semicontinuous;

(H_6) $\mathbb{R}^k \times \mathbb{R}^k \ni (y, z) \rightsquigarrow F(x, y, z)$ is lower semicontinuous for all $x \in \Omega$.

Remark 8.2 If $F : \Omega \times \mathbb{R}^k \times \mathbb{R}^k \rightsquigarrow \mathbb{R}$ is a closed-valued Carathéodory map, i.e., for any $(y, z) \in \mathbb{R}^k \times \mathbb{R}^k$, $x \in \Omega \rightsquigarrow F(x, y, z)$ is measurable and for any $x \in \Omega$, $(y, z) \in \mathbb{R}^k \times \mathbb{R}^k \rightsquigarrow F(x, y, z)$ is continuous, then the hypotheses (H_1) and (H_6) hold automatically.

The aim of this section is to study the following *hemivariational inclusion problem*:
(HI) Find $u \in K$ such that

$$\sigma(\mathcal{A}(u), v - u) + G(u, v - u) + \int_\Omega F(x, Tu(x), Tv(x) - Tu(x))dx \subseteq \mathbb{R}_+, \ \forall v \in K.$$

$$(8.1)$$

The main result of this section is the following.

Theorem 8.1 ([10]) *Let K be a nonempty compact convex subset of a Banach space X. Suppose $F : \Omega \times \mathbb{R}^k \times \mathbb{R}^k \rightsquigarrow \mathbb{R}$ and $G : K \times X \rightsquigarrow \mathbb{R}$ are two set-valued maps satisfying $(H_1) - (H_6)$ and F is closed valued. If $\mathcal{A} : K \rightsquigarrow X^*$ is upper demicontinuous on K with bounded values, then (HI) has at least one solution.*

Proof For any $v \in K$ we set

$$S_v := \left\{ \begin{array}{c} u \in K : \mathbb{R}_+ \supseteq \sigma(\mathcal{A}(u), v - u) + G(u, v - u) \\ + \int_\Omega F(x, Tu(x), Tv(x) - Tu(x))dx \end{array} \right\}$$

First, we prove that S_v is closed set for all $v \in K$. Fix a $v \in K$. Of course, $S_v \neq \emptyset$, since $v \in S_v$, due to (H_4). Now, let $\{u_n\}$ be a sequence in S_v which converges to $u \in X$. We prove that $u \in S_v$. Since $T : X \to L^p(\Omega, \mathbb{R}^k)$ is continuous, it follows that $Tu_n \to Tu$ in $L^p(\Omega, \mathbb{R}^k)$ as $n \to \infty$. Using Proposition 8.1 we get the existence of a subsequence $\{u_m\}$ of $\{u_n\}$, such that

$$\limsup_{n\to\infty} \sigma(\mathcal{A}(u_n), v - u_n) = \lim_{m\to\infty} \sigma(\mathcal{A}(u_m), v - u_m). \tag{8.2}$$

Moreover, there exists a subsequence $\{Tu_l\}$ of $\{Tu_m\}$ and $g \in L^p(\Omega, \mathbb{R}_+)$ such that

$$|Tu_l(x)| \leq g(x), \quad Tu_l(x) \to Tu(x) \quad \text{for a.e. } x \in \Omega. \tag{8.3}$$

In the relation

$$\sigma(\mathcal{A}(u), v - u) + G(u, v - u) + \int_\Omega F(x, Tu(x), Tv(x) - Tu(x))dx \subseteq \mathbb{R}_+$$

taking the lower limit and using Lemma B.1 with $X := \mathbb{R}$ we obtain

$$\mathbb{R}_+ = \liminf_{l\to\infty} \mathbb{R}_+ \supseteq \liminf_{l\to\infty} \sigma(\mathcal{A}(u_l), v - u_l) + \liminf_{l\to\infty} G(u_l, v - u_l) \tag{8.4}$$

$$+ \liminf_{l\to\infty} \int_\Omega F(x, Tu_l(x), Tv(x) - Tu_l(x))dx .$$

Using Proposition B.2, relation (8.2) and Proposition 8.1, we obtain

$$\liminf_{l\to\infty} \sigma(\mathcal{A}(u_l), v - u_l) = \lim_{l\to\infty} \sigma(\mathcal{A}(u_l), v - u_l) \tag{8.5}$$

$$= \limsup_{n\to\infty} \sigma(\mathcal{A}(u_n), v - u_n) \leq \sigma(\mathcal{A}(u), v - u).$$

From (H_5) and the characterization of lower semicontinuity of set-valued function we obtain

$$G(u, v - u) \subseteq \liminf_{l\to\infty} G(u_l, v - u_l). \tag{8.6}$$

Let $F_l := F(\cdot, Tu_l(\cdot), Tv(\cdot) - Tu_l(\cdot))$. From (H_1) follows that, F_l is measurable, for any l. The function $x \in \Omega \mapsto \sup_l \text{dist}(0, F_l(x))$ is integrable. Indeed, from (H_2) and relation (8.3) we have

$$\text{dist}(0, F(x)) \leq (h_1(x) + h_2(x)[g(x)]^{p-1})|Tv(x) - Tu_l(x)|$$

$$\leq (h_1(x) + h_2(x)[g(x)]^{p-1})(|Tv(x)| + g(x)),$$

a.e. $x \in \Omega$. Let $h(x) := (h_1(x) + h_2(x)[g(x)]^{p-1})(|Tv(x)| + g(x))$. Hölder's inequality and condition for h_1 and h_2 ensure that $h \in L^1(\Omega, \mathbb{R})$. Therefore the function $x \in \Omega \mapsto \sup_l \text{dist}(0, F_l(x))$ is integrable. Applying the Lebesque dominated convergence theorem for set-valued map, on hase

$$\int_\Omega \liminf_{l \to \infty} F_l(x)dx \subseteq \liminf_{i \to \infty} \int_\Omega F_l(x)dx. \tag{8.7}$$

Of course $\int_\Omega \liminf_{l \to \infty} F_l(x)dx = \int_\Omega \liminf_{l \to \infty} F(x, Tu_l(x), Tv(x) - Tu_l(x))$ is measurable (see, e.g., Aubin-Frankowska [1, Theorem 8.6.7]). Using hypothesis (H_6) and the characterization of lower semicontinuity of set-valued maps with sequences and (8.3), one has

$$F(x, Tu(x), Tv(x) - Tu(x)) \subseteq \liminf_{l \to \infty} F(x, Tu_l(x), Tv(x) - Tu_l(x))$$

for a.e. $x \in \Omega$. Using the elementary property of the set-valued integral and (8.7) we obtain

$$\int_\Omega F(x, Tu(x), Tv(x) - Tu(x))dx \subseteq \liminf_{l \to \infty} \int_\Omega F_l(x)dx \tag{8.8}$$

Therefore, from (8.5), (8.6), (8.8) and (8.4) we obtain

$$\sigma(\mathcal{A}(u), v - u) + G(u, v - u) + \int_\Omega F(x, Tu(x), Tv(x) - Tu(x))dx \subseteq \mathbb{R}_+,$$

i.e. $u \in S_v$.

Finally we prove that $S : K \rightsquigarrow K$ is a KKM-map. To this end, let $\{v_1, \ldots, v_n\}$ be an arbitrary finite subset of K. We prove that $\text{co}\{v_1, \ldots, v_n\} \subseteq \bigcup_{i=1}^n S_{v_i}$. Assuming the contrary, there exists $\lambda_i \geq 0$ $(i \in \{1, \ldots, n\})$ such that $\sum_{i=1}^n \lambda_i = 1$ and $\bar{v} = \sum_{i=1}^n \lambda_i v_i \notin S_{v_i}$ for all $i \in \{1, \ldots, n\}$. The above relation means that for all $i \in \{1, \ldots, n\}$

$$\left[\sigma(\mathcal{A}(\bar{v}), v_i - \bar{v}) + G(\bar{v}, v_i - \bar{v} + \int_\Omega F(x, T\bar{v}(x), Tv_i(x) - T\bar{v}(x))dx)\right] \cap \mathbb{R}_-^* \neq \varnothing.$$

Let $I := \{i \in \{1, \ldots, n\} : \lambda_i > 0\}$. From the above obtain

$$\varnothing \neq \left\{\sum_{i \in I} \lambda_i \sigma(\mathcal{A}(\bar{v}), v_i - \bar{v}) + G(\bar{v}, v_i - \bar{v})\right.$$

$$\left. + \int_\Omega F(x, T\bar{v}(x), Tv_i(x) - T\bar{v}(x))dx\right\} \cap \mathbb{R}_-^*.$$

Using the sublinearity of the function $h \in X \mapsto \sigma(\mathcal{A}(\bar{v}), h)$, $(H3)$ the linearity of T and $(H4)$, we obtain

$$
\begin{aligned}
\emptyset \neq &\left\{ \sigma\left(\mathcal{A}(\bar{v}), \sum_{i \in I} \lambda_i v_i - \sum_{i \in I} \lambda_i \bar{v}\right) + \sum_{i \in I} \lambda_i G(\bar{v}, v_i - \bar{v}) \right. \\
&\left. + \left\{ \sum_{i \in I} \lambda_i \int_{\Omega} F(x, T\bar{v}(x), Tv_i(x) - T\bar{v}(x)) dx \right\} \cap \mathbb{R}_-^* \right. \\
\subseteq &\left\{ \sigma(\mathcal{A}(\bar{v}), 0) + G\left(\bar{v}, \sum_{i \in I} \lambda_i v_i - \sum_{i \in I} \lambda_i \bar{v}\right) \right. \\
&\left. + \int_{\Omega} \overline{\sum_{i \in I} \lambda_i F(x, T\bar{v}(x), Tv_i(x) - T\bar{v}(x))} dx \right\} \cap \mathbb{R}_-^* \\
\subseteq &\left\{ G(\bar{v}, 0) \int_{\Omega} F\left(x, T\bar{v}(x), \sum_{i \in I} Tv_i(x) - \sum_{i \in I} T\bar{v}(x)\right) dx \right\} \cap \mathbb{R}_-^* \\
= &\left\{ G(\bar{v}, 0) \int_{\Omega} F(x, T\bar{v}(x), 0) dx \right\} \cap \mathbb{R}_-^* \subseteq \left\{ \mathbb{R}_+ + \int_{\Omega} \mathbb{R}_+ \right\} \cap \mathbb{R}_-^* \\
= &\emptyset,
\end{aligned}
$$

which is contradiction. This means that S is a KKM-map. Since K is compact, applying Corollary D.1, we obtain that $\bigcap_{v \in K} S_v \neq \emptyset$,, i.e., (8.1) has at least a solution. $\qquad \square$

When K is not compact, we can state the following result, using the coercivity assumption.

Theorem 8.2 ([10]) *Let K be a nonempty closed convex subset of a Banach space X. Let $\mathcal{A}$, G and F as in Theorem 8.1. In addition, suppose that there exists a compact subset K_0 of K and an element $w_0 \in K_0$ such that*

$$
\left\{ \sigma(\mathcal{A}(u), u_0 - u) + \int_{\Omega} F(x, Tu, Tw_0 - Tu) dx + G(u, w_0 - u) \right\} \cap \mathbb{R}_-^* \neq \emptyset \quad (8.9)
$$

for all $u \in K \setminus K_0$. Then (HI) has at least a solution.

Proof We define the map S as in Theorem 8.1. Clearly, S is a KKM-map and S_v is closed for all $v \in K$, as seen above. Moreover, $S_{w_0} \subset K_0$. Indeed, assuming the contrary, there exists an element $u \in S_{w_0} \subseteq K$ such that $u \notin K_0$. But this contradicts (8.9). Since K_0 is compact, the set S_{w_0} is also compact. Applying again Corollary D.1, we obtain a solution for (HI). $\qquad \square$

8.2   Variational-Hemivariational Inequalities with Lack of Compactness

In this section we prove the existence of at least one solution for a variational-hemivariational inequality on a closed and convex set using the well-known theorem of Knaster-Kuratowski-Mazurkiewicz due to Ky Fan, i.e., Corollary D.1. The theoretical results can be applied to Schrödinger type problems and for problems with radially symmetric functions.

Let $(X, \| \cdot \|)$ be a Banach space and X^* its topological dual, $\langle \cdot, \cdot \rangle$ denotes the duality pairing between X^* and X. Let $\Omega \subseteq \mathbb{R}^n$ be an unbounded domain, let p be such that $1 < p < n$ and we denote $p^* := \frac{np}{n-p}$. Assume the following conditions hold.

(X) Assume that for $s \in [p, p^*)$ the embedding $X \hookrightarrow L^s(\Omega)$ is compact;

(A_1) Let $A : X \to X^*$ be an operator with the following property: for any sequence $\{u_n\}_n$ in X which converges weakly to $u \in X$ it holds

$$\langle Au, u - w \rangle \leq \liminf_{n \to \infty} \langle Au_n, u_n - w \rangle, \quad \forall w \in X;$$

(A_2) There exists $\lambda := \inf_{u \in X \setminus \{0\}} \frac{\langle Au, u \rangle}{\|u\|^p} > 0$.

Remark 8.3 Let $A : X \to X^*$ be a linear and continuous operator, which is *positive*, i.e., $\langle Au, u \rangle \geq 0$, for all $u \in X$. These assumptions imply that A is weakly sequentially continuous and that (A_1) is satisfied.

If $a : X \times X \to \mathbb{R}$ is a bilinear form, which is compact, i.e., for any sequences $\{u_n\}_n$ and $\{v_n\}_n$ from X such that $u_n \rightharpoonup u$ and $v_n \rightharpoonup v$ $(u, v \in X)$ it follows that $a(u_n, v_n) \to a(u, v)$, then the operator $A : X \to X^*$ defined by

$$\langle Au, v \rangle := a(u, v), \quad \forall u, v \in X$$

satisfies assumption (A_1).

We continue with the assumptions for our problem.

(f_1) Let $f : \Omega \times \mathbb{R} \to \mathbb{R}$ be a Carathéodory function, such that for some $\alpha > 0$ it holds

$$|f(x, y)| \leq \alpha |y|^{p-1} + \beta(x),$$

for a.e. $x \in \Omega$ and all $y \in \mathbb{R}$, where $\beta \in L^{\frac{p}{p-1}}(\Omega)$;

(f_2) we assume that the constants from (f_1) and (A_2) satisfy $\alpha C_p^p < \lambda$.

(j_1) Assume that $j : \Omega \times \mathbb{R} \to \mathbb{R}$ is a Carathéodory function, which is locally Lipschitz with respect to the second variable, and there exists $c > 0, r \in [p, p^*)$ such that

$$|\xi| \leq c(|y|^{p-1} + |y|^{r-1})$$

for a.e. $x \in \Omega$, all $y \in \mathbb{R}$ and all $\xi \in \partial_C^2 j(x, y)$;

(j_2) there exists $k \in L^{\frac{p}{p-1}}(\Omega)$ such that

$$|j^0(x, y; -y)| \leq k(x)|y|, \quad \forall x \in \Omega, \ \forall y \in \mathbb{R},$$

where $j^0(x, y; z)$ denotes the generalized directional derivative of $j(x, \cdot)$ at the point $y \in X$ in the direction $z \in X$.

Let $K \subseteq X$. In this paper we investigate the existence of at least one solution for the following variational-hemivariational inequality:

(VHI) Find $u \in K$ such that

$$\langle Au, v - u \rangle + \int_\Omega f(x, u(x))(v(x) - u(x))\mathrm{d}x + \int_\Omega j^0(x, u(x); v(x) - u(x))\mathrm{d}x \geq 0,$$

for all $v \in K$.

Lemma 8.1 *Suppose that X is a Banach space.*

1. *Assume that (j_1) is satisfied and X_1 and X_2 are nonempty subsets of X.*
   (a) *If the embedding $X \hookrightarrow L^s(\Omega)$ is continuous for each $s \in [p, p^*]$, then the mapping*

$$X_1 \times X_2 \ni (u, v) \mapsto \int_\Omega j^0(x, u(x); v(x))\mathrm{d}x \in \mathbb{R}$$

   is upper semicontinuous;
   (b) *Moreover, if $X \hookrightarrow L^s(\Omega)$ is compact for $s \in [p, p^*)$, then the above mapping is weakly upper semicontinuous;*
2. *Assume that (f_1) holds and that $X \hookrightarrow L^p(\Omega)$ is compact. Then, for each $v \in X$ the mapping*

$$X \ni u \mapsto \int_\Omega f(x, u(x))(v(x) - u(x))\mathrm{d}x \in \mathbb{R}$$

is weakly sequentially continuous.

Proof

(1a) Let $\{(u_n, v_n)\}_n \subset X_1 \times X_2$ be a sequence converging to $(u, v) \in X_1 \times X_2$. Since $X \hookrightarrow L^p(\Omega)$, $X \hookrightarrow L^r(\Omega)$ are continuous, it follows that

$$u_n \to u, \ v_n \to v \text{ in } L^p(\Omega) \text{ and in } L^r(\Omega) \text{ as } n \to \infty.$$

There exists a subsequence $\{(u_{n_k}, v_{n_k})\}_k$ of $\{(u_n, v_n)\}_n$ such that

$$\limsup_{n\to\infty} \int_\Omega j^0(x, u_n(x); v_n(x)) dx = \lim_{k\to\infty} \int_\Omega j^0(x, u_{n_k}(x); v_{n_k}(x)) dx. \qquad (8.10)$$

By Theorem 4.9 in [3] it follows that there exists $\bar{u}, \bar{v} \in L^p(\Omega)$ and $\hat{u}, \hat{v} \in L^r(\Omega)$ and two subsequences $\{u_{n_i}\}_i$ and $\{v_{n_i}\}_i$ of $\{u_{n_k}\}_k$ and $\{v_{n_k}\}_k$ such that for a.e. $x \in \Omega$ it hold

$$u_{n_i}(x) \to u(x), \ \text{and } v_{n_i}(x) \to v(x) \text{ as } i \to \infty \qquad (8.11)$$

and

$$|u_{n_i}(x)| \le \bar{u}(x), \ |u_{n_i}(x)| \le \hat{u}(x) \text{ and } |v_{n_i}(x)| \le \bar{v}(x), |v_{n_i}(x)| \le \hat{v}(x) \text{ for all } i \in \mathbb{N}.$$

By assumption (j_1) and the properties of j^0, see Proposition 2.3, it follows that for all $i \in \mathbb{N}$ and for a.e. $x \in \Omega$ it holds

$$|j^0(x, u_{n_i}(x); v_{n_i}(x))| \le |\xi_i||v_{n_i}(x)| \le c|\bar{u}(x)|^{p-1}|\bar{v}(x)| + c|\hat{u}(x)|^{r-1}|\hat{v}(x)|,$$

where $\xi_i \in \partial_C^2 j(x, u_{n_i}(x); v_{n_i}(x))$. By using $\bar{u}, \bar{v} \in L^p(\Omega)$, $\hat{u}, \hat{v} \in L^r(\Omega)$ and Hölder's inequality we have $|\bar{u}|^{p-1}|\bar{v}| + |\hat{u}|^{r-1}|\hat{v}| \in L^1(\Omega)$. The Fatou-Lebesgue Theorem implies

$$\lim_{i\to\infty} \int_\Omega j^0(x, u_{n_i}(x); v_{n_i}(x)) dx \le \int_\Omega \limsup_{i\to\infty} j^0(x, u_{n_i}(x); v_{n_i}(x)) dx. \qquad (8.12)$$

The mapping $j^0(x, \cdot; \cdot)$ is upper-semicontinuous, see Proposition 2.3 and by (8.11) we obtain

$$\limsup_{i\to\infty} j^0(x, u_{n_i}(x); v_{n_i}(x)) \le j^0(x, u(x); v(x)). \qquad (8.13)$$

We use (8.10), (8.12) and (8.13) to get

$$\limsup_{n\to\infty} \int_\Omega j^0(x, u_n(x); v_n(x)) dx \le \int_\Omega j^0(x, u(x); v(x)) dx.$$

(1*b*) Let $\{(u_n, v_n)\}_n \subset X_1 \times X_2$ be a sequence converging weakly to $(u, v) \in X_1 \times X_2$. Since $X \hookrightarrow L^s(\Omega)$ is compact for $s \in [p, p^*)$, it follows that

$$u_n \to u, \, v_n \to v \text{ in in } L^p(\Omega) \text{ and in } L^r(\Omega) \text{ as } n \to \infty.$$

From now on the proof is similar to the case (1a).

(2) Let $\{u_n\}_n \subset X$ be a sequence converging weakly to $u \in X$. Since $X \hookrightarrow L^p(\Omega)$ is compact, it follows that

$$u_n \to u \text{ in in } L^p(\Omega) \text{ as } n \to \infty.$$

By Theorem 4.9 in [3] it follows that there exists $\bar{u} \in L^p(\Omega)$ and a subsequences $\{u_{n_i}\}_i$ of $\{u_n\}_n$ such that for a.e. $x \in \Omega$ it holds

$$u_{n_i}(x) \to u(x) \text{ as } i \to \infty \tag{8.14}$$

and

$$|u_{n_i}(x)| \leq |\bar{u}(x)|, \text{ for all } i \in \mathbb{N}.$$

By assumption (f1) it follows that for all $i \in \mathbb{N}$ and for a.e. $x \in \Omega$ it hold

$$|f(x, u_{n_i}(x))(u_{n_i}(x) - v(x))| \leq (\alpha|\bar{u}(x)|^{p-1} + \beta(x))(|\bar{u}(x)| + |v(x)|).$$

By using $\bar{u} \in L^p(\Omega)$, $\beta \in L^{\frac{p}{p-1}}(\Omega)$ and Hölder's inequality we have $(\alpha|\bar{u}|^{p-1} + \beta)(|\bar{u}| + |v|) \in L^1(\Omega)$. Since f is a Carathéodory function, it follows by the Dominated Convergence Theorem and by (8.14) that

$$\lim_{i \to \infty} \int_\Omega f(x, u_{n_i}(x))(u_{n_i}(x) - v(x))dx = \int_\Omega \lim_{i \to \infty} f(x, u_{n_i}(x))(u_{n_i}(x) - v(x))dx$$

$$= \int_\Omega f(x, u(x))(u(x) - v(x))dx.$$

Hence, every subsequence admits a subsequence which converges to the same limit, and we get

$$\lim_{n \to \infty} \int_\Omega f(x, u_n(x))(u_n(x) - v(x))dx = \int_\Omega f(x, u(x))(u(x) - v(x))dx.$$

$\square$

Theorem 8.3 ([10]) *Suppose that X is a reflexive Banach space and that $K \subseteq X$ is a nonempty, closed, convex and bounded set and that the hypotheses (X), (A_1), (f_1), $(j1)$, are fulfilled. Then, (VHI) has at least one solution.*

Proof Let $G : K \rightsquigarrow X$ be the multivalued mapping defined by

$$G(v) := \left\{ \begin{array}{l} u \in K : \langle Au, v - u \rangle + \int_\Omega f(x, u(x))(v(x) - u(x))dx \\ \quad + \int_\Omega j^0(x, u(x); v(x) - u(x))dx \geq 0 \end{array} \right\}$$

Note that for each $v \in K$ one has $G(v) \neq \varnothing$ as $v \in G(v)$. We verify the assumptions of Corollary D.1 are fulfilled the weak topology.

STEP 1.   *For $v \in K$ the set $G(v)$ is weakly closed.*
Let $\{u_n\}_n \subset G(v)$ such that $u_n \rightharpoonup u$ in the space X. By Lemma 8.1 and (A_1) it follows that

$$0 \leq \limsup_{n \to \infty} \left(\langle Au_n, v - u_n \rangle + \int_\Omega f(x, u_n(x))(v(x) - u_n(x))dx \right.$$

$$\left. + \int_\Omega j^0(x, u_n(x); v(x) - u_n(x))dx \right)$$

$$\leq \langle Au, v - u \rangle + \int_\Omega f(x, u(x))(v(x) - u(x))dx + \int_\Omega j^0(x, u(x); v(x) - u(x))dx.$$

Hence $u \in G(v)$.
STEP 2.   *G is a KKM mapping.*
We argue by contradiction, let $v_1, \ldots, v_n \in K$ and $u \in \mathrm{co}\{v_1, \ldots, v_n\}$ such that $u \notin \bigcup_{i=1}^n G(v_i)$. This implies that for all $i \in \{1, \ldots, n\}$ we have

$$\langle Au, v - u \rangle + \int_\Omega f(x, u(x))(v(x) - u(x))dx + \int_\Omega j^0(x, u(x); v(x) - u(x))dx < 0.$$

We denote by

$$C := \left\{ \begin{array}{l} v \in K : \langle Au, v - u \rangle + \int_\Omega f(x, u(x))(v(x) - u(x))dx \\ \quad + \int_\Omega j^0(x, u(x); v(x) - u(x))dx < 0 \end{array} \right\}$$

Observe that $u_1, \ldots, u_n \in C$ and that C is a convex set, since $j^0(x, u(x); \cdot)$ is positively homogeneous and subadditive. This implies $u \in C$, which is a contradiction.
STEP 3.   *For each $w \in K$ the set $G(w)$ is weakly compact.*
$G(v)$ is a bounded set (since K is bounded) and it is weakly closed (by STEP 1). The Eberlein-Šmulian Theorem implies that $G(v)$ is weakly compact.

The above steps ensure that the conditions of Corollary D.1 are satisfied (for the weak topology), therefore $\bigcap_{v \in K} G(v) \neq \emptyset$, i.e., the solution set of (VHI) is nonempty. $\square$

Theorem 8.4 ([10]) *Suppose that X is a reflexive Banach space and $K \subseteq X$ is a nonempty, closed, and convex set and that the hypotheses (X), (A_1), (A_2), (f_1), (f_2), (j_1), (j_2) are fulfilled. Then, (VHI) has at least one solution.*

Proof Without loss of generality we assume that $0 \in K$. For any positive integer n, set

$$K_n := \{v \in K : \|v\| \leq n\}.$$

Thus, $0 \in K_n$ for all $n \in \mathbb{N}$.

Let $n \in \mathbb{N}$. Applying Theorem 8.3 there exists $u_n \in K_n$ such that for all $v \in K_n$ it holds

$$\langle Au_n, v - u_n \rangle + \int_\Omega f(x, u_n(x))(v(x) - u_n(x))dx \tag{8.15}$$

$$+ \int_\Omega j^0(x, u_n(x); v(x) - u_n(x))dx \geq 0.$$

We prove that $\{u_n\}_n$ is a bounded sequence in X. In (8.15) we take $v = 0$ and get

$$\langle Au_n, u_n \rangle + \int_\Omega f(x, u_n(x))u_n(x)dx \leq \int_\Omega j^0(x, u_n(x); -u_n(x))dx. \tag{8.16}$$

By the assumption (j_2), (f_1) and by the continuity of the embedding $X \hookrightarrow L^p(\Omega)$, it follows that

$$\int_\Omega j^0(x, u_n(x); -u_n(x))dx \leq \int_\Omega k(x)|u_n(x)|dx \leq \|k\|_{L^{p'}(\Omega)} C_p \|u_n\|, \tag{8.17}$$

and

$$\left| \int_\Omega f(x, u_n(x))u_n(x)dx \right| \leq \alpha C_p^p \|u_n\|^p + \|\beta\|_{L^{p'}(\Omega)} C_p \|u_n\|,$$

where C_p denotes the embedding constant. Using (f_2) we obtain

$$(\lambda - \alpha C_p^p) \|u_n\|^p - C_p \|\beta\|_{L^{p'}(\Omega)} \|u_n\| \leq \langle Au_n, u_n \rangle + \int_\Omega f(x, u_n(x))u_n(x)dx.$$

Since $p > 1$, it follows by (f_2), (8.16) and (8.17) that $\{u_n\}_n$ is a bounded sequence in X.

This property and the closedness of K, implies that there exist $u \in K$ and a subsequence, which we denote also $\{u_n\}_n$, such that $u_n \rightharpoonup u$ in X.

By (A_1), Lemma 8.1 it follows that

$$\limsup_{n\to\infty} \int_\Omega j^0(x, u_n(x); v_n(x))dx \le \int_\Omega j^0(x, u(x); v(x))dx, \tag{8.18}$$

$$\lim_{n\to\infty} \int_\Omega f(x, u_n(x))(u_n(x) - v(x))dx = \int_\Omega f(x, u(x))(u(x) - v(x))dx, \tag{8.19}$$

$$\limsup_{n\to\infty}\langle Au_n, v - u_n \rangle \le \langle Au, v - u \rangle. \tag{8.20}$$

Let $v \in K$ be fixed. Then there exists $n_0 \in \mathbb{N}$ such that $v \in K_n$ for all $n \ge n_0$. We pass to lim sup as $n \to \infty$ in (8.15), use (8.18), (8.19) and (8.20) and obtain that $u \in K$ is a solution of (VHI). □

Example 8.1 This is an example of a Schrödinger type problem. Let $n > 2$ and $V : \mathbb{R}^n \to \mathbb{R}$ be a continuous function such that

$$\inf_{x\in\mathbb{R}^n} V(x) > 0 \text{ and for every } M > 0 \text{ meas}\Big(\{x \in \mathbb{R}^n : V(x) \le M\}\Big) < \infty.$$

The space

$$X := \Big\{u \in W^{1,2}(\mathbb{R}^n) : \int_{\mathbb{R}^n} |\nabla u(x)|^2 + V(x)u^2(x)dx < \infty\Big\}$$

equipped with the inner product

$$(u, v)_X := \int_{\mathbb{R}^n} \nabla u(x)\nabla v(x) + V(x)u(x)v(x)dx$$

is a Hilbert space. It is known that $X \hookrightarrow L^s(\mathbb{R}^n)$ is continuous for $s \in \left[2, \frac{2n}{n-2}\right]$, since $W^{1,2}(\mathbb{R}^n) \hookrightarrow L^s(\mathbb{R}^n)$ is continuous for $s \in \left[2, \frac{2n}{n-2}\right]$. Bartsch and Wang proved in [2] that $X \hookrightarrow L^s(\mathbb{R}^n)$ is compact for $s \in \left[2, \frac{2n}{n-2}\right)$. Hence, assumption (X) is satisfied for $p = 2$. We consider $A : X \to X$ to be defined by

$$\langle Au, v \rangle := (u, v)_X.$$

By the properties of the norm and of the weak convergence, it follows that (A_1) and (A_2) are satisfied. Theorem 8.3 can be applied assuming that f and j satisfy (f_1) and (j_1), respectively, and that $K \subset X$ is a nonempty, closed, convex and bounded set. If f and j

satisfy (f_1), (f_2) and (j_1), (j_2), respectively, and if $K \subseteq X$ is a nonempty, closed, and convex set, then by Theorem 8.4, it follows that (VHI) has at least one solution.

Example 8.2 Another Schrödinger type problem can be analogously formulated, if we consider for $n > 2$ the Hilbert space

$$X := \left\{ u \in L^2(\mathbb{R}^n) : \int_{\mathbb{R}^n} |\nabla u(x)|^2 + |x|^2 u^2(x) dx < \infty \right\}$$

equipped with the inner product

$$(u, v)_X := \int_{\mathbb{R}^n} \nabla u(x) \nabla v(x) + |x|^2 u(x) v(x) dx.$$

Note, that $X \hookrightarrow L^s(\mathbb{R}^n)$ is compact for $s \in \left[2, \frac{2n}{n-2} \right)$ (see Kavian, [8]). Similarly, as in Example 8.1, Theorems 8.3 and 8.4 can be applied.

In Theorems 8.3 and 8.4 it is very important that the conditions (X), (j_1) and (j_2) are satisfied. In this example we modify the conditions (j_1), (j_2) and prove that Theorems 8.3 and 8.4 still hold.

Let $a : \mathbb{R}^L \times \mathbb{R}^M \to \mathbb{R}$ ($L \geq 2$) be a nonnegative continuous function satisfying the following assumptions:

(a_1) $a(x, y) \geq a_0 > 0$ if $|(x, y)| \geq R$ for a large $R > 0$;
(a_2) $a(x, y) \to +\infty$, when $|y| \to +\infty$ uniformly for $x \in \mathbb{R}^L$;
(a_3) $a(x, y) = a(x', y)$ for all $x, x' \in \mathbb{R}^L$ with $|x| = |x'|$ and all $y \in \mathbb{R}^M$.

Consider the following subspaces of $W^{1,p}(\mathbb{R}^L \times \mathbb{R}^M)$

$$\tilde{E} := \left\{ u \in W^{1,p}(\mathbb{R}^L \times \mathbb{R}^M) : u(s, t) = u(s', t) \, \forall s, s' \in \mathbb{R}^L, |s| = |s'|, \forall t \in \mathbb{R}^M \right\},$$

$$E := \left\{ u \in W^{1,p}(\mathbb{R}^L \times \mathbb{R}^M) : \int_{\mathbb{R}^{L+M}} a(x)|u(x)|^p dx < \infty \right\},$$

$$X := \tilde{E} \cap E = \left\{ u \in \tilde{E} : \int_{\mathbb{R}^{L+M}} a(x)|u(x)|^p dx < \infty \right\}$$

endowed with the norm

$$\|u\|^p = \int_{\mathbb{R}^{L+M}} |\nabla u(x)|^p dx + \int_{\mathbb{R}^{L+M}} a(x)|u(x)|^p dx.$$

Morais Filho, Souto and Marcos Do proved in [6] the following result: X *is continuously embedded in* $L^s(\mathbb{R}^L \times \mathbb{R}^M)$ *if* $s \in [p, p^*]$, *and compactly embedded if* $s \in (p, p^*)$.
Let

$$G := \left\{ g : E \to E : g(v) = v \circ \begin{pmatrix} R & 0 \\ 0 & Id_{\mathbb{R}^M} \end{pmatrix}, R \in O(\mathbb{R}^L) \right\},$$

where $O(\mathbb{R}^L)$ is the set of all rotations on $\mathbb{R}^L$ and $Id_{\mathbb{R}^M}$ denotes the $M \times M$ identity matrix. The elements of G leave $\mathbb{R}^{L+M}$ invariant, i.e. $g(\mathbb{R}^{L+M}) = \mathbb{R}^{L+M}$ for all $g \in G$.
The action of G over E is defined by

$$(gu)(x) = u(g^{-1}x), \quad g \in G, \ u \in E, \text{ a.e. } x \in \mathbb{R}^{L+M}.$$

As usual we write gu in place of $\pi(g)u$.
A function u defined on $\mathbb{R}^{L+M}$ is said to be G-invariant if

$$u(gx) = u(x), \quad \forall g \in G, \text{ a.e. } x \in \mathbb{R}^{L+M}.$$

Then $u \in E$ is G-invariant if and only if $u \in X$. We observe that the norm $\| \cdot \|$ is G-invariant.
We assume that $j : \mathbb{R}^L \times \mathbb{R}^M \times \mathbb{R} \to \mathbb{R}$ is a Carathéodory function, which is locally Lipschitz in the second variable (the real variable) and satisfies the following conditions:

(j_1') $j(x, 0) = 0$, and there exist $c > 0$ and $r \in (p, p^*)$ such that

$$|\xi| \le c(|y|^{p-1} + |y|^{r-1}), \forall \xi \in \partial_C^2 j(x, y), \ (x, y) \in \mathbb{R}^{L+M} \times \mathbb{R};$$

(j_3) $\lim_{y \to 0} \frac{\max\{|\xi| : \xi \in \partial_C^2 j(x,y)\}}{|y|^{p-1}} = 0$ uniformly for every $x \in \mathbb{R}^{L+M}$;
(j_4) $j(\cdot, y)$ is G-invariant for all $y \in \mathbb{R}$.

To derive the results of Theorem 8.3 we use the following result proved in [12, Proposition 5.1] instead of Lemma 8.1:
If $j : \mathbb{R}^L \times \mathbb{R}^M \times \mathbb{R} \to \mathbb{R}$ *verifies the conditions* (j_1'), (j_3) *and* (j_4) *then*

$$u \in X \mapsto \int_{\mathbb{R}^{L+M}} j(x, u(x)) dx$$

is weakly sequentially continuous.

In the same way as in Theorems 8.3 and 8.4 we can prove the following existence result:

Theorem 8.5

(i) *Let $K \subset X$ be a nonempty, closed, convex and bounded set. Let $A : E \to E^*$ be an operator satisfying (A_1). Assume that j satisfies (j_1'), (j_3) and (j_4). Then, there exists $u \in K$ such that*

$$\langle Au, v - u \rangle + \int_{\mathbb{R}^{L+M}} j^0(x, u(x); v(x) - u(x)) \mathrm{d}x \geq 0, \quad \forall v \in K. \tag{8.21}$$

(ii) *Moreover, if $K \subset X$ is a nonempty, closed and convex set and $A : X \to X^*$ is an operator satisfying (A_1), (A_2) and if we assume that j satisfies (j_1'), (j_2), (j_3) and (j_4). Then, there exists $u \in K$ such that (8.21) holds.*

8.3 Nonlinear Hemivariational Inequalities

This section is dedicated to the study of the following *nonlinear hemivariational inequality*
(NHI) Find $u \in K$ such that

$$\Lambda(u, v) + \int_{\Omega} j^0(x, \hat{u}(x); \hat{v}(x) - \hat{u}(x)) \mathrm{d}x \geq \langle f, v - u \rangle, \quad \forall v \in K,$$

where X is a real Banach space, $\emptyset \neq K \subseteq X$, $\Lambda : K \times K \to \mathbb{R}$ is a given function and $T : X \to L^p(\Omega; \mathbb{R}^k)$ is a linear and continuous operator, where $1 < p < \infty$, $k \geq 1$, and Ω is a bounded open set in $\mathbb{R}^N$. We shall denote $Tu := \hat{u}$ and by p' the conjugated exponent of p. Let K be a subset of X and $j : \Omega \times \mathbb{R}^k \to \mathbb{R}$ is a function such that the mapping

$$j(\cdot, y) : \Omega \to \mathbb{R} \text{ is measurable } \forall y \in \mathbb{R}^k. \tag{H_j^1}$$

We assume that at least one of the following conditions holds true: either there exist $l \in L^{p'}(\Omega; \mathbb{R})$ such that

$$|j(x, y_1) - j(x, y_2)| \leq l(x)|y_1 - y_2|, \quad \forall x \in \Omega, \forall y_1, y_2 \in \mathbb{R}^k, \tag{H_j^2}$$

or

$$\text{the mapping } j(x, \cdot) \text{ is locally Lipschitz} \forall x \in \Omega, \tag{H_j^3}$$

and there exist $C > 0$ such that

$$|\zeta| \le C(1 + |y|^{p-1}), \quad \forall x \in \Omega, \forall \zeta \in \partial_C^2 j(x, y). \qquad (H_j^4)$$

Regarding $\Lambda : K \times K \to \mathbb{R}$ we assume

(H_Λ^1) $\Lambda(u, u) = 0$, for all $u \in K$;
(H_Λ^2) $v \mapsto \Lambda(u, v)$ is convex, for all $u \in K$;
(H_Λ^3) $u \mapsto \Lambda(u, v)$ is upper semicontinuous, for all $v \in K$;

We point out the fact that the study of inequality problems involving nonlinear terms has captured special attention in the last few years. We just refer to the prototype problem of finding $u \in K$ such that

$$\Lambda(u, v) \ge \langle f, v - u \rangle, \quad \forall v \in K. \qquad (8.22)$$

Nonlinear inequality problems of the type (8.22) model some equilibrium problems drawn from operations research, as well as some unilateral boundary value problems stemming from mathematical physics and were introduced by Gwinner [7] who investigated the existence theory and abstract stability analysis in the setting of reflexive Banach spaces.

The main object of this section is to establish existence results for the nonlinear hemivariational inequality (NHI) for general maps, without monotonicity assumptions. As a consequence to our theorems, we will derive some existence results for hemivariational inequalities that have been studied in [13, 14] and [15] as it will be seen at the end of this section.

Theorem 8.6 ([4]) *Let K be a nonempty, closed and convex subset of X and assume j satisfies the conditions (H_j^1) and (H_j^2) or $(H_j^3) - (H_j^4)$, $T : X \to L^p(\Omega; \mathbb{R}^k)$ is linear and continuous and Λ satisfies $(H_\Lambda^1) - (H_\Lambda^3)$. If K is not compact assume in addition*

(H_K) *The set K possesses a nonempty compact convex subset K_1 with the property that for each $u \in K \setminus K_1$ there exists $v \in K_1$ such that*

$$\Lambda(u, v) + \int_\Omega j^0(x, \hat{u}(x); \hat{v}(x) - \hat{u}(x)) dx < \langle f, v - u \rangle.$$

Then for each $f \in X^*$ problem (NHI) has at least one solution in K.

Proof For each $v \in K$ we define the set

$$N(v) := \left\{ u \in K : \Lambda(u, v) + \int_\Omega j^0(x, \hat{u}(x); \hat{v}(x) - \hat{u}(x)) dx \ge \langle f, v - u \rangle \right\}.$$

We point out the fact that the solution set of (NHI) is $S := \bigcap_{v \in K} N(v)$.

First we prove that for each $v \in K$ the set $N(v)$ is closed. Let $\{u_n\} \subset N(v)$ be a sequence which converges to u as $n \to \infty$. We show that $u \in N(v)$. Since j satisfies the conditions $(H^1_j) - (H^2_j)$ or (H^1_j), (H^3_j), (H^4_j) the application

$$(u, v) \mapsto \int_\Omega j^0(x, \hat{u}(x); \hat{v}(x) - \hat{u}(x)) dx$$

is upper semicontinuous (see Panagiotopoulos et al. [15, Lemma 1]). Since T is linear and continuous, $\hat{u}_n \to \hat{u}$ and by the fact that $u_n \in N(v)$ for each n, we have

$$\langle f, v - u \rangle = \limsup_{n \to \infty} \langle f, v - u_n \rangle \leq \limsup_{n \to \infty} \left[\Lambda(u_n, v) + \int_\Omega j^0(x, \hat{u}_n(x); \hat{v}(x) - \hat{u}_n(x)) dx \right]$$

$$\leq \limsup_{n \to \infty} \Lambda(u_n, v) + \limsup_{n \to \infty} \int_\Omega j^0(x, \hat{u}_n(x); \hat{v}(x) - \hat{u}_n(x)) dx$$

$$\leq \Lambda(u, v) + \int_\Omega j^0(x, \hat{u}(x); \hat{v}(x) - \hat{u}(x)) dx.$$

This is equivalent to $u \in N(v)$.

Arguing by contradiction, suppose that $S = \varnothing$. Then for each $u \in K$ there exists $v \in K$ such that

$$\Lambda(u, v) + \int_\Omega j^0(x, \hat{u}(x); \hat{v}(x) - \hat{u}(x)) \, dx < \langle f, v - u \rangle. \tag{8.23}$$

We define the set valued map $F : K \rightsquigarrow K$ by

$$F(u) := \left\{ v \in K : \Lambda(u, v) + \int_\Omega j^0(x, \hat{u}(x); \hat{v}(x) - \hat{u}(x)) dx < \langle f, v - u \rangle \right\}.$$

Taking (8.23) into account we deduce that $F(u)$ is nonempty for each $u \in K$. Using the fact that Λ is convex with respect to the second variable, T is linear and the application $\hat{v} \mapsto j^0(x, \hat{u}; \hat{v})$ is also convex, we obtain that $F(u)$ is a convex set.

Now, for each $v \in K$, the set

$$F^{-1}(v) := \{u \in K : v \in F(u)\}$$

$$= \left\{ u \in K : \Lambda(u, v) + \int_\Omega j^0(x, \hat{u}(x); \hat{v}(x) - \hat{u}(x)) dx < \langle f, v - u \rangle \right\}$$

$$= \left\{ u \in K : \Lambda(u, v) + \int_\Omega j^0(x, \hat{u}(x); \hat{v}(x) - \hat{u}(x)) dx \geq \langle f, v - u \rangle \right\}^c$$

$$= [N(v)]^c =: O_v$$

is open in K. We claim next that $\bigcup_{v \in K} O_v = K$. To prove that, let $u \in K$. As $F(u)$ is nonempty it follows that there exists $v \in F(u)$ which implies $u \in F^{-1}(v)$. Thus $K \subseteq \bigcup_{v \in K} O_v$, the converse inclusion being obvious.

Finally, (H_K) ensures that $u \notin N(v)$. This implies that the set $D := \bigcap_{v \in K_1} O_v^c = \bigcap_{v \in K_1} N(v) \subset K_1$ is empty or compact as a closed subset of the compact set K_1. Taking $K_0 = K_1$ we have proved that the set valued map F satisfies the conditions of Theorem D.3, hence there exists $u_0 \in K$ such that $u_0 \in F(u_0)$, that is,

$$0 = \Lambda(u_0, u_0) + \int_\Omega j^0(x, \hat{u}_0(x); \hat{u}_0(x) - \hat{u}_0(x))dx < \langle f, u_0 - u_0 \rangle = 0,$$

which is a contradiction. Hence the solution set S of problem (NHI) is nonempty. $\qquad \square$

Remark 8.4 If X is reflexive, K is bounded, closed and convex, the operator $T : X \to L^p(\Omega; \mathbb{R}^k)$ is linear and compact and Λ is weakly upper semicontinuous with respect to the first variable (instead of being upper semicontinuous), then condition (H_K) in Theorem 8.6 can be dropped, because in these conditions,

$$(u, v) \mapsto \int_\Omega j^0(x, \hat{u}(x); \hat{v}(x) - \hat{u}(x))dx$$

is weakly upper semicontinuous. The proof is identical to that of Theorem 8.6, but the conditions of Theorem D.3 are satisfied for the weak topology.

Lemma 8.2 *If K is a nonempty, bounded, closed and convex subset of a real reflexive Banach space X and $f \in X^*$ be fixed. Consider a Banach space Y and let $L : X \to Y$ be linear and compact and $J : Y \to \mathbb{R}$ be a locally Lipschitz functional. Suppose in addition that $\Lambda : X \times X \to \mathbb{R}$ is a function which satisfies (H_Λ^1) and*

(H_Λ^4) $\Lambda(u, v) + \Lambda(v, u) \geq 0$, *for all $u, v \in X$;*
(H_Λ^5) $u \mapsto \Lambda(u, v)$ *is weakly upper semicontinuous and concave.*

Then there exists $u \in K$ such that

$$\Lambda(u, v) + J^0(Lu; Lv - Lu) \geq \langle f, v - u \rangle, \quad \forall v \in K.$$

Proof Set

$$g(v, u) := -\Lambda(u, v) - \langle f, u - v \rangle - J^0(Lu; Lv - Lu)$$

and

$$h(v, u) := \Lambda(v, u) - \langle f, u - v \rangle - J^0(Lu; Lv - Lu).$$

By condition (H_Λ^4) we have

$$g(v, u) - h(v, u) = -[\Lambda(u, v) + \Lambda(v, u)] \le 0, \quad \forall u, v \in X.$$

The mapping $u \mapsto g(v, u)$ is weakly lower semicontinuous for each $v \in X$, while the mapping $v \mapsto h(v, u)$ is concave for each $u \in X$. We apply Mosco's Alternative (Theorem D.8) with $\lambda := 0$ and $\phi := I_K$, where I_K denotes the indicator function of the set K. Clearly I_K is proper, convex and lower semicontinuous since K is nonempty, convex and closed. We obtain that exists $u \in K$ satisfying

$$g(v, u) + I_K(u) - I_K(v) \le 0, \quad \forall v \in X;$$

A simple computation yields that there exists $u \in K$ such that

$$\Lambda(u, v) + J^0(Lu; Lv - Lu) \ge \langle f, v - u \rangle, \quad \forall v \in K,$$

which exactly the desired conclusion. $\qquad\square$

The second existence result concerning the nonlinear hemivariational inequality problem can now be stated as follows:

Theorem 8.7 ([4]) *Let K be a bounded, closed and convex subset of a real reflexive Banach space X and assume j satisfies (H_j^1) and (H_j^2) or (H_j^1), (H_j^3) and (H_j^4). If T is linear and compact and Λ satisfies (H_Λ^1), (H_Λ^4) and (H_Λ^5), then (NHI) has at least one solution.*

Proof Apply Lemma 8.2 with $Y := L^p(\Omega; \mathbb{R}^k)$, $L := T$ and $J : L^p(\Omega; \mathbb{R}^k) \to \mathbb{R}$ $J(u) := \int_\Omega j(x, u(x))dx$. $\qquad\square$

8.4 Systems of Nonlinear Hemivariational Inequalities

In the last section of this chapter we take a step further and study a system of nonlinear hemivariational inequalities. We use a fixed-point for multivalued functions due to Lin [11] to establish several existence results including some sufficient coercivity conditions for the case of unbounded subsets. Such results come in handy in Contact Mechanics when describing the contact between a piezoelectric body and a foundation (see Part IV) or in the study of Nash-type equilibrium points (see, e.g., [5, 9, 16, 17]).

Let n be a positive integer, let $X_1, \ldots, X_n$ be real reflexive Banach spaces and $Y_1, \ldots, Y_n$ be real Banach spaces such that X_k is compactly embedded into Y_k, for each $k \in \{1, \ldots, n\}$. We denote by $i_k : X_k \to Y_k$ the embedding operator and $\hat{u}_k := i_k(u_k)$.

Throughout this section we investigate the existence of solutions for the following *system of nonlinear hemivariational inequalities:*

$(SNHI)$ Find $(u_1, \ldots, u_n) \in K_1 \times \ldots \times K_n$ such that

$$
\begin{cases}
\psi_1(u_1, \ldots, u_n, v_1) + J^0_{,1}(\hat{u}_1, \ldots, \hat{u}_n; \hat{v}_1 - \hat{u}_1) \geq \langle F_1(u_1, \ldots, u_n), v_1 - u_1 \rangle_{X_1} \\
\ldots \quad \ldots \quad \ldots \quad \ldots \quad \ldots \quad \ldots \quad \ldots \quad \ldots \quad \ldots \quad \ldots \quad \ldots \quad \ldots \\
\psi_n(u_1, \ldots, u_n, v_n) + J^0_{,n}(\hat{u}_1, \ldots, \hat{u}_n; \hat{v}_n - \hat{u}_n) \geq \langle F_n(u_1, \ldots, u_n), v_n - u_n \rangle_{X_n},
\end{cases}
$$

for all $(v_1, \ldots, v_n) \in K_1 \times \ldots \times K_n$.

Here and hereafter, for each $k \in \{1, \ldots, n\}$, we assume

- $K_k \subseteq X_k$ is nonempty closed and convex;
- $\psi_k : X_1 \times \ldots \times X_k \times \ldots \times X_n \times X_k \to \mathbb{R}$ is a nonlinear functional;
- $J : Y_1 \times \ldots \times Y_n \to \mathbb{R}$ is a regular locally Lipschitz functional;
- $F_k : X_1 \times \ldots \times X_k \times \ldots \times X_n \to X^*_k$ is a nonlinear operator.

In order to establish the existence of at least one solution for problem $(SNHI)$ we assume:

(H_1) For each $k \in \{1, \ldots, n\}$, the functional $\psi_k : X_1 \times \ldots \times X_k \times \ldots \times X_n \times X_k \to \mathbb{R}$ satisfies

    (i) $\psi_k(u_1, \ldots, u_k, \ldots, u_n, u_k) = 0$, for all $u_k \in X_k$;

    (ii) For each $v_k \in X_k$ the mapping $(u_1, \ldots, u_n) \mapsto \psi_k(u_1, \ldots, u_n, v_k)$ is weakly upper semicontinuous;

    (iii) For each $(u_1, \ldots, u_n) \in X_1 \times \ldots \times X_n$ the mapping $v_k \mapsto \psi_k(u_1, \ldots, u_n, v_k)$ is convex.

(H_2) For each $k \in \{1, \ldots, n\}$, $F_k : X_1 \times \ldots \times X_k \times \ldots \times X_n \to X^*_k$ is a nonlinear operator such that

$$
\liminf_{m \to \infty} \langle F_k\left(u^m_1, \ldots, u^m_n\right), v_k - u^m_k \rangle_{X_k} \geq \langle F_k(u_1, \ldots, u_n), v_k - u_k \rangle_{X_k}
$$

whenever $\left(u^m_1, \ldots, u^m_n\right) \rightharpoonup (u_1, \ldots, u_n)$ as $m \to \infty$ and $v_k \in X_k$ is fixed.

The first existence result of this section refers to the case when the sets K_k are bounded, closed and convex and it is given by the following theorem.

Theorem 8.8 ([5]) *For each $k \in \{1, \ldots, n\}$ let $K_k \subset X_k$ be a nonempty, bounded, closed and convex set and assume that conditions $(H_1) - (H_2)$ hold. Then $(SNHI)$ has at least one solution.*

The existence of solutions for our system will be a direct consequence of the fact that a *vector hemivariational inequality* admits solutions. Let us introduce the following notations:

- $X := X_1 \times \ldots \times X_n$, $K := K_1 \times \ldots \times K_n$ and $Y := Y_1 \times \ldots \times Y_n$;
- $u := (u_1, \ldots, u_n)$ and $\hat{u} := (\hat{u}_1, \ldots, \hat{u}_n)$;
- $\Psi : X \times X \to \mathbb{R}$, $\Psi(u, v) := \sum_{k=1}^{n} \Psi_k(u_1, \ldots, u_k, \ldots, u_n, v_k)$;
- $F : X \to X^*$, $\langle Fu, v \rangle_X := \sum_{k=1}^{n} \langle F_k(u_1, \ldots, u_n), v_k \rangle_{X_k}$.

and formulate the following vector hemivariational inequality
(VHI) Find $u \in K$ such that

$$\Psi(u, v) + J^0(\hat{u}; \hat{v} - \hat{u}) \geq \langle Fu, v - u \rangle_X, \quad \forall v \in K.$$

Remark 8.5 If $(H_1) - (i)$ holds, then any solution $u^0 := (u_1^0, \ldots, u_n^0) \in K_1 \times \ldots \times K_n$ of the vector hemivariational inequality (VHI) is also a solution of the system $(SNHI)$.

Indeed, if for a $k \in \{1, \ldots, n\}$ we fix $v_k \in K_k$ and for $j \neq k$ we consider $v_j := u_j^0$, using Proposition 2.3 and the fact that u^0 solves (VHI) we obtain

$$\left\langle F_k\left(u_1^0, \ldots, u_n^0\right), v_k - u_k^0 \right\rangle_{X_k} = \sum_{j=1}^{n} \left\langle F_j\left(u_1^0, \ldots, u_n^0\right), v_j - u_j^0 \right\rangle_{X_j}$$

$$= \left\langle Fu^0, v - u^0 \right\rangle_X \leq \Psi\left(u^0, v\right) + J^0\left(\hat{u}^0; \hat{v} - \hat{u}^0\right)$$

$$\leq \sum_{j=1}^{n} \Psi_j\left(u_1^0, \ldots, u_j^0, \ldots, u_n^0, v_j\right) + \sum_{j=1}^{n} J_{,j}^0\left(\hat{u}_1^0, \ldots, \hat{u}_n^0; \hat{v}_j - \hat{u}_j^0\right)$$

$$= \Psi_k\left(u_1^0, \ldots, u_k^0, \ldots, u_n^0, v_k\right) + J_{,k}^0\left(\hat{u}_1^0, \ldots, \hat{u}_n^0; \hat{v}_k - \hat{u}_k^0\right).$$

As $k \in \{1, \ldots, n\}$ and $v_k \in K_k$ were arbitrarily fixed, we conclude that $(u_1^0, \ldots, u_n^0) \in K_1 \times \ldots \times K_n$ is a solution of our system $(SNHI)$.

Proof of Theorem 8.8 According to Remark 8.5 it suffices to prove that problem (VHI) possesses a solution. With this end in view we consider the set $\mathcal{A} \subset K \times K$ as follows

$$\mathcal{A} := \left\{(v, u) \in K \times K : \Psi(u, v) + J^0(\hat{u}; \hat{v} - \hat{u}) - \langle Fu, v - u \rangle_X \geq 0 \right\}.$$

The following steps ensure that the set $\mathcal{A}$ satisfies the conditions required in Theorem D.2 for the weak topology of the space X.

STEP 1.   *For each $v \in K$ the set $N(v) := \{u \in K : (v, u) \in \mathcal{A}\}$ is weakly closed.*
In order to prove the above assertion, for a fixed $v \in K$ we consider the functional $\alpha : K \to \mathbb{R}$ defined by

$$\alpha(u) := \Psi(u, v) + J^0(\hat{u}; \hat{v} - \hat{u}) - \langle Fu, v - u \rangle_X$$

and we shall prove that it is weakly upper semicontinuous. Let us consider a sequence $\{u^m\} \subset K$ such that $u^m \rightharpoonup u$ as $m \to \infty$. Taking into account that i_k is compact for each $k \in \{1, \ldots, n\}$ we deduce that $\hat{u}^m \to \hat{u}$ as $m \to \infty$. Using $(H_1) - (ii)$ we obtain

$$\limsup_{m \to \infty} \Psi(u^m, v) = \limsup_{m \to \infty} \sum_{k=1}^{n} \psi_k(u_1^m, \ldots, u_n^m, v_k) \leq \sum_{k=1}^{n} \limsup_{m \to \infty} \psi_k(u_1^m, \ldots, u_n^m, v_k)$$

$$\leq \sum_{k=1}^{n} \psi_k(u_1, \ldots, u_n, v_k) = \Psi(u, v).$$

On the other hand, using Proposition 2.3 we deduce that

$$\limsup_{m \to \infty} J^0(\hat{u}^m; \hat{v} - \hat{u}^m) \leq J^0(\hat{u}; \hat{v} - \hat{u}).$$

Finally, using $(H2)$ we have

$$\limsup_{m \to \infty}[-\langle Fu^m, v - u^m \rangle_X] = -\liminf_{m \to \infty} \langle Fu^m, v - u^m \rangle_X$$

$$= -\liminf_{m \to \infty} \sum_{k=1}^{n} \langle F_k(u_1^m, \ldots, u_n^m), v_k - u_k^m \rangle_{X_k}$$

$$\leq -\sum_{k=1}^{n} \langle F_k(u_1, \ldots, u_n), v_k - u_k \rangle_{X_k} = -\langle Fu, v - u \rangle_X$$

It is clear from the above relations that the functional α is weakly upper semicontinuous, therefore the set

$$[\alpha \geq \lambda] := \{u \in K : \alpha(u) \geq \lambda\}$$

is weakly closed for any $\lambda \in \mathbb{R}$. Taking $\lambda = 0$ we obtain that the set $N(v)$ is weakly closed.

STEP 2.   *For each $u \in K$ the set $M(u) := \{v \in K : (v, u) \notin \mathcal{A}\}$ is either convex or empty.*

Let us fix $u \in K$ and assume that $M(u)$ is nonempty. Let v^1, v^2 be two elements of $M(u)$, $t \in (0, 1)$ and $v^t := tv^1 + (1 - t)v^2$. Using (H1)-(iii) we obtain

$$\Psi(u, v^t) = \sum_{k=1}^{n} \psi_k \left(u_1, \ldots, u_n, tv_k^1 + (1-t)v_k^2\right)$$

$$\leq t \sum_{k=1}^{n} \psi_k \left(u_1, \ldots, u_n, v_k^1\right) + (1-t) \sum_{k=1}^{n} \psi_k \left(u_1, \ldots, u_n, v_k^2\right)$$

$$= t\Psi(u, v^1) + (1-t)\Psi(u, v^2),$$

which shows that the mapping $v \mapsto \Psi(u, v)$ is convex. On the other hand Proposition 2.3 ensures that the mapping $\hat{v} \mapsto J^0(\hat{u}; \hat{v} - \hat{u})$ is convex. Using the fact that the mapping $v \mapsto \langle Fu, v - u \rangle_X$ is affine we are led to

$$\Psi(u, v^t) + J^0(\hat{u}; \hat{v}^t - \hat{u}) - \langle Fu, v^t - u \rangle_X \leq t \left[\Psi(u, v^1) + J^0(\hat{u}; \hat{v}^1 - \hat{u}) - \langle Fu, v^1 - u \rangle_X \right]$$

$$+ (1-t) \left[\Psi(u, v^2) + J^0(\hat{u}; \hat{v}^2 - \hat{u}) - \langle Fu, v^2 - u \rangle_X \right] < 0,$$

which means that $v^t \in M(u)$, therefore $M(u)$ is a convex set.

STEP 3. $(u, u) \in \mathcal{A}$ for each $u \in K$.

Let $u \in K$ be fixed. Using $(H_1) - (i)$ we obtain

$$\Psi(u, u) + J^0(\hat{u}; \hat{u} - \hat{u}) - \langle Fu, u - u \rangle_X = \sum_{k=1}^{n} \psi_k(u_1, \ldots, u_k, \ldots, u_n, u_k) = 0,$$

and this is equivalent to $(u, u) \in \mathcal{A}$.

STEP 4. *The set $B := \{u \in K : (v, u) \in \mathcal{A}$ for all $v \in K\}$ is compact.*

First we observe that K is a weakly compact subset of the reflexive space X as it is bounded, closed and convex. Then, we observe that the set B can be rewritten in the following way

$$B = \bigcap_{v \in K} \mathcal{N}(v).$$

This shows that B is also a weakly compact set as it is an intersection of weakly closed subsets of K.

We are now able to apply Lin's theorem (see Theorem D.2) and conclude that there exists $u^0 \in B \subseteq K$ such that $K \times \{u^0\} \subset \mathcal{A}$. This means that

$$\Psi(u^0, v) + J^0(\hat{u}^0; \hat{v} - \hat{u}^0) \geq \langle Fu^0, v - u^0 \rangle_X, \quad \text{for all } v \in K,$$

therefore u^0 solves problem (VHI) and, according to Remark 8.5, it is a solution of $(SNHI)$.

We will show next that if we change the hypotheses on the nonlinear functionals ψ_k we are still able to prove the existence of at least one solution for our system. Let us consider that instead of (H_1) we have the following set of hypotheses

(H_3) For each $k \in \{1, \ldots, n\}$, the functional $\psi_k : X_1 \times \ldots \times X_k \times \ldots \times X_n \times X_k \to \mathbb{R}$ satisfies

(i) $\psi_k(u_1, \ldots, u_k, \ldots, u_n, u_k) = 0$ for all $u_k \in X_k$;
(ii) For each $k \in \{1, \ldots, n\}$ and any pair $(u_1, \ldots, u_k, \ldots, u_n)$,
       $(v_1, \ldots, v_k, \ldots, v_n) \in X_1 \times \ldots \times X_k \times \ldots \times X_n$ we have

$$\psi_k(u_1, \ldots, u_k, \ldots, u_n, v_k) + \psi_k(v_1, \ldots, v_k, \ldots, v_n, u_k) \geq 0;$$

(iii) For each $(u_1, \ldots, u_n) \in X_1 \times \ldots \times X_n$ the mapping $v_k \mapsto \psi_k(u_1, \ldots, u_n, v_k)$ is weakly lower semicontinuous;
(iv) For each $v_k \in X_k$ the mapping $(u_1, \ldots, u_n) \mapsto \psi_k(u_1, \ldots, u_n, v_k)$ is concave. □

Theorem 8.9 ([5]) *For each $k \in \{1, \ldots, n\}$ let $K_k \subset X_k$ be a nonempty, bounded, closed and convex set and let us assume that conditions $(H_2) - (H_3)$ are fulfilled. Then $(SNHI)$ has at least one solution.*

In order to prove Theorem 8.9 we will need the following lemma.

Lemma 8.3 *Assume that $(H3)$ holds. Then*

(a) $\Psi(u, v) + \Psi(v, u) \geq 0$ *for all $u, v \in X$;*
(b) *For each $v \in X$ the mapping $u \mapsto -\Psi(v, u)$ is weakly upper semicontinuous;*
(c) *For each $u \in X$ the mapping $v \mapsto -\Psi(v, u)$ is convex.*

Proof

(a) Taking into account $(H_3) - (ii)$ and the way the functional $\Psi : X \times X \to \mathbb{R}$ was defined we find

$$\Psi(u, v) + \Psi(v, u) = \sum_{k=1}^{n} [\psi_k(u_1, \ldots, u_k, \ldots, u_n, v_k) + \psi_k(v_1, \ldots, v_k, \ldots, v_n, u_k)] \geq 0.$$

(b) Let $v \in X$ be fixed and let $\{u^m\} \subset X$ be a sequence which converges weakly to some $u \in X$. Using $(H3) - (iii)$ and the fact that $u^m \to u$ we obtain

$$\limsup_{m \to \infty} \left[-\Psi(v, u^m) \right] = -\liminf_{m \to \infty} \Psi(v, u^m) = -\liminf_{m \to \infty} \sum_{k=1}^{n} \psi_k(v_1, \ldots, v_n, u_k^m)$$

$$\leq -\sum_{k=1}^{n} \liminf_{m \to \infty} \psi_k(v_1, \ldots, v_n, u_k^m) \leq -\sum_{k=1}^{n} \psi_k(v_1, \ldots, v_n, u_k)$$

$$= -\Psi(v, u).$$

(c) Let $u, v^1, v^2 \in X$ and $t \in (0, 1)$. Keeping $(H3) - (iv)$ in mind we deduce that

$$\Psi\left(tv^1 + (1-t)v^2, u\right) = \sum_{k=1}^{n} \psi_k\left(tv_1^1 + (1-t)v_1^2, \ldots, tv_n^1 + (1-t)v_n^2, u_k\right)$$

$$\geq \sum_{k=1}^{n} t\psi_k\left(v_1^1, \ldots, v_n^1, u_k\right) + (1-t)\psi_k\left(v_1^2, \ldots, v_n^2, u_k\right)$$

$$= t\Psi(v^1, u) + (1-t)\Psi(v^2, u).$$

We have prove that the mapping $v \mapsto \Psi(v, u)$ is concave, hence the application $v \mapsto -\Psi(v, u)$ must be convex. $\qquad \square$

Proof of Theorem 8.9 Let us consider the set $\mathcal{A} \subset K \times K$ defined by

$$\mathcal{A} := \{(v, u) \in K \times K : -\Psi(v, u) + J^0(\hat{u}; \hat{v} - \hat{u}) - \langle Fu, v - u \rangle_X \geq 0\}.$$

Lemma 8.3 ensures that we can follow the same steps as in the proof of Theorem 8.8 to conclude that the conditions required in Lin's theorem are fulfilled. Thus we get the existence of an element $u^0 \in K$ such that $K \times \{u^0\} \subset \mathcal{A}$ which is equivalent to

$$-\Psi(v, u^0) + J^0(\hat{u}^0; \hat{v} - \hat{u}^0) \geq \langle Fu^0, v - u^0 \rangle_X \quad \text{for all } v \in K. \tag{8.24}$$

On the other hand Lemma 8.3 tells us that

$$\Psi(u^0, v) + \Psi(v, u^0) \geq 0, \quad \text{for all } v \in K. \tag{8.25}$$

Combining relations (8.24) and (8.25) we deduce that u^0 solves problem (VHI), therefore it is a solution of problem $(SNHI)$.

Let us consider now the case when at least one of the subsets K_k is unbounded and either conditions $(H_1) - (H_2)$ or $(H_2) - (H_3)$ hold. Let $R > 0$ be such that the set $K_{k,R} := K_k \cap \bar{B}_{X_k}(0; R)$ is nonempty for every $k \in \{1, \ldots, n\}$. Then, for each $k \in \{1, \ldots, n\}$ the set $K_{k,R}$ is nonempty, bounded, closed and convex and according to Theorem 8.8 or Theorem 8.9 the following problem possesses at least one solution.

(SR) Find $(u_1, \ldots, u_n) \in K_{1,R} \times \ldots \times K_{n,R}$ such that

$$\begin{cases} \psi_1(u_1, \ldots, u_n, v_1) + J_{,1}^0(\hat{u}_1, \ldots, \hat{u}_n; \hat{v}_1 - \hat{u}_1) \geq \langle F_1(u_1, \ldots, u_n), v_1 - u_1 \rangle_{X_1} \\ \ldots \quad \ldots \quad \ldots \quad \ldots \quad \ldots \quad \ldots \quad \ldots \quad \ldots \quad \ldots \quad \ldots \quad \ldots \\ \psi_n(u_1, \ldots, u_n, v_n) + J_{,n}^0(\hat{u}_1, \ldots, \hat{u}_n; \hat{v}_n - \hat{u}_n) \geq \langle F_n(u_1, \ldots, u_n), v_n - u_n \rangle_{X_n}, \end{cases}$$

for all $(v_1, \ldots, v_n) \in K_{1,R} \times \ldots \times K_{n,R}$.

We have the following existence result concerning the case of at least one unbounded subset.   □

Theorem 8.10 ([5]) *For each $k \in \{1, \ldots, n\}$ let $K_k \subset X_k$ be a nonempty, closed and convex set and assume that there exists at least one index $k_0 \in \{1, \ldots, n\}$ such that K_{k_0} is unbounded. Assume in addition that either $(H_1) - (H_2)$ or $(H_2) - (H_3)$ hold. Then $(SNHI)$ possesses at least one solution if and only if the following condition holds:*

(H_4) there exists $R > 0$ such that $K_{k,R}$ is nonempty for every $k \in \{1, \ldots, n\}$ and at least one solution $(u_1^0, \ldots, u_n^0)$ of problem (SR) satisfies

$$u_k^0 \in B_{X_k}(0; R), \quad \forall k \in \{1, \ldots, n\}.$$

Proof The necessity is obvious.

In order to prove the sufficiency for each $k \in \{1, \ldots, n\}$ let us fix $v_k \in K_k$ and define the scalar

$$\lambda_k := \begin{cases} \dfrac{1}{2}, & \text{if } u_k^0 = v_k \\ \min\left\{\dfrac{1}{2}; \dfrac{R - \|u_k^0\|_{X_k}}{\|v_k - u_k^0\|_{X_k}}\right\}, & \text{otherwise.} \end{cases}$$

Condition (H_4) ensures that $\lambda_k \in (0, 1)$, therefore $w_{\lambda_k} := u_k^0 + \lambda_k(v_k - u_k^0)$ is an element of $K_{k,R}$ due to the convexity of the set K_k.

CASE 1. $(H_1) - (H_2)$ *hold.*

Using the fact $(u_1^0, \ldots, u_n^0)$ is a solution of (SR) for each $k \in \{1, \ldots, n\}$ we have

$$\psi_k(u_1^0, \ldots, u_n^0, w_{\lambda_k}) + J_{,k}^0(\hat{u}_1^0, \ldots, \hat{u}_n^0; \hat{w}_{\lambda_k} - \hat{u}_k^0) \geq \langle F_k(u_1^0, \ldots, u_n^0), w_{\lambda_k} - u_k^0 \rangle_{X_k}$$

$$(8.26)$$

In this case relation (8.26) leads to

$$\lambda_k \langle F_k(u_1^0, \ldots, u_n^0), v_k - u_k^0 \rangle_{X_k} = \langle F_k(u_1^0, \ldots, u_n^0), w_{\lambda_k} - u_k^0 \rangle_{X_k}$$

$$\leq \lambda_k \psi_k(u_1^0, \ldots, u_n^0, v_k) + (1 - \lambda_k) \psi_k(u_1^0, \ldots, u_n^0, u_k^0) + \lambda_k J_{,k}^0(\hat{u}_1^0, \ldots, \hat{u}_n^0; \hat{v}_k - \hat{u}_k^0)$$

$$= \lambda_k \left[\psi_k(u_1^0, \ldots, u_n^0, v_k) + J_{,k}^0(\hat{u}_1^0, \ldots, \hat{u}_n^0; \hat{v}_k - \hat{u}_k^0) \right].$$

Dividing by λ_k the above inequality and taking into account that $v_k \in K_k$ was arbitrary fixed we conclude that $(u_1^0, \ldots, u_n^0)$ is a solution of $(SNHI)$.

CASE 2. $(H_2) - (H_3)$ hold.

Theorem 8.9 ensures that (see (8.24))

$$-\Psi(w, u^0) + J^0(\hat{u}^0; \hat{w} - u^0) \geq \langle Fu^0, w - u^0 \rangle, \quad \forall w \in K_R = K_{1,R} \times \ldots \times K_{n,R}.$$

Choosing $w_k := w_{\lambda_k}$ and $w_j = u_j^0$ for $j \neq k$ in the above relation we obtain

$$\lambda_k \left\langle F_k(u_1^0, \ldots, u_n^0), v_k - u_k^0 \right\rangle_{X_k} = \left\langle F_k(u_1^0, \ldots, u_n^0), w_{\lambda_k} - u_k^0 \right\rangle_{X_k}$$

$$= \sum_{j=1}^n \left\langle F_k(u_1^0, \ldots, u_n^0), w_j - u_j^0 \right\rangle_{X_k} = \langle Fu^0, w - u^0 \rangle_X$$

$$\leq -\Psi(w, u^0) + J^0(\hat{u}^0; \hat{w} - \hat{u}^0)$$

$$= -\sum_{j=1}^n \psi_j(w_1, \ldots, w_j, \ldots, w_n, u_j^0) + \sum_{j=1}^n J_{,j}^0(\hat{u}_1^0, \ldots, \hat{u}_n^0; \hat{w}_j - \hat{u}_j^0)$$

$$= -\psi_k(u_1^0, \ldots, w_{\lambda_k}, \ldots, u_n^0, u_k^0) + J_{,k}^0(\hat{u}_1^0, \ldots, \hat{u}_n^0; \hat{w}_{\lambda_k} - \hat{u}_k^0)$$

$$\leq -\lambda_k \psi_k(u_1^0, \ldots, v_k, \ldots, u_n^0, u_k^0) - (1 - \lambda_k) \psi_k(u_1^0, \ldots, u_k^0, \ldots, u_n^0, u_k^0)$$

$$+ \lambda_k J_{,k}^0(\hat{u}_1^0, \ldots, \hat{u}_n^0; \hat{v}_k - \hat{u}_k^0)$$

$$\leq \lambda_k \left[-\psi_k(u_1^0, \ldots, v_k, \ldots, u_n^0, u_k^0) + J_{,k}^0(\hat{u}_1^0, \ldots, \hat{u}_n^0; \hat{v}_k - \hat{u}_k^0) \right]$$

Dividing by λ_k we obtain that

$$-\psi_k(u_1^0, \ldots, v_k, \ldots, u_n^0, u_k^0) + J_{,k}^0(\hat{u}_1^0, \ldots, \hat{u}_n^0; \hat{v}_k - \hat{u}_k^0) \geq \left\langle F_k(u_1^0, \ldots, u_n^0), v_k - u_k^0 \right\rangle_{X_k}.$$

Combining the above inequality and $(H_3) - (ii)$ we deduce the for each $k \in \{1, \ldots, n\}$ the following inequality takes place

$$\psi_k(u_1^0, \ldots, u_k^0, \ldots, u_n^0, v_k) + J_{,k}^0(\hat{u}_1^0, \ldots, \hat{u}_n^0; \hat{v}_k - \hat{u}_k^0) \geq \left\langle F_k(u_1^0, \ldots, u_n^0), v_k - u_k^0 \right\rangle_{X_k},$$

which means that $(u_1^0, \ldots, u_n^0)$ is a solution of $(SNHI)$, since $v_k \in K_k$ was arbitrary fixed.   □

Corollary 8.1 *For each $k \in \{1, \ldots, n\}$ let $K_k \subset X_k$ be a nonempty, closed and convex set and assume that there exists at least one index $k_0 \in \{1, \ldots, n\}$ such that K_{k_0} is unbounded. Assume in addition that either $(H_1)-(H_2)$ or $(H_2)-(H_3)$ hold. Then a sufficient condition for $(SNHI)$ to possess solution is*

(H_5) *there exists $R_0 > 0$ such that K_{k,R_0} is nonempty for every $k \in \{1, \ldots, n\}$ and for each $(u_1, \ldots, u_n) \in K_1 \times \ldots \times K_n \setminus K_{1,R_0} \times \ldots \times K_{n,R_0}$ there exists $(v_1^0, \ldots, v_n^0) \in K_{1,R_0} \times \ldots \times K_{n,R_0}$ such that*

$$\psi_k(u_1, \ldots, u_n, v_k^0) + J_{,k}^0(\hat{u}_1, \ldots, \hat{u}_n; \hat{v}_k^0 - \hat{u}_k) < \langle F_k(u_1, \ldots, u_n), v_k^0 - u_k \rangle_{X_k}, \tag{8.27}$$

for all $k \in \{1, \ldots, n\}$.

Proof Let us fix $R > R_0$. According to Theorem 8.8 or Theorem 8.9 problem (SR) has at least one solution. Let $(u_1, \ldots, u_n) \in K_{1,R} \times \ldots \times K_{n,R}$ be a solution of (SR). We shall prove that $(u_1, \ldots, u_n)$ also solves $(SNHI)$.

CASE 1. $u_k \in B_{X_k}(0, R)$ for all $k \in \{1, \ldots, n\}$.
   In this case we have nothing to prove as Theorem 8.10 ensures that $(u_1, \ldots, u_n)$ is a solution of $(SNHI)$.

CASE 2.   *There exists at least one index $j_0 \in \{1, \ldots, n\}$ such that $u_{j_0} \notin B_{X_{j_0}}(0, R)$.*
   In this case $\|u_{j_0}\|_{X_{j_0}} = R > R_0$, therefore $(u_1, \ldots, u_n) \notin K_{1,R_0} \times \ldots \times K_{n,R_0}$ and according to (H_5) there exist $(v_1^0, \ldots, v_n^0) \in K_{1,R_0} \times \ldots \times K_{n,R_0}$ such that 8.27 holds. For each $k \in \{1, \ldots, n\}$ let us fix $v_k \in K_k$ and define the scalar

$$\lambda_k := \begin{cases} \dfrac{1}{2} & \text{if } v_k = v_k^0 \\[2mm] \min\left\{\dfrac{1}{2}, \dfrac{R-R_0}{\|v_k - v_k^0\|_{X_k}}\right\} & \text{otherwise.} \end{cases}$$

Obviously $\lambda_k \in (0, 1)$ and $w_{\lambda_k} = v_k^0 + \lambda_k(v_k - v_k^0) \in K_{k,R}$. Furthermore, we observe that

$$w_{\lambda_k} - u_k = v_k^0 - u_k + \lambda_k v_k - \lambda_k v_k^0 + \lambda_k u_k - \lambda_k u_k = \lambda_k(v_k - u_k) + (1 - \lambda_k)(v_k^0 - u_k).$$

CASE 2.1 $(H_1) - (H_2)$ *hold.*

Using the fact that $(u_1, \ldots, u_n)$ solves (SR) we obtain the following estimates

$$\langle F_k(u_1, \ldots, u_n), w_{\lambda_k} - u_k \rangle = \lambda_k \langle F_k(u_1, \ldots, u_n), v_k - u_k \rangle_{X_k}$$

$$+ (1 - \lambda_k)\langle F_k(u_1, \ldots, u_n), v_k^0 - u_k \rangle_{X_k}$$

$$\leq \psi_k(u_1, \ldots, u_n, w_{\lambda_k}) + J_{,k}^0(\hat{u}_1, \ldots, \hat{u}_n; \hat{w}_{\lambda_k} - \hat{u}_k)$$

$$\leq \lambda_k \left[\psi_k(u_1, \ldots, u_n, v_k) + J_{,k}^0(\hat{u}_1, \ldots, \hat{u}_n; \hat{v}_k - \hat{u}_k) \right]$$

$$+ (1 - \lambda_k) \left[\psi_k(u_1, \ldots, u_n, v_k^0) + J_{,k}^0(\hat{u}_1, \ldots, \hat{u}_n; \hat{v}_k^0 - \hat{u}_k) \right].$$

Combining the above relation and (8.27) we obtain that

$$F_k(u_1, \ldots, u_n), v_k - u_k \rangle_{X_k} \leq \psi_k(u_1, \ldots, u_n, v_k) + J_{,k}^0(\hat{u}_1, \ldots, \hat{u}_n; v_k - u_k)$$

for all $k \in \{1, \ldots, n\}$, which means that $(u_1, \ldots, u_n)$ is a solution of $(SNHI)$.

CASE 2.2. $(H_2) - (H_3)$ *hold.*

The fact that $(u_1, \ldots, u_n)$ solves (SR) and relation 8.24 allow us to conclude that

$$-\Psi(w, u) + J^0(\hat{u}, \hat{w} - \hat{u}) \geq \langle Fu, w - u \rangle_X, \quad \forall w \in K_R := K_{1,R} \times \ldots \times K_{n,R}.$$

Choosing $w_k := w_{\lambda_k}$ and $w_j = u_j$ for $j \neq k$ in the above relation and using $(H_3) - (iv)$ we obtain

$$\langle F_k(u_1, \ldots, u_n), w_{\lambda_k} - u_k \rangle_{X_k} = \lambda_k \langle F_k(u_1, \ldots, u_n), v_k - u_k \rangle_{X_k}$$

$$+ (1 - \lambda_k) \left\langle F_k(u_1, \ldots, u_n), v_k^0 - u_k \right\rangle_{X_k}$$

$$= \sum_{j=1}^n \langle F_k(u_1, \ldots, u_n), w_j - u_j \rangle_{X_k} = \langle Fu, w - u \rangle$$

$$\leq -\Psi(w, u) + J^0(\hat{u}; \hat{w} - \hat{u})$$

$$= -\sum_{j=1}^n \psi_j(w_1, \ldots, w_j, \ldots, w_n, u_j) + \sum_{j=1}^n J_{,j}^0(\hat{u}_1, \ldots, \hat{u}_n; \hat{w}_j - \hat{u}_j)$$

$$= -\psi_k(u_1, \ldots, w_{\lambda_k}, \ldots, u_n, u_k)) + J_{,k}^0(\hat{u}_1, \ldots, \hat{u}_n; \hat{w}_{\lambda_k} - \hat{u}_k)$$

$$\leq -\lambda_k \psi_k(u_1, \ldots, v_k, \ldots, u_n, u_k) - (1 - \lambda_k)\psi_k(u_1, \ldots, v_k^0, \ldots, u_n, u_k)$$

$$+ \lambda_k J_{,k}^0(\hat{u}_1, \ldots, \hat{u}_n; \hat{v}_k - \hat{u}_k) + (1 - \lambda_k)J^0(\hat{u}_1, \ldots, \hat{u}_n; \hat{v}_k^0 - \hat{u}_k).$$

Using $(H_3) - (ii)$ and (8.27) we deduce that

$$F_k(u_1, \ldots, u_n), v_k - u_k)_{X_k} \leq \psi_k(u_1, \ldots, u_n, v_k) + J^0_{,k}(\hat{u}_1, \ldots, \hat{u}_n; v_k - u_k)$$

for all $k \in \{1, \ldots, n\}$ which means that $(u_1, \ldots, u_n)$ is a solution of $(SNHI)$.   □

In order to simplify some computations let us assume next that $0 \in K_k$ for each $k \in \{1, \ldots, n\}$. In this case $K_{k,R} \neq \varnothing$ for every $k \in \{1, \ldots, n\}$ and every $R > 0$.

Corollary 8.2 *For each $k \in \{1, \ldots, n\}$ let $K_k \subset X_k$ be a nonempty, closed and convex set and assume that there exists at least one index $k_0 \in \{1, \ldots, n\}$ such that K_{k_0} is unbounded and either $(H_1) - (H_2)$ or $(H_2) - (H_3)$ hold. Assume in addition that for each $k \in \{1, \ldots, n\}$ the following conditions hold*

(H_6) *There exists a function $c : \mathbb{R}_+ \to \mathbb{R}_+$ with the property that $\lim_{t \to \infty} c(t) = +\infty$ such that*

$$-\sum_{k=1}^{n} \psi_k(u_1, \ldots, u_k, \ldots, u_n, 0) \geq c(\|u\|_X)\|u\|_X,$$

for all $(u_1, \ldots, u_n) \in X_1 \times \ldots \times X_n$, where $\|u\|_X := \left(\sum_{k=1}^{n} \|u_k\|^2_{X_k}\right)^{1/2}$;
(H_7) *There exists $M_k > 0$ such that*

$$J^0_{,k}(w_1, \ldots, w_k, \ldots, w_n; -w_k) \leq M_k\|w_k\|_{Y_k}, \quad \forall(w_1, \ldots, w_n) \in Y_1 \times \ldots \times Y_n;$$

(H_8) *There exists $m_k > 0$ such that*

$$\|F_k(u_1, \ldots, u_k, \ldots, u_n)\|_{X^*_k} \leq m_k, \quad \forall(u_1, \ldots, u_n) \in X_1 \times \ldots \times X_n.$$

Then the system $(SNHI)$ has at least one solution.

Proof For each $R > 0$ Theorem 8.8 (or Theorem 8.9) enables us to conclude that there exists a solution $(u_{1R}, \ldots, u_{nR}) \in K_{1,R} \times \ldots \times K_{n,R}$ of problem (SR). We shall prove that there exists $R_0 > 0$ such that

$$u_{kR_0} \in \text{int } B_{X_k}(0; R_0), \quad \text{for all } k \in \{1, \ldots, n\},$$

which, according to Theorem 8.10, means that $(u_{1R_0}, \ldots, u_{nR_0})$ is a solution of the system $(SNHI)$.

Arguing by contradiction let us assume that for each $R > 0$ there exists at least one index $j_0 \in \{1, \ldots, n\}$ such that $u_{j_0R} \notin B_{X_{j_0}}(0, R)$, therefore $\|u_{j_0R}\|_{X_{j_0}} = R$. Using the fact that $(u_{1R}, \ldots, u_{nR})$ solves (SR) we conclude that for each $k \in \{1, \ldots, n\}$ the following inequality holds

$$\psi_k(u_{1R}, \ldots, u_{nR}, v_k) + J^0_{,k}(\hat{u}_{1R}, \ldots, \hat{u}_{nR}; \hat{v}_k - \hat{u}_{kR}) \geq \langle F_k(u_{1R}, \ldots, u_{nR}), v_k - u_{kR} \rangle_{X_k},$$
(8.28)

for all $v_k \in K_{k,R}$.

Taking $v_k = 0$ in (8.28), summing and using $(H_6) - (H_8)$ we have

$$c(\|u\|_X)\|u\|_X \leq -\sum_{k=1}^{n} \psi_k(u_{1R}, \ldots, u_{j_0R}, \ldots, u_{nR}, 0)$$

$$\leq \sum_{k=1}^{n} \left[\langle F_k(u_{1R}, \ldots, u_{nR}), u_{kR} \rangle_{X_k} + J^0_{,k}(\hat{u}_{1R}, \ldots, \hat{u}_k, \ldots, \hat{u}_{nR}; -\hat{u}_k) \right]$$

$$\leq \sum_{k=1}^{n} \left(\|F_k(u_{1R}, \ldots, u_{nR})\|_{X_k^*} \|u_k\|_{X_k} + M_k \|\hat{u}_{kR}\|_{Y_k} \right)$$

$$\leq \sum_{k=1}^{n} \left[(m_k + M_k \|T_k\|) \|u_{kR}\|_{X_k} \right]$$

$$\leq C\|u\|_X.$$

Dividing by $\|u\|_X$ and letting $R \to +\infty$ we obtain a contradiction since the left-hand term of the inequality is unbounded while the right-hand term remains bounded. □

References

1. J.-P. Aubin, H. Frankowska, *Set-Valued Analysis* (Birkhäuser, Boston, 1990)
2. T. Bartsch, Z.-Q. Wang, Existence and multiplicity results for some superlinear elliptic problems in $\mathbb{R}^N$. Commun. Partial Differ. Equ. **20**, 1725–1741 (1995)
3. H. Brezis, *Functional Analysis, Sobolev Spaces and Partial Differential Equations* (Springer, Berlin, 2011)
4. N. Costea, V. Rădulescu, Hartman–Stampacchia results for stably pseudomonotone operators and non-linear hemivariational inequalities. Appl. Anal. **89**, 165–188 (2010)
5. N. Costea, C. Varga, Systems of nonlinear hemivariational inequalities and applications. Topol. Methods Nonlinear Anal. **41**, 39–65 (2013)
6. D.C. de Morais Filho, M.A.S. Souto, J.M.D.O, A compactness embedding lemma, a principle of symmetric criticality and applications to elliptic problems, in *Proyecciones*, vol. 19 (Universidad Catolica del Norte, Antofagasta, 2000), pp. 1–17
7. J. Gwinner, Stability of monotone variational inequalities with various applications, in *Variational Inequalities and Network Equilibrium Problems*, ed. by F. Gianessi, A. Maugeri (Plenum Press, New York, 1995), pp. 123–142
8. O. Kavian, *Introduction à la Théorie des Point Critique et Applications aux Proble'emes Elliptique* (Springer, Berlin, 1995)

9. A. Kristály, Location of Nash equilibria: a Riemannian approach. Proc. Am. Math. Soc. **138**, 1803–1810 (2010)
10. A. Kristály, C. Varga, A set-valued approach to hemivariational inequalities. Topol. Methods Nonlinear Anal. **24**, 297–307 (2004)
11. T.-C. Lin, Convex sets, fixed points, variational and minimax inequalities. Bull. Aust. Math. Soc. **34**, 107–117 (1986)
12. H. Lisei, C. Varga, Some applications to variational-hemivariational inequalities of the principle of symmetric criticality for Motreanu-Panagiotopoulos type functionals. J. Global Optim. **36**, 283–305 (2006)
13. D. Motreanu, V. Rădulescu, Existence results for inequality problems with lack of convexity. Numer. Funct. Anal. Optim. **21**, 869–884 (2000)
14. Z. Naniewicz, P.D. Panagiotopoulos, *Mathematical Theory of Hemivariational Inequalities and Applications* (Marcel Dekker, New York, 1995)
15. P.D. Panagiotopoulos, M. Fundo, V. Rădulescu, Existence theorems of Hartman-Stampacchia type for hemivariational inequalities and applications. J. Global Optim. **15**, 41–54 (1999)
16. R. Precup, Nash-type equilibria for systems of Szulkin functionals. Set-Valued Var. Anal. **24**, 471–482 (2016)
17. D. Repovš, C. Varga, A Nash-type solution for hemivariational inequality systems. Nonlinear Anal. **74**, 5585–5590 (2011)

Nonsmooth Nash Equilibria on Smooth Manifolds 9

9.1 Nash Equilibria on Curved Spaces

After the seminal paper of Nash [14] there has been considerable interest in the theory of Nash equilibria due to its applicability in various real-life phenomena (game theory, price theory, networks, etc). The Nash equilibrium problem involves n players such that each player know the equilibrium strategies of the partners, but moving away from his/her own strategy alone a player has nothing to gain. Formally, if the sets K_i denote the strategies of the players and $f_i : K_1 \times \ldots \times K_n \to \mathbf{R}$ are their loss-functions, $i \in \{1, \ldots, n\}$, the objective is to find an n-tuple $\mathbf{p} = (p_1, \ldots, p_n) \in \mathbf{K} = K_1 \times \ldots \times K_n$ such that

$$f_i(\mathbf{p}) = f_i(p_i, \mathbf{p}_{-i}) \leq f_i(q_i, \mathbf{p}_{-i}) \text{ for every } q_i \in K_i \text{ and } i \in \{1, \ldots, n\},$$

where $(q_i, \mathbf{p}_{-i}) = (p_1, \ldots, p_{i-1}, q_i, p_{i+1}, \ldots, p_n) \in \mathbf{K}$. Such point $\mathbf{p}$ is called a *Nash equilibrium point for* $(\mathbf{f}, \mathbf{K}) = (f_1, \ldots, f_n; K_1, \ldots, K_n)$, the set of these points being denoted by $\mathcal{S}_{NE}(\mathbf{f}, \mathbf{K})$. The starting point of our analysis is the following result of Nash [14, 15]:

Theorem 9.1 (Nash [14, 15]) *Let $K_1, \ldots, K_n$ be nonempty, compact, convex subsets of Hausdorff topological vector spaces and $f_i : K_1 \times \ldots \times K_n \to \mathbb{R}$ ($i = 1, \ldots, n$) be continuous functions such that $K_i \ni q_i \mapsto f_i(q_i, \mathbf{p}_{-i})$ is quasiconvex for all fixed $\mathbf{p} \in \mathbf{K}$. Then $\mathcal{S}_{NE}(\mathbf{f}, \mathbf{K}) \neq \emptyset$.*

The original proof of Theorem 9.1 is based on the Kakutani fixed point theorem, and we postpone it since a more general result will be provided in the sequel.

While most of the known developments in the Nash equilibrium theory deeply exploit the usual convexity of the sets K_i together with the vector space structure of their ambient spaces M_i (i.e., $K_i \subset M_i$), it is nevertheless true that these results are in large part *geometrical* in nature. The main purpose of the present section is to enhance those geometrical and analytical structures which serve as a basis of a systematic study of location of Nash-type equilibria in a general setting as presently possible. In the light of these facts our contribution to the Nash equilibrium theory should be considered intrinsical and analytical rather than game-theoretical, based on nonsmooth analysis on manifolds. In fact, we assume *a priori* that the strategy sets of the players are *geodesic convex* subsets of certain finite-dimensional Riemannian manifolds. This approach can be widely applied when the strategy sets are 'curved'. We also notice that the choice of such Riemannian structures does not influence the Nash equilibrium points.

Let $K_1, \ldots, K_n$ ($n \geq 2$) be non-empty sets, corresponding to the strategies of n players and $f_i : K_1 \times \ldots \times K_n \to \mathbb{R}$ ($i \in \{1, \ldots, n\}$) be the payoff functions, respectively. Throughout this section, the following notations/conventions are used:

- $\mathbf{K} = K_1 \times \ldots \times K_n$; $\mathbf{f} = (f_1, \ldots, f_n)$; $(\mathbf{f}, \mathbf{K}) = (f_1, \ldots, f_n; K_1, \ldots, K_n)$;
- $\mathbf{p} = (p_1, \ldots, p_n)$;
- $\mathbf{p}_{-i}$ is a strategy profile of all players except for player i;
  $(q_i, \mathbf{p}_{-i}) = (p_1, \ldots, p_{i-1}, q_i, p_{i+1}, \ldots, p_n)$; in particular, $(p_i, \mathbf{p}_{-i}) = \mathbf{p}$;
- $\mathbf{K}_{-i}$ is the strategy set profile of all players except for player i;
  $(U_i, \mathbf{K}_{-i}) = K_1 \times \ldots \times K_{i-1} \times U_i \times K_{i+1} \times \ldots \times K_n$ for some $U_i \supset K_i$.

We recall that a set $K \subset M$ is *geodesic convex* if every two points $q_1, q_2 \in K$ can be joined by a unique minimizing geodesic whose image belongs to K. Note that (2.14) is also valid for every $q_1, q_2 \in K$ in a geodesic convex set K since $\exp_{q_i}^{-1}$ is well-defined on $K, i \in \{1, 2\}$. The function $f : K \to \mathbb{R}$ is *convex*, if $f \circ \gamma : [0, 1] \to \mathbb{R}$ is convex in the usual sense for every geodesic $\gamma : [0, 1] \to K$ provided that $K \subset M$ is a geodesic convex set.

An immediate extension of Theorem 9.1 reads as follows:

Theorem 9.2 ([6]) *Let* (M_i, g_i) *be finite-dimensional Riemannian manifolds, $K_i \subset M_i$ be non-empty, compact, geodesic convex sets, and $f_i : \mathbf{K} \to \mathbb{R}$ be continuous functions such that $K_i \ni q_i \mapsto f_i(q_i, \mathbf{p}_{-i})$ is convex on K_i for every $\mathbf{p}_{-i} \in \mathbf{K}_{-i}, i \in \{1, \ldots, n\}$. Then there exists at least one Nash equilibrium point for* $(\mathbf{f}, \mathbf{K})$, *i.e.,* $S_{NE}(\mathbf{f}, \mathbf{K}) \neq \emptyset$.

In order to prove Theorem 9.2 we first start with the following expected result.

Proposition 9.1 *Any geodesic convex set $K \subset M$ is contractible.*

Proof Let us fix $p \in K$ arbitrarily. Since K is geodesic convex, every point $q \in K$ can be connected to p uniquely by the geodesic segment $\gamma_q : [0, 1] \to K$, i.e., $\gamma_q(0) = p$, $\gamma_q(1) = q$. Moreover, the map $K \ni q \mapsto \exp_p^{-1}(q) \in T_p M$ is well-defined and continuous. Note that $\gamma_q(t) = \exp_p(t \exp_p^{-1}(q))$. We define the map $F : [0, 1] \times K \to K$ by $F(t, q) = \gamma_q(t)$; it is clear that F is continuous, $F(1, q) = q$ and $F(0, q) = p$ for all $q \in K$, i.e., the identity map id_K is homotopic to the constant map p. □

Proof of Theorem 9.2 Let $X = \mathbf{K} = \Pi_{i=1}^n K_i$ and $h : X \times X \to \mathbb{R}$ defined by $h(\mathbf{q}, \mathbf{p}) = \sum_{i=1}^n [f_i(q_i, \mathbf{p}_{-i}) - f_i(\mathbf{p})]$. First of all, note that the sets K_i are ANRs, due to Hanner's theorem, see Bessaga and Pełczyński [2, Theorem 5.1]. Moreover, since a product of a finite family of ANRs is an ANR, see [2, Corollary 5.5], it follows that X is an ANR. Due to Proposition 9.1, X is contractible, thus acyclic.

We notice that the function h is continuous, and $h(\mathbf{p}, \mathbf{p}) = 0$ for every $\mathbf{p} \in X$. Consequently, the set $\{(\mathbf{q}, \mathbf{p}) \in X \times X : 0 > h(\mathbf{q}, \mathbf{p})\}$ is open.

It remains to prove that $S_{\mathbf{p}} = \{\mathbf{q} \in X : 0 > h(\mathbf{q}, \mathbf{p})\}$ is contractible or empty for all $\mathbf{p} \in X$. Assume that $S_{\mathbf{p}} \neq \emptyset$ for some $\mathbf{p} \in X$. Then, there exists $i_0 \in \{1, \dots, n\}$ such that $f_{i_0}(q_{i_0}, \mathbf{p}_{-i_0}) - f_{i_0}(\mathbf{p}) < 0$ for some $q_{i_0} \in K_{i_0}$. Therefore, $\mathbf{q} = (q_{i_0}, \mathbf{p}_{-i_0}) \in S_{\mathbf{p}}$, i.e., $\mathrm{pr}_i S_{\mathbf{p}} \neq \emptyset$ for every $i \in \{1, \dots, n\}$. Now, we fix $\mathbf{q}^j = (q_1^j, \dots, q_n^j) \in S_{\mathbf{p}}$, $j \in \{1, 2\}$ and let $\gamma_i : [0, 1] \to K_i$ be the unique geodesic joining the points $q_i^1 \in K_i$ and $q_i^2 \in K_i$ (note that K_i is geodesic convex), $i \in \{1, \dots, n\}$. Let $\gamma : [0, 1] \to \mathbf{K}$ be defined by $\gamma(t) = (\gamma_1(t), \dots, \gamma_n(t))$. Due to the convexity of the function $K_i \ni q_i \mapsto f_i(q_i, \mathbf{p}_{-i})$, for every $t \in [0, 1]$, we have

$$
h(\gamma(t), \mathbf{p}) = \sum_{i=1}^n [f_i(\gamma_i(t), \mathbf{p}_{-i}) - f_i(\mathbf{p})]
$$

$$
\leq \sum_{i=1}^n [t f_i(\gamma_i(1), \mathbf{p}_{-i}) + (1 - t) f_i(\gamma_i(0), \mathbf{p}_{-i}) - f_i(\mathbf{p})]
$$

$$
= t h(\mathbf{q}^2, \mathbf{p}) + (1 - t) h(\mathbf{q}^1, \mathbf{p})
$$

$$
< 0.
$$

Consequently, $\gamma(t) \in S_{\mathbf{p}}$ for every $t \in [0, 1]$, i.e., $S_{\mathbf{p}}$ is a geodesic convex set in the product manifold $\mathbf{M} = \Pi_{i=1}^n M_i$ endowed with its natural (warped-)product metric (with the constant weight functions 1), see O'Neill [17, p. 208]. Now, Proposition 9.1 implies that $S_{\mathbf{p}}$ is contractible.

We are in the position to apply Theorem D.9. Therefore, there exists $\mathbf{p} \in \mathbf{K}$ such that $0 = h(\mathbf{p}, \mathbf{p}) \leq h(\mathbf{q}, \mathbf{p})$ for every $\mathbf{q} \in \mathbf{K}$. In particular, putting $\mathbf{q} = (q_i, \mathbf{p}_{-i})$, $q_i \in K_i$ fixed, we obtain that $f_i(q_i, \mathbf{p}_{-i}) - f_i(\mathbf{p}) \geq 0$ for every $i \in \{1, \dots, n\}$, i.e., $\mathbf{p}$ is a Nash equilibrium point for $(\mathbf{f}, \mathbf{K})$. □

9.2   Comparison of Nash-Type Equilibria on Manifolds

Similarly to Theorem 9.2, let us assume that for every $i \in \{1, \ldots, n\}$, one can find a finite-dimensional Riemannian manifold (M_i, g_i) such that the strategy set K_i is closed and geodesic convex in (M_i, g_i). Let $\mathbf{M} = M_1 \times \ldots \times M_n$ be the product manifold with its standard Riemannian product metric

$$\mathbf{g}(\mathbf{V}, \mathbf{W}) = \sum_{i=1}^{n} g_i(V_i, W_i) \tag{9.1}$$

for every $\mathbf{V} = (V_1, \ldots, V_n), \mathbf{W} = (W_1, \ldots, W_n) \in T_{p_1}M_1 \times \ldots \times T_{p_n}M_n = T_{\mathbf{p}}\mathbf{M}$. Let $\mathbf{U} = U_1 \times \ldots \times U_n \subset \mathbf{M}$ be an open set such that $\mathbf{K} \subset \mathbf{U}$; we always mean that U_i inherits the Riemannian structure of (M_i, g_i). Let

$$\mathcal{L}_{(\mathbf{K}, \mathbf{U}, \mathbf{M})} = \{\mathbf{f} = (f_1, \ldots, f_n) \in C^0(\mathbf{K}, \mathbb{R}^n) : f_i : (U_i, \mathbf{K}_{-i}) \to \mathbb{R} \text{ is continuous and}$$

$$f_i(\cdot, \mathbf{p}_{-i}) \text{ is locally Lipschitz on } (U_i, g_i)$$

$$\text{for all } \mathbf{p}_{-i} \in \mathbf{K}_{-i}, \ i \in \{1, \ldots, n\}\}.$$

Definition 9.1 [6] Let $\mathbf{f} \in \mathcal{L}_{(\mathbf{K}, \mathbf{U}, \mathbf{M})}$. The set of *Nash-Clarke equilibrium points for* $(\mathbf{f}, \mathbf{K})$ is

$$\mathcal{S}_{NC}(\mathbf{f}, \mathbf{K}) = \{\mathbf{p} \in \mathbf{K} : f_i^0(\mathbf{p}; \exp_{p_i}^{-1}(q_i)) \geq 0 \text{ for all } q_i \in K_i, \ i \in \{1, \ldots, n\}\}.$$

Here, $f_i^0(\mathbf{p}; \exp_{p_i}^{-1}(q_i))$ denotes the Clarke generalized derivative of $f_i(\cdot, \mathbf{p}_{-i})$ at point $p_i \in K_i$ in direction $\exp_{p_i}^{-1}(q_i) \in T_{p_i}M_i$. More precisely,

$$f_i^0(\mathbf{p}; \exp_{p_i}^{-1}(q_i)) = \limsup_{q \to p_i, q \in U_i, \ t \to 0^+} \frac{f_i(\sigma_{q, \exp_{p_i}^{-1}(q_i)}(t), \mathbf{p}_{-i}) - f_i(q, \mathbf{p}_{-i})}{t}, \tag{9.2}$$

where $\sigma_{q,v}(t) = \exp_q(tw)$, and $w = d(\exp_q^{-1} \circ \exp_{p_i})_{\exp_{p_i}^{-1}(q)} v$ for $v \in T_{p_i}M_i$, and $t > 0$ is small enough.

The following existence result is available concerning the Nash-Clarke points for $(\mathbf{f}, \mathbf{K})$.

Theorem 9.3 ([6]) *Let* (M_i, g_i) *be complete finite-dimensional Riemannian manifolds,* $K_i \subset M_i$ *be non-empty, compact, geodesic convex sets, and* $\mathbf{f} \in \mathcal{L}_{(\mathbf{K}, \mathbf{U}, \mathbf{M})}$ *such that for every* $\mathbf{p} \in \mathbf{K}$, $i \in \{1, \ldots, n\}$, $K_i \ni q_i \mapsto f_i^0(\mathbf{p}; \exp_{p_i}^{-1}(q_i))$ *is convex and* f_i^0 *is upper semicontinuous on its domain of definition. Then* $\mathcal{S}_{NC}(\mathbf{f}, \mathbf{K}) \neq \emptyset$.

Proof The proof is similar to that of Theorem 9.2; we show only the differences. Let $X = \mathbf{K} = \Pi_{i=1}^n K_i$ and $h : X \times X \to \mathbb{R}$ defined by $h(\mathbf{q}, \mathbf{p}) = \sum_{i=1}^n f_i^0(\mathbf{p}; \exp_{p_i}^{-1}(q_i))$. It is clear that $h(\mathbf{p}, \mathbf{p}) = 0$ for every $\mathbf{p} \in X$.

First of all, the upper-semicontinuity of $h(\cdot, \cdot)$ on $X \times X$ implies the fact that the set $\{(\mathbf{q}, \mathbf{p}) \in X \times X : 0 > h(\mathbf{q}, \mathbf{p})\}$ is open.

Now, let $\mathbf{p} \in X$ such that $S_{\mathbf{p}} = \{\mathbf{q} \in X : 0 > h(\mathbf{q}, \mathbf{p})\}$ is not empty. Then, there exists $i_0 \in \{1, \dots, n\}$ such that $f_i^0(\mathbf{p}; \exp_{p_i}^{-1}(q_{i_0})) < 0$ for some $q_{i_0} \in K_{i_0}$. Consequently, $\mathbf{q} = (q_{i_0}, \mathbf{p}_{-1}) \in S_{\mathbf{p}}$, i.e., $\mathrm{pr}_i S_{\mathbf{p}} \neq \emptyset$ for every $i \in \{1, \dots, n\}$. Now, we fix $\mathbf{q}^j = (q_1^j, \dots, q_n^j) \in S_{\mathbf{p}}$, $j \in \{1, 2\}$, and let $\gamma_i : [0, 1] \to K_i$ be the unique geodesic joining the points $q_i^1 \in K_i$ and $q_i^2 \in K_i$. Let also $\gamma : [0, 1] \to \mathbf{K}$ defined by $\gamma(t) = (\gamma_1(t), \dots, \gamma_n(t))$. Due to the convexity assumption on the function $K_i \ni q_i \mapsto f_i^0(\mathbf{p}; \exp_{p_i}^{-1}(q_i))$ for every $\mathbf{p} \in \mathbf{K}$, $i \in \{1, \dots, n\}$, the convexity of the function $[0, 1] \ni t \mapsto h(\gamma(t), \mathbf{p})$, $t \in [0, 1]$ easily follows. Therefore, $\gamma(t) \in S_{\mathbf{p}}$ for every $t \in [0, 1]$, i.e., $S_{\mathbf{p}}$ is a geodesic convex set, thus contractible.

Lemma D.9 implies the existence of $\mathbf{p} \in \mathbf{K}$ such that $0 = h(\mathbf{p}, \mathbf{p}) \leq h(\mathbf{q}, \mathbf{p})$ for every $\mathbf{q} \in \mathbf{K}$. In particular, if $\mathbf{q} = (q_i, \mathbf{p}_{-i})$, $q_i \in K_i$ fixed, we obtain that $f_i^0(\mathbf{p}; \exp_{p_i}^{-1}(q_i)) \geq 0$ for every $i \in \{1, \dots, n\}$, i.e., $\mathbf{p}$ is a Nash-Clarke equilibrium point for $(\mathbf{f}, \mathbf{K})$. □

Remark 9.1 Although Theorem 9.3 gives a possible approach to locate Nash equilibria on Riemannian manifolds, its applicability is quite reduced. Indeed, $f_i^0(\mathbf{p}; \exp_{p_i}^{-1}(\cdot))$ has no convexity property in general, unless we are in the Euclidean setting or the set K_i is a geodesic segment. For instance, if $\mathbb{H}^2$ is the standard Poincaré upper-plane with the metric $g_{\mathbb{H}} = (\frac{\delta_{ij}}{y^2})$ and we consider the function $f : \mathbb{H}^2 \times \mathbb{R} \to \mathbb{R}$, $f((x, y), r) = rx$ and the geodesic segment $\gamma(t) = (1, e^t)$ in $\mathbb{H}^2$, $t \in [0, 1]$, the function

$$t \mapsto f_1^0(((2, 1), r); \exp_{(2,1)}^{-1}(\gamma(t))) = r \left(e^{2t} \frac{\sinh 2}{2} + e^t \cosh 1 \sqrt{e^{2t}(\cosh 1)^2 - 1} \right)^{-1}$$

is not convex.

The limited applicability of Theorem 9.3 comes from the involved form of the set $S_{NC}(\mathbf{f}, \mathbf{K})$ which motivates the introduction and study of the following concept which plays the central role in the present section.

Definition 9.2 [7] Let $\mathbf{f} \in \mathcal{L}_{(\mathbf{K}, \mathbf{U}, \mathbf{M})}$. The set of *Nash-Stampacchia equilibrium points for* $(\mathbf{f}, \mathbf{K})$ is

$$S_{NS}(\mathbf{f}, \mathbf{K}) = \{\mathbf{p} \in \mathbf{K} : \exists \xi_C^i \in \partial_C^i f_i(\mathbf{p}) \text{ such that } \langle \xi_C^i, \exp_{p_i}^{-1}(q_i) \rangle_{g_i} \geq 0,$$

$$\text{for all } q_i \in K_i, \ i \in \{1, \dots, n\}\}.$$

Here, $\partial_C^i f_i(\mathbf{p})$ denotes the Clarke subdifferential of the function $f_i(\cdot, \mathbf{p}_{-i})$ at point $p_i \in K_i$, i.e., $\partial_C f_i(\cdot, \mathbf{p}_{-i})(p_i) = \mathrm{co}(\partial_L f_i(\cdot, \mathbf{p}_{-i})(p_i))$.

Our first aim is to compare the three Nash-type equilibrium points. Before to do that, we introduce another two classes of functions. If $U_i \subset M_i$ is geodesic convex for every $i \in \{1, \ldots, n\}$, we may define

$$\mathcal{K}_{(\mathbf{K},\mathbf{U},\mathbf{M})} = \{\mathbf{f} \in C^0(\mathbf{K}, \mathbb{R}^n) : f_i : (U_i, \mathbf{K}_{-i}) \to \mathbb{R} \text{ is continuous and } f_i(\cdot, \mathbf{p}_{-i}) \text{ is}$$
$$\text{convex on } (U_i, g_i) \text{ for all } \mathbf{p}_{-i} \in \mathbf{K}_{-i}, \ i \in \{1, \ldots, n\}\},$$

and

$$C_{(\mathbf{K},\mathbf{U},\mathbf{M})} = \{\mathbf{f} \in C^0(\mathbf{K}, \mathbb{R}^n) : f_i : (U_i, \mathbf{K}_{-i}) \to \mathbb{R} \text{ is continuous and } f_i(\cdot, \mathbf{p}_{-i}) \text{ is of}$$
$$\text{class } C^1 \text{ on } (U_i, g_i) \text{ for all } \mathbf{p}_{-i} \in \mathbf{K}_{-i}, \ i \in \{1, \ldots, n\}\}.$$

Remark 9.2 Due to Azagra, Ferrera and López-Mesas [1, Proposition 5.2], one has that $\mathcal{K}_{(\mathbf{K},\mathbf{U},\mathbf{M})} \subset \mathcal{L}_{(\mathbf{K},\mathbf{U},\mathbf{M})}$. Moreover, it is clear that $C_{(\mathbf{K},\mathbf{U},\mathbf{M})} \subset \mathcal{L}_{(\mathbf{K},\mathbf{U},\mathbf{M})}$.

The main result of this subsection reads as follows.

Theorem 9.4 ([7]) *Let (M_i, g_i) be finite-dimensional Riemannian manifolds, $K_i \subset M_i$ be non-empty, closed, geodesic convex sets, $U_i \subset M_i$ be open sets containing K_i, and $f_i : \mathbf{K} \to \mathbb{R}$ be some functions, $i \in \{1, \ldots, n\}$. Then, we have*

(i) $S_{NE}(\mathbf{f}, \mathbf{K}) \subset S_{NS}(\mathbf{f}, \mathbf{K}) = S_{NC}(\mathbf{f}, \mathbf{K})$ *whenever* $\mathbf{f} \in \mathcal{L}_{(\mathbf{K},\mathbf{U},\mathbf{M})}$;

(ii) $S_{NE}(\mathbf{f}, \mathbf{K}) = S_{NS}(\mathbf{f}, \mathbf{K}) = S_{NC}(\mathbf{f}, \mathbf{K})$ *whenever* $\mathbf{f} \in \mathcal{K}_{(\mathbf{K},\mathbf{U},\mathbf{M})}$;

Proof (i) First, we prove that $S_{NE}(\mathbf{f}, \mathbf{K}) \subset S_{NS}(\mathbf{f}, \mathbf{K})$. Indeed, we have
$\mathbf{p} \in S_{NE}(\mathbf{f}, \mathbf{K}) \Leftrightarrow$

$\Leftrightarrow f_i(q_i, \mathbf{p}_{-i}) \geq f_i(\mathbf{p}) \text{ for all } q_i \in K_i, \ i \in \{1, \ldots, n\}$

$\Leftrightarrow 0 \in \partial_{cl}(f_i(\cdot, \mathbf{p}_{-i}) + \delta_{K_i})(p_i), \ i \in \{1, \ldots, n\}$

$\Rightarrow 0 \in \partial_L(f_i(\cdot, \mathbf{p}_{-i}) + \delta_{K_i})(p_i), \ i \in \{1, \ldots, n\} \quad \text{(cf. Theorem 2.4)}$

$\Rightarrow 0 \in \partial_L f_i(\cdot, \mathbf{p}_{-i})(p_i) + \partial_L \delta_{K_i}(p_i), \ i \in \{1, \ldots, n\} \quad \text{(cf. Propositions 2.12 and 2.13)}$

$\Rightarrow 0 \in \partial_C f_i(\cdot, \mathbf{p}_{-i})(p_i) + \partial_L \delta_{K_i}(p_i), \ i \in \{1, \ldots, n\}$

$\Leftrightarrow 0 \in \partial_C^i f_i(\mathbf{p}) + N_L(p_i; K_i), \ i \in \{1, \ldots, n\}$

$\Leftrightarrow \exists \xi_C^i \in \partial_C^i f_i(\mathbf{p}) \text{ such that } \langle \xi_C^i, \exp_{p_i}^{-1}(q_i) \rangle_{g_i} \geq 0 \text{ for all } q_i \in K_i, i \in \{1, \ldots, n\}$

$$\text{(cf. Corollary 2.1)}$$

$\Leftrightarrow \mathbf{p} \in S_{NS}(\mathbf{f}, \mathbf{K}).$

Now, we prove $S_{NS}(\mathbf{f}, \mathbf{K}) \subset S_{NC}(\mathbf{f}, \mathbf{K})$; more precisely, we have
$\mathbf{p} \in S_{NS}(\mathbf{f}, \mathbf{K}) \Leftrightarrow$

$\Leftrightarrow 0 \in \partial_C^i f_i(\mathbf{p}) + N_L(p_i; K_i), \ i \in \{1, \ldots, n\}$

$\Leftrightarrow 0 \in \partial_C^i f_i(\mathbf{p}) + \partial_{cl}\delta_{K_i}(p_i), \ i \in \{1, \ldots, n\}$ (cf. Corollary 2.1)

$\Leftrightarrow 0 \in \partial_{cl}(f_i^0(\mathbf{p}; \exp_{p_i}^{-1}(\cdot)))(p_i) + \partial_{cl}\delta_{K_i}(p_i), \ i \in \{1, \ldots, n\}$ (cf. Theorem 2.5)

$\Rightarrow 0 \in \partial_{cl}(f_i^0(\mathbf{p}; \exp_{p_i}^{-1}(\cdot)) + \delta_{K_i})(p_i), \ i \in \{1, \ldots, n\}$ (cf. Proposition 2.14)

$\Leftrightarrow f_i^0(\mathbf{p}; \exp_{p_i}^{-1}(q_i)) \geq 0 \ \text{for all } q_i \in K_i, \ i \in \{1, \ldots, n\}$

$\Leftrightarrow \mathbf{p} \in S_{NC}(\mathbf{f}, \mathbf{K}).$

In order to prove $S_{NC}(\mathbf{f}, \mathbf{K}) \subset S_{NS}(\mathbf{f}, \mathbf{K})$, we recall that $f_i^0(\mathbf{p}; \exp_{p_i}^{-1}(\cdot))$ is locally Lipschitz in a neighborhood of p_i. Thus, we have
$\mathbf{p} \in S_{NC}(\mathbf{f}, \mathbf{K}) \Leftrightarrow$

$\Leftrightarrow 0 \in \partial_{cl}(f_i^0(\mathbf{p}; \exp_{p_i}^{-1}(\cdot)) + \delta_{K_i})(p_i), \ i \in \{1, \ldots, n\}$

$\Rightarrow 0 \in \partial_L(f_i^0(\mathbf{p}; \exp_{p_i}^{-1}(\cdot)) + \delta_{K_i})(p_i), \ i \in \{1, \ldots, n\}$ (cf. Theorem 2.1)

$\Rightarrow 0 \in \partial_L(f_i^0(\mathbf{p}; \exp_{p_i}^{-1}(\cdot)))(p_i) + \partial_L\delta_{K_i}(p_i), \ i \in \{1, \ldots, n\}$

$$\text{(cf. Propositions 2.12 and 2.13)}$$

$\Leftrightarrow 0 \in \partial_C(f_i(\cdot, \mathbf{p_{-i}}))(p_i) + \partial_L\delta_{K_i}(p_i), \ i \in \{1, \ldots, n\}$ (cf. Theorem 2.5)

$\Leftrightarrow 0 \in \partial_C^i(f_i(\mathbf{p})) + N_L(p_i; K_i), \ i \in \{1, \ldots, n\}$

$\Leftrightarrow \mathbf{p} \in S_{NS}(\mathbf{f}, \mathbf{K}).$

(ii) Due to (i) and Remark 9.2, it is enough to prove that $S_{NC}(\mathbf{f}, \mathbf{K}) \subset S_{NE}(\mathbf{f}, \mathbf{K})$. Let $\mathbf{p} \in S_{NC}(\mathbf{f}, \mathbf{K})$, i.e., for every $i \in \{1, \ldots, n\}$ and $q_i \in K_i$,

$$f_i^0(\mathbf{p}; \exp_{p_i}^{-1}(q_i)) \geq 0. \tag{9.3}$$

Fix $i \in \{1, \ldots, n\}$ and $q_i \in K_i$ arbitrary. Since $f_i(\cdot, \mathbf{p_{-i}})$ is convex on (U_i, g_i), on account of (2.18), we have

$$f_i^0(\mathbf{p}; \exp_{p_i}^{-1}(q_i)) = \lim_{t \to 0^+} \frac{f_i(\exp_{p_i}(t \exp_{p_i}^{-1}(q_i)), \mathbf{p_{-i}}) - f_i(\mathbf{p})}{t}. \tag{9.4}$$

Note that the function

$$R(t) = \frac{f_i(\exp_{p_i}(t \exp_{p_i}^{-1}(q_i)), \mathbf{p_{-i}}) - f_i(\mathbf{p})}{t}$$

is well-defined on the whole interval $(0, 1]$; indeed, $t \mapsto \exp_{p_i}(t \exp_{p_i}^{-1}(q_i))$ is the minimal geodesic joining the points $p_i \in K_i$ and $q_i \in K_i$ which belongs to $K_i \subset U_i$. Moreover, it is well-known that $t \mapsto R(t)$ is non-decreasing on $(0, 1]$. Consequently,

$$f_i(q_i, \mathbf{p}_{-i}) - f_i(\mathbf{p}) = f_i(\exp_{p_i}(\exp_{p_i}^{-1}(q_i)), \mathbf{p}_{-i}) - f_i(\mathbf{p}) = R(1) \geq \lim_{t \to 0^+} R(t).$$

Now, (9.3) and (9.4) give that $\lim_{t \to 0^+} R(t) \geq 0$, which concludes the proof.   □

Remark 9.3

(a) As we can see, the key tool in the proof of $\mathcal{S}_{NS}(\mathbf{f}, \mathbf{K}) = \mathcal{S}_{NC}(\mathbf{f}, \mathbf{K})$ is the locally Lipschitz property of the function $f_i^0(\mathbf{p}; \exp_{p_i}^{-1}(\cdot))$ near p_i.

(b) In [6] only the sets $\mathcal{S}_{NE}(\mathbf{f}, \mathbf{K})$ and $\mathcal{S}_{NC}(\mathbf{f}, \mathbf{K})$ have been considered. Note however that the Nash-Stampacchia concept is more appropriate to find Nash equilibrium points in general contexts, see also the applications in Sect. 9.4 for both compact and non-compact cases. Moreover, via $\mathcal{S}_{NS}(\mathbf{f}, \mathbf{K})$ we realize that the optimal geometrical framework to develop this study is the class of Hadamard manifolds. In the next sections we develop this approach.

9.3   Nash-Stampacchia Equilibria on Hadamard Manifolds

Before providing the main results on this subsection, we recall some basic notions from the theory of metric projections on *Hadamard manifolds*.

Let (M, g) be an m-dimensional Riemannian manifold $(m \geq 2)$, $K \subset M$ be a non-empty set. Let

$$P_K(q) = \{p \in K : d_g(q, p) = \inf_{z \in K} d_g(q, z)\}$$

be the set of *metric projections* of the point $q \in M$ to the set K. Due to the theorem of Hopf-Rinow, if (M, g) is complete, then any closed set $K \subset M$ is *proximinal*, i.e., $P_K(q) \neq \emptyset$ for all $q \in M$. In general, P_K is a set-valued map. When $P_K(q)$ is a singleton for every $q \in M$, we say that K is a *Chebyshev set*. The map P_K is *non-expansive* if

$$d_g(p_1, p_2) \leq d_g(q_1, q_2) \text{ for all } q_1, q_2 \in M \text{ and } p_1 \in P_K(q_1), p_2 \in P_K(q_2).$$

In particular, K is a Chebyshev set whenever the map P_K is non-expansive.

A non-empty closed set $K \subset M$ verifies the *obtuse-angle property* if for fixed $q \in M$ and $p \in K$ the following two statements are equivalent:

(OA_1) $p \in P_K(q)$;

(OA_2) If $\gamma : [0, 1] \to M$ is the unique minimal geodesic from $\gamma(0) = p \in K$ to $\gamma(1) = q$, then for every geodesic $\sigma : [0, \delta] \to K$ ($\delta \geq 0$) emanating from the point p, we have $g(\dot{\gamma}(0), \dot{\sigma}(0)) \leq 0$.

Remark 9.4

(a) In the Euclidean case $(\mathbf{R}^m, \langle \cdot, \cdot \rangle_{\mathbf{R}^m})$, (here, $\langle \cdot, \cdot \rangle_{\mathbf{R}^m}$ is the standard inner product in $\mathbf{R}^m$), every non-empty closed convex set $K \subset \mathbf{R}^m$ verifies the obtuse-angle property, see Moskovitz-Dines [13], which reduces to the well-known geometric form:

$$p \in P_K(q) \Leftrightarrow \langle q - p, z - p \rangle_{\mathbf{R}^m} \leq 0 \text{ for all } z \in K.$$

(b) The first variational formula of Riemannian geometry shows that (OA_1) implies (OA_2) for every closed set $K \subset M$ in a complete Riemannian manifold (M, g). However, the converse does not hold in general; for a detailed discussion, see Kristály, Rădulescu and Varga [8].

A Riemannian manifold (M, g) is a *Hadamard manifold* if it is complete, simply connected and its sectional curvature is non-positive. It is easy to check that on a Hadamard manifold (M, g) every geodesic convex set is a Chebyshev set. Moreover, we have

Proposition 9.2 *Let (M, g) be a finite-dimensional Hadamard manifold, $K \subset M$ be a closed set. The following statements hold true:*

(i) *(Walter [19]) If $K \subset M$ is geodesic convex, it verifies the obtuse-angle property;*

(ii) *(Grognet [5]) P_K is non-expansive if and only if $K \subset M$ is geodesic convex.*

We also recall that on a Hadamard manifold (M, g), if $h(p) = d_g^2(p, p_0)$, $p_0 \in M$ is fixed, then

$$\mathrm{grad}h(p) = -2\exp_p^{-1}(p_0). \tag{9.5}$$

In the sequel, let (M_i, g_i) be finite-dimensional Hadamard manifolds, $i \in \{1, \ldots, n\}$, and $\mathbf{M} = M_1 \times \ldots \times M_n$ be the product manifold with its Riemannian product metric from (9.1) Standard arguments show that $(\mathbf{M}, \mathbf{g})$ is also a Hadamard manifold, see O'Neill [17, Lemma 40, p. 209]. Moreover, on account of the characterization of (warped) product geodesics, see O'Neill [17, Proposition 38, p. 208], if $\exp_\mathbf{p}$ denotes the usual exponential map on $(\mathbf{M}, \mathbf{g})$ at $\mathbf{p} \in \mathbf{M}$, then for every $\mathbf{V} = (V_1, \ldots, V_n) \in T_\mathbf{p}\mathbf{M}$, we have

$$\exp_\mathbf{p}(\mathbf{V}) = (\exp_{p_1}(V_1), \ldots, \exp_{p_n}(V_n)).$$

We consider that $K_i \subset M_i$ are non-empty, closed, geodesic convex sets and $U_i \subset M_i$ are open sets containing K_i, $i \in \{1, \ldots, n\}$.

Let $\mathbf{f} \in \mathcal{L}_{(\mathbf{K},\mathbf{U},\mathbf{M})}$. The *diagonal Clarke subdifferential of* $\mathbf{f} = (f_1, \ldots, f_n)$ at $\mathbf{p} \in \mathbf{K}$ is

$$\partial_C^\Delta \mathbf{f}(\mathbf{p}) = (\partial_C^1 f_1(\mathbf{p}), \ldots, \partial_C^n f_n(\mathbf{p})).$$

From the definition of the metric $\mathbf{g}$, for every $\mathbf{p} \in \mathbf{K}$ and $\mathbf{q} \in \mathbf{M}$ it turns out that

$$\langle \xi_C^\Delta, \exp_\mathbf{p}^{-1}(\mathbf{q}) \rangle_\mathbf{g} = \sum_{i=1}^n \langle \xi_C^i, \exp_{p_i}^{-1}(q_i) \rangle_{g_i}, \quad \xi_C^\Delta = (\xi_C^1, \ldots, \xi_C^n) \in \partial_C^\Delta \mathbf{f}(\mathbf{p}). \tag{9.6}$$

9.3.1 Fixed Point Characterization of Nash-Stampacchia Equilibria

For each $\alpha > 0$ and $\mathbf{f} \in \mathcal{L}_{(\mathbf{K},\mathbf{U},\mathbf{M})}$, we define the set-valued map $A_\alpha^\mathbf{f} : \mathbf{K} \to 2^\mathbf{K}$ by

$$A_\alpha^\mathbf{f}(\mathbf{p}) = P_\mathbf{K}(\exp_\mathbf{p}(-\alpha \partial_C^\Delta \mathbf{f}(\mathbf{p}))), \quad \mathbf{p} \in \mathbf{K}.$$

Note that for each $\mathbf{p} \in \mathbf{K}$, the set $A_\alpha^\mathbf{f}(\mathbf{p})$ is non-empty and compact. The following result plays a crucial role in our further investigations.

Theorem 9.5 ([7]) *Let* (M_i, g_i) *be finite-dimensional Hadamard manifolds,* $K_i \subset M_i$ *be non-empty, closed, geodesic convex sets,* $U_i \subset M_i$ *be open sets containing* K_i, $i \in \{1, \ldots, n\}$, *and* $\mathbf{f} \in \mathcal{L}_{(\mathbf{K},\mathbf{U},\mathbf{M})}$. *Then the following statements are equivalent:*

 (*i*) $\mathbf{p} \in \mathcal{S}_{NS}(\mathbf{f}, \mathbf{K})$;
 (*ii*) $\mathbf{p} \in A_\alpha^\mathbf{f}(\mathbf{p})$ *for all* $\alpha > 0$;
 (*iii*) $\mathbf{p} \in A_\alpha^\mathbf{f}(\mathbf{p})$ *for some* $\alpha > 0$.

Proof In view of relation (9.6) and the identification between $T_\mathbf{p}\mathbf{M}$ and $T_\mathbf{p}^*\mathbf{M}$, see (2.11), we have that

$$\mathbf{p} \in \mathcal{S}_{NS}(\mathbf{f}, \mathbf{K}) \Leftrightarrow \exists \xi_C^\Delta = (\xi_C^1, \ldots, \xi_C^n) \in \partial_C^\Delta \mathbf{f}(\mathbf{p}) \text{ such that} \tag{9.7}$$

$$\langle \xi_C^\Delta, \exp_\mathbf{p}^{-1}(\mathbf{q}) \rangle_\mathbf{g} \geq 0 \text{ for all } \mathbf{q} \in \mathbf{K}$$

$$\Leftrightarrow \exists \xi_C^\Delta = (\xi_C^1, \ldots, \xi_C^n) \in \partial_C^\Delta \mathbf{f}(\mathbf{p}) \text{ such that}$$

$$\mathbf{g}(-\alpha \xi_C^\Delta, \exp_\mathbf{p}^{-1}(\mathbf{q})) \leq 0 \text{ for all } \mathbf{q} \in \mathbf{K} \text{ and}$$

for all/some $\alpha > 0$.

On the other hand, let $\gamma, \sigma : [0, 1] \rightarrow \mathbf{M}$ be the unique minimal geodesics defined by $\gamma(t) = \exp_{\mathbf{p}}(-t\alpha\xi_C^\Delta)$ and $\sigma(t) = \exp_{\mathbf{p}}(t \exp_{\mathbf{p}}^{-1}(\mathbf{q}))$ for any fixed $\alpha > 0$ and $\mathbf{q} \in \mathbf{K}$. Since $\mathbf{K}$ is geodesic convex in $(\mathbf{M}, \mathbf{g})$, then $\operatorname{Im}\sigma \subset \mathbf{K}$ and

$$\mathbf{g}(\dot{\gamma}(0), \dot{\sigma}(0)) = \mathbf{g}(-\alpha\xi_C^\Delta, \exp_{\mathbf{p}}^{-1}(\mathbf{q})). \tag{9.8}$$

Taking into account relation (9.8) and Proposition 9.2-(i), i.e., the validity of the obtuse-angle property on the Hadamard manifold $(\mathbf{M}, \mathbf{g})$, (9.7) is equivalent to

$$\mathbf{p} = \gamma(0) = P_\mathbf{K}(\gamma(1)) = P_\mathbf{K}(\exp_{\mathbf{p}}(-\alpha\xi_C^\Delta)),$$

which is nothing but $\mathbf{p} \in A_\alpha^\mathbf{f}(\mathbf{p})$.   □

Remark 9.5 Note that the implications $(ii) \Rightarrow (i)$ and $(iii) \Rightarrow (i)$ hold for arbitrarily Riemannian manifolds, see Remark 9.4. These implications are enough to find Nash-Stampacchia equilibrium points for $(\mathbf{f}, \mathbf{K})$ via fixed points of the map $A_\alpha^\mathbf{f}$. However, in the sequel we exploit further aspects of the Hadamard manifolds as non-expansiveness of the projection operator of geodesic convex sets and a Rauch-type comparison property. Moreover, in the spirit of Nash's original idea that Nash equilibria appear exactly as fixed points of a specific map, Theorem 9.5 provides a full characterization of Nash-Stampacchia equilibrium points for $(\mathbf{f}, \mathbf{K})$ via the fixed points of the set-valued map $A_\alpha^\mathbf{f}$ whenever (M_i, g_i) are Hadamard manifolds.

In the sequel, two cases will be considered to guarantee Nash-Stampacchia equilibrium points for $(\mathbf{f}, \mathbf{K})$, depending on the compactness of the strategy sets K_i.

9.3.2 Nash-Stampacchia Equilibrium Points: Compact Case

Our first result guarantees the existence of a Nash-Stampacchia equilibrium point for $(\mathbf{f}, \mathbf{K})$ whenever the sets K_i are compact, $i \in \{1, \ldots, n\}$.

Theorem 9.6 ([7]) *Let (M_i, g_i) be finite-dimensional Hadamard manifolds, $K_i \subset M_i$ be non-empty, compact, geodesic convex sets, and $U_i \subset M_i$ be open sets containing K_i, $i \in \{1, \ldots, n\}$. Assume that $\mathbf{f} \in \mathcal{L}_{(\mathbf{K}, \mathbf{U}, \mathbf{M})}$ and $\mathbf{K} \ni \mathbf{p} \mapsto \partial_C^\Delta \mathbf{f}(\mathbf{p})$ is upper semicontinuous. Then there exists at least one Nash-Stampacchia equilibrium point for $(\mathbf{f}, \mathbf{K})$, i.e., $S_{NS}(\mathbf{f}, \mathbf{K}) \neq \emptyset$.*

Proof Fix $\alpha > 0$ arbitrary. We prove that the set-valued map $A_\alpha^\mathbf{f}$ has closed graph. Let $(\mathbf{p}, \mathbf{q}) \in \mathbf{K} \times \mathbf{K}$ and the sequences $\{\mathbf{p}_k\}, \{\mathbf{q}_k\} \subset \mathbf{K}$ such that $\mathbf{q}_k \in A_\alpha^\mathbf{f}(\mathbf{p}_k)$ and $(\mathbf{p}_k, \mathbf{q}_k) \rightarrow (\mathbf{p}, \mathbf{q})$ as $k \rightarrow \infty$. Then, for every $k \in \mathbb{N}$, there exists $\xi_{C,k}^\Delta \in \partial_C^\Delta \mathbf{f}(\mathbf{p}_k)$ such that $\mathbf{q}_k = P_\mathbf{K}(\exp_{\mathbf{p}_k}(-\alpha\xi_{C,k}^\Delta))$. On account of Proposition 2.13 $(i) \Leftrightarrow (ii)$, the sequence $\{\xi_{C,k}^\Delta\}$ is bounded on the cotangent bundle $T^*\mathbf{M}$. Using the identification between elements of the tangent and cotangent fibers, up to a subsequence, we may assume that $\{\xi_{C,k}^\Delta\}$ converges to an element $\xi_C^\Delta \in T_\mathbf{p}^*\mathbf{M}$. Since the set-valued map $\partial_C^\Delta \mathbf{f}$ is upper semicontinuous on $\mathbf{K}$ and $\mathbf{p}_k \to \mathbf{p}$ as $k \to \infty$, we have that $\xi_C^\Delta \in \partial_C^\Delta \mathbf{f}(\mathbf{p})$. The non-expansiveness of $P_\mathbf{K}$ (see Proposition 9.2-(ii)) gives that

$$\mathbf{d_g}(\mathbf{q}, P_\mathbf{K}(\exp_\mathbf{p}(-\alpha\xi_C^\Delta))) \le \mathbf{d_g}(\mathbf{q}, \mathbf{q}_k) + \mathbf{d_g}(\mathbf{q}_k, P_\mathbf{K}(\exp_\mathbf{p}(-\alpha\xi_C^\Delta)))$$

$$= \mathbf{d_g}(\mathbf{q}, \mathbf{q}_k) + \mathbf{d_g}(P_\mathbf{K}(\exp_{\mathbf{p}_k}(-\alpha\xi_{C,k}^\Delta)), P_\mathbf{K}(\exp_\mathbf{p}(-\alpha\xi_C^\Delta)))$$

$$\le \mathbf{d_g}(\mathbf{q}, \mathbf{q}_k) + \mathbf{d_g}(\exp_{\mathbf{p}_k}(-\alpha\xi_{C,k}^\Delta), \exp_\mathbf{p}(-\alpha\xi_C^\Delta))$$

Letting $k \to \infty$, both terms in the last expression tend to zero. Indeed, the former follows from the fact that $\mathbf{q}_k \to \mathbf{q}$ as $k \to \infty$, while the latter is a simple consequence of the local behaviour of the exponential map. Thus,

$$\mathbf{q} = P_\mathbf{K}(\exp_\mathbf{p}(-\alpha\xi_C^\Delta)) \in P_\mathbf{K}(\exp_\mathbf{p}(-\alpha\partial_C^\Delta \mathbf{f}(\mathbf{p}))) = A_\alpha^\mathbf{f}(\mathbf{p}),$$

i.e., the graph of $A_\alpha^\mathbf{f}$ is closed.

By definition, for each $\mathbf{p} \in \mathbf{K}$ the set $\partial_C^\Delta \mathbf{f}(\mathbf{p})$ is convex, so contractible. Since both $P_\mathbf{K}$ and the exponential map are continuous, $A_\alpha^\mathbf{f}(\mathbf{p})$ is contractible as well for each $\mathbf{p} \in \mathbf{K}$, so acyclic (see [12]).

Now, we are in the position to apply Begle's fixed point theorem, equivalent to Lemma D.9, see e.g. McClendon [12, Proposition 1.1]. Consequently, there exists $\mathbf{p} \in \mathbf{K}$ such that $\mathbf{p} \in A_\alpha^\mathbf{f}(\mathbf{p})$. On account of Theorem 9.5, $\mathbf{p} \in \mathcal{S}_{NS}(\mathbf{f}, \mathbf{K})$. $\square$

Remark 9.6

(a) Since $\mathbf{f} \in \mathcal{L}_{(\mathbf{K},\mathbf{U},\mathbf{M})}$ in Theorem 9.6, the partial Clarke gradients $q \mapsto \partial_C f_i(\cdot, \mathbf{p}_{-i})(q)$ are upper semicontinuous, $i \in \{1, \ldots, n\}$. However, in general, the diagonal Clarke subdifferential $\partial_C^\Delta \mathbf{f}(\cdot)$ does not inherit this regularity property.
(b) Two applications to Theorem 9.6 will be given in Examples 9.1 and 9.2; the first on the Poincaré disc, the second on the manifold of positive definite, symmetric matrices.

9.3.3 Nash-Stampacchia Equilibrium Points: Non-compact Case

In the sequel, we are focusing to the location of Nash-Stampacchia equilibrium points for $(\mathbf{f}, \mathbf{K})$ in the case when K_i are *not* necessarily compact on the Hadamard manifolds (M_i, g_i). Simple examples show that even the C^∞-smoothness of the payoff functions are not enough to guarantee the existence of Nash(-Stampacchia) equilibria.

Indeed, if $f_1, f_2 : \mathbb{R}^2 \to \mathbb{R}$ are defined as $f_1(x, y) = f_2(x, y) = e^{-x-y}$, and $K_1 = K_2 = [0, \infty)$, then $\mathcal{S}_{NS}(\mathbf{f}, \mathbf{K}) = \mathcal{S}_{NE}(\mathbf{f}, \mathbf{K}) = \emptyset$. Therefore, in order to prove existence/location of Nash(-Stampacchia) equilibria on not necessarily compact strategy sets, one needs to require more specific assumptions on $\mathbf{f} = (f_1, \ldots, f_n)$. Two such possible ways are described in the sequel.

The first existence result is based on a suitable coercivity assumption and Theorem 9.6. For a fixed $\mathbf{p}_0 \in \mathbf{K}$, we introduce the hypothesis:

$(H_{\mathbf{p}_0})$ There exists $\xi_C^0 \in \partial_C^\triangle \mathbf{f}(\mathbf{p}_0)$ such that

$$L_{\mathbf{p}_0} = \limsup_{d_\mathbf{g}(\mathbf{p},\mathbf{p}_0)\to\infty} \frac{\sup_{\xi_C \in \partial_C^\triangle \mathbf{f}(\mathbf{p})} \langle \xi_C, \exp_\mathbf{p}^{-1}(\mathbf{p}_0)\rangle_\mathbf{g} + \langle \xi_C^0, \exp_{\mathbf{p}_0}^{-1}(\mathbf{p})\rangle_\mathbf{g}}{d_\mathbf{g}(\mathbf{p}, \mathbf{p}_0)} < -\|\xi_C^0\|_\mathbf{g}, \ \mathbf{p} \in \mathbf{K}.$$

Remark 9.7

(a) A similar assumption to hypothesis $(H_{\mathbf{p}_0})$ can be found in Németh [16] in the context of variational inequalities.

(b) Note that for the above numerical example, $(H_{\mathbf{p}_0})$ is not satisfied for any $\mathbf{p}_0 = (x_0, y_0) \in [0, \infty) \times [0, \infty)$. Indeed, one has $L_{(x_0,y_0)} = -e^{x_0+y_0}$, and $\|\xi_C^0\|_\mathbf{g} = e^{x_0+y_0}\sqrt{2}$. Therefore, the facts that $S_{NS}(\mathbf{f}, \mathbf{K}) = S_{NE}(\mathbf{f}, \mathbf{K}) = \emptyset$ are not unexpected.

The precise statement of the existence result is as follows.

Theorem 9.7 ([7]) *Let (M_i, g_i) be finite-dimensional Hadamard manifolds, $K_i \subset M_i$ be non-empty, closed, geodesic convex sets, and $U_i \subset M_i$ be open sets containing K_i, $i \in \{1, \ldots, n\}$. Assume that $\mathbf{f} \in \mathcal{L}_{(\mathbf{K},\mathbf{U},\mathbf{M})}$, the map $\mathbf{K} \ni \mathbf{p} \mapsto \partial_C^\triangle \mathbf{f}(\mathbf{p})$ is upper semicontinuous, and hypothesis $(H_{\mathbf{p}_0})$ holds for some $\mathbf{p}_0 \in \mathbf{K}$. Then there exists at least one Nash-Stampacchia equilibrium point for $(\mathbf{f}, \mathbf{K})$, i.e., $S_{NS}(\mathbf{f}, \mathbf{K}) \neq \emptyset$.*

Proof Let $E_0 \in \mathbb{R}$ such that $L_{\mathbf{p}_0} < -E_0 < -\|\xi_C^0\|_\mathbf{g}$. On account of hypothesis $(H_{\mathbf{p}_0})$ there exists $R > 0$ large enough such that for every $\mathbf{p} \in \mathbf{K}$ with $d_\mathbf{g}(\mathbf{p}, \mathbf{p}_0) \geq R$, we have

$$\sup_{\xi_C \in \partial_C^\triangle \mathbf{f}(\mathbf{p})} \langle \xi_C, \exp_\mathbf{p}^{-1}(\mathbf{p}_0)\rangle_\mathbf{g} + \langle \xi_C^0, \exp_{\mathbf{p}_0}^{-1}(\mathbf{p})\rangle_\mathbf{g} \leq -E_0 d_\mathbf{g}(\mathbf{p}, \mathbf{p}_0).$$

It is clear that $\mathbf{K} \cap \overline{B}_\mathbf{g}(\mathbf{p}_0, R) \neq \emptyset$, where $\overline{B}_\mathbf{g}(\mathbf{p}_0, R)$ denotes the closed geodesic ball in $(\mathbf{M}, \mathbf{g})$ with center $\mathbf{p}_0$ and radius R. In particular, from (2.14) and (2.12), for every $\mathbf{p} \in \mathbf{K}$ with $d_\mathbf{g}(\mathbf{p}, \mathbf{p}_0) \geq R$, the above relation yields

$$\sup_{\xi_C \in \partial_C^\triangle \mathbf{f}(\mathbf{p})} \langle \xi_C, \exp_\mathbf{p}^{-1}(\mathbf{p}_0)\rangle_\mathbf{g} \leq -E_0 d_\mathbf{g}(\mathbf{p}, \mathbf{p}_0) + \|\xi_C^0\|_\mathbf{g}\| \exp_{\mathbf{p}_0}^{-1}(\mathbf{p})\|_\mathbf{g} \qquad (9.9)$$

$$= (-E_0 + \|\xi_C^0\|_\mathbf{g})d_\mathbf{g}(\mathbf{p}, \mathbf{p}_0)$$

$$< 0.$$

Let $\mathbf{K}_R = \mathbf{K} \cap \overline{B}_\mathbf{g}(\mathbf{p}_0, R)$. It is clear that $\mathbf{K}_R$ is a geodesic convex, compact subset of $\mathbf{M}$. By applying Theorem 9.6, we immediately have that $\tilde{\mathbf{p}} \in \mathcal{S}_{NS}(\mathbf{f}, \mathbf{K}_R) \neq \emptyset$, i.e., there exists $\tilde{\xi}_C \in \partial_C^\Delta \mathbf{f}(\tilde{\mathbf{p}})$ such that

$$\langle \tilde{\xi}_C, \exp_{\tilde{\mathbf{p}}}^{-1}(\mathbf{p})\rangle_\mathbf{g} \geq 0 \text{ for all } \mathbf{p} \in \mathbf{K}_R. \tag{9.10}$$

It is also clear that $\mathbf{d}_\mathbf{g}(\tilde{\mathbf{p}}, \mathbf{p}_0) < R$. Indeed, assuming the contrary, we obtain from (9.9) that $\langle \tilde{\xi}_C, \exp_{\tilde{\mathbf{p}}}^{-1}(\mathbf{p}_0)\rangle_\mathbf{g} < 0$, which contradicts relation (9.10). Now, fix $\mathbf{q} \in \mathbf{K}$ arbitrarily. Thus, for $\varepsilon > 0$ small enough, the element $\mathbf{p} = \exp_{\tilde{\mathbf{p}}}(\varepsilon \exp_{\tilde{\mathbf{p}}}^{-1}(\mathbf{q}))$ belongs both to $\mathbf{K}$ and $\overline{B}_\mathbf{g}(\mathbf{p}_0, R)$, so $\mathbf{K}_R$. By substituting $\mathbf{p}$ into (9.10), we obtain that $\langle \tilde{\xi}_C, \exp_{\tilde{\mathbf{p}}}^{-1}(\mathbf{q})\rangle_\mathbf{g} \geq 0$. The arbitrariness of $\mathbf{q} \in \mathbf{K}$ shows that $\tilde{\mathbf{p}} \in \mathbf{K}$ is actually a Nash-Stampacchia equilibrium point for $(\mathbf{f}, \mathbf{K})$, which ends the proof.   □

The second result in the non-compact case is based on a suitable Lipschitz-type assumption. In order to avoid technicalities in our further calculations, we will consider that $\mathbf{f} \in C_{(\mathbf{K}, \mathbf{U}, \mathbf{M})}$. In this case, $\partial_C^\Delta \mathbf{f}(\mathbf{p})$ and $A_\alpha^\mathbf{f}(\mathbf{p})$ are singletons for every $\mathbf{p} \in \mathbf{K}$ and $\alpha > 0$.

For $\mathbf{f} \in C_{(\mathbf{K}, \mathbf{U}, \mathbf{M})}$, $\alpha > 0$ and $0 < \rho < 1$ we introduce the hypothesis:

$(H_\mathbf{K}^{\alpha, \rho})$ $\mathbf{d}_\mathbf{g}(\exp_\mathbf{p}(-\alpha \partial_C^\Delta \mathbf{f}(\mathbf{p})), \exp_\mathbf{q}(-\alpha \partial_C^\Delta \mathbf{f}(\mathbf{q}))) \leq (1 - \rho)\mathbf{d}_\mathbf{g}(\mathbf{p}, \mathbf{q})$ for all $\mathbf{p}, \mathbf{q} \in \mathbf{K}$.

Remark 9.8 One can show that $(H_\mathbf{K}^{\alpha, \rho})$ implies $(H_{\mathbf{p}_0})$ for every $\mathbf{p}_0 \in \mathbf{K}$ whenever (M_i, g_i) are Euclidean spaces. However, it is not clear if the same holds for Hadamard manifolds.

Finding fixed points for $A_\alpha^\mathbf{f}$, one could expect to apply dynamical systems; we consider both *discrete* and *continuous* ones. First, for some $\alpha > 0$ and $\mathbf{p}_0 \in \mathbf{M}$ fixed, we consider the discrete dynamical system

$$\mathbf{p}_{k+1} = A_\alpha^\mathbf{f}(P_\mathbf{K}(\mathbf{p}_k)). \tag{$(DDS)_\alpha$}$$

Second, according to Theorem 9.5, we clearly have that

$$\mathbf{p} \in \mathcal{S}_{NS}(\mathbf{f}, \mathbf{K}) \Leftrightarrow 0 = \exp_\mathbf{p}^{-1}(A_\alpha^\mathbf{f}(\mathbf{p})) \text{ for all/some } \alpha > 0.$$

Consequently, for some $\alpha > 0$ and $\mathbf{p}_0 \in \mathbf{M}$ fixed, the above equivalence motivates the study of the continuous dynamical system

$$\begin{cases} \dot{\eta}(t) = \exp_{\eta(t)}^{-1}(A_\alpha^\mathbf{f}(P_\mathbf{K}(\eta(t)))) \\ \eta(0) = \mathbf{p}_0. \end{cases} \tag{$(CDS)_\alpha$}$$

The following result describes the exponential stability of the orbits in both cases.

Theorem 9.8 ([7]) *Let (M_i, g_i) be finite-dimensional Hadamard manifolds, $K_i \subset M_i$ be non-empty, closed geodesics convex sets, $U_i \subset M_i$ be open sets containing K_i, and $f_i : \mathbf{K} \to \mathbb{R}$ be functions, $i \in \{1, \dots, n\}$ such that $\mathbf{f} \in C_{(\mathbf{K}, \mathbf{U}, \mathbf{M})}$. Assume that $(H_{\mathbf{K}}^{\alpha, \rho})$ holds true for some $\alpha > 0$ and $0 < \rho < 1$. Then the set of Nash-Stampacchia equilibrium points for $(\mathbf{f}, \mathbf{K})$ is a singleton, i.e., $\mathcal{S}_{NS}(\mathbf{f}, \mathbf{K}) = \{\tilde{\mathbf{p}}\}$. Moreover, for each $\mathbf{p}_0 \in \mathbf{M}$, we have*

(i) the orbit $\{\mathbf{p}_k\}$ of $(DDS)_\alpha$ converges exponentially to $\tilde{\mathbf{p}} \in \mathbf{K}$ and

$$\mathbf{d_g}(\mathbf{p}_k, \tilde{\mathbf{p}}) \le \frac{(1-\rho)^k}{\rho} \mathbf{d_g}(\mathbf{p}_1, \mathbf{p}_0) \ for \ all \ k \in \mathbb{N};$$

(ii) the orbit η of $(CDS)_\alpha$ is globally defined on $[0, \infty)$ and it converges exponentially to $\tilde{\mathbf{p}} \in \mathbf{K}$ and

$$\mathbf{d_g}(\eta(t), \tilde{\mathbf{p}}) \le e^{-\rho t} \mathbf{d_g}(\mathbf{p}_0, \tilde{\mathbf{p}}) \ for \ all \ t \ge 0.$$

Proof Let $\mathbf{p}, \mathbf{q} \in \mathbf{M}$ be arbitrarily fixed. On account of the non-expansiveness of the projection $P_{\mathbf{K}}$ (see Proposition 9.2-(ii)) and hypothesis $(H_{\mathbf{K}}^{\alpha, \rho})$, we have that $\mathbf{d_g}((A_\alpha^{\mathbf{f}} \circ P_{\mathbf{K}})(\mathbf{p}), (A_\alpha^{\mathbf{f}} \circ P_{\mathbf{K}})(\mathbf{q}))$

$$= \mathbf{d_g}(P_{\mathbf{K}}(\exp_{P_{\mathbf{K}}(\mathbf{p})}(-\alpha \partial_C^\Delta \mathbf{f}(P_{\mathbf{K}}(\mathbf{p})))), P_{\mathbf{K}}(\exp_{P_{\mathbf{K}}(\mathbf{q})}(-\alpha \partial_C^\Delta \mathbf{f}(P_{\mathbf{K}}(\mathbf{q})))))$$

$$\le \mathbf{d_g}(\exp_{P_{\mathbf{K}}(\mathbf{p})}(-\alpha \partial_C^\Delta \mathbf{f}(P_{\mathbf{K}}(\mathbf{p}))), \exp_{P_{\mathbf{K}}(\mathbf{q})}(-\alpha \partial_C^\Delta \mathbf{f}(P_{\mathbf{K}}(\mathbf{q}))))$$

$$\le (1-\rho) \mathbf{d_g}(P_{\mathbf{K}}(\mathbf{p}), P_{\mathbf{K}}(\mathbf{q}))$$

$$\le (1-\rho) \mathbf{d_g}(\mathbf{p}, \mathbf{q}),$$

which means that the map $A_\alpha^{\mathbf{f}} \circ P_{\mathbf{K}} : \mathbf{M} \to \mathbf{M}$ is a $(1-\rho)$-contraction on $\mathbf{M}$.

(i) Since $(\mathbf{M}, \mathbf{d_g})$ is a complete metric space, the standard Banach fixed point argument shows that $A_\alpha^{\mathbf{f}} \circ P_{\mathbf{K}}$ has a unique fixed point $\tilde{\mathbf{p}} \in M$. Since $\mathrm{Im} A_\alpha^{\mathbf{f}} \subset \mathbf{K}$, then $\tilde{\mathbf{p}} \in \mathbf{K}$. Therefore, we have that $A_\alpha^{\mathbf{f}}(\tilde{\mathbf{p}}) = \tilde{\mathbf{p}}$. Due to Theorem 9.5, $\mathcal{S}_{NS}(\mathbf{f}, \mathbf{K}) = \{\tilde{\mathbf{p}}\}$ and the estimate for $\mathbf{d_g}(\mathbf{p}_k, \tilde{\mathbf{p}})$ yields in a usual manner.

(ii) Since $A_\alpha^{\mathbf{f}} \circ P_{\mathbf{K}} : \mathbf{M} \to \mathbf{M}$ is a $(1-\rho)$-contraction on $\mathbf{M}$ (thus locally Lipschitz in particular), the map $\mathbf{M} \ni \mathbf{p} \mapsto G(\mathbf{p}) := \exp_{\mathbf{p}}^{-1}(A_\alpha^{\mathbf{f}}(P_{\mathbf{K}}(\mathbf{p})))$ is of class C^{1-0}. Now, we may guarantee the existence of a unique maximal orbit $\eta : [0, T_{\max}) \to \mathbf{M}$ of $(CDS)_\alpha$.

We assume that $T_{\max} < \infty$. Let us consider the Lyapunov function $h : [0, T_{\max}) \to \mathbb{R}$ defined by

$$h(t) = \frac{1}{2} \mathbf{d_g}^2(\eta(t), \tilde{\mathbf{p}}).$$

The function h is differentiable for a.e. $t \in [0, T_{\max})$ and in the differentiable points of η we have

$$h'(t) = -\mathbf{g}(\dot{\eta}(t), \exp^{-1}_{\eta(t)}(\tilde{\mathbf{p}}))$$

$$= -\mathbf{g}(\exp^{-1}_{\eta(t)}(A^{\mathbf{f}}_{\alpha}(P_{\mathbf{K}}(\eta(t)))), \exp^{-1}_{\eta(t)}(\tilde{\mathbf{p}})) \qquad\qquad (cf.\ (CDS)_{\alpha})$$

$$= -\mathbf{g}(\exp^{-1}_{\eta(t)}(A^{\mathbf{f}}_{\alpha}(P_{\mathbf{K}}(\eta(t)))) - \exp^{-1}_{\eta(t)}(\tilde{\mathbf{p}}), \exp^{-1}_{\eta(t)}(\tilde{\mathbf{p}}))$$

$$\quad - \mathbf{g}(\exp^{-1}_{\eta(t)}(\tilde{\mathbf{p}}), \exp^{-1}_{\eta(t)}(\tilde{\mathbf{p}}))$$

$$\leq \| \exp^{-1}_{\eta(t)}(A^{\mathbf{f}}_{\alpha}(P_{\mathbf{K}}(\eta(t)))) - \exp^{-1}_{\eta(t)}(\tilde{\mathbf{p}})\|_{\mathbf{g}} \cdot \| \exp^{-1}_{\eta(t)}(\tilde{\mathbf{p}})\|_{\mathbf{g}} - \| \exp^{-1}_{\eta(t)}(\tilde{\mathbf{p}})\|^2_{\mathbf{g}}.$$

In the last estimate we used the Cauchy-Schwartz inequality (2.12). From (2.14) we have that

$$\| \exp^{-1}_{\eta(t)}(\tilde{\mathbf{p}}))\|_{\mathbf{g}} = \mathbf{d}_{\mathbf{g}}(\eta(t), \tilde{\mathbf{p}}). \qquad\qquad (9.11)$$

We claim that for every $t \in [0, T_{\max})$ one has

$$\| \exp^{-1}_{\eta(t)}(A^{\mathbf{f}}_{\alpha}(P_{\mathbf{K}}(\eta(t)))) - \exp^{-1}_{\eta(t)}(\tilde{\mathbf{p}})\|_{\mathbf{g}} \leq \mathbf{d}_{\mathbf{g}}(A^{\mathbf{f}}_{\alpha}(P_{\mathbf{K}}(\eta(t))), \tilde{\mathbf{p}}). \qquad (9.12)$$

To see this, fix a point $t \in [0, T_{\max})$ where η is differentiable, and let $\gamma : [0, 1] \to \mathbf{M}$, $\tilde{\gamma} : [0, 1] \to T_{\eta(t)}\mathbf{M}$ and $\overline{\gamma} : [0, 1] \to T_{\eta(t)}\mathbf{M}$ be three curves such that

- γ is the unique minimal geodesic joining the two points $\gamma(0) = \tilde{\mathbf{p}} \in \mathbf{K}$ and $\gamma(1) = A^{\mathbf{f}}_{\alpha}(P_{\mathbf{K}}(\eta(t)))$;
- $\tilde{\gamma}(s) = \exp^{-1}_{\eta(t)}(\gamma(s))$, $s \in [0, 1]$;
- $\overline{\gamma}(s) = (1 - s) \exp^{-1}_{\eta(t)}(\tilde{\mathbf{p}}) + s \exp^{-1}_{\eta(t)}(A^{\mathbf{f}}_{\alpha}(P_{\mathbf{K}}(\eta(t))))$, $s \in [0, 1]$.

By the definition of γ, we have that

$$L_{\mathbf{g}}(\gamma) = \mathbf{d}_{\mathbf{g}}(A^{\mathbf{f}}_{\alpha}(P_{\mathbf{K}}(\eta(t))), \tilde{\mathbf{p}}). \qquad\qquad (9.13)$$

Moreover, since $\overline{\gamma}$ is a segment of the straight line in $T_{\eta(t)}\mathbf{M}$ that joins the endpoints of $\tilde{\gamma}$, we have that

$$l(\overline{\gamma}) \leq l(\tilde{\gamma}); \qquad\qquad (9.14)$$

here, l denotes the length function on $T_{\eta(t)}\mathbf{M}$. Moreover, since the curvature of $(\mathbf{M}, \mathbf{g})$ is non-positive, we may apply a Rauch-type comparison result for the lengths of γ and $\tilde{\gamma}$, see do Carmo [4, Proposition 2.5, p.218], obtaining that

$$l(\tilde{\gamma}) \leq L_{\mathbf{g}}(\gamma). \qquad\qquad (9.15)$$

Combining relations (9.13), (9.14) and (9.15) with the fact that

$$l(\overline{\gamma}) = \| \exp_{\eta(t)}^{-1}(A_\alpha^{\mathbf{f}}(P_{\mathbf{K}}(\eta(t)))) - \exp_{\eta(t)}^{-1}(\tilde{\mathbf{p}})\|_{\mathbf{g}},$$

relation (9.12) holds true.

Coming back to $h'(t)$, in view of (9.11) and (9.12), it turns out that

$$h'(t) \le \mathbf{d_g}(A_\alpha^{\mathbf{f}}(P_{\mathbf{K}}(\eta(t))), \tilde{\mathbf{p}}) \cdot \mathbf{d_g}(\eta(t), \tilde{\mathbf{p}}) - \mathbf{d}_{\mathbf{g}}^2(\eta(t), \tilde{\mathbf{p}}). \tag{9.16}$$

On the other hand, note that $\tilde{\mathbf{p}} \in \mathcal{S}_{NS}(\mathbf{f}, \mathbf{K})$, i.e., $A_\alpha^{\mathbf{f}}(\tilde{\mathbf{p}}) = \tilde{\mathbf{p}}$. By exploiting the non-expansiveness of the projection operator $P_{\mathbf{K}}$, see Proposition 9.2-(ii), and $(H_{\mathbf{K}}^{\alpha,\rho})$, we have that

$$\begin{aligned}
\mathbf{d_g}(A_\alpha^{\mathbf{f}}(P_{\mathbf{K}}(\eta(t))), \tilde{\mathbf{p}}) &= \mathbf{d_g}(A_\alpha^{\mathbf{f}}(P_{\mathbf{K}}(\eta(t))), A_\alpha^{\mathbf{f}}(\tilde{\mathbf{p}})) \\
&= \mathbf{d_g}(P_{\mathbf{K}}(\exp_{P_{\mathbf{K}}(\eta(t))}(-\alpha\partial_C^{\Delta}\mathbf{f}(P_{\mathbf{K}}(\eta(t))))), P_{\mathbf{K}}(\exp_{\tilde{\mathbf{p}}}(-\alpha\partial_C^{\Delta}\mathbf{f}(\tilde{\mathbf{p}})))) \\
&\le \mathbf{d_g}(\exp_{P_{\mathbf{K}}(\eta(t))}(-\alpha\partial_C^{\Delta}\mathbf{f}(P_{\mathbf{K}}(\eta(t)))), \exp_{\tilde{\mathbf{p}}}(-\alpha\partial_C^{\Delta}\mathbf{f}(\tilde{\mathbf{p}}))) \\
&\le (1-\rho)\mathbf{d_g}(P_{\mathbf{K}}(\eta(t)), \tilde{\mathbf{p}}) \\
&= (1-\rho)\mathbf{d_g}(P_{\mathbf{K}}(\eta(t)), P_{\mathbf{K}}(\tilde{\mathbf{p}})) \\
&\le (1-\rho)\mathbf{d_g}(\eta(t), \tilde{\mathbf{p}}).
\end{aligned}$$

Combining the above relation with (9.16), for a.e. $t \in [0, T_{\max})$ it yields

$$h'(t) \le (1-\rho)\mathbf{d}_{\mathbf{g}}^2(\eta(t), \tilde{\mathbf{p}}) - \mathbf{d}_{\mathbf{g}}^2(\eta(t), \tilde{\mathbf{p}}) = -\rho\mathbf{d}_{\mathbf{g}}^2(\eta(t), \tilde{\mathbf{p}}),$$

which is nothing but

$$h'(t) \le -2\rho h(t) \quad \text{for a.e. } t \in [0, T_{\max}).$$

Due to the latter inequality, we have that

$$\frac{\mathrm{d}}{\mathrm{d}t}[h(t)e^{2\rho t}] = [h'(t) + 2\rho h(t)]e^{2\rho t} \le 0 \quad \text{for a.e. } t \in [0, T_{\max}).$$

After integration, one gets

$$h(t)e^{2\rho t} \le h(0) \quad \text{for all } t \in [0, T_{\max}). \tag{9.17}$$

According to (9.17), the function h is bounded on $[0, T_{\max})$; thus, there exists $\overline{\mathbf{p}} \in \mathbf{M}$ such that $\lim_{t \nearrow T_{\max}} \eta(t) = \overline{\mathbf{p}}$. The last limit means that η can be extended toward the value $T_{\max}$, which contradicts the maximality of $T_{\max}$. Thus, $T_{\max} = \infty$.

Now, relation (9.17) leads to the required estimate; indeed, we have

$$\mathbf{d_g}(\eta(t), \tilde{\mathbf{p}}) \le e^{-\rho t} \mathbf{d_g}(\eta(0), \tilde{\mathbf{p}}) = e^{-\rho t} \mathbf{d_g}(\mathbf{p_0}, \tilde{\mathbf{p}}) \quad \text{for all } t \in [0, \infty),$$

which concludes the proof of (ii). □

9.3.4 Curvature Rigidity

The obtuse-angle property and the non-expansiveness of $P_{\mathbf{K}}$ for the closed, geodesic convex set $\mathbf{K} \subset \mathbf{M}$ played indispensable roles in the proof of Theorems 9.5–9.8, which are well-known features of Hadamard manifolds (see Proposition 9.2). In Sect. 9.3 the product manifold $(\mathbf{M}, \mathbf{g})$ is considered to be a Hadamard one due to the fact that (M_i, g_i) are Hadamard manifolds themselves for each $i \in \{1, \ldots, n\}$. We actually have the following characterization which is also of geometric interests in its own right and entitles us to assert that Hadamard manifolds are the natural framework to develop the theory of Nash-Stampacchia equilibria on manifolds.

Theorem 9.9 ([7]) *Let (M_i, g_i) be complete, simply connected Riemannian manifolds, $i \in \{1, \ldots, n\}$, and $(\mathbf{M}, \mathbf{g})$ their product manifold. The following statements are equivalent:*

(i) *Any non-empty, closed, geodesic convex set $\mathbf{K} \subset \mathbf{M}$ verifies the obtuse-angle property and $P_{\mathbf{K}}$ is non-expansive;*

(ii) *(M_i, g_i) are Hadamard manifolds for every $i \in \{1, \ldots, n\}$.*

Proof $(ii) \Rightarrow (i)$. As mentioned before, if (M_i, g_i) are Hadamard manifolds for every $i \in \{1, \ldots, n\}$, then $(\mathbf{M}, \mathbf{g})$ is also a Hadamard manifold, see O'Neill [17, Lemma 40, p. 209]. It remains to apply Proposition 9.2 for the Hadamard manifold $(\mathbf{M}, \mathbf{g})$.

$(i) \Rightarrow (ii)$. We first prove that $(\mathbf{M}, \mathbf{g})$ is a Hadamard manifold. Since (M_i, g_i) are complete and simply connected Riemannian manifolds for every $i \in \{1, \ldots, n\}$, the same is true for $(\mathbf{M}, \mathbf{g})$. We now show that the sectional curvature of $(\mathbf{M}, \mathbf{g})$ is non-positive. To see this, let $\mathbf{p} \in \mathbf{M}$ and $\mathbf{W_0}, \mathbf{V_0} \in T_{\mathbf{p}}\mathbf{M} \setminus \{\mathbf{0}\}$. We claim that the sectional curvature of the two-dimensional subspace $S = \text{span}\{\mathbf{W_0}, \mathbf{V_0}\} \subset T_{\mathbf{p}}\mathbf{M}$ at the point $\mathbf{p}$ is non-positive, i.e., $K_{\mathbf{p}}(S) \le 0$. We assume without loosing the generality that $\mathbf{V_0}$ and $\mathbf{W_0}$ are $\mathbf{g}$-perpendicular, i.e., $\mathbf{g}(\mathbf{W_0}, \mathbf{V_0}) = 0$.

Let us fix $r_{\mathbf{p}} > 0$ and $\delta > 0$ such that $B_{\mathbf{g}}(\mathbf{p}, r_{\mathbf{p}})$ is a totally normal ball of $\mathbf{p}$ and

$$\delta \left(\|\mathbf{W_0}\|_{\mathbf{g}} + 2\|\mathbf{V_0}\|_{\mathbf{g}} \right) < r_{\mathbf{p}}. \tag{9.18}$$

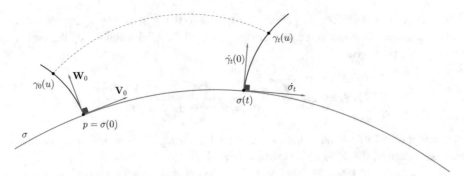

Fig. 9.1 The construction of parallel transport along the geodesic segment σ

Let $\sigma : [-\delta, 2\delta] \to \mathbf{M}$ be the geodesic segment $\sigma(t) = \exp_{\mathbf{p}}(t\mathbf{V}_0)$ and $\mathbf{W}$ be the unique parallel vector field along σ with the initial data $\mathbf{W}(0) = \mathbf{W}_0$. For any $t \in [0, \delta]$, let $\gamma_t : [0, \delta] \to \mathbf{M}$ be the geodesic segment $\gamma_t(u) = \exp_{\sigma(t)}(u\mathbf{W}(t))$.

Let us fix $t, u \in [0, \delta]$ arbitrarily, $u \neq 0$. Due to (9.18), the geodesic segment $\gamma_t|_{[0,u]}$ belongs to the totally normal ball $B_{\mathbf{g}}(\mathbf{p}, r_{\mathbf{p}})$ of $\mathbf{p}$; thus, $\gamma_t|_{[0,u]}$ is the unique minimal geodesic joining the point $\gamma_t(0) = \sigma(t)$ to $\gamma_t(u)$. Moreover, since $\mathbf{W}$ is the parallel transport of $\mathbf{W}(0) = \mathbf{W}_0$ along σ, we have $\mathbf{g}(\mathbf{W}(t), \dot{\sigma}(t)) = \mathbf{g}(\mathbf{W}(0), \dot{\sigma}(0)) = \mathbf{g}(\mathbf{W}_0, \mathbf{V}_0) = 0$; therefore,

$$\mathbf{g}(\dot{\gamma}_t(0), \dot{\sigma}(t)) = \mathbf{g}(\mathbf{W}(t), \dot{\sigma}(t)) = 0,$$

see Fig. 9.1.

Consequently, the minimal geodesic segment $\gamma_t|_{[0,u]}$ joining $\gamma_t(0) = \sigma(t)$ to $\gamma_t(u)$, and the set $\mathbf{K} = \mathrm{Im}\sigma = \{\sigma(t) : t \in [-\delta, 2\delta]\}$ fulfill hypothesis (OA_2). Note that $\mathrm{Im}\sigma$ is a closed, geodesic convex set in $\mathbf{M}$; thus, from hypothesis (i) it follows that $\mathrm{Im}\sigma$ verifies the obtuse-angle property and $P_{\mathrm{Im}\sigma}$ is non-expansive. Thus, (OA_2) implies (OA_1), i.e., for every $t, u \in [0, \delta]$, we have $\sigma(t) \in P_{\mathrm{Im}\sigma}(\gamma_t(u))$. Since $\mathrm{Im}\sigma$ is a Chebyshev set (cf. the non-expansiveness of $P_{\mathrm{Im}\sigma}$), for every $t, u \in [0, \delta]$, we have

$$P_{\mathrm{Im}\sigma}(\gamma_t(u)) = \{\sigma(t)\}. \tag{9.19}$$

Thus, for every $t, u \in [0, \delta]$, relation (9.19) and the non-expansiveness of $P_{\mathrm{Im}\sigma}$ imply

$$\mathbf{d}_{\mathbf{g}}(\mathbf{p}, \sigma(t)) = \mathbf{d}_{\mathbf{g}}(\sigma(0), \sigma(t)) = \mathbf{d}_{\mathbf{g}}(P_{\mathrm{Im}\sigma}(\gamma_0(u)), P_{\mathrm{Im}\sigma}(\gamma_t(u))) \tag{9.20}$$

$$\leq \mathbf{d}_{\mathbf{g}}(\gamma_0(u), \gamma_t(u)).$$

The above construction (i.e., the parallel transport of $\mathbf{W}(0) = \mathbf{W}_0$ along σ) and the formula of the sectional curvature in the parallelogramoid of Levi-Civita defined by the points p, $\sigma(t)$, $\gamma_0(u)$, $\gamma_t(u)$ give

$$K_{\mathbf{p}}(S) = \lim_{u,t\to 0} \frac{\mathbf{d}_g^2(\mathbf{p}, \sigma(t)) - \mathbf{d}_g^{mathrm2}(\gamma_0(u), \gamma_t(u))}{\mathbf{d}_g(\mathbf{p}, \gamma_0(u)) \cdot \mathbf{d}_g(\mathbf{p}, \sigma(t))}.$$

According to (9.20), the latter limit is non-positive, so $K_{\mathbf{p}}(S) \leq 0$, which concludes the first part, namely, $(\mathbf{M}, \mathbf{g})$ is a Hadamard manifold.

Now, a result of Chen [3, Theorem 1] implies that the metric spaces (M_i, d_{g_i}) are Aleksandrov NPC spaces for every $i \in \{1, \ldots, n\}$. Consequently, for each $i \in \{1, \ldots, n\}$, the Riemannian manifolds (M_i, g_i) have non-positive sectional curvature, thus they are Hadamard manifolds. The proof is complete.   ◇

Remark 9.9 The obtuse-angle property and the non-expansiveness of the metric projection are also key tools behind the theory of monotone vector fields, proximal point algorithms and variational inequalities developed on Hadamard manifolds; see Li, López and Martín-Márquez [10, 11], and Németh [16]. Within the class of Riemannian manifolds, Theorem 9.9 shows in particular that Hadamard manifolds are indeed the appropriate frameworks for developing successfully the approaches in [10, 11], and [16] and further related works.

9.4   Examples of Nash Equilibria on Curved Settings

In this subsection we present various examples where our main results for Nash equilibria can be efficiently applied; for convenience, we give all the details in our calculations by keeping also the notations from the previous subsections.

Example 9.1 Let

$$K_1 = \{(x_1, x_2) \in \mathbb{R}_+^2 : x_1^2 + x_2^2 \leq 4 \leq (x_1 - 1)^2 + x_2^2\}, \quad K_2 = [-1, 1],$$

and the functions $f_1, f_2 : K_1 \times K_2 \to \mathbb{R}$ defined for $(x_1, x_2) \in K_1$ and $y \in K_2$ by

$$f_1((x_1, x_2), y) = y(x_1^3 + y(1 - x_2)^3); \quad f_2((x_1, x_2), y) = -y^2 x_2 + 4|y|(x_1 + 1).$$

It is clear that $K_1 \subset \mathbb{R}^2$ is not convex in the usual sense while $K_2 \subset \mathbb{R}$ is. However, if we consider the Poincaré upper-plane model $(\mathbb{H}^2, g_{\mathbb{H}})$, the set $K_1 \subset \mathbb{H}^2$ is geodesic convex (and compact) with respect to the metric $g_{\mathbb{H}} = (\frac{\delta_{ij}}{x_2^2})$. Therefore, we embed the set K_1 into the Hadamard manifold $(\mathbb{H}^2, g_{\mathbb{H}})$, and K_2 into the standard Euclidean space $(\mathbb{R}, g_0)$. After natural extensions of $f_1(\cdot, y)$ and $f_2((x_1, x_2), \cdot)$ to the whole $U_1 = \mathbb{H}^2$

and $U_2 = \mathbb{R}$, respectively, we clearly have that $f_1(\cdot, y)$ is a C^1 function on $\mathbb{H}^2$ for every $y \in K_2$, while $f_2((x_1, x_2), \cdot)$ is a locally Lipschitz function on $\mathbb{R}$ for every $(x_1, x_2) \in K_1$. Thus, $\mathbf{f} = (f_1, f_2) \in \mathcal{L}_{(K_1 \times K_2, \mathbb{H}^2 \times \mathbb{R}, \mathbb{H}^2 \times \mathbb{R})}$ and for every $((x_1, x_2), y) \in \mathbf{K} = K_1 \times K_2$, we have

$$\partial_C^1 f_1((x_1, x_2), y) = \operatorname{grad} f_1(\cdot, y)(x_1, x_2) = \left(g_{\mathbb{H}}^{ij} \frac{\partial f_1(\cdot, y)}{\partial x_j} \right)_i = 3yx_2^2(x_1^2, -y(1-x_2)^2);$$

$$\partial_C^2 f_2((x_1, x_2), y) = \begin{cases} -2yx_2 - 4(x_1 + 1) & \text{if } y < 0, \\ 4(x_1 + 1)[-1, 1] & \text{if } y = 0, \\ -2yx_2 + 4(x_1 + 1) & \text{if } y > 0. \end{cases}$$

It is now clear that the map $\mathbf{K} \ni ((x_1, x_2), y) \mapsto \partial_C^{\Delta} \mathbf{f}(((x_1, x_2), y))$ is upper semicontinuous. Consequently, on account of Theorem 9.6, $\mathcal{S}_{NS}(\mathbf{f}, \mathbf{K}) \neq \emptyset$, and its elements are precisely the solutions $((\tilde{x}_1, \tilde{x}_2), \tilde{y}) \in \mathbf{K}$ of the system

$$\begin{cases} \langle \partial_C^1 f_1((\tilde{x}_1, \tilde{x}_2), \tilde{y}), \exp_{(\tilde{x}_1, \tilde{x}_2)}^{-1}(q_1, q_2) \rangle_{g_{\mathbb{H}}} \geq 0 & \text{for all } (q_1, q_2) \in K_1, \\ \xi_C^2(q - \tilde{y}) \geq 0 \text{ for some } \xi_C^2 \in \partial_C^2 f_2((\tilde{x}_1, \tilde{x}_2), \tilde{y}) \text{ for all } q \in K_2. \end{cases} \tag{(S_1)}$$

In order to solve (S_1) we first observe that

$$K_1 \subset \{(x_1, x_2) \in \mathbb{R}^2 : \sqrt{3} \leq x_2 \leq 2(x_1 + 1)\}. \tag{9.21}$$

We distinguish four cases:

(a) If $\tilde{y} = 0$ then both inequalities of (S_1) hold for every $(\tilde{x}_1, \tilde{x}_2) \in K_1$ by choosing $\xi_C^2 = 0 \in \partial_C^2 f_2((\tilde{x}_1, \tilde{x}_2), 0)$ in the second relation. Thus, $((\tilde{x}_1, \tilde{x}_2), 0) \in \mathcal{S}_{NS}(\mathbf{f}, \mathbf{K})$ for every $(\tilde{x}_1, \tilde{x}_2) \in \mathbf{K}$.

(b) Let $0 < \tilde{y} < 1$. The second inequality of (S_1) gives that $-2\tilde{y}\tilde{x}_2 + 4(\tilde{x}_1 + 1) = 0$; together with (9.21) it yields $0 = \tilde{y}\tilde{x}_2 - 2(\tilde{x}_1 + 1) < \tilde{x}_2 - 2(\tilde{x}_1 + 1) \leq 0$, a contradiction.

(c) Let $\tilde{y} = 1$. The second inequality of (S_1) is true if and only if $-2\tilde{x}_2 + 4(\tilde{x}_1 + 1) \leq 0$. Due to (9.21), we necessarily have $\tilde{x}_2 = 2(\tilde{x}_1 + 1)$; this Euclidean line intersects the set K_1 in the unique point $(\tilde{x}_1, \tilde{x}_2) = (0, 2) \in K_1$. By the geometrical meaning of the exponential map one can conclude that

$$\{t \exp_{(0,2)}^{-1}(q_1, q_2) : (q_1, q_2) \in K_1, t \geq 0\} = \{(x, -y) \in \mathbb{R}^2 : x, y \geq 0\}.$$

Taking into account this relation and $\partial_C^1 f_1((0, 2), 1) = (0, -12)$, the first inequality of (S_1) holds true as well. Therefore, $((0, 2), 1) \in \mathcal{S}_{NS}(\mathbf{f}, \mathbf{K})$.

(d) Similar reason as in (b) (for $-1 < \tilde{y} < 0$) and (c) (for $\tilde{y} = -1$) gives that $((0, 2), -1) \in \mathcal{S}_{NS}(\mathbf{f}, \mathbf{K})$.

Thus, from (a)–(d) we have that

$$\mathcal{S}_{NS}(\mathbf{f}, \mathbf{K}) = (K_1 \times \{0\}) \cup \{((0, 2), 1), ((0, 2), -1)\}.$$

Now, on account of Theorem 9.4 (i) we may easily select the Nash equilibrium points for $(\mathbf{f}, \mathbf{K})$ among the elements of $\mathcal{S}_{NS}(\mathbf{f}, \mathbf{K})$ obtaining that $\mathcal{S}_{NE}(\mathbf{f}, \mathbf{K}) = K_1 \times \{0\}$.

In the rest of this subsection we deal with some applications involving matrices; thus, we recall some basic notions from the matrix-calculus. Fix $n \geq 2$. Let $M_n(\mathbb{R})$ be the set of symmetric $n \times n$ matrices with real values, and $M_n^+(\mathbb{R}) \subset M_n(\mathbb{R})$ be the cone of symmetric positive definite matrices. The standard inner product on $M_n(\mathbb{R})$ is defined as

$$\langle U, V \rangle = \text{tr}(UV). \tag{9.22}$$

Here, $\text{tr}(Y)$ denotes the trace of $Y \in M_n(\mathbb{R})$. It is well-known that $(M_n(\mathbb{R}), \langle \cdot, \cdot \rangle)$ is an Euclidean space, the unique geodesic between $X, Y \in M_n(\mathbb{R})$ is

$$\gamma_{X,Y}^E(s) = (1 - s)X + sY, \quad s \in [0, 1]. \tag{9.23}$$

The set $M_n^+(\mathbb{R})$ will be endowed with the Killing form

$$\langle\langle U, V \rangle\rangle_X = \text{tr}(X^{-1}VX^{-1}U), \quad X \in M_n^+(\mathbb{R}), \ U, V \in T_X(M_n^+(\mathbb{R})). \tag{9.24}$$

Note that the pair $(M_n^+(\mathbb{R}), \langle\langle \cdot, \cdot \rangle\rangle)$ is a Hadamard manifold, see Lang [9, Chapter XII], and $T_X(M_n^+(\mathbb{R})) \simeq M_n(\mathbb{R})$. The unique geodesic segment connecting $X, Y \in M_n^+(\mathbb{R})$ is defined by

$$\gamma_{X,Y}^H(s) = X^{1/2}(X^{-1/2}YX^{-1/2})^s X^{1/2}, \quad s \in [0, 1]. \tag{9.25}$$

In particular, $\frac{d}{ds}\gamma_{X,Y}^H(s)|_{s=0} = X^{1/2} \ln(X^{-1/2}YX^{-1/2})X^{1/2}$; consequently, for each $X, Y \in M_n^+(\mathbb{R})$, we have

$$\exp_X^{-1} Y = X^{1/2} \ln(X^{-1/2}YX^{-1/2})X^{1/2}.$$

Moreover, the metric function on $M_n^+(\mathbb{R})$ is given by

$$d_H^2(X, Y) = \langle\langle \exp_X^{-1} Y, \exp_X^{-1} Y \rangle\rangle_X = \text{tr}(\ln^2(X^{-1/2}YX^{-1/2})). \tag{9.26}$$

Example 9.2 Let

$$K_1 = [0, 2], \ K_2 = \{X \in M_n^+(\mathbb{R}) : \text{tr}(\ln^2 X) \leq 1 \leq \det X \leq 2\},$$

and the functions $f_1, f_2 : K_1 \times K_2 \to \mathbb{R}$ defined by

$$f_1(t, X) = (\max(t, 1))^{n-1} \text{tr}^2(X) - 4n \ln(t+1) S_2(X), \tag{9.27}$$

$$f_2(t, X) = g(t) \left(\text{tr}(X^{-1}) + 1 \right)^{t+1} + h(t) \ln \det X. \tag{9.28}$$

Here, $S_2(Y)$ denotes the second elementary symmetric function of the eigenvalues $\lambda_1, \dots, \lambda_n$ of Y, i.e.,

$$S_2(Y) = \sum_{1 \le i_1 < i_2 \le n} \lambda_{i_1} \lambda_{i_2}, \tag{9.29}$$

and $g, h : K_1 \to \mathbb{R}$ are two continuous functions such that

$$h(t) \ge 2(n+1)g(t) \ge 0 \text{ for all } t \in K_1. \tag{9.30}$$

The elements of $\mathcal{S}_{NE}(\mathbf{f}, \mathbf{K})$ are the solutions $(\tilde{t}, \tilde{X}) \in \mathbf{K}$ of the system

$$\begin{cases} \left[(\max(t, 1))^{n-1} - (\max(\tilde{t}, 1))^{n-1} \right] \text{tr}^2(\tilde{X}) \ge 4n S_2(\tilde{X}) \ln \frac{t+1}{\tilde{t}+1}, & \forall t \in K_1, \\ g(\tilde{t}) \left[(\text{tr}(Y^{-1}) + 1)^{\tilde{t}+1} - (\text{tr}(\tilde{X}^{-1}) + 1)^{\tilde{t}+1} \right] + h(\tilde{t}) \ln \frac{\det Y}{\det \tilde{X}} \ge 0, \forall Y \in K_2. \end{cases} \tag{S_2}$$

The involved forms in (S_2) suggest an approach via the Nash-Stampacchia equilibria for $(\mathbf{f}, \mathbf{K})$; first of all, we have to find the appropriate context where the machinery described in Sect. 9.3 works efficiently.

At first glance, the natural geometric framework seems to be $M_n(\mathbb{R})$ with the inner product $\langle \cdot, \cdot \rangle$ defined in (9.22). Note however that the set K_2 is not geodesic convex with respect to $\langle \cdot, \cdot \rangle$. Indeed, let $X = \text{diag}(2, 1, \dots, 1) \in K_2$ and $Y = \text{diag}(1, 2, \dots, 1) \in K_2$ and $\gamma_{X,Y}^E$ be the Euclidean geodesic connecting them, see (9.23); although $\gamma_{X,Y}^E(s) \in M_n^+(\mathbb{R})$ and $\text{tr}(\ln^2(\gamma_{X,Y}^E(s))) = \ln^2(2-s) + \ln^2(1+s) \le \ln^2 2$ for every $s \in [0, 1]$, we have that $\det(\gamma_{X,Y}^E(s)) > 2$ for every $0 < s < 1$. Consequently, a more appropriate metric is needed to provide some sort of geodesic convexity for K_2. To complete this fact, we restrict our attention to the cone of symmetric positive definite matrices $M_n^+(\mathbb{R})$ with the metric introduced in (9.24).

Let $I_n \in M_n^+(\mathbb{R})$ be the identity matrix, and $\overline{B}_H(I_n, 1)$ be the closed geodesic ball in $M_n^+(\mathbb{R})$ with center I_n and radius 1. Note that

$$K_2 = \overline{B}_H(I_n, 1) \cap \{ X \in M_n^+(\mathbb{R}) : 1 \le \det X \le 2 \}.$$

Indeed, for every $X \in M_n^+(\mathbb{R})$, we have

$$d_H^2(I_n, X) = \text{tr}(\ln^2 X). \tag{9.31}$$

Since K_2 is bounded and closed, on account of the Hopf-Rinow theorem, K_2 is compact. Moreover, as a geodesic ball in the Hadamard manifold $(M_n^+(\mathbb{R}), \langle\langle\cdot, \cdot\rangle\rangle)$, the set $\overline{B}_H(I_n, 1)$ is geodesic convex. Keeping the notation from (9.25), if $X, Y \in K_2$, one has for every $s \in [0, 1]$ that

$$\det(\gamma_{X,Y}^H(s)) = (\det X)^{1-s}(\det Y)^s \in [1, 2],$$

which shows the geodesic convexity of K_2 in $(M_n^+(\mathbb{R}), \langle\langle\cdot, \cdot\rangle\rangle)$.

After naturally extending the functions $f_1(\cdot, X)$ and $f_2(t, \cdot)$ to $U_1 = (-\frac{1}{2}, \infty)$ and $U_2 = M_n^+(\mathbb{R})$ by the same expressions (see (9.27) and (9.28)), we clearly have that $\mathbf{f} = (f_1, f_2) \in \mathcal{L}_{(\mathbf{K}, \mathbf{U}, \mathbf{M})}$, where $\mathbf{U} = U_1 \times U_2$, and $\mathbf{M} = \mathbb{R} \times M_n^+(\mathbb{R})$. A standard computation shows that for every $(t, X) \in U_1 \times K_2$, we have

$$\partial_C^1 f_1(t, X) = -\frac{4n S_2(X)}{t+1} + \mathrm{tr}^2(X) \cdot \begin{cases} 0 & \text{if } -1/2 < t < 1, \\ [0, n-1] & \text{if } t = 1, \\ (n-1)t^{n-2} & \text{if } 1 < t. \end{cases}$$

For every $t \in K_1$, the Euclidean gradient of $f_2(t, \cdot)$ at $X \in U_2 = M_n^+(\mathbb{R})$ is

$$f_2'(t, \cdot)(X) = -g(t)(t+1)\left(\mathrm{tr}(X^{-1}) + 1\right)^t X^{-2} + h(t)X^{-1},$$

thus the Riemannian derivative has the form

$$\partial_C^2 f_2(t, X) = \mathrm{grad}\, f_2(t, \cdot)(X) = X f_2'(t, \cdot)(X)X$$
$$= -g(t)(t+1)\left(\mathrm{tr}(X^{-1}) + 1\right)^t I_n + h(t)X.$$

The above expressions show that $\mathbf{K} \ni (t, X) \mapsto \partial_C^\Delta \mathbf{f}(t, X)$ is upper semicontinuous. Therefore, Theorem 9.6 implies that $\mathcal{S}_{NS}(\mathbf{f}, \mathbf{K}) \neq \emptyset$, and its elements $(\tilde{t}, \tilde{X}) \in \mathbf{K}$ are precisely the solutions of the system

$$\begin{cases} \xi_1(t - \tilde{t}) \geq 0 \text{ for some } \xi_1 \in \partial_C^1 f_1(\tilde{t}, \tilde{X}) \text{ for all } t \in K_1, \\ \langle\langle\partial_C^2 f_2(\tilde{t}, \tilde{X}), \exp_{\tilde{X}}^{-1} Y\rangle\rangle_{\tilde{X}} \geq 0 \qquad \text{for all } Y \in K_2, \end{cases} \qquad ((S_2'))$$

We notice that the solutions of (S_2') and (S_2) coincide. In fact, we may show that $\mathbf{f} \in \mathcal{K}_{(\mathbf{K}, \mathbf{U}, \mathbf{M})}$; thus from Theorem 9.4 (ii) we have that $\mathcal{S}_{NE}(\mathbf{f}, \mathbf{K}) = \mathcal{S}_{NS}(\mathbf{f}, \mathbf{K}) = \mathcal{S}_{NC}(\mathbf{f}, \mathbf{K})$. It is clear that the map $t \mapsto f_1(t, X)$ is convex on U_1 for every $X \in K_2$. Moreover, $X \mapsto f_2(t, X)$ is also a convex function on $U_2 = M_n^+(\mathbb{R})$ for every $t \in K_1$. Indeed, fix

$X, Y \in K_2$ and let $\gamma_{X,Y}^H : [0, 1] \to K_2$ be the unique geodesic segment connecting X and Y, see (9.25). For every $s \in [0, 1]$, we have that

$$\ln \det(\gamma_{X,Y}^H(s)) = \ln((\det X)^{1-s}(\det Y)^s)$$
$$= (1 - s) \ln \det X + s \ln \det Y$$
$$= (1 - s) \ln \det(\gamma_{X,Y}^H(0)) + s \ln \det(\gamma_{X,Y}^H(1)).$$

The Riemannian Hessian of $X \mapsto \mathrm{tr}(X^{-1})$ with respect to $\langle\langle \cdot, \cdot \rangle\rangle$ is

$$\mathrm{Hess}(\mathrm{tr}(X^{-1}))(V, V) = \mathrm{tr}(X^{-2}VX^{-1}V) = |X^{-1}VX^{-1/2}|_F^2 \geq 0,$$

where $| \cdot |_F$ denotes the standard Fröbenius norm. Thus, $X \mapsto \mathrm{tr}(X^{-1})$ is convex (see Udriște [18, §3.6]), so $X \mapsto (\mathrm{tr}(X^{-1}) + 1)^{t+1}$. Combining the above facts with the non-negativity of g and h (see (9.30)), it yields that $\mathbf{f} \in \mathcal{K}_{(\mathbf{K,U,M})}$ as we claimed.

By recalling the notation from (9.29), the inequality of Newton has the form

$$S_2(Y) \leq \frac{n - 1}{2n}\mathrm{tr}^2(Y) \quad \text{for all } Y \in M_n(\mathbb{R}). \tag{9.32}$$

The possible cases are as follow:

(a) Let $0 \leq \tilde{t} < 1$. Then the first relation from (S_2') implies $-\frac{4n S_2(\tilde{X})}{\tilde{t}+1} \geq 0$, a contradiction.

(b) If $1 < \tilde{t} < 2$, the first inequality from (S_2') holds if and only if

$$S_2(\tilde{X}) = \frac{n - 1}{4n}\tilde{t}^{n-2}(\tilde{t} + 1)\mathrm{tr}^2(\tilde{X}),$$

which contradicts Newton's inequality (9.32).

(c) If $\tilde{t} = 2$, from the first inequality of (S_2') it follows that

$$3(n - 1)2^{n-4}\mathrm{tr}^2(\tilde{X}) \leq n S_2(\tilde{X}),$$

contradicting again (9.32).

(d) Let $\tilde{t} = 1$. From the first relation of (S_2') we necessarily have that $0 = \xi_1 \in \partial_C^1 f_1(1, \tilde{X})$. This fact is equivalent to

$$\frac{2n S_2(\tilde{X})}{\mathrm{tr}^2(\tilde{X})} \in [0, n - 1],$$

which holds true, see (9.32). In this case, the second relation from (S_2') becomes

$$-2g(1)\left(\mathrm{tr}(\tilde{X}^{-1})+1\right)\langle\langle I_n, \exp_{\tilde{X}}^{-1} Y\rangle\rangle_{\tilde{X}} + h(1)\langle\langle \tilde{X}, \exp_{\tilde{X}}^{-1} Y\rangle\rangle_{\tilde{X}} \geq 0, \ \forall Y \in K_2.$$

By using (9.24) and the well-known formula $e^{\mathrm{tr}(\ln X)} = \det X$, the above inequality reduces to

$$-2g(1)(\mathrm{tr}(\tilde{X}^{-1})+1)\mathrm{tr}(\tilde{X}^{-1}\ln(\tilde{X}^{-1/2}Y\tilde{X}^{-1/2})) + h(1)\ln\frac{\det Y}{\det \tilde{X}} \geq 0, \ \forall Y \in K_2. \quad (9.33)$$

We also distinguish three cases:

(d1) If $g(1) = h(1) = 0$, then $S_{NE}(\mathbf{f}, \mathbf{K}) = S_{NS}(\mathbf{f}, \mathbf{K}) = \{1\} \times K_2$.
(d2) If $g(1) = 0$ and $h(1) > 0$, then (9.33) implies that $S_{NE}(\mathbf{f}, \mathbf{K}) = S_{NS}(\mathbf{f}, \mathbf{K}) = \{(1, \tilde{X}) \in \mathbf{K} : \det \tilde{X} = 1\}$.
(d3) If $g(1) > 0$, then (9.30) implies that $(1, I_n) \in S_{NE}(\mathbf{f}, \mathbf{K}) = S_{NS}(\mathbf{f}, \mathbf{K})$.   ◇

Remark 9.10 We easily observed in the case (d3) that $\tilde{X} = I_n$ solves (9.33). Note that the same is not evident at all for the second inequality in (S_2). We also notice that the determination of the whole set $S_{NS}(\mathbf{f}, \mathbf{K})$ in (d3) is quite difficult; indeed, after a simple matrix-calculus we realize that (9.33) is equivalent to the equation

$$\tilde{X} = P_{K_2}\left(e^{-\frac{h(1)}{2g(1)(\mathrm{tr}(\tilde{X}^{-1})+1)}} \tilde{X} e^{\tilde{X}^{-1}}\right),$$

where P_{K_2} is the metric projection with respect to the metric d_H.

Example 9.3

(a) Assume that K_i is closed and convex in the Euclidean space $(M_i, g_i) = (\mathbb{R}^{m_i}, \langle\cdot,\cdot\rangle_{\mathbb{R}^{m_i}})$, $i \in \{1, \ldots, n\}$, and let $\mathbf{f} \in C_{(\mathbf{K},\mathbf{U},\mathbb{R}^m)}$ where $m = \sum_{i=1}^n m_i$. If $\partial_C^\Delta \mathbf{f}$ is L−globally Lipschitz and κ-strictly monotone on $\mathbf{K} \subset \mathbb{R}^m$, then the function $\mathbf{f}$ verifies $(H_{\mathbf{K}}^{\alpha,\rho})$ with $\alpha = \frac{\kappa}{L^2}$ and $\rho = \frac{\kappa^2}{2L^2}$. (Note that the above facts imply that $\kappa \leq L$, thus $0 < \rho < 1$.) Indeed, for every $\mathbf{p}, \mathbf{q} \in \mathbf{K}$ we have that

$$d_g^2(\exp_\mathbf{p}(-\alpha\partial_C^\Delta \mathbf{f}(\mathbf{p})), \exp_\mathbf{q}(-\alpha\partial_C^\Delta \mathbf{f}(\mathbf{q})))$$

$$= \|\mathbf{p} - \alpha\partial_C^\Delta \mathbf{f}(\mathbf{p}) - (\mathbf{q} - \alpha\partial_C^\Delta \mathbf{f}(\mathbf{q}))\|_{\mathbb{R}^m}^2 = \|\mathbf{p} - \mathbf{q} - (\alpha\partial_C^\Delta \mathbf{f}(\mathbf{p}) - \alpha\partial_C^\Delta \mathbf{f}(\mathbf{q}))\|_{\mathbb{R}^m}^2$$

$$= \|\mathbf{p} - \mathbf{q}\|_{\mathbb{R}^m}^2 - 2\alpha\langle \mathbf{p} - \mathbf{q}, \partial_C^\Delta \mathbf{f}(\mathbf{p}) - \partial_C^\Delta \mathbf{f}(\mathbf{q})\rangle_{\mathbb{R}^m} + \alpha^2\|\partial_C^\Delta \mathbf{f}(\mathbf{p}) - \alpha\partial_C^\Delta \mathbf{f}(\mathbf{q})\|_{\mathbb{R}^m}^2$$

$$\leq (1 - 2\alpha\kappa + \alpha^2 L^2)\|\mathbf{p} - \mathbf{q}\|_{\mathbb{R}^m}^2 = \left(1 - \frac{\kappa^2}{L^2}\right)d_g^2(\mathbf{p}, \mathbf{q})$$

$$\leq (1 - \rho)^2 d_g^2(\mathbf{p}, \mathbf{q}).$$

(b) Let

$$K_1 = [0, \infty), \quad K_2 = \{X \in M_n(\mathbb{R}) : \text{tr}(X) \geq 1\},$$

and the functions $f_1, f_2 : K_1 \times K_2 \to \mathbb{R}$ defined by

$$f_1(t, X) = g(t) - ct\,\text{tr}(X), \quad f_2(t, X) = \text{tr}((X - h(t)A)^2).$$

Here, $g, h : K_1 \to \mathbb{R}$ are two functions such that g is of class C^2 verifying

$$0 < \inf_{K_1} g'' \leq \sup_{K_1} g'' < \infty, \tag{9.34}$$

h is L_h-globally Lipschitz, while $A \in M_n(\mathbb{R})$ and $c > 0$ are fixed such that

$$c + L_h\sqrt{\text{tr}(A^2)} < 2 \inf_{K_1} g'' \quad \text{and} \quad cn + 2L_h\sqrt{\text{tr}(A^2)} < 4. \tag{9.35}$$

Now, we consider the space $M_n(\mathbb{R})$ endowed with the inner product defined in (9.22). We observe that K_2 is geodesic convex but not compact in $(M_n(\mathbb{R}), \langle \cdot, \cdot \rangle)$. After a natural extension of functions $f_1(\cdot, X)$ to $\mathbb{R}$ and $f_2(t, \cdot)$ to the whole $M_n(\mathbb{R})$, we can state that $\mathbf{f} = (f_1, f_2) \in C_{(\mathbf{K}, \mathbf{U}, \mathbf{M})}$, where $\mathbf{U} = \mathbf{M} = \mathbb{R} \times M_n(\mathbb{R})$. On account of (9.34), after a computation it follows that the map

$$\partial_C^A \mathbf{f}(t, X) = (g'(t) - c\,\text{tr}(X), 2(X - h(t)A))$$

is L-globally Lipschitz and κ-strictly monotone on $\mathbf{K}$ with

$$L = \max\left((2 \sup_{K_1} g'' + 8L_h\text{tr}(A^2))^{1/2}, \left(2c^2n + 8\right)^{1/2}\right) > 0,$$

$$\kappa = \min\left(\inf_{K_1} g'' - \frac{c}{2} - \frac{L_h\sqrt{\text{tr}(A^2)}}{2}, 1 - \frac{cn}{4} - \frac{L_h\sqrt{\text{tr}(A^2)}}{2}\right) > 0.$$

According to (a), $\mathbf{f}$ verifies $(H_{\mathbf{K}}^{\alpha,\rho})$ with $\alpha = \frac{\kappa}{L^2}$ and $\rho = \frac{\kappa^2}{2L^2}$. On account of Theorem 9.8, the set of Nash-Stampacchia equilibrium points for $(\mathbf{f}, \mathbf{K})$ contains exactly one point $(\tilde{t}, \tilde{X}) \in \mathbf{K}$ and the orbits of both dynamical systems $(DDS)_\alpha$ and $(CDS)_\alpha$ exponentially converge to $(\tilde{t}, \tilde{X})$. Moreover, one also has that $\mathbf{f} \in \mathcal{K}_{(\mathbf{K}, \mathbf{U}, \mathbf{M})}$; thus, due to Theorem 9.4-(ii) we have that $\mathcal{S}_{NE}(\mathbf{f}, \mathbf{K}) = \mathcal{S}_{NS}(\mathbf{f}, \mathbf{K}) = \{(\tilde{t}, \tilde{X})\}$.

References

1. D. Azagra, J. Ferrera, F. López-Mesas, Nonsmooth analysis and Hamilton-Jacobi equations on Riemannian manifolds. J. Funct. Anal. **220**, 304–361 (2005)
2. C.A. Bessaga, A. Peł czyński, *Selected Topics in Infinite-Dimensional Topology* (PWN— Polish Scientific Publishers, Warsaw, 1975). Monografie Matematyczne, Tom 58. [Mathematical Monographs, Vol. 58]
3. C.-H. Chen, Warped products of metric spaces of curvature bounded from above. Trans. Am. Math. Soc. **351**, 4727–4740 (1999)
4. M. P. do Carmo, in *Riemannian Geometry*. Mathematics: Theory & Applications (Birkhäuser, Boston, 1992). Translated from the second Portuguese edition by Francis Flaherty.
5. S. Grognet, Théorème de Motzkin en courbure négative. Geom. Dedicata **79**, 219–227 (2000)
6. A. Kristály, Location of Nash equilibria: a Riemannian approach. Proc. Am. Math. Soc. **138**, 1803–1810 (2010)
7. A. Kristály, Nash-type equilibria on Riemannian manifolds: a variational approach. J. Math. Pures Appl. (9) **101**, 660–688 (2014)
8. A. Kristály, V. Rădulescu, C. Varga, in *Variational Principles in Mathematical Physics, Geometry, and Economics: Qualitative Analysis of Nonlinear Equations and Unilateral Problems*, vol. 136 of Encyclopedia of Mathematics and its Applications (Cambridge University Press, Cambridge, 2010)
9. S. Lang, in *Fundamentals of Differential Geometry*, vol. 191 of Graduate Texts in Mathematics (Springer, New York, 1999)
10. C. Li, G. López, V. Martín-Márquez, Monotone vector fields and the proximal point algorithm on Hadamard manifolds. J. Lond. Math. Soc. (2) **79**, 663–683 (2009)
11. C. Li, G. López, V. Martín-Márquez, Iterative algorithms for nonexpansive mappings on Hadamard manifolds. Taiwan. J. Math. **14**, 541–559 (2010)
12. J. F. McClendon, Minimax and variational inequalities for compact spaces. Proc. Am. Math. Soc. **89**, 717–721 (1983)
13. D. Moskovitz, L.L. Dines, Convexity in a linear space with an inner product. Duke Math. J. **5**, 520–534 (1939)
14. J. Nash, Equilibrium points in n-person games. Proc. Natl. Acad. Sci. U.S.A. **36**, 48–49 (1950)
15. J. Nash, Non-cooperative games. Ann. Math. **54**, 286–295 (1951)
16. S. Z. Németh, Variational inequalities on Hadamard manifolds. Nonlinear Anal. **52**, 1491–1498 (2003)
17. B. O'Neill, in *Semi-Riemannian Geometry*, vol. 103 of Pure and Applied Mathematics (Academic Press, Inc. [Harcourt Brace Jovanovich, Publishers], New York, 1983). With applications to relativity.
18. C. Udriște, in *Convex Functions and Optimization Methods on Riemannian Manifolds*, vol. 297 of Mathematics and its Applications (Kluwer Academic Publishers Group, Dordrecht, 1994)
19. R. Walter, On the metric projection onto convex sets in Riemannian spaces. Arch. Math. (Basel) **25**, 91–98 (1974)

Inequality Problems Governed by Set-valued Maps of Monotone Type

10.1 Variational-Hemivariational Inequalities

Throughout this section X will denote a real reflexive Banach space with its dual space X^* and $T : X \to L^p(\Omega; \mathbb{R}^k)$ will be a linear and compact operator where $1 < p < \infty$ and Ω is a bounded and open subset of $\mathbb{R}^N$. We shall denote $\hat{u} := Tu$ and by p' the conjugated exponent of p. Let $j = j(x, y) : \Omega \times \mathbb{R}^k \to \mathbb{R}$ be a Carathéodory function, locally Lipschitz with respect to the second variable which satisfies the following condition:
(H_j) there exist $C > 0$ such that

$$|\zeta| \leq C(1 + |y|^{p-1}) \tag{10.1}$$

for a.e. $x \in \Omega$, all $y \in \mathbb{R}^k$ and all $\zeta \in \partial_C^2 j(x, y)$.

Let K be a nonempty closed, convex subset of X and $\phi : X \to (-\infty, +\infty]$ a convex and lower semicontinuous functional such that

$$K_\phi := D(\phi) \cap K \neq \emptyset. \tag{10.2}$$

Assuming A is a set valued mapping from K into X^*, with $D(A) = K$ our aim is to study the following *multivalued variational-hemivariational inequality*:

$(MVHI)$ Find $u \in K$ and $u^* \in A(u)$ such that

$$\langle u^*, v - u \rangle + \phi(v) - \phi(u) + \int_\Omega j_{,2}^0(x, \hat{u}(x); \hat{v}(x) - \hat{u}(x))dx \geq 0, \ \forall v \in K. \tag{10.3}$$

N. Costea et al., *Variational and Monotonicity Methods in Nonsmooth Analysis*,
Frontiers in Mathematics, https://doi.org/10.1007/978-3-030-81671-1_10

As it will be seen, this problem closely links to the *dual variational-hemivariational inequality*:

$(DVHI)$ Find $u \in K$ such that

$$\sup_{v^* \in A(v)} \langle v^*, u - v \rangle \leq \phi(v) - \phi(u) + \int_\Omega j^0_{,2}(x, \hat{u}(x); \hat{v}(x) - \hat{u}(x))dx, \quad \forall v \in K. \quad (10.4)$$

Definition 10.1 A set valued mapping $A : K \rightsquigarrow X^*$ is said to be *lower hemicontinuous* on K if the restriction of A to every line segment of K is lower semicontinuous from $s - X$ into $w^* - X^*$.

We denote by S and $S^\star$ the solutions sets of problem $(MVHI)$ and problem $(DHVI)$, respectively. The following result, due to Costea and Lupu [1], highlights the relationship between the two problems.

Theorem 10.1 *Let K be a nonempty closed and convex subset of the real reflexive Banach space X. If $A : K \rightsquigarrow X^*$ is monotone, then $S \subseteq S^\star$. In addition, if A is lower hemicontinuous, then $S^\star = S$.*

Proof Let $u \in S$ and $v \in K$ be arbitrary fixed. Then it exists $u^* \in A(u)$ such that (10.3) holds. Since A is monotone we have

$$\langle v^* - u^*, v - u \rangle \geq 0, \quad \forall v^* \in A(v). \quad (10.5)$$

Hence, adding (10.3) and (10.5) we have

$$\langle v^*, v - u \rangle + \phi(v) - \phi(u) + \int_\Omega j^0_{,2}(x, \hat{u}(x); \hat{v}(x) - \hat{u}(x)) \, dx \geq 0, \quad \forall v^* \in A(v). \quad (10.6)$$

This is equivalent to

$$\sup_{v^* \in A(v)} \langle v^*, u - v \rangle \leq \phi(v) - \phi(u) + \int_\Omega j^0_{,2}(x, \hat{u}(x); \hat{v}(x) - \hat{u}(x)) \, dx. \quad (10.7)$$

Since v has been arbitrary chosen, it follows that (10.7) holds for all $v \in K$ which implies that $u \in S^\star$.

In addition if A is lower hemicontinuous, we will show that $S = S^\star$. Suppose $u \in S^\star$ and let $v \in K$ be arbitrary fixed. We define the sequence $\{u_n\}_{n \geq 1}$ by $u_n := u + \frac{1}{n}(v - u)$. Clearly $\{u_n\} \subset K$ by the convexity of K. For any $u^* \in A(u)$, using the lower hemicontinuity of A, a sequence $u_n^* \in A(u_n)$ can be determined such that $u_n^* \rightharpoonup u^*$. Taking into account that $u \in S^\star$, for each $n \geq 1$ we have

$$\langle u_n^*, u - u_n \rangle \leq \phi(u_n) - \phi(u) + \int_\Omega j_{,2}^0(x, \hat{u}(x); \hat{u}_n(x) - \hat{u}(x)) \, \mathrm{d}x \qquad (10.8)$$

But ϕ is convex and $j_{,2}^0(x, \hat{u}; \lambda \hat{v}) = \lambda j_{,2}^0(x, \hat{u}; \hat{v})$ for all $\lambda > 0$. Therefore (10.8) may be written, equivalently

$$0 \leq \left\langle u_n^*, \frac{1}{n}(v - u) \right\rangle + \phi \left(\frac{1}{n}v + \frac{n-1}{n}u \right) - \phi(u) + \int_\Omega j_{,2}^0 \left(x, \hat{u}(x); \frac{1}{n}(\hat{v}(x) - \hat{u}(x)) \right) \, \mathrm{d}x$$

$$\leq \frac{1}{n} \left[\langle u_n^*, v - u \rangle + \phi(v) - \phi(u) + \int_\Omega j_{,2}^0(x, \hat{u}(x); \hat{v}(x) - \hat{u}(x)) \, \mathrm{d}x \right].$$

Multiplying the last relation by n and passing to the limits as $n \to \infty$ we obtain the $u \in S$. □

We are now in position to establish the existence of solutions when the constraint set K is bounded. More precisely we have the following result.

Theorem 10.2 ([1]) *Let K be a nonempty, bounded, closed and convex subset of the real reflexive Banach space X and $A : K \rightsquigarrow X^*$ a set valued mapping which is monotone and lower hemicontinuous on K. If $T : X \to L^p(\Omega; \mathbb{R}^k)$ is linear and compact and j satisfies the condition (10.1) then problem $(MVHI)$ possesses at least one solution.*

Proof For any $v \in K_\phi$ define two set valued mappings $F, G : K \cap D(\phi) \rightsquigarrow X$ as follows:

$$F(v) := \left\{ u \in K_\phi : \begin{array}{l} \exists u^* \in A(u) \text{ s.t. } \langle u^*, v - u \rangle + \phi(v) - \phi(u) \\ + \int_\Omega j_{,2}^0(x, \hat{u}(x); \hat{v}(x) - \hat{u}(x)) \mathrm{d}x \geq 0 \end{array} \right\}$$

and

$$G(v) := \left\{ u \in K_\phi : \begin{array}{l} \sup_{v^* \in A(v)} \langle v^*, u - v \rangle \leq \phi(v) - \phi(u) \\ + \int_\Omega j_{,2}^0(x, \hat{u}(x); \hat{v}(x) - \hat{u}(x)) \mathrm{d}x \end{array} \right\}.$$

We divide the proof into several steps as follows.

STEP 1. *F is a KKM mapping.*
    If F is not a KKM mapping, then there exists $\{v_1, \ldots, v_n\} \subset K_\phi$ such that

$$\mathrm{co}\{v_1, \ldots, v_n\} \not\subset \bigcup_{i=1}^n F(v_i),$$

i.e., there exists a $v_0 \in co\{v_1, \ldots, v_n\}$, $v_0 := \sum_{i=1}^{n} \lambda_i v_i$, where $\lambda_i \in [0, 1]$, $i \in \overline{1, n}$, $\sum_{i=1}^{n} \lambda_i = 1$, but $v_0 \notin \bigcup_{i=1}^{n} F(v_i)$. By the definition of F, we have

$$\langle v_0^*, v_i - v_0 \rangle + \phi(v_i) - \phi(v_0) + \int_\Omega j_{,2}^0(x, \hat{v}_0(x); \hat{v}_i(x) - \hat{v}_0(x))dx < 0, \quad \forall v_0^* \in A(v_0)$$

for $i \in \overline{1, n}$. It follows from the convexity of $\hat{v} \longmapsto j_{,2}^0(x, \hat{u}; \hat{v})$ and the convexity of ϕ that for each $v_0^* \in A(v_0)$ we have

$$0 = \langle v_0^*, v_0 - v_0 \rangle + \phi(v_0) - \phi(v_0) + \int_\Omega j_{,2}^0(x, \hat{v}_0(x); \hat{v}_0(x) - \hat{v}_0(x))dx$$

$$= \left\langle v_0^*, \sum_{i=1}^{n} \lambda_i v_i - v_0 \right\rangle + \phi\left(\sum_{i=1}^{n} \lambda_i v_i\right) - \phi(v_0)$$

$$+ \int_\Omega j_{,2}^0\left(x, \hat{v}_0(x); \sum_{i=1}^{n} \lambda_i \hat{v}_i(x) - \hat{v}_0(x)\right) dx$$

$$\leq \sum_{i=1}^{n} \lambda_i \left[\langle v_0^*, v_i - v_0 \rangle + \phi(v_i) - \phi(v_0) + \int_\Omega j_{,2}^0(x, \hat{v}_0(x); \hat{v}_i(x) - \hat{v}_0(x))dx\right]$$

$$< 0.$$

which is a contradiction. This implies that F is a KKM mapping.

STEP 2. $F(v) \subseteq G(v)$ for all $v \in K_\phi$.

For a given $v \in K_\phi$, let $u \in F(v)$. Then, there exists $u^* \in A(u)$ such that

$$\langle u^*, v - u \rangle + \phi(v) - \phi(u) + \int_\Omega j_{,2}^0(x, \hat{u}(x); \hat{v}(x) - \hat{u}(x)) \, dx \geq 0.$$

Since A is monotone, we have

$$\langle v^* - u^*, v - u \rangle \geq 0, \quad \forall v^* \in A(v).$$

It follows from the last two relations that

$$\langle v^*, v - u \rangle + \phi(v) - \phi(u) + \int_\Omega j_{,2}^0(x, \hat{u}(x); \hat{v}(x) - \hat{u}(x))dx \geq 0, \quad \forall v^* A(v)$$

which may be equivalently rewritten

$$\sup_{v^* \in A(v)} \langle v^*, u - v \rangle \leq \phi(v) - \phi(v) + \int_\Omega j_{,2}^0(x, \hat{u}(x); \hat{v}(x) - \hat{u}(x))dx$$

and so $u \in G(v)$. In particular, this implies that G is also a KKM mapping.

STEP 3. $G(v)$*is weakly closed for each* $v \in K_\phi$.

Let $\{u_n\} \subset G(v)$ be a sequence which converges weakly to u as $n \to \infty$. We must prove that $u \in G(v)$. Since $u_n \in G(v)$ for all $n \geq 1$ and ϕ is weakly lower semicontinuous, for each $v^* \in A(v)$ we have

$$
0 \leq \limsup_{n\to\infty} \left[\langle v^*, v - u_n \rangle + \phi(v) - \phi(u_n) + \int_\Omega j_{,2}^0(x, \hat{u}_n(x); \hat{v}(x) - \hat{u}_n(x))dx \right]
$$

$$
\leq \lim_{n\to\infty} \langle v^*, v - u_n \rangle + \phi(v) - \liminf_{n\to\infty} \phi(u_n) + \limsup_{n\to\infty} \int_\Omega j_{,2}^0(x, \hat{u}_n(x); \hat{v}(x) - \hat{u}_n(x))dx
$$

$$
\leq \langle v^*, v - u \rangle + \phi(v) - \phi(u) + \int_\Omega j_{,2}^0(x, \hat{u}(x); \hat{v}(x) - \hat{u}(x))dx.
$$

This is equivalent to $u \in G(v)$.

STEP 4. $G(v)$*is weakly compact for all* $v \in K_\phi$.

Indeed, since K is bounded, closed and convex, we know that K is weakly compact, and so $G(v)$ is weakly compact for each $v \in K \cap D(\phi)$, as it is a weakly closed subset of an weakly compact set.

Therefore conditions of Corollary D.1 are satisfied in the weak topology. It follows that

$$
\bigcap_{v \in K_\phi} G(v) \neq \emptyset.
$$

This yields that there exists an element $u \in K_\phi$ such that, for any $v \in K_\phi$

$$
\sup_{v^* \in A(v)} \langle v^*, v - u \rangle \leq \phi(v) - \phi(u) + \int_\Omega j_{,2}^0(x, \hat{u}(x); \hat{v}(x) - \hat{u}(x))dx.
$$

This inequality is trivially satisfied for any $v \notin D(\phi)$ which means that the inequality problem $(DVHI)$ has at least one solution. Theorem 10.1 enables us to claim that inequality problem $(MVHI)$ also possesses a solution.   □

We present below some coercivity conditions ensuring the existence of solutions for unbounded constraint sets. Without loss of generality we may assume that $0 \in K_\phi$ and let us consider the sets $K_n := \{u \in K : \|u\| \leq n\}$ for $n \geq 1$.

If K is nonempty, unbounded, closed and convex subset of X and $A : K \rightsquigarrow X^*$ is monotone and lower hemicontinuous, then by Theorem 10.2, for every $n \geq 1$ there exists $u_n \in K_n$ and $u_n^* \in A(u_n)$ such that

$$
\langle u_n^*, v - u_n \rangle + \phi(v) - \phi(u_n) + \int_\Omega j_{,2}^0(x, \hat{u}_n(x); \hat{v}(x) - \hat{u}_n(x)) \, dx \geq 0, \ \forall v \in K_n, \quad (10.9)
$$

Theorem 10.3 ([1]) *Assume that the same hypotheses as in Theorem 10.2 hold without the assumption of boundedness of K and let $u_n \in K_n$ and $u_n^* \in A(u_n)$ be two sequences such that (10.9) is satisfied for every $n \geq 1$. Then each of the following condition is sufficient for the problem $(MVHI)$ to possess a solution:*

(C_1) *There exists a positive integer n_0 such that $\|u_{n_0}\| < n_0$;*

(C_2) *There exists a positive integer n_0 such that*

$$\langle u_{n_0}^*, -u_{n_0} \rangle + \phi(0) - \phi(u_{n_0}) + \int_\Omega j_{,2}^0(x, \hat{u}_{n_0}(x); -\hat{u}_{n_0}(x)) dx \leq 0;$$

(C_3) *There exists $u_0 \in K_\phi$ and $q \geq p$ such that for any unbounded sequence $\{w_n\} \subset K$ one has*

$$\frac{\langle w_n^*, w_n - u_0 \rangle}{\|w_n\|^q} \to \infty, \qquad as\ n \to \infty$$

for every $w_n^* \in A(w_n)$.

Proof Let $v \in K$ be arbitrary fixed.

Assume (C_1) holds and take $t > 0$ small enough such that $w := u_{n_0} + t(v - u_{n_0})$ satisfies $w \in K_{n_0}$ (it suffices to take $t = 1$ if $v := u_{n_0}$ and $t < (n_0 - \|u_{n_0}\|)/\|v - u_{n_0}\|$ otherwise). By (10.9) we have

$$0 \leq \langle u_{n_0}^*, w - u_{n_0} \rangle + \phi(w) - \phi(u_{n_0}) + \int_\Omega j_{,2}^0(x, \hat{u}_{n_0}(x); \hat{w}(x) - \hat{u}_{n_0}(x)) dx$$

$$\leq t \left[\langle u_{n_0}^*, v - u_{n_0} \rangle + \phi(v) - \phi(u_{n_0}) + \int_\Omega j_{,2}^0(x, \hat{u}_{n_0}(x); \hat{v}(x) - \hat{u}_{n_0}(x)) dx \right].$$

Dividing by t the last relation we observe that u_{n_0} is a solution of $(MVHI)$.

Now, let us assume that (C_2) is fulfilled. In this case, some $t \in (0, 1)$ can be found such that $tv \in K_{n_0}$. Taking (10.9) into account

$$0 \leq \langle u_{n_0}^*, tv - u_{n_0} \rangle + \phi(tv) - \phi(u_{n_0}) + \int_\Omega j_{,2}^0(x, \hat{u}_{n_0}(x); t\hat{v}(x) - \hat{u}_{n_0}(x)) dx$$

$$= \langle u_{n_0}^*, t(v - u_{n_0}) + (1 - t)(-u_{n_0}) \rangle + \phi(tv + (1 - t)0) - \phi(u_{n_0})$$

$$+ \int_\Omega j_{,2}^0(x, \hat{u}_{n_0}(x); t(\hat{v}(x) - \hat{u}_{n_0}(x)) + (1 - t)(-\hat{u}_{n_0}(x))) dx$$

$$\leq t \left[\langle u_{n_0}^*, v - u_{n_0} \rangle + \phi(v) - \phi(u_{n_0}) + \int_\Omega j_{,2}^0(x, \hat{u}_{n_0}(x); \hat{v}(x) - \hat{u}_{n_0}(x)) dx \right]$$

$$+(1-t)\left[\langle u_{n_0}^*, -u_{n_0}\rangle + \phi(0) - \phi(u_{n_0}) + \int_\Omega j_{,2}^0(x, \hat{u}_{n_0}(x); -\hat{u}_{n_0}(x))dx\right]$$

$$\leq t\left[\langle u_{n_0}^*, v - u_{n_0}\rangle + \phi(v) - \phi(u_{n_0}) + \int_\Omega j_{,2}^0(x, \hat{u}_{n_0}(x); \hat{v}(x) - \hat{u}_{n_0}(x))dx\right].$$

Dividing again by t the conclusion follows.

Assuming that (C_3) holds we observe that there exists $n_0 > 0$ such that $u_0 \in K_n$ for all $n \geq n_0$. We claim that the sequence $\{u_n\}$ is bounded. Suppose by contradiction that up to a subsequence $\|u_n\| \to \infty$. Since $w_n := u_n/\|u_n\|$ is bounded, passing eventually to a subsequence (still denoted w_n for the sake of simplicity), we may assume that $w_n \rightharpoonup w$. The function ϕ being convex and lower semicontinuous, it is bounded from below by an affine and continuous function (see Theorem 1.3), which means that for some $\zeta \in X^*$ and some $\alpha \in \mathbb{R}$ we have

$$\langle \zeta, u\rangle + \alpha \leq \phi(u), \quad \forall u \in X.$$

This leads to

$$-\phi(u) \leq \|\zeta\| \cdot \|u\| - \alpha, \quad \forall u \in X. \tag{10.10}$$

On the other hand, for any $y, h \in \mathbb{R}^k$ there exists $\xi \in \partial_C j(x, y)$ such that

$$j_{,2}^0(x, y; h) = \xi \cdot h = \max\left\{\eta \cdot h : \eta \in \partial_C^2 j(x, y)\right\}.$$

It follows from (10.1) that

$$\left|j_{,2}^0(x, \hat{u}(x); \hat{v}(x))\right| \leq C\left(1 + |\hat{u}(x)|^{p-1}\right)|\hat{v}(x)|$$

and using Hölder's inequality we obtain that

$$\left|\int_\Omega j_{,2}^0(x, \hat{u}(x); \hat{v}(x))dx\right| \leq C\left((\text{meas}(\Omega))^{\frac{p-1}{p}}\|\hat{v}\|_p + \|\hat{u}\|_p^{p-1}\|\hat{v}\|_p\right)$$

$$\leq C_1\|v\| + C_2\|u\|^{p-1}\|v\| \tag{10.11}$$

for some suitable constants $C_1, C_2 > 0$. Relations (10.9), (10.10), and (10.11) show that

$$\langle u_n^*, u_n - u_0\rangle \leq \phi(u_0) - \phi(u_n) + \int_\Omega j_{,2}^0(x, \hat{u}_n(x); \hat{u}_0(x) - \hat{u}_n(x))dx$$

$$\leq \phi(u_0) + \|\zeta\| \cdot \|u_n\| - \alpha + C_1\|u_n - u_0\| + C_2\|u_n\|^{p-1}\|u_n - u_0\|.$$

Thus

$$\frac{\langle u_n^*, u_n - u_0 \rangle}{\|u_n\|^q} \leq \frac{\phi(u_0) - \alpha}{\|u_n\|^q} + \frac{\|\zeta\|}{\|u_n\|^{q-1}} + C_1 \left\| \frac{w_n}{\|u_n\|^{q-1}} - \frac{u_0}{\|u_n\|^q} \right\| + $$
$$C_2 \left\| \frac{w_n}{\|u_n\|^{q-p}} - \frac{u_0}{\|u_n\|^{q-p+1}} \right\|$$

and passing to the limit as $n \to \infty$ we reach a contradiction, since $1 < p \leq q$.

Since $\{u_n\}$ is bounded, a $n_0 \geq 1$ can be found such that $\|u_{n_0}\| < n_0$ and by (C_1) the corresponding solution of (10.9) u_{n_0} solves $(MHVI)$.   □

10.2   Quasi-Hemivariational Inequalities

Let $(X, \| \cdot \|)$ be a real Banach space which is continuously embedded in $L^p(\Omega; \mathbb{R}^n)$, for some $1 < p < +\infty$ and $n \geq 1$, where Ω is a bounded domain in $\mathbb{R}^m$, $m \geq 1$. Let i be the canonical injection of X into $L^p(\Omega; \mathbb{R}^n)$ and denote by $i^* : L^q(\Omega; \mathbb{R}^n) \to X^*$ the adjoint operator of i $(1/p + 1/q = 1)$.

Throughout this section $A : X \rightsquigarrow X^*$ is a nonlinear set-valued mapping, $F : X \to X^*$ is a nonlinear operator and $J : L^p(\Omega; \mathbb{R}^n) \to \mathbb{R}$ is a locally Lipschitz functional. We also assume that $h : X \to \mathbb{R}$ is a given nonnegative functional.

The aim is to study the existence of solutions for the following *multivalued quasi-hemivariational inequality*:

$(MQHI)$ Find $u \in X$ and $u^* \in A(u)$ such that

$$\langle u^*, v \rangle + h(u) J^0(iu; iv) \geq \langle Fu, v \rangle, \quad \forall v \in X.$$

The above problem is called a quasi-hemivariational inequality because, in general, we cannot determine a function G such that $\partial_C G(u) = h(u)\partial J(u)$.

As we will see next problem $(MQHI)$ can be rewritten equivalently as an *inclusion* in the following way:

(IP) Find $u \in X$ such that

$$Fu \in A(u) + h(u) i^* \partial_C J(iu), \quad \text{in } X^*.$$

An element $u \in X$ is called a solution of (IP) if there exist $u^* \in A(u)$ and $\zeta \in \partial_C J(iu)$ such that

$$\langle u^*, v \rangle + h(u)\langle i^* \zeta, v \rangle = \langle Fu, v \rangle, \quad \forall v \in X. \tag{10.12}$$

Proposition 10.1 *An element $u \in X$ is a solution of problem (IP) if and only if it solves problem $(MQHI)$.*

Proof

$(MQHI) \Rightarrow (IP)$ Let $u \in X$ be a solution of $(MQHI)$. Then, by Proposition 2.4, there exists $\zeta_u \in \partial_C J(iu)$ such that for all $w \in L^p(\Omega; \mathbb{R}^n)$ we have

$$J^0(iu; w) = \langle \zeta_u, w \rangle_{L^q \times L^p} = \max\{\langle \zeta, w \rangle_{L^q \times L^p} : \zeta \in \partial_C J(iu)\}.$$

Taking $w := iv$ and using the fact that u is a solution of $(MQHI)$ we obtain

$$\langle u^*, v \rangle + h(u)\langle i^* \zeta_u, v \rangle \geq \langle Fu, v \rangle, \quad \forall v \in X,$$

for some $u^* \in A(u)$. Taking $-v$ instead of v in the above relation we deduce that (10.12) holds therefore u is a solution of problem (IP).

$(IP) \Rightarrow (MQHI)$ Let $u \in X$ be a solution of (IP). Then, there exist $u^* \in A(u)$ and $\zeta \in \partial_C J(iu)$ such that (10.12) takes place. As $\zeta \in \partial_C J(iu)$ we obtain that

$$\langle \zeta, w \rangle_{L^q \times L^p} \leq J^0(iu; w), \quad \forall w \in L^p(\Omega; \mathbb{R}^n).$$

For a fixed $v \in X$ we define $w := iv$ and taking into account that h is nonnegative we get

$$h(u)\langle i^* \zeta, v \rangle = h(u)\langle \zeta, iv \rangle_{L^q \times L^p} \leq h(u)J^0(iu; iv) \tag{10.13}$$

Combining (10.12) and (10.13) we obtain that u solves inequality problem $(MQHI)$.   □

Sometimes, due to some technical reasons, it is useful to study hemivariational inequalities of the type $(MQHI)$ whose solution is sought in a nonempty, closed and convex subset K of X, the so-called set of constraints. This leads us to the study of the following inequality problem:

(P_K) Find $u \in K$ and $u^* \in A(u)$ such that

$$\langle u^*, v - u \rangle + h(u)J^0(iu; iv - iu) \geq \langle Fu, v - u \rangle, \quad \forall v \in K.$$

The first main result of this section is given by the following theorem.

Theorem 10.4 ([2]) *Let K be a nonempty compact convex subset of the real Banach space X. Assume that:*

(H_1) $A : X \to X^*$ *is l.s.c. from $s - X$ into $w^* - X^*$;*
(H_2) $h : X \to \mathbb{R}$ *is a continuous nonnegative functional;*
(H_3) $F : X \to X^*$ *satisfies* $\limsup\limits_{n \to \infty} \langle Fu_n, v - u_n \rangle \geq \langle Fu, v - u \rangle$, *whenever $u_n \to u$.*

Then the inequality problem (P_K) has at least one solution.

Proof Arguing by contradiction, let us assume that problem (P_K) has no solution. Then, for each $u \in K$, there exists $v \in K$ such that

$$\sup_{u^* \in A(u)} \langle u^*, v - u \rangle + h(u) J^0(iu; iv - iu) < \langle Fu, v - u \rangle. \tag{10.14}$$

We introduce the set-valued mapping $\Lambda : K \rightsquigarrow K$ defined by

$$\Lambda(v) := \left\{ u \in K : \inf_{u^* \in A(u)} \langle u^*, v - u \rangle + h(u) J^0(iu; iv - iu) \geq \langle Fu, v - u \rangle \right\}.$$

We claim that the set-valued map Λ has nonempty closed values.

The fact that $\Lambda(v)$ is nonempty is obvious as $v \in \Lambda(v)$ for each $v \in K$.

In order to prove the above claim let us fix $v \in K$ and consider a sequence $\{u_n\}_{n \geq 1} \subset \Lambda(v)$ which converges to some $u \in K$. We shall prove that $u \in \Lambda(v)$. As $u_n \in \Lambda(v)$, for each $n \geq 1$ we get that

$$\langle u_n^*, v - u_n \rangle + h(u_n) J^0(iu_n; iv - iu_n) \geq \langle Fu_n, v - u_n \rangle, \ \forall u_n^* \in A(u_n). \tag{10.15}$$

Let $u^* \in A(u)$ be fixed and let $\bar{u}_n^* \in A(u_n)$ such that $\bar{u}_n^* \rightharpoonup u^*$ in X^* (the existence of such a sequence is ensured by the fact that A is l.s.c. with respect to the weak* topology of X^*). On the other hand, using the continuous embedding of X into $L^p(\Omega; \mathbb{R}^n)$ we obtain that $iu_n \to iu$ in $L^p(\Omega; \mathbb{R}^n)$. Passing to lim sup as $n \to \infty$ in (10.15) we obtain the following estimates:

$$\langle Fu, v - u \rangle \leq \limsup_{n \to \infty} \langle Fu_n, v - u_n \rangle \leq \limsup_{n \to \infty} \left[\langle \bar{u}_n^*, v - u_n \rangle + h(u_n) J^0(iu_n; iv - iu_n) \right]$$

$$\leq \limsup_{n \to \infty} \langle \bar{u}_n^*, v - u_n \rangle + \limsup_{n \to \infty} [h(u_n) - h(u)] J^0(iu_n; iv - iu_n)$$

$$+ \limsup_{n \to \infty} h(u) J^0(iu_n; iv - iu_n)$$

$$\leq \langle u^*, v - u \rangle + h(u) J^0(iu; iv - iu).$$

This shows that $u \in \Lambda(v)$ hence Λ has closed values.

According to (10.14) for each $u \in K$ there exists $v \in K$ such that $u \in [\Lambda(v)]^c :=$ $X - \Lambda(v)$. This means that the family $\{[\Lambda(v)]^c\}_{v \in K}$ is an open covering of the compact set K. Therefore there exists a finite subset $\{v_1, \ldots, v_N\}$ of K such that $\{[\Lambda(v_j)]^c\}_{1 \leq j \leq N}$ is a finite subcover of K. For each $j \in \{1, \ldots, N\}$ let $\delta_j(u)$ be the distance between u and the set $\Lambda(v_j)$ and define $\beta_j : K \to \mathbb{R}$ as follows:

$$\beta_j(u) := \frac{\delta_j(u)}{\sum\limits_{k=1}^{N} \delta_k(u)}.$$

Clearly, for each $j \in \{1, \ldots, N\}$, β_j is a Lipschitz continuous function that vanishes on $\Lambda(v_j)$ and $0 \leq \beta_j(u) \leq 1$, for all $u \in K$. Moreover, $\sum_{j=1}^{N} \beta_j(u) = 1$. Let us consider next the operator $S : K \to K$ defined by

$$S(u) := \sum_{j=1}^{N} \beta_j(u)v_j.$$

We shall prove that S is a completely continuous operator. We have

$$\|Su_1 - Su_2\| = \left\| \sum_{j=1}^{N}(\beta(u_1) - \beta(u_2))v_j \right\| \leq \sum_{j=1}^{N} \|v_j\| \, \|\beta(u_1) - \beta(u_2)\|$$

$$\leq \sum_{j=1}^{N} \|v_j\| \, L_j \, \|u_1 - u_2\| \leq L \, \|u_1 - u_2\|,$$

which shows that S is Lipschitz continuous hence continuous.

Let M be a bounded subset of K. As $\overline{S(M)}$ is a closed subset of the compact set K we conclude that $S(M)$ is relatively compact, hence S maps bounded sets into relatively compact sets which shows that S is a compact map. Thus, by Schauder's fixed point theorem, there exists $u_0 \in K$ such that $S(u_0) = u_0$.

Let us define next the functional $g : K \to \mathbb{R}$

$$g(u) := \inf_{u^* \in A(u)} \langle u^*, S(u) - u \rangle + h(u)J^0(iu, iS(u) - iu) - \langle Fu, S(u) - u \rangle.$$

Taking into account the way the operator S was constructed, for each $u \in K$, we have:

$$g(u) = \inf_{u^* \in A(u)} \left\langle u^*, \sum_{j=1}^{N} \beta_j(u)(v_j - u) \right\rangle + h(u) J^0 \left(iu, \sum_{j=1}^{N} \beta_j(u)(iv_j - iu) \right)$$

$$- \left\langle Fu, \sum_{j=1}^{N} \beta_j(u)(v_j - u) \right\rangle$$

$$\leq \sum_{j=1}^{N} \beta_j(u) \left[\inf_{u^* \in A(u)} \langle u^*, v_j - u \rangle + h(u) J^0(iu, iv_j iu) - \langle Fu, v_j - u \rangle \right].$$

Let $u \in K$ be arbitrary fixed. For each index $j \in \{1, \ldots, N\}$ we distinguish the following possibilities:

CASE 1. $u \in \left[\Lambda(v_j) \right]^c$.
   In this case we have

$$\beta_j(u) > 0$$

and

$$\inf_{u^* \in A(u)} \langle u^*, v_j - u \rangle + h(u) J^0(iu, iv_j - iu) - \langle Fu, v_j - u \rangle < 0.$$

CASE 2. $u \in \Lambda(v_j)$.
   In this case we have

$$\beta_j(u) = 0$$

and

$$\inf_{u^* \in A(u)} \langle u^*, v_j - u \rangle + h(u) J^0(iu, iv_j - iu) - \langle Fu, v_j - u \rangle \geq 0.$$

Taking into account that $K \subseteq \bigcup_{j=1}^{N} \left[\Lambda(v_j) \right]^c$ we deduce that there exists at least one index $j_0 \in \{1, \ldots, N\}$ such that $u \in \left[\Lambda(v_{j_0}) \right]^c$. This shows that $g(u) < 0$ for all $u \in K$.
   On the other hand, $g(u_0) = 0$ and thus we have obtained a contradiction that completes the proof.   □

We point out the fact that in the above case when K is a compact convex subset of X we do not impose any monotonicity conditions on A, nor we assume X to be a reflexive space. However, in applications, most problems lead to an inequality whose solution is sought in a closed and convex subset of the space X. Weakening the hypotheses on K by assuming that K is only bounded, closed and convex, we need to impose certain monotonicity properties on A and assume in addition that X is reflexive.

Theorem 10.5 ([2]) *Let K be a nonempty, bounded, closed and convex subset of the real reflexive Banach space X which is compactly embedded in $L^p(\Omega; \mathbb{R}^n)$. Assume that:*

(H_4) $A : X \to X^*$ *is l.s.c. from $s - X$ into $w - X^*$ and relaxed α monotone;*

(H_5) $\alpha : X \to \mathbb{R}$ *is a functional such that $\limsup_{n\to\infty} \alpha(u_n) \geq \alpha(u)$ whenever $u_n \rightharpoonup u$ and*
$$\lim_{t\downarrow 0} \frac{\alpha(tu)}{t} = 0;$$

(H_6) $h : X \to \mathbb{R}$ *is a nonnegative sequentially weakly continuous functional;*

(H_7) $F : X \to X^*$ *is an operator such that $u \mapsto \langle Fu, v - u\rangle$ is weakly lower semicontinuous.*

Then the inequality problem (P_K) has at least one solution in K.

Proof Let us define the set-valued mapping $\Theta : K \rightsquigarrow K$

$$\Theta(v) := \left\{ \begin{array}{l} u \in K : \alpha(v - u) \leq \inf_{v^* \in A(v)} \langle v^*, v - u\rangle + h(u)J^0(iu; iv - iu) \\ \qquad\qquad\qquad -\langle Fu, v - u\rangle \end{array} \right\}.$$

We show first that Θ has weakly closed values. Let us fix $v \in K$ and consider a sequence $\{u_n\}_{n\geq 1} \subset \Theta(v)$ such that $u_n \rightharpoonup u$ in X. We must prove that $u \in \Theta(v)$. First we observe that the compactness of the embedding operator i implies that the sequence $\{iu_n\}_{n\geq 1}$ converges strongly to iu in $L^p(\Omega, \mathbb{R}^n)$.

For each $v^* \in A(v)$ we have

$$\alpha(v - u) \leq \limsup_{n\to\infty} \left[\langle v^*, v - u_n\rangle + h(u_n)J^0(iu_n; iv - iu_n) - \langle Fu_n, v - u_n\rangle \right]$$

$$\leq \langle v^*, v - u\rangle + h(u)J^0(iu, iv - iu) - \langle Fu, v - u\rangle,$$

which shows that $u \in \Theta(v)$ and thus the proof of the claim is complete.

Let us prove next that Θ is a KKM mapping. Arguing by contradiction, assume there exists a finite subset $\{v_1, \ldots, v_N\} \subset K$ and $u_0 := \sum_{j=1}^N \lambda_j v_j$, with $\lambda_j \in [0, 1]$ and $\sum_{j=1}^N \lambda_j = 1$ such that $u_0 \notin \bigcup_{j=1}^N \Theta(v_j)$. This is equivalent to

$$\inf_{v_j^* \in A(v_j)} \langle v_j^*, v_j - u_0\rangle + h(u_0)J^0(iu_0; iv_j - iu_0) - \langle Fu_0, v_j - u_0\rangle < \alpha(v_j - u_0), \quad (10.16)$$

for all $j \in \{1, \ldots, N\}$.

On the other hand, A is a relaxed α monotone operator and thus, for each $j \in \{1, \ldots, N\}$ we have

$$\langle u_0^* - v_j^*, v_j - u_0 \rangle \leq -\alpha(v_j - u_0), \quad \forall u_0^* \in A(u_0), \ \forall v_j^* \in A(v_j). \tag{10.17}$$

Combining (10.16) and (10.17) we are led to

$$\langle u_0^*, v_j - u_0 \rangle + h(u_0) J^0(iu_0; iv_j - iu_0) - \langle Fu_0, v_j - u_0 \rangle < 0, \quad \forall u_0^* \in A(u_0). \tag{10.18}$$

Using (10.18) and the fact that $J^0(iu_0; \cdot)$ is subadditive, for fixed $u_0^* \in A(u_0)$ we have

$$0 = \langle u_0^*, u_0 - u_0 \rangle + h(u_0) J^0(iu_0; iu_0 - iu_0) - \langle Fu_0, u_0 - u_0 \rangle$$

$$= \left\langle u_0^*, \sum_{j=1}^N \lambda_j(v_j - u_0) \right\rangle + h(u_0) J^0 \left(iu_0; \sum_{j=1}^N \lambda_j(iv_j - iu_0) \right) - \left\langle Fu_0, \sum_{j=1}^N \lambda_j(v_j - u_0) \right\rangle$$

$$\leq \sum_{j=1}^N \lambda_j \left[\langle u_0^*, v_j - u_0 \rangle + h(u_0) J^0(iu_0; iv_j - iu_0) - \langle Fu_0, v_j - u_0 \rangle \right]$$

$$< 0,$$

which obviously is a contradiction and thus the proof of the claim is complete.

Since $\Theta(v)$ is a weakly closed subset of K and K is weakly compact set as it is a bounded, closed and convex subset of the real reflexive Banach space X, it follows that $\Theta(v)$ it is weakly compact for each $v \in K$. Thus we can apply Corollary D.1 to conclude that $\bigcap_{v \in K} \Theta(v) \neq \emptyset$.

Let $u_0 \in \bigcap_{v \in K} \Theta(v)$. This implies that for each $w \in K$ we have

$$\inf_{w^* \in A(w)} \langle w^*, w - u_0 \rangle + h(u_0) J^0(iu_0; iw - iu_0) - \langle Fu_0, w - u_0 \rangle \geq \alpha(w - u_0).$$

Let $v \in K$ be fixed and define $w_\lambda := u_0 + \lambda(v - u_0)$, $\lambda \in (0, 1)$. Using the fact that $w_\lambda \in K$ and taking into account the above relation we deduce that

$$\langle w_\lambda^*, v - u_0 \rangle + h(u_0) J^0(iu_0, iv - iu_0) - \langle Fu_0, v - u_0 \rangle \geq \frac{\alpha(\lambda(v - u_0))}{\lambda}, \ \forall w_\lambda^* \in A(w_\lambda).$$

Letting $\lambda \to 0$ and using the l.s.c. of A we obtain that u_0 solves problem (P_K).   □

As we have seen above the boundedness of the set K played a key role in proving that problem (P_K) admits at least one solution. In the case when K is the whole space X, assuming that the same hypotheses as in Theorem 10.5 hold, we shall need an extra condition to overcome the lack of boundedness. For each real number $R > 0$ taking $K := \bar{B}(0; R) = \{u \in X : \|u\| \leq R\}$ we know from Theorem 10.5 that problem (P_R) Find $u_R \in \bar{B}(0; R)$ and $u_R^* \in A(u_R)$ such that

$$\langle u_R^*, v - u_R \rangle + h(u_R) J^0(i u_R; i v - i u_R) \geq \langle F u_R, v - u_R \rangle, \quad \forall v \in \bar{B}(0; R),$$

admits at least one solution.

Theorem 10.6 ([2]) *Assume that the same hypotheses as in Theorem 10.5 hold in the case $K := X$. Then problem $(MQHI)$ admits at least one solution if and only if the following condition holds true:*

(H_8) *There exists $R > 0$ such that at least one solution u_R of problem (P_R) satisfies $u_R \in B(0; R)$.*

Proof The necessity is obvious.

In order to prove the sufficiency fix $v \in X$. We shall prove that u_R is a solution of $(MQHI)$. First we define

$$\lambda := \begin{cases} 1, & \text{if } u_R = v \\ \frac{R - \|u_R\|}{\|v - u_R\|}, & \text{otherwise} . \end{cases}$$

Since $u_R \in B(0; R)$ we conclude that $\lambda > 0$ and that $w_\lambda := u_R + \lambda(v - u_R) \in \bar{B}(0; R)$. Using that u_R solves problem (P_R) we find

$$\langle F u_R, \lambda(v - u_R) \rangle = \langle F u_R, w_\lambda - u_R \rangle \leq \langle u_R^*, w_\lambda - u_R \rangle + h(u_R) J^0(i u_R; i w_\lambda - i u_R)$$
$$= \langle u_R^*, \lambda(v - u_R) \rangle + h(u_R) J^0(i u_R; \lambda(i v - i u_R))$$
$$= \lambda \left[\langle u_R^*, v - u_R \rangle + h(u_R) J^0(i u_R; i v - i u_R) \right].$$

Dividing by $\lambda > 0$ we conclude that u_R solves problem $(MQHI)$. $\square$

Corollary 10.1 *Let us assume that the same hypotheses as in Theorem 10.5 hold in the case $K := X$. Then a sufficient condition for problem $(MQHI)$ to posses a solution is:*

(H_9) *There exists $R_0 > 0$ such that for each $u \in X \setminus \bar{B}(0; R_0)$ there exists $v \in B(0; R_0)$ with the property that*

$$\sup_{u^* \in A(u)} \langle u^*, v - u \rangle + h(u) J^0(i u; i v - i u) < \langle F u, v - u \rangle.$$

Proof Let us fix $R > R_0$. According to Theorem 10.5 there exists $u_R \in \bar{B}(0, R)$ and $\bar{u}_R^* \in A(u_R)$ such that

$$\langle \bar{u}_R^*, v - u_R \rangle + h(u_R)J^0(iu_R; iv - iu_R) \geq \langle Fu_R, v - u_R \rangle, \quad \forall v \in \bar{B}(0; R). \quad (10.19)$$

CASE 1. $u_R \in B(0; R)$.

Then we have nothing to prove, Theorem 10.6 showing that u_R is a solution of problem $(MQHI)$.

CASE 2. $u_R \in \partial \bar{B}(0; R)$.

In this case $\|u_R\| = R > R_0$ and thus $u_R \in X \setminus \bar{B}(0; R_0)$. According to our hypothesis there exists $\bar{v} \in B(0; R_0)$ such that

$$\sup_{u_R^* \in A(u_R)} \langle u_R^*, \bar{v} - u_R \rangle + h(u_R)J^0(iu_R; i\bar{v} - iu_R) < \langle Fu_R, \bar{v} - u_R \rangle. \quad (10.20)$$

Let us fix $v \in X$. Defining

$$\lambda := \begin{cases} 1, & \text{if } v = \bar{v} \\ \frac{R - R_0}{\|v - \bar{v}\|}, & \text{otherwise,} \end{cases}$$

we observe that $w_\lambda := \bar{v} + \lambda(v - \bar{v}) \in \bar{B}(0; R)$. On the other hand we observe that

$$w_\lambda - u_R = \bar{v} - u_R + \lambda(v - \bar{v}) + \lambda u_R - \lambda u_R = \lambda(v - u_R) + (1 - \lambda)(\bar{v} - u_R).$$

Taking w_λ instead of v in (10.19) and using (10.20) we are led to the following estimates

$$\langle Fu_R, \lambda(v - u_R) + (1 - \lambda)(\bar{v} - u_R) \rangle = \langle Fu_R, w_\lambda - u_R \rangle$$
$$\leq \langle \bar{u}_R^*, w_\lambda - u_R \rangle + h(u_R)J^0(iu_R; iw_\lambda - iu_R)$$
$$\leq \lambda \left[\langle \bar{u}_R^*, v - u_R \rangle + h(u_R)J^0(iu_R; iv - iu_R) \right]$$
$$+ (1 - \lambda) \left[\langle \bar{u}_R^*, \bar{v} - u_R \rangle + h(u_R)J^0(iu_R; i\bar{v} - iu_R) \right]$$
$$\leq \lambda \left[\langle \bar{u}_R^*, v - u_R \rangle + h(u_R)J^0(iu_R; iv - iu_R) \right] + (1 - \lambda)\langle Fu_R, \bar{v} - u_R \rangle.$$

This shows that

$$\langle \bar{u}_R^*, v - u_R \rangle + h(u_R)J^0(iu_R; iv - iu_R) \geq \langle Fu_R; v - u_R \rangle, \quad \forall v \in X,$$

which means that u_R solves problem $(MQHI)$ and thus the proof is complete.   □

Corollary 10.2 *Let us assume that the same hypotheses as in Theorem 10.5 hold in the case $K := X$. Assume in addition that:*

(H_{10}) *A is coercive, i.e. there exists a function $c : \mathbb{R}_+ \to \mathbb{R}_+$ with the property that $\lim_{r \to \infty} c(r) = +\infty$ such that*

$$\inf_{u^* \in A(u)} \langle u^*, u \rangle \geq c(\|u\|)\|u\|;$$

(H_{11}) *there exists a constant $k > 0$ such that $h(v)J^0(iv; -iv) \leq k\|v\|$ for all $v \in X$;*
(H_{12}) *there exists a constant $m > 0$ such that $\|Fu\|_{X^*} \leq m$ for all $u \in X$.*

Then problem $(MQHI)$ has at least one solution.

Proof For each $R > 0$ Theorem 10.5 guarantees that there exist $u_R \in X$ and $u_R^* \in A(u_R)$ such that

$$\langle u_R^*, v - u_R \rangle + h(u_R)J^0(iu; iv - iu_R) \geq \langle Fu_R, v - u_R \rangle, \quad \forall v \in \bar{B}(0; R). \quad (10.21)$$

We shall prove that there exists $R_0 > 0$ such that $u_{R_0} \in B(0; R_0)$ which according to Theorem 10.6 is equivalent to the fact that u_R is a solution of problem $(MQHI)$. Arguing by contradiction, assume that $u_R \in \partial \bar{B}(0; R)$ for all $R > 0$. Taking $v = 0$ in (10.21) we have

$$c(R)R = c(\|u_R\|)\|u_R\| \leq \langle u_R^*, u_R \rangle \leq \langle Fu_R, u_R \rangle + h(u_R)J^0(iu_R; -iu_R)$$

$$\leq \|Fu_R\|_{X^*}\|u_R\| + k\|u_R\| \leq (m + k)R.$$

Dividing by $R > 0$ we obtain that $c : \mathbb{R}_+ \to \mathbb{R}_+$ is bounded from above which contradicts the fact that $\lim_{R \to \infty} c(R) = +\infty$. $\qquad \square$

10.3 Variational-Like Inequalities

In 1989 Parida, Sahoo, and Kumar [4] introduced a new type of inequality problem of variational type which had the form:
Find $u \in K$ such that

$$\langle A(u), \eta(v, u) \rangle \geq 0, \quad \forall v \in K, \quad (10.22)$$

where $K \subseteq \mathbb{R}^n$ is a nonempty closed and convex set and $A : K \to \mathbb{R}^n$, $\eta : K \times K \to \mathbb{R}^n$ are two continuous maps. The authors called (10.22) *variational-like inequality problem* and showed that this kind of inequalities can be related to some mathematical programming problems.

In this section the goal is to extend the results obtained in [4] to the following setting: X is a Banach space (not necessarily reflexive) with X^* and $X^{**} = (X^*)^*$ its dual and bidual, respectively, K is a nonempty closed and convex subset X^{**} and $A : K \to X^*$ is a set-valued map. More precisely, we are interested in finding solutions for the following inequality problems:

$(QVLI)$ Find $u \in K_\phi$ such that

$$\exists u^* \in A(u): \quad \langle u^*, \eta(v, u) \rangle + \phi(v) - \phi(u) \geq 0, \quad \forall v \in K, \tag{10.23}$$

(VLI) Find $u \in K$ such that

$$\exists u^* \in A(u): \quad \langle u^*, \eta(v, u) \rangle \geq 0, \quad \forall v \in K, \tag{10.24}$$

where $K \subseteq X^{**}$ is nonempty closed and convex, $\eta : K \times K \to X^{**}$, $A : K \to X^*$ is a set-valued map, $\phi : X^{**} \to \mathbb{R} \cup \{+\infty\}$ is a proper convex and lower semicontinuous functional such that $K_\phi := K \cap D(\phi) \neq \varnothing$, with $D(\phi)$ the effective domain of the functional ϕ. We call these problems *quasi-variational-like inequality* and *variational-like inequality*, respectively. Note that if ϕ is the indicator function of the set K, then $(QVLI)$ reduces to (VLI).

Definition 10.2 A solution $u_0 \in K_\phi$ of inequality problem (10.23) is called *strong* if $\langle u^*, \eta(v, u_0) \rangle + \phi(v) - \phi(u_0) \geq 0$ holds for all $v \in K$ and all $u^* \in A(u_0)$.

It is clear from the above definition that if A is a single-valued operator, then the concepts of solution and strong solution are one and the same.

First we consider the case of non-reflexive Banach spaces. Before stating the results concerning the existence of solutions for problem (10.23) we indicate below some hypotheses that will be needed in the sequel.

$(\mathcal{H}_A^1)$ $A : K \leadsto X^*$ is l.s.c. from $s - X$ into $w^* - X^*$ and has nonempty values;
$(\mathcal{H}_A^2)$ $A : K \to X^*$ is u.s.c. from $s - X$ into $w^* - X^*$ has nonempty w^*-compact values;
$(\mathcal{H}_\phi)$ $\phi : X^{**} \to \mathbb{R} \cup \{+\infty\}$ is a proper convex l.s.c. functional;
$(\mathcal{H}_\eta)$ $\eta : K \times K \to X^{**}$ is such that
    (i) for all $v \in K$ the map $u \mapsto \eta(v, u)$ is continuous;
    (ii) for all $u, v, w \in K$ and all $w^* \in A(w)$, the map $v \mapsto \langle w^*, \eta(v, u) \rangle$ is convex and $\langle w^*, \eta(u, u) \rangle \geq 0$;

Theorem 10.7 ([3]) *Let X be a nonreflexive Banach space and $K \subseteq X^{**}$ nonempty closed and convex. Assume that $(\mathcal{H}_\phi)$, $(\mathcal{H}_\eta)$ and either $(\mathcal{H}_A^1)$ or $(\mathcal{H}_A^2)$ hold. If the set*

K_ϕ *is not compact we assume in addition that for some nonempty compact convex subset* C *of* K_ϕ *the following condition holds*

(H_C) *for each* $u \in K_\phi \setminus C$ *there exist* $u_0^* \in A(u)$ *and* $\bar{v} \in C$ *with the property that*

$$\langle u_0^*, \eta(\bar{v}, u) \rangle + \phi(\bar{v}) - \phi(u) < 0.$$

Then $(QVLI)$ *has at least one strong solution.*

Proof Arguing by contradiction let us assume that (10.23) has no strong solution. Then, for each $u \in K_\phi$ there exist $\bar{u}^* \in A(u)$ and $v = v(u, \bar{u}^*) \in K$ such that

$$\langle \bar{u}^*, \eta(v, u) \rangle + \phi(v) - \phi(u) < 0. \tag{10.25}$$

It is clear that the element v for which (10.25) takes place satisfies $v \in \mathcal{D}(\phi)$, therefore $v \in K_\phi$. We consider next the set-valued map $F : K_\phi \to 2^{K_\phi}$ defined by

$$F(u) := \left\{ v \in K_\phi : \langle \bar{u}^*, \eta(v, u) \rangle + \phi(v) - \phi(u) < 0 \right\},$$

where $\bar{u}^* \in A(u)$ is given in (10.25).

STEP 1.   *For each* $u \in K_\phi$ *the set* $F(u)$ *is nonempty and convex.*

Let $u \in K_\phi$ be arbitrarily fixed. Then (10.25) implies that $F(u)$ is nonempty. Let $v_1, v_2 \in F(u)$, $\lambda \in (0, 1)$ and define $w = \lambda v_1 + (1 - \lambda)v_2$. We have

$$\langle \bar{u}^*, \eta(w, u) \rangle + \phi(w) - \phi(u) \leq \lambda \left[\langle \bar{u}^*, \eta(v_1, u) \rangle + \phi(v_1) - \phi(u) \right]$$
$$+ (1 - \lambda) \left[\langle \bar{u}^*, \eta(v_2, u) \rangle + \phi(v_2) - \phi(u) \right] < 0,$$

which shows that $w \in F(u)$, therefore $F(u)$ in a convex subset of K_ϕ.

STEP 2.   *For each* $v \in K_\phi$ *the set* $F^{-1}(v) := \{u \in K_\phi : v \in F(u)\}$ *is open.*

Let us fix $v \in K_\phi$. Taking into account that

$$F^{-1}(v) = \left\{ u \in K_\phi : \exists \bar{u}^* \in A(u) \text{ s.t. } \langle \bar{u}^*, \eta(v, u) \rangle + \phi(v) - \phi(u) < 0 \right\}$$

we shall prove that

$$\left[F^{-1}(v) \right]^c = \left\{ u \in K_\phi : \langle u^*, \eta(v, u) \rangle + \phi(v) - \phi(u) \geq 0, \text{ for all } u^* \in A(u) \right\}$$

is a closed subset of K_ϕ. Let $\{u_\lambda\}_{\lambda \in I} \subset \left[F^{-1}(v) \right]^c$ be a net converging to some $u \in K_\phi$. Then for each $\lambda \in I$ we have

$$\langle u_\lambda^*, \eta(v, u_\lambda) \rangle + \phi(v) - \phi(u_\lambda) \geq 0, \quad \text{for all } u_\lambda^* \in A(u_\lambda). \tag{10.26}$$

Taking into account that $\eta(\cdot, \cdot)$ is continuous with respect to the second variable we obtain that

$$\eta(v, u_\lambda) \to \eta(v, u). \tag{10.27}$$

CASE 1. $(\mathcal{H}_A^1)$ holds.

We fix $u^* \in A(u)$ and for each $\lambda \in I$ we can determine $u_\lambda^* \in A(u_\lambda)$ such that

$$u_\lambda^* \rightharpoonup u^* \text{ in } X^*,$$

since A is l.s.c. from K endowed with the strong topology into X^* endowed with the w^*-topology, which combined with (10.27) shows that $\langle u_\lambda^*, \eta(v, u_\lambda) \rangle \to \langle u^*, \eta(v, u) \rangle$.

CASE 2. $(\mathcal{H}_A^2)$ holds.

We define the compact set $D := \{u_\lambda : \lambda \in I\} \cup \{u\}$ and apply Proposition B.9 to conclude that $A(D)$ is a w^*-compact set, which means that $\{u_\lambda^*\}_{\lambda \in I}$ admits a subnet $\{u_\lambda^*\}_{\lambda \in J}$ such that $u_\lambda^* \rightharpoonup u^*$ for some $u^* \in X^*$. But, A is u.s.c. and thus $u^* \in A(u)$. Since $u_\lambda^* \rightharpoonup u^*$ and $\eta(v, u_\lambda) \to \eta(v, u)$ we deduce that $\langle u_\lambda^*, \eta(v, u_\lambda) \rangle \to \langle u^*, \eta(v, u) \rangle$.

Using (10.26) we get

$$0 \leq \limsup \left[\langle u_\lambda^*, \eta(v, u_\lambda) \rangle + \phi(v) - \phi(u_\lambda) \right]$$
$$\leq \limsup \langle u_\lambda^*, \eta(v, u_\lambda) \rangle + \phi(v) - \liminf \phi(u_\lambda)$$
$$\leq \langle u^*, \eta(v, u) \rangle + \phi(v) - \phi(u),$$

which means that $u \in \left[F^{-1}(v) \right]^c$, therefore $\left[F^{-1}(v) \right]^c$ is a closed subset of K_ϕ.

STEP 3. $K_\phi = \bigcup_{v \in K_\phi} \text{int}_{K_\phi} F^{-1}(v)$.

We only need to prove that $K_\phi \subseteq \bigcup_{v \in K_\phi} \text{int}_{K_\phi} F^{-1}(v)$ as the converse inclusion is satisfied since $F^{-1}(v)$ is a subset K_ϕ for all $v \in K_\phi$. For each $u \in K_\phi$ there exist $v \in K_\phi$ such that $v \in F(u)$ (such a v exists since $F(u)$ is nonempty) and thus $u \in F^{-1}(v) \subseteq \bigcup_{v \in K_\phi} F^{-1}(v) = \bigcup_{v \in K_\phi} \text{int}_{K_\phi} F^{-1}(v)$.

If the K_ϕ is not compact then the last condition of our theorem implies that for each $u \in K_\phi \setminus C$ there exists $\bar{v} \in C$ such that $u \in F^{-1}(\bar{v}) = \text{int}_{K_\phi} F^{-1}(\bar{v})$. This observation and the above Claims ensure that all the conditions of Theorem D.4 are satisfied for $S = T = F$ and we deduce that the set-valued map $F : K_\phi \to 2^{K_\phi}$ admits a fixed point $u_0 \in K_\phi$, i.e. $u_0 \in F(u_0)$. This can be rewritten equivalently as

$$0 \leq \langle \bar{u}_0^*, \eta(u_0, u_0) \rangle + \phi(u_0) - \phi(u_0) < 0.$$

We have reached thus a contradiction which completes the proof. $\square$

We consider next the case of reflexive Banach spaces. In order to prove our existence results, throughout this subsection, we shall use some of the following hypotheses:

(H_A^1) $A : K \rightsquigarrow X^*$ is l.s.c. from $s - X$ into $w - X^*$ and has nonempty values;

(H_A^2) $A : K \rightsquigarrow X^*$ is u.s.c. from $s - X$ into $w - X^*$ has nonempty w-compact values;

(H_ϕ) $\phi : X \to \mathbb{R} \cup \{+\infty\}$ is a proper convex l.s.c. functional such that $K_\phi \neq \emptyset$;

(H_η^1) $\eta : K \times K \to X$ is such that

> (i) for all $v \in K$ the map $u \mapsto \eta(v, u)$ is continuous;
> (ii) for all $u, v, w \in K$ and all $w^* \in A(w)$ the map $v \mapsto \langle w^*, \eta(v, u) \rangle$ is convex and $\langle w^*, \eta(u, u) \rangle \geq 0$;

(H_η^2) $\eta : K \times K \to X$ is such that

> (i) $\eta(u, v) + \eta(v, u) = 0$ for all $u, v \in K$;
> (ii) for all $u, v, w \in K$ and all $w^* \in A(w)$, $v \mapsto \langle w^*, \eta(v, u) \rangle$ is convex and l.s.c.;

(H_α^1) $\alpha : X \to \mathbb{R}$ is weakly l.s.c. and $\limsup\limits_{\lambda \downarrow 0} \frac{\alpha(\lambda v)}{\lambda} \geq 0$ for all $v \in X$;

(H_α^2) $\alpha : X \to \mathbb{R}$ is a such that

> (i) $\alpha(0) = 0$;
> (ii) $\limsup_{\lambda \downarrow 0} \frac{\alpha(\lambda v)}{\lambda} \geq 0$, for all $v \in X$;
> (iii) $\alpha(u) \leq \limsup \alpha(u_\lambda)$, whenever $u_\lambda \rightharpoonup u$ in X;

The following theorem is a variant of Theorem 10.7 in the framework of reflexive Banach spaces.

Theorem 10.8 ([3]) *Let X be a real reflexive Banach space and $K \subseteq X$ nonempty compact and convex. Assume that (H_ϕ), (H_η^1) and either (H_A^1) or (H_A^2) hold. Then $(QVLI)$ has at least one strong solution.*

The proof of Theorem 10.8 follows basically the same steps as the proof of Theorem 10.7, therefore we shall omit it.

We point out the fact that in the above case when K is a compact convex subset of X we do not impose any monotonicity conditions on the set-valued operator A. However, in applications, most problems lead to an inequality whose solution is sought in a closed and convex subset of the space X. Weakening the hypotheses on K by assuming that K is only bounded, closed and convex, we need to impose certain monotonicity properties on A.

Theorem 10.9 ([3]) *Let K be a nonempty bounded closed and convex subset of the real reflexive Banach space X. Let $A : K \to X^*$ be a relaxed $\eta - \alpha$ monotone map and assume that (H_ϕ), (H_η^2), and (H_α^1) hold. If in addition*

(H_A^1) *holds, then $(QVLI)$ has at least one strong solution;*

(H_A^2) *holds, then $(QVLI)$ has at least one solution.*

Proof We shall apply Mosco's Alternative for the weak topology of X. First we note that K is weakly compact as it is a bounded closed and convex subset of the real reflexive space X and $\phi : X \rightarrow \mathbb{R} \cup \{+\infty\}$ is weakly lower semicontinuous as it is convex and lower semicontinuous. We define $f, g : X \times X \rightarrow \mathbb{R}$ as follows

$$f(v, u) := - \inf_{v^* \in A(v)} \langle v^*, \eta(v, u) \rangle + \alpha(v - u)$$

and

$$g(v, u) := \sup_{u^* \in A(u)} \langle u^*, \eta(u, v) \rangle.$$

Let us fix $u, v \in X$ and choose $\bar{v}^* \in A(v)$ such that $\langle \bar{v}^*, \eta(v, u) \rangle = \inf_{v^* \in A(v)} \langle v^*, \eta(v, u) \rangle$. For and arbitrary fixed $u^* \in A(u)$ we have

$$g(v, u) - f(v, u) = \sup_{u^* \in A(u)} \langle u^*, \eta(u, v) \rangle + \inf_{v^* \in A(v)} \langle v^*, \eta(v, u) \rangle - \alpha(v - u)$$

$$\geq \langle u^*, \eta(u, v) \rangle + \langle \bar{v}^*, \eta(v, u) \rangle - \alpha(v - u)$$

$$= \langle \bar{v}^*, \eta(v, u) \rangle - \langle u^*, \eta(v, u) \rangle - \alpha(v - u) \geq 0.$$

It is easy to check that conditions imposed on η and α ensure that the map $u \mapsto f(v, u)$ is weakly lower semicontinuous, while the map $v \mapsto g(v, u)$ is concave. Applying Mosco's Alternative for $\mu := 0$ we conclude that there exists $u_0 \in K_\phi$ such that

$$f(v, u_0) + \phi(u_0) - \phi(v) \leq 0, \quad \forall v \in X,$$

since $g(v, v) = 0$ for all $v \in X$. A simple computation shows that for each $w \in K$ we have

$$\langle w^*, \eta(w, u_0) \rangle + \phi(w) - \phi(u_0) \geq \alpha(w - u_0), \quad \forall w^* \in A(w). \tag{10.28}$$

Let us fix $v \in K$ and define $w_\lambda := u_0 + \lambda(v - u_0)$, with $\lambda \in (0, 1)$. Then for a fixed $w_\lambda^* \in A(w_\lambda)$ from (10.28) we have

$$\alpha(\lambda(v - u_0)) \leq \langle w_\lambda^*, \eta(w_\lambda, u_0) \rangle + \phi(w_\lambda) - \phi(u_0) \leq \lambda \langle w_\lambda^*, \eta(v, u_0) \rangle$$

$$+ (1 - \lambda)\langle w_\lambda^*, \eta(u_0, u_0) \rangle + \lambda\phi(v) + (1 - \lambda)\phi(u_0) - \phi(u_0)$$

$$= \lambda \left[\langle w_\lambda^*, \eta(v, u_0) \rangle + \phi(v) - \phi(u_0) \right],$$

which leads to

$$\frac{\alpha(\lambda(v - u_0))}{\lambda} \leq \langle w_\lambda^*, \eta(v, u_0)\rangle + \phi(v) - \phi(u_0). \tag{10.29}$$

CASE 1. (H_A^1) holds.

We shall prove next that u_0 is a strong solution of inequality problem (10.23). Let $u_0^* \in A(u_0)$ be arbitrarily fixed. Combining the fact that $w_\lambda \to u_0$ as $\lambda \downarrow 0$ with the fact that A is l.s.c. from K endowed with the strong topology into X^* endowed with the w−topology we deduce that for each $\lambda \in (0, 1)$ we can find $w_\lambda^* \in A(w_\lambda)$ such that $w_\lambda^* \rightharpoonup u_0^*$ as $\lambda \downarrow 0$. Taking the superior limit in (10.29) as $\lambda \downarrow 0$ and keeping in mind (H_α^1) we get

$$0 \leq \limsup_{\lambda \downarrow 0} \frac{\alpha(\lambda(v - u_0))}{\lambda} \leq \limsup_{\lambda \downarrow 0} \left[\langle w_\lambda^*, \eta(v, u_0)\rangle + \phi(v) - \phi(u_0)\right]$$

$$= \langle u_0^*, \eta(v, u_0)\rangle + \phi(v) - \phi(u_0),$$

which shows that u_0 is a strong solution of (10.23), since $v \in K$ and $u_0^* \in A(u_0)$ were arbitrarily fixed.

CASE 2. (H_A^2) holds.

We shall prove in this case that u_0 is a solution of (10.23). Reasoning as in the proof of Theorem 10.7-CASE 2 we infer that there exists $\bar{u}_0^* \in A(u_0)$ and a subnet $\{w_\lambda^*\}_{\lambda \in J}$ of $\{w_\lambda^*\}_{\lambda \in (0,1)}$ such that $w_\lambda^* \rightharpoonup \bar{u}_0^*$ as $\lambda \downarrow 0$. Combining this with relation (10.29) and hypothesis (H_α^1) we conclude that

$$0 \leq \limsup_{\lambda \downarrow 0} \frac{\alpha(\lambda(v - u_0))}{\lambda} \leq \limsup_{\lambda \downarrow 0} \left[\langle w_\lambda^*, \eta(v, u_0)\rangle + \phi(v) - \phi(u_0)\right]$$

$$= \langle \bar{u}_0^*, \eta(v, u_0)\rangle + \phi(v) - \phi(u_0),$$

which shows that u_0 is a solution of (10.23), since $v \in K$ was arbitrarily fixed. $\square$

Weakening even more the hypotheses by assuming that the set-valued map $A : K \to X^*$ is *relaxed $\eta - \alpha$ quasimonotone* instead of being *relaxed $\eta - \alpha$ monotone* the existence of solutions for inequality problem (10.23) is an open problem in the case when K is nonempty bounded closed and convex. However, in this case we can prove the following existence result concerning inequality (VLI).

Theorem 10.10 ([3]) *Let K be a nonempty bounded closed and convex subset of the real reflexive Banach space X. Let $A : K \to X^*$ be a relaxed $\eta - \alpha$ quasimonotone map and assume that (H_η^2) and (H_α^2) hold. If in addition*

(H_A^1) holds, then (VLI) possesses at least one strong solution;
(H_A^2) holds, then (VLI) possesses at least one solution.

Proof Define $G : K \rightsquigarrow X$ in the following way:

$$G(v) := \left\{u \in K : \langle v^*, \eta(v, u)\rangle \ge \alpha(v - u), \ \forall v^* \in A(v)\right\}.$$

First of all, note that $v \in G(v)$ for all $v \in K$ hence $G(v)$ is nonempty for all $v \in K$. Now, we prove that $G(v)$ is weakly closed for all $v \in K$. Let $\{u_\lambda\}_{\lambda \in I} \subset G(v)$ be a net such that u_λ converges weakly to some $u \in K$. Then, we have

$$\alpha(v - u) \le \limsup \alpha(v - u_\lambda) \le \limsup \langle v^*, \eta(v, u_\lambda)\rangle = \limsup \left[-\langle v^*, \eta(u_\lambda, v)\rangle\right]$$
$$= -\liminf \langle v^*, \eta(u_\lambda, v)\rangle \le -\langle v^*, \eta(u, v)\rangle = \langle v^*, \eta(v, u)\rangle,$$

for all $v^* \in A(v)$. It follows that $u \in G(v)$, so $G(v)$ is weakly closed.

CASE 1.   *G is a KKM map.*

Since K is bounded closed and convex in X which is reflexive, it follows that K is weakly compact and thus $G(v)$ is weakly compact for all $v \in K$ as it is a weakly closed subset of K. Applying Corollary D.1, we have $\bigcap_{v \in K} G(v) \ne \emptyset$ and the set of solutions of (VLI) is nonempty. In order to see that let $u_0 \in \bigcap_{v \in K} G(v)$. This implies that for each $w \in K$ we have

$$\langle w^*, \eta(w, u)\rangle \ge \alpha(w - u), \quad \text{for all } w^* \in A(w).$$

Let v be fixed in K and for $\lambda \in (0, 1)$ define $w_\lambda = u_0 + \lambda(v - u_0)$. We infer that

$$\alpha(\lambda(v - u_0)) \le \langle w_\lambda^*, \eta(w_\lambda, u_0)\rangle \le \lambda\langle w_\lambda^*, \eta(v, u_0)\rangle + (1 - \lambda)\langle w_\lambda^*, \eta(u_0, u_0)\rangle$$
$$= \lambda\langle w_\lambda^*, \eta(v, u_0)\rangle$$

for all $w_\lambda^* \in A(w_\lambda)$.

Applying the same arguments as in the previous proof we conclude that u_0 is a strong solution of inequality problem (10.24) if (H_A^1) holds, while if (H_A^2) holds then u_0 is a solution of inequality problem (10.24).

CASE 2.   *G is not a KKM map.*

Consider $\{v_1, v_2, \ldots, v_N\} \subseteq K$ and $u_0 = \sum_{j=1}^N \lambda_j v_j$ with $\lambda_j \in [0, 1]$ and $\sum_{j=1}^N \lambda_j = 1$ such that $u_0 \notin \bigcup_{j=1}^N G(v_j)$. The existence of such u_0 is guaranteed by the fact that

G is not a KKM map. This implies that for all $j \in \{1, \ldots, N\}$ there exists $\bar{v}_j^* \in A(v_j)$ such that

$$\langle \bar{v}_j^*, \eta(v_j, u_0) \rangle < \alpha(v_j - u_0) \tag{10.30}$$

Now, we claim that there exists a neighborhood U of u_0 such that (10.30) takes place for all $w \in U \cap K$, that is

$$\langle \bar{v}_j^*, \eta(v_j, w) \rangle < \alpha(v_j - w), \quad \forall w \in U \cap K.$$

Arguing by contradiction let us assume that for any neighborhood U of u_0 there exists an index $j_0 \in \{1, \ldots, N\}$ and an element $w_0 \in U \cap K$ such that

$$\langle v_{j_0}^*, \eta(v_{j_0}, w_0) \rangle \geq \alpha(v_{j_0} - w_0), \quad \forall v_{j_0}^* \in A(v_{j_0}). \tag{10.31}$$

Choose $U = \bar{B}_X(u_0; \lambda)$ and for each $\lambda > 0$ one can find a $j_0 \in \{1, \ldots, N\}$ and $w_\lambda \in \bar{B}_X(u_0; \lambda) \cap K$ such that

$$\langle v_{j_0}^*, \eta(v_{j_0}, w_\lambda) \rangle \geq \alpha(v_j - w_\lambda), \quad \forall v_{j_0}^* \in A(v_{j_0}).$$

Let us fix $v_{j_0}^* \in A(v_{j_0})$. Using the fact that $w_\lambda \to u_0$ as $\lambda \downarrow 0$ and taking the superior limit in the above relation, we obtain

$$\alpha(v_{j_0} - u_0) \leq \limsup_{\lambda \downarrow 0} \alpha(v_{j_0} - w_\lambda) \leq \limsup_{\lambda \downarrow 0} \langle v_{j_0}^*, \eta(v_{j_0}, w_\lambda) \rangle$$

$$= -\liminf_{\lambda \downarrow 0} \langle v_{j_0}^*, \eta(w_\lambda, v_{j_0}) \rangle \leq -\langle v_{j_0}^*, \eta(u_0, v_{j_0}) \rangle$$

$$= \langle v_{j_0}^*, \eta(v_{j_0}, u_0) \rangle,$$

which contradicts with relation (10.30) and this contradiction completes the proof of the claim. Now, using the fact that A is relaxed $\eta - \alpha$ quasimonotne map, we prove that

$$\langle w^*, \eta(v_j, w) \rangle \leq 0, \; \forall w \in K \cap U, \; \forall w^* \in A(w), \; \forall j \in \{1, \ldots, N\}. \tag{10.32}$$

In order to prove (10.32) assume by contradiction there exists $w_0 \in K \cap U, \; w_0^* \in A(w_0)$ and $j_0 \in \{1, \ldots, N\}$ such that $\langle w_0^*, \eta(v_{j_0}, w_0) \rangle > 0$. From the fact that A is relaxed $\eta - \alpha$ quasimonotone it follows that

$$\langle v_{j_0}^*, \eta(v_{j_0}, w_0) \rangle \geq \alpha(v_{j_0} - w_0), \quad \forall v_{j_0}^* \in A(v_{j_0}),$$

which contradicts the fact that (10.30) holds for all $w \in U \cap K$ and all $j \in \{1, \dots, N\}$. On the other hand, for arbitrary fixed $w \in K \cap U$ and $\bar{w}^* \in A(w)$ we have

$$\langle \bar{w}^*, \eta(u_0, w) \rangle = \left\langle \bar{w}^*, \eta \left(\sum_{j=1}^N \lambda_j v_j, w \right) \right\rangle \le \sum_{j=1}^N \lambda_j \langle \bar{w}^*, \eta(v_j, w) \rangle \le 0.$$

Thus, we obtain

$$0 \le \langle \bar{w}^*, -\eta(u_0, w) \rangle = \langle \bar{w}^*, \eta(w, u_0) \rangle.$$

But $\bar{w}^* \in A(w)$ was choosen arbitrary but fixed and thus for each $w \in U \cap K$ we have

$$\langle w^*, \eta(w, u_0) \rangle \ge 0, \qquad \text{for all } w^* \in A(w) \tag{10.33}$$

We shall prove next that u_0 solves inequality problem (10.24). Consider $v \in K$ to be arbitrary fixed.

CASE 2.1 $v \in U$.

In this case the entire line segment

$$(u_0, v) := \{ u_0 + \lambda(v - u_0) : \lambda \in (0, 1) \}$$

is contained in $U \cap K$ and, according to (10.33), for each $w_\lambda \in (u_0, v)$ and each $w_\lambda^* \in A(w_\lambda)$ we have

$$0 \le \langle w_\lambda^*, \eta(w_\lambda, u_0) \rangle \le \lambda \langle w_\lambda^*, \eta(v, u_0) \rangle + (1 - \lambda) \langle w_\lambda^*, \eta(u_0, u_0) \rangle = \lambda \langle w_\lambda^*, \eta(v, u_0) \rangle$$

Let us assume that (H_A^1) and fix $u^* \in A(u)$. Then for each $\lambda \in (0, 1)$ we can determine $\bar{w}_\lambda^* \in A(w_\lambda)$ such that $\bar{w}_\lambda^* \rightharpoonup u^*$ as $\lambda \downarrow 0$.

If (H_A^2) holds, then there exists $\bar{u}_0^* \in A(u_0)$ for which we can determine a subnet $\{w_\lambda^*\}_{\lambda \in J}$ of $\{w_\lambda^*\}_{\lambda \in (0,1)}$ such that $w_\lambda^* \rightharpoonup \bar{u}_0^*$ in X^* as $\lambda \downarrow 0$.

Dividing by $\lambda > 0$ the above relation and taking into account the previous observation we conclude (after passing to the limit as $\lambda \downarrow 0$) that u_0 is a strong solution of problem (10.24) if (H_A^1) holds (u_0 is a solution of problem (10.24) if (H_A^2) holds).

CASE 2.2 $v \in K \setminus U$.

Since K is convex and $u_0, v \in K$, then we have that $(u_0, v) \subseteq K$. From $v \notin U$ there exists $\lambda_0 \in (0, 1)$ such that $v_0 = u_0 + \lambda_0(v - u_0) \in (u_0, v)$ and has the property that the entire line segment (u_0, v_0) is contained in $U \cap K$. Thus, for each $\lambda \in (0, 1)$ the element $w_\lambda = u_0 + \lambda(v_0 - u_0) \in K \cap V$, but $v_0 = u_0 + \lambda_0(v - u_0)$, hence $w_\lambda = u_0 + \lambda_0\lambda(v - u) \in K \cap V$ and $w_\lambda \to u_0$ as $\lambda \downarrow 0$. Applying the same arguments as in CASE 2.1 we infer that u_0 is a strong solution of problem (10.24) if (H_A^1) holds (u_0 is a solution of problem (10.24) if (H_A^2) holds) and this completes the proof. □

Let us turn our attention towards the case when K is a unbounded closed and convex subset of X. We shall establish next some sufficient conditions for the existence of solutions of problems $(QVLI)$ and (VLI). For every $r > 0$ we define

$$K_r := \{u \in K : \|u\| \leq r\} \quad \text{and} \quad K_r^- := \{u \in K : \|u\| < r\},$$

and consider the problems
Find $u_r \in K_r \cap D(\phi)$ such that

$$\exists u_r^* \in A(u_r) : \quad \langle u_r^*, \eta(v, u_r) \rangle + \phi(v) - \phi(u_r) \geq 0, \quad \forall v \in K_r, \tag{10.34}$$

and
Find $u_r \in K_r$ such that

$$\exists u_r^* \in A(u_r) : \quad \langle u_r^*, \eta(v, u_r) \rangle \geq 0, \quad \forall v \in K_r. \tag{10.35}$$

It is clear from above that the solution sets of problems (10.34) and (10.35) are nonempty. We have the following characterization for the existence of solutions in the case of unbounded closed and convex subsets.

Theorem 10.11 ([3]) *Assume that the same hypotheses as in Theorem 10.9 hold without the assumption of boundedness of K. Then each of the following conditions is sufficient for inequality problem $(QVLI)$ to admit at least one strong solution (solution):*

(C_1) *there exists $r_0 > 0$ and $u_0 \in K_{r_0}^-$ such that u_{r_0} solves (10.34).*
(C_2) *there exists $r_0 > 0$ such that for each $u \in K \setminus K_{r_0}$ we can find $\bar{v} \in K_{r_0}$ such that*

$$\langle u^*, \eta(\bar{v}, u) \rangle + \phi(\bar{v}) - \phi(u) \leq 0, \quad \forall u^* \in A(u).$$

(C_3) *there exists $\bar{u} \in K$ and a function $c : \mathbb{R}_+ \to \mathbb{R}_+$ with the property that $\lim\limits_{r \to +\infty} c(r) = +\infty$ such that*

$$\inf_{u^* \in A(u)} \langle u^*, \eta(u, \bar{u}) \rangle \geq c(\|u\|) \|u\|, \quad \forall u \in K.$$

Proof Let $v \in K$ be arbitrary fixed.

Assume (C_1) holds.

We define

$$s_0 := \begin{cases} \frac{1}{2}, & \text{if } v = u_{r_0} \\ \min\left\{\frac{1}{2}; \frac{r_0 - \|u_{r_0}\|}{\|v - u_{r_0}\|}\right\}, & \text{otherwise,} \end{cases}$$

and observe that $w_{s_0} := u_{r_0} + s_0(v - u_{r_0})$ belongs to K_{r_0} as $0 < s_0 < 1$ and K_{r_0} is convex. Assuming that (H_A^1) holds and using the fact that u_{r_0} is a strong solution of problem (10.34) we deduce that for each $u_{r_0}^* \in A(u_{r_0})$ we have

$$0 \leq \langle u_{r_0}^*, \eta(w_{s_0}, u_{r_0}) \rangle + \phi(w_{s_0}) - \phi(u_{r_0})$$

$$\leq s_0 \langle u_{r_0}^*, \eta(v, u_{r_0}) \rangle + (1 - s_0) \langle u_{r_0}^*, \eta(u_{r_0}, u_{r_0}) \rangle$$

$$+ s_0 \phi(v) + (1 - s_0)\phi(u_{r_0}) + \phi(u_{r_0})$$

$$= s_0 \left[\langle u_{r_0}^*, \eta(v, u_{r_0}) \rangle + \phi(v) - \phi(u_{r_0}) \rangle \right].$$

Dividing by $s_0 > 0$ we obtain that u_{r_0} is a strong solution of $(QVLI)$ as $v \in K$ was chosen arbitrary.

In a similar way we prove that u_{r_0} is a solution of inequality problem $(QVLI)$ if (H_A^2) holds.

Assume (C_2) holds.

Let us fix $r > r_0$. Then problem (10.34) admits at one solution $u_r \in K_r$. We observe that we only need to study the case when $\|u_r\| = r$. Indeed, if $\|u_r\| < r$, then $u_r \in K_r^-$ and by condition $(\mathcal{H}_1)$ u_r solves problem (10.23). The fact that $\|u_r\| = r$ implies that $u_r \in K \setminus K_{r_0}$ and thus we have

$$\langle u_r^*, \eta(\bar{v}, u_r) \rangle + \phi(\bar{v}) - \phi(u_r) \leq 0, \quad \forall u_r^* \in A(u_r). \tag{10.36}$$

We define

$$s_1 := \begin{cases} \frac{1}{2}, & \text{if } v = \bar{v} \\ \min\left\{\frac{1}{2}, \frac{r - r_0}{\|v - \bar{v}\|}\right\}, & \text{otherwise,} \end{cases}$$

and observe that $w_{s_1} := \bar{v} + s_1(v - \bar{v})$ belongs to K_r and

$$w_{s_1} - u_r = s_1(v - u_r) + (1 - s_1)(\bar{v} - u_r).$$

Let us assume that (H_A^1) holds and u_r is a strong solution of inequality (10.34). Then for each $u_r^* \in A(u_r)$ we have

$$0 \le \langle u_r^*, \eta(w_{s_1}, u_r) \rangle + \phi(w_{s_1}) - \phi(u_r)$$

$$\le s_1 \langle u_r^*, \eta(v, u_r) \rangle + (1 - s_1) \langle u_r^*, \eta(\bar{v}, u_r) \rangle + s_0 \phi(v) + (1 - s_1) \phi(\bar{v}) - \phi(u_r),$$

which leads to

$$0 \le s_1 \left[\langle u_r^*, \eta(v, u_r) \rangle + \phi(v) - \phi(u_r) \right] + (1 - s_1) \left[\langle u_r^*, \eta(\bar{v}, u_r) \rangle + \phi(\bar{v}) - \phi(u_r) \right],$$
$$(10.37)$$

for all $u_r^* \in A(u_r)$. Combining (10.36) and (10.37) we infer that u_r is a strong solution of $(QVLI)$.

In a similar way we prove that u_{r_0} is a solution of inequality problem $(QVLI)$ if (H_A^2) holds.

Assume (C_3) holds.

For each $r > 0$ problem (10.34) admits at least a solution $u_r \in K_r$. We shall prove that there exists r_0 such that $u_{r_0} \in K_{r_0}^-$, which according to $(\mathcal{H}_1)$ means that u_{r_0} solves $(QVLI)$. Arguing by contradiction, let us assume that $\|u_r\| = r$ for all $r > 0$. First we observe that the function ϕ is bounded from below by an affine and continuous function as it is convex and lower semicontinuous, therefore there exists $\xi \in X^*$ and $\beta \in \mathbb{R}$ such that

$$\phi(u) \ge \langle \xi, u \rangle + \beta, \quad \forall u \in X.$$

Taking $v := \bar{u}$ in (10.34) we obtain:

$$c(r)r = c(\|u_r\|)\|u_r\| \le \langle u_r^*, \eta(u_r, \bar{u}) \rangle \le \phi(\bar{u}) - \phi(u_r) \le \phi(\bar{u}) - \beta - \langle \xi, u_r \rangle$$

$$\le r \|\xi\|_* + \phi(\bar{u}) - \beta.$$

Dividing by $r > 0$ and then letting $r \to +\infty$ we obtain a contradiction since the left-hand side term of the inequality diverges, while the right-hand side term remains bounded.   □

Using the same arguments as above we are also able to prove the following characterization for the existence of solution of inequality problem (10.24) in the case of unbounded closed and convex subsets.

Theorem 10.12 ([3]) *Assume that the same hypotheses as in Theorem 10.10 hold without the assumption of boundedness of K. Then each of the following conditions is sufficient for inequality problem (VLI) to admit at least one strong solution (solution):*

(C_1') *there exists $r_0 > 0$ and $u_0 \in K_{r_0}^-$ such that u_{r_0} solves (10.35).*

(C_2') *there exists $r_0 > 0$ such that for each $u \in K \setminus K_{r_0}$ we can find $\bar{v} \in K_{r_0}$ such that*

$$\langle u^*, \eta(\bar{v}, u) \rangle \leq 0, \quad \forall u^* \in A(u).$$

(C_3') *there exists $\bar{u} \in K$ and a function $c : \mathbb{R}_+ \to \mathbb{R}_+$ with the property that $\lim\limits_{r \to +\infty} c(r) = +\infty$ such that*

$$\inf\limits_{u^* \in A(u)} \langle u^*, \eta(u, \bar{u}) \rangle \geq c(\|u\|)\|u\|, \quad \forall u \in K.$$

References

1. N. Costea, C. Lupu, On a class of variational-hemivariational inequalities involving set-valued mappings. Adv. Pure Appl. Math. **1**, 233–246 (2010)
2. N. Costea, V. Rădulescu, Inequality problems of quasi-hemivariational type involving set-valued operators and a nonlinear term. J. Global Optim. **52**, 743–756 (2012)
3. N. Costea, D.A. Ion, C. Lupu, Variational-like inequality problems involving set-valued maps and generalized monotonicity. J. Optim. Theory Appl. **155**, 79–99 (2012)
4. J. Parida, M. Sahoo, A. Kumar, A variational-like inequality problem. Bull. Aust. Math. Soc. **39**, 225–231 (1989)

Part IV

Applications to Nonsmooth Mechanics

Antiplane Shear Deformation of Elastic Cylinders in Contact with a Rigid Foundation

11

11.1 The Antiplane Model and Formulation of the Problem

Let us consider a deformable body $\mathcal{B}$ that we refer to a cartesian system $Ox_1x_2x_3$. Assume $\mathcal{B}$ is a cylinder with generators parallel to the x_3-axes and the cross section is a regular domain Ω in the plane Ox_1x_2. Furthermore, the generators are sufficiently long so the end effects in the axial direction are negligible. Thus, we can consider that $\mathcal{B} := \Omega \times (-\infty, \infty)$. We denote by $\partial\Omega =: \Gamma$ the boundary of Ω and we assume that Γ_1, Γ_2, Γ_3 are three open measurable parts that form a partition of Γ (i.e., $\Gamma = \overline{\Gamma}_1 \cup \overline{\Gamma}_2 \cup \overline{\Gamma}_3$; $\Gamma_i \cap \Gamma_j = \varnothing \ \forall i, j \in \{1, 2, 3\}$, $i \neq j$) such that $meas(\Gamma_1) > 0$. Suppose $\mathcal{B}$ is clamped on $\Gamma_1 \times (-\infty, \infty)$ and it is in frictional contact over $\Gamma_3 \times (-\infty, \infty)$ with a rigid foundation. In addition, the cylindrical body is subjected to volume forces of density f_0 in $\Omega \times (-\infty, \infty)$ and to surface tractions of density f_2 on $\Gamma_2 \times (-\infty, \infty)$.

Let $\mathbb{S}_3$ be the linear space of second order symmetric tensors in $\mathbb{R}^3$ (or, equivalently, the space of symmetric matrices of order 3), while "$\cdot$", "$:$" and $\|\cdot\|$ stand for the inner products and the Euclidean norms on $\mathbb{R}^3$ and $\mathbb{S}_3$, respectively. We have:

$$u \cdot v = u_i v_i, \quad \|v\| = (v \cdot v)^{1/2}, \quad \forall u := (u_i), \ v := (v_i) \in \mathbb{R}^3,$$

$$\sigma : \tau = \sigma_{ij} \tau_{ij}, \quad \|\tau\| = (\tau : \tau)^{1/2}, \quad \forall \sigma := (\sigma_{ij}), \ \tau := (\tau_{ij}) \in \mathbb{S}_3.$$

Here and below, the indices i and j run between 1 and 3 and the summation convention of the repeated indices is adopted.

© The Author(s), under exclusive license to Springer Nature Switzerland AG 2021
N. Costea et al., *Variational and Monotonicity Methods in Nonsmooth Analysis*,
Frontiers in Mathematics, https://doi.org/10.1007/978-3-030-81671-1_11

Loading the body in the following particular way,

$$f_0 := (0, 0, f_0) \quad \text{with} \quad f_0 := f_0(x_1, x_2) : \Omega \to \mathbb{R}, \tag{11.1}$$

$$f_2 := (0, 0, f_2) \quad \text{with} \quad f_2 := f_2(x_1, x_2) : \Gamma_2 \to \mathbb{R}, \tag{11.2}$$

we get a displacement field of the form

$$u := (0, 0, u) \quad \text{with} \quad u := u(x_1, x_2) : \Omega \to \mathbb{R}. \tag{11.3}$$

Concerning the unit outward normal to Γ, we have to write

$$n := (v_1, v_2, 0), \quad v_i := v_i(x_1, x_2) : \Gamma \to \mathbb{R}, \ i \in \{1, 2\}. \tag{11.4}$$

The infinitesimal strain tensor becomes

$$\varepsilon(u) = \begin{pmatrix} 0 & 0 & \frac{1}{2} u_{,1} \\ 0 & 0 & \frac{1}{2} u_{,2} \\ \frac{1}{2} u_{,1} & \frac{1}{2} u_{,2} & 0 \end{pmatrix}, \tag{11.5}$$

where $u_{,i} := \partial u / \partial x_i, \ i \in \{1, 2\}$.

Let $\sigma := (\sigma_{ij})$ denote the stress field and recall that, for the stationary processes, the equilibrium equation

$$\text{Div } \sigma + f_0 = 0_{\mathbb{R}^3}, \quad \text{in } \Omega \times (-\infty, \infty) \tag{11.6}$$

takes place, where $\text{Div } \sigma := (\sigma_{ij,j}), \ i \in \{1, 2, 3\}$.

Let us assume that the stress field σ has the following form

$$\sigma(x) := \begin{pmatrix} 0 & 0 & a_1(x, \nabla u) \\ 0 & 0 & a_2(x, \nabla u) \\ a_1(x, \nabla u) & a_2(x, \nabla u) & 0 \end{pmatrix} \tag{11.7}$$

where $x := (x_1, x_2) \in \overline{\Omega} \subset \mathbb{R}^2$ and $a(x, y) := (a_1(x, y), a_2(x, y)) : \overline{\Omega} \times \mathbb{R}^2 \to \mathbb{R}^2$.

Taking into account (11.1), (11.3), and (11.6), it follows that the equilibrium equation reduces to the following scalar equation

$$\text{div}(a(x, \nabla u)) + f_0 = 0, \quad \text{in } \Omega. \tag{11.8}$$

To complete the model, the boundary conditions must be specified. According to the physical setting,

$$u = 0_{\mathbb{R}^3}, \text{ on } \Gamma_1 \times (-\infty, \infty),$$

and

$$\sigma n = f_2, \text{ on } \Gamma_2 \times (-\infty, \infty).$$

Taking into account (11.2), (11.3), and (11.7), the previous vectorial boundary conditions reduce to the following *scalar conditions*

$$u = 0, \text{ on } \Gamma_1 \tag{11.9}$$

and

$$a(x, \nabla u) \cdot \nu = f_2, \text{ on } \Gamma_2, \tag{11.10}$$

where $\nu := (\nu_1, \nu_2)$, i.e, the 2-dimensional vector comprising only the first two components of the unit outward normal to Γ.

For a vector w we denote by w_n and w_T its *normal* and *tangential* components on the boundary, that is

$$w_n := w \cdot n, \quad w_T := w - w_n n. \tag{11.11}$$

Similarly, for a regular tensor field σ, we define its *normal* and *tangential* components to be the normal and the tangential components of the *Cauchy vector* σn, that is,

$$\sigma_n := (\sigma n) \cdot n, \quad \sigma_T := \sigma n - \sigma_n n. \tag{11.12}$$

Let us describe the frictional contact on $\Gamma_3 \times (-\infty, \infty)$. Taking into account (11.3) and (11.4) we conclude that the normal displacement vanishes, which shows that the contact is *bilateral*, i.e., the contact is kept during the process. From (11.3), (11.4), (11.7), (11.11), and (11.12) we deduce

$$u_T = (0, 0, u), \quad \sigma_T = (0, 0, \sigma_\tau), \text{ with } \sigma_\tau(x) := a(x, \nabla u(x)) \cdot \nu(x). \tag{11.13}$$

We model the frictional contact by the following boundary condition,

$$-\sigma_\tau(x) \in h(x, u(x)) \, \partial_C^2 j(x, u(x)), \text{ on } \Gamma_3, \tag{11.14}$$

where h and j are given functions which depend on the variable $x := (x_1, x_2)$ and do not depend on x_3 and, the notation $\partial_C^2 j(x, t)$ denotes the Clarke's generalized gradient of the mapping $t \mapsto j(x, t)$.

Putting together equations and conditions (11.8), (11.9), (11.10), and (11.14) we obtain a mathematical model which describes the *antiplane shear deformation of an elastic cylinder in frictional contact with a rigid foundation*:

Find a displacement $u : \overline{\Omega} \to \mathbb{R}$ such that

$$
\begin{cases}
\operatorname{div}(a(x, \nabla u)) + f_0 = 0, & \text{in } \Omega \\
u = 0, & \text{on } \Gamma_1 \\
a(x, \nabla u) \cdot v = f_2, & \text{on } \Gamma_2 \\
-a(x, \nabla u) \cdot v \in h(x, u) \, \partial_C^2 j(x, u), & \text{on } \Gamma_3.
\end{cases} \qquad ((P):)
$$

Once the displacement field u is determined, the stress tensor σ can be obtained via relation (11.7).

11.2   Weak Formulation and Solvability of the Problem

We assume that Ω is an open, connected, bounded subset of $\mathbb{R}^2$, with Lipshitz continuous boundary. In addition, we admit the following hypotheses:

(H_f) $f_0 \in L^2(\Omega)$ and $f_2 \in L^2(\Gamma_2)$.

(H_h) $h : \Gamma_3 \times \mathbb{R} \to \mathbb{R}$ is a Carathéodory function. Moreover, there exists a positive constant h_0 such that $0 \le h(x, t) \le h_0$, for all $t \in \mathbb{R}$, a.e. $x \in \Gamma_3$.

(H_j) $j : \Gamma_3 \times \mathbb{R} \to \mathbb{R}$ is a function which is measurable with respect to the first variable, and there exists $k \in L^2(\Gamma_3)$ such that, for all $x \in \Gamma_3$ and all $t_1, t_2 \in \mathbb{R}$, we have

$$
|j(x, t_1) - j(x, t_2)| \le k(x)|t_1 - t_2|.
$$

(H_a) $a : \overline{\Omega} \times \mathbb{R}^2 \to \mathbb{R}^2$ is a Carathéodory function which satisfies:

(i) there exist $\alpha > 0$ and $b \in L^2(\Omega)$ such that for a.e. $x \in \Omega$ and all $y \in \mathbb{R}^2$

$$
\|a(x, y)\| \le \alpha(b(x) + \|y\|);
$$

(ii) There exists $m > 0$ such that $a(x, y) \cdot y \ge m\|y\|^2$ for all $y \in \mathbb{R}^2$ and a.e. $x \in \Omega$;

(iii) $[a(x, y_1) - a(x, y_2)] \cdot (y_1 - y_2) \ge 0$, for all $y_1, y_2 \in \mathbb{R}^2$ and a.e. $x \in \Omega$.

Let us consider the functional space

$$V := \{v \in H^1(\Omega) : \gamma v = 0 \text{ a.e. on } \Gamma_1\},$$

where $\gamma : H^1(\Omega) \to L^2(\Gamma)$ denotes the Sobolev trace operator. For simplicity, everywhere below, we will omit to write γ to indicate the Sobolev trace on the boundary, writing v instead of γv. Since $meas(\Gamma_1) > 0$, it is well known that V is a Hilbert space endowed with the inner product

$$\langle u, v \rangle_V := \int_\Omega \nabla u \cdot \nabla v dx, \quad \forall u, v \in V,$$

and the associated norm is

$$\|v\|_V := \left(\int_\Omega \|\nabla v\|^2 dx \right)^{1/2},$$

which is equivalent with the usual norm on $H^1(\Omega)$. Using Sobolev's trace theorem we deduce that there exists $C > 0$ such that

$$\|v\|_{L^2(\Gamma_3)} \leq C \|v\|_V, \quad \forall v \in V.$$

Next, we define the operator $A : V \to V$ by

$$\langle Au, v \rangle_V := \int_\Omega a(x, \nabla u) \cdot \nabla v dx, \quad \forall u, v \in V. \tag{11.15}$$

Remark 11.1 It is easy to check that, if hypotheses (H_a) are fulfilled, then

(i) the operator A is well defined;
(ii) $\langle Au_n, v \rangle_V \to \langle Au, v \rangle_V$, for each $v \in V$, whenever $u_n \to u$ in V as $n \to \infty$;
(iii) $\langle Av, v \rangle_V \geq m \|v\|_V^2$, for all $v \in V$;
(iv) $\langle Av - Au, v - u \rangle_V \geq 0$, for all $u, v \in V$.

We are now able to provide a variational formulation for problem (P). To this end, consider $v \in V$ to be a test function and we multiply the first line of the problem (P) by $v - u$. To simplify the notation, we will not indicate explicitly the dependence on x. Assuming that the functions involved in the writing of the problem (P) are regular enough, after integration by parts, we obtain

$$\int_\Omega a(x, \nabla u) \cdot (\nabla v - \nabla u) dx = \int_\Gamma a(x, \nabla u) \cdot v(v - u) d\Gamma + \int_\Omega f_0(v - u) dx.$$

Taking into account the boundary conditions, we see that

$$\int_\Gamma a(x, \nabla u) \cdot v(v - u) d\Gamma = \int_{\Gamma_3} a(x, \nabla u) \cdot v(v - u) d\Gamma + \int_{\Gamma_2} f_2(v - u) d\Gamma.$$

On the other hand, from the definition of Clarke's generalized gradient, combined with the last line of problem (P), we have

$$-a(x, \nabla u) \cdot v(v - u) \le h(x, u) j^0(x, u; v - u), \quad \text{a.e. on } \Gamma_3$$

which implies

$$\int_{\Gamma_3} a(x, \nabla u) \cdot v(v - u) d\Gamma \ge - \int_{\Gamma_3} h(x, u) j^0(x, u; v - u) d\Gamma.$$

Thus, we arrive to the following variational formulation of the problem (P).
(P_V) Find $u \in V$ such that

$$\langle Au - g, v - u \rangle_V + \int_{\Gamma_3} h(x, u) j^0(x, u; v - u) d\Gamma \ge 0, \quad \forall v \in V, \qquad (11.16)$$

where g is the element of V given by the Riesz's representation theorem as follows,

$$\langle g, v \rangle_V = \int_\Omega f_0 v dx + \int_{\Gamma_2} f_2 v d\Gamma, \quad \forall v \in V.$$

Any function $u \in V$ which satisfies (11.16) is called a *weak solution* of problem (P).

Next we focus on the weak solvability of the problem (P). More precisely, we prove the following existence result.

Theorem 11.1 ([2, Theorem 4.1]) *Assume conditions (H_f), (H_h), (H_j), and (H_a) are fulfilled. Then, there exists at least one solution for problem (P_V).*

In order to prove Theorem 11.1 we need several auxiliary results.

Lemma 11.1 *Let K be a nonempty, closed and convex subset of V. Under hypotheses (H_h), (H_j), and (H_a) the set of the solutions for the problem*

(P_1) *Find $u \in K$ such that*

$$\langle Au - g, v - u \rangle_V + \int_{\Gamma_3} h(x, u) j^0(x, u; v - u) d\Gamma \ge 0, \; \forall v \in K, \qquad (11.17)$$

coincides with the set of the solutions for the problem

(P_2) *Find $u \in K$ such that*

$$\langle Av - g, v - u \rangle_V + \int_{\Gamma_3} h(x, u) j^0(x, u; v - u) d\Gamma \geq 0, \ \forall v \in K. \qquad (11.18)$$

Proof Let $u \in K$ be a solution of (P_1). By Remark 11.1-(iv) we have

$$\langle Av - Au, v - u \rangle_V \geq 0, \ \forall v \in K.$$

Summing the last relation and (11.17) we conclude that u is a solution of (P_2). Conversely, let us assume that u is a solution of (P_2). Fix $v \in K$ and define

$$w := u + t(v - u), \ t \in (0, 1).$$

We have

$$\langle Aw - g, w - u \rangle_V + \int_{\Gamma_3} h(x, u) j^0(x, u; w - u) d\Gamma \geq 0.$$

Using the positive homogeneity of the map $j^0(x, u; \cdot)$ it follows that

$$t \langle Aw - g, v - u \rangle_V + t \int_{\Gamma_3} h(x, u) j^0(x, u; v - u) d\Gamma \geq 0.$$

Keeping in mind Remark 11.1-(ii), we divide by $t > 0$ and pass to the limit as $t \to 0$. Thus, we get (11.18). Therefore, $u \in K$ is a solution of (P_1). $\qquad \square$

Lemma 11.2 *Let K be a nonempty, bounded, closed and convex subset of V. Under hypotheses (H_h), (H_j), and (H_a), there exists at least one solution for (P_1).*

Proof For each $v \in K$, we define two set valued mappings $G, H : K \rightsquigarrow K$ as follows:

$$G(v) := \left\{ u \in K : \langle Au - g, v - u \rangle_V + \int_{\Gamma_3} h(x, u) j^0(x, u; v - u) d\Gamma \geq 0 \right\},$$

$$H(v) := \left\{ u \in K : \langle Av - g, v - u \rangle_V + \int_{\Gamma_3} h(x, u) j^0(x, u; v - u) d\Gamma \geq 0 \right\}.$$

STEP 1. *G is a KKM mapping.*

If G is not a KKM mapping, then there exists $\{v_1 \ldots, v_N\} \subset K$ such that

$$\text{co}\{v_1, \ldots, v_N\} \not\subset \bigcup_{i=1}^N G(v_i),$$

i.e., there exists $v_0 \in \text{co}\{v_1, \ldots, v_N\}$, $v_0 := \sum_{i=1}^{N} \lambda_i v_i$, with $\lambda_i \geq 0$, $i \in \{1, \ldots, N\}$ and $\sum_{i=1}^{N} \lambda_i = 1$, such that $v_0 \notin \bigcup_{i=1}^{N} G(v_i)$. By the definition of G, we have

$$\langle Av_0 - g, v_i - v_0 \rangle_V + \int_{\Gamma_3} h(x, v_0) j^0(x, v_0; v_i - v_0) d\Gamma < 0,$$

for each $i \in \{1, \ldots, N\}$. It follows that,

$$0 = \langle Av_0 - g, v_0 - v_0 \rangle_V + \int_{\Gamma_3} h(x, v_0) j^0(x, v_0; v_0 - v_0) d\Gamma$$

$$= \left\langle Av_0 - g, \sum_{i=1}^{N} \lambda_i v_i - v_0 \right\rangle_V + \int_{\Gamma_3} h(x, v_0) j^0 \left(x, v_0; \sum_{i=1}^{N} \lambda_i v_i - v_0 \right) d\Gamma$$

$$\leq \sum_{i=1}^{N} \lambda_i \left[\langle Av_0 - g, v_i - v_0 \rangle_V + \int_{\Gamma_3} h(x, v_0) j^0(x, v_0; v_i - v_0) d\Gamma \right] < 0,$$

which is a contradiction.

STEP 2. $G(v) \subseteq H(v)$ for all $v \in K$.

For a given $v \in K$, arbitrarily fixed, let $u \in G(v)$. This implies by the definition of G that

$$\langle Au - g, v - u \rangle_V + \int_{\Gamma_3} h(x, u) j^0(x, u; v - u) d\Gamma \geq 0.$$

On the other hand, we recall that

$$\langle Av - Au, v - u \rangle_V \geq 0,$$

and summing the last two relations it follows that $u \in H(v)$.

Thus $G(v) \subseteq H(v)$, which implies that H is also a KKM mapping.

STEP 3. $H(v)$ is weakly closed for all $v \in K$.

For a fixed $v \in K$ let us consider the sequence $\{u_n\}_n \subset H(v)$ such that $u_n \rightharpoonup u$ in V. We will prove that $u \in H(v)$. We have

$$0 \leq \limsup_{n \to \infty} \left[\langle Av - g, v - u_n \rangle_V + \int_{\Gamma_3} h(x, u_n) j^0(x, u_n; v - u_n) d\Gamma \right]$$

$$\leq \langle Av - g, v - u \rangle_V + \limsup_{n \to \infty} \int_{\Gamma_3} h(x, u_n) j^0(x, u_n; v - u_n) d\Gamma.$$

Using Sobolev's trace theorem we conclude that

$$u_n \to u \text{ in } L^2(\Gamma_3)$$

and passing eventually to a subsequence we get

$$u_n(x) \to u(x) \text{ a.e. on } \Gamma_3.$$

On the other hand, (H_j) enables us to conclude that

$$|j^0(x, u(x); v(x))| \le k(x)|v(x)| \text{ a.e. } x \in \Gamma_3. \tag{11.19}$$

Next, using Fatou's lemma, we have

$$\limsup_{n \to \infty} \int_{\Gamma_3} h(x, u_n(x)) j^0(x, u_n(x); v(x) - u_n(x)) d\Gamma \le$$

$$\int_{\Gamma_3} \limsup_{n \to \infty} |h(x, u_n(x)) - h(x, u(x))| k(x)|u_n(x) - v(x)| d\Gamma$$

$$+ \int_{\Gamma_3} h(x, u(x)) \limsup_{n \to \infty} j^0(x, u_n(x); v(x) - u_n(x)) d\Gamma \le$$

$$\int_{\Gamma_3} h(x, u(x)) j^0(x, u(x); v(x) - u(x)) d\Gamma.$$

We can conclude that

$$0 \le \langle Av - g, v - u \rangle_V + \int_{\Gamma_3} h(x, u) j^0(x, u; v - u) d\Gamma,$$

which is equivalent to $u \in H(v)$.

Since K is bounded, closed and convex, we know that K is weakly compact. So, $H(v)$ is weakly compact for each $v \in K$ as it is a closed subset of a compact set (in the weak topology). Therefore, the conditions of Corollary D.1 are satisfied in the weak topology. It follows that

$$\bigcap_{v \in K} H(v) \ne \varnothing,$$

and, from Lemma 11.1, we get

$$\bigcap_{v \in K} G(v) = \bigcap_{v \in K} H(v) \ne \varnothing.$$

Hence, there exists at least one solution of problem (P_1).   □

Proof of Theorem 11.1 For each $n \geq 1$, set $K_n := \{u \in V : \|u\|_V \leq n\}$. Lemma 11.2 guarantees the existence of a sequence $\{u_n\}_n$ such that for all $v \in K_n$ one has

$$\langle Au_n - g, v - u_n \rangle_V + \int_{\Gamma_3} h(x, u_n) j^0(x, u_n; v - u_n) d\Gamma \geq 0. \tag{11.20}$$

STEP 1.   *There exists a positive integer n_0 such that $\|u_{n_0}\|_V < n_0$.*
   Arguing by contradiction let us suppose that $\|u_n\|_V = n$ for each $n \geq 1$. Taking $v := 0_V$ in (11.20), we have

$$\langle Au_n, u_n \rangle_V \leq \langle g, u_n \rangle_V + \int_{\Gamma_3} h(x, u_n) j^0(x, u_n; -u_n) d\Gamma.$$

Taking into account (11.19) and using $(H_a) - (iii)$, we get,

$$\langle Au_n, u_n \rangle_V \leq \|g\|_V \|u_n\|_V + \int_{\Gamma_3} h_0 k(x) |u_n(x)| d\Gamma$$
$$\leq \|g\|_V \|u_n\|_V + h_0 \|k\|_{L^2(\Gamma_3)} \|u_n\|_{L^2(\Gamma_3)}$$
$$\leq \|g\|_V \|u_n\|_V + h_0 \|k\|_{L^2(\Gamma_3)} C \|u_n\|_V.$$

Thus,

$$\frac{\langle Au_n, u_n \rangle_V}{\|u_n\|_V} \leq \|g\|_V + h_0 C \|k\|_{L^2(\Gamma_3)} < \infty,$$

which contradicts the fact that

$$\frac{\langle Au_n, u_n \rangle_V}{\|u_n\|_V} \geq \frac{m\|u_n\|_V^2}{\|u_n\|_V} = m\|u_n\|_V \to \infty.$$

STEP 2.   *u_{n_0} solves problem (P_V).*
   Since $\|u_{n_0}\|_V < n_0$, for each $v \in V$ we can choose $t > 0$ such that $w := u_{n_0} + t(v - u_{n_0}) \in K_{n_0}$ (it suffices to take $t := 1$ if $v = u_{n_0}$ and $t < (n_0 - \|u_{n_0}\|_V)/\|v - u_{n_0}\|_V$ otherwise). It follows from (11.20) and the positive homogeneity of the map $v \mapsto j^0(x, u; v)$ that

$$0 \leq \langle Au_{n_0} - g, w - u_{n_0} \rangle_V + \int_{\Gamma_3} h(x, u_{n_0}) j^0(x, u_{n_0}; w - u_{n_0}) d\Gamma$$

$$= t\langle Au_{n_0} - g, v - u_{n_0} \rangle_V + t \int_{\Gamma_3} h(x, u_{n_0}) j^0(x, u_{n_0}; v - u_{n_0}) d\Gamma.$$

Dividing by $t > 0$ the conclusion follows.   □

11.3 Examples of Constitutive Laws

In this section we present examples of elastic constitutive laws which lead to the particular form of the stress field σ considered in (11.7).

Example 11.1 (Linear Constitutive Law) We can describe the behavior of the material with the constitutive law

$$\sigma := \lambda(tr\ \varepsilon(u))I_3 + 2\mu\varepsilon(u), \tag{11.21}$$

where λ and μ are Lamé's coefficients, $tr\ \varepsilon(u) = \varepsilon_{kk}(u)$ and I_3 is the unit tensor.

Using (11.21) and (11.5) we obtain that, in the antiplane context, the stress field has the following form

$$\sigma = \begin{pmatrix} 0 & 0 & \mu u_{,1} \\ 0 & 0 & \mu u_{,2} \\ \mu u_{,1} & \mu u_{,2} & 0 \end{pmatrix}.$$

We assume that μ depends on the variable $x := (x_1, x_2)$ and it is independent on x_3. Furthermore, we assume that μ satisfies

(H_μ) $\mu \in L^\infty(\Omega)$ and there exists $\mu^* \in \mathbb{R}$ such that $\mu(x) \geq \mu^* > 0$ a.e. $x \in \Omega$.

We take $a : \overline{\Omega} \times \mathbb{R}^2 \to \mathbb{R}^2$, $a(x, y) := \mu(x)y$ and point out the fact that under (H_μ), the hypotheses (H_a) are fulfilled.

Example 11.2 (Piecewise Linear Constitutive Law) We can consider the following constitutive law, see for example Han and Sofonea [3],

$$\sigma := \lambda(tr\ \varepsilon(u))I_3 + 2\mu\varepsilon(u) + 2\beta(\varepsilon(u) - P_\mathcal{K}\varepsilon(u)) \tag{11.22}$$

where λ, μ, $\beta > 0$ are the coefficients of the material, $tr\ \varepsilon := \varepsilon_{kk}$, I_3 is the identity tensor, $\mathcal{K}$ is the nonempty, closed and convex *von Mises set*

$$\mathcal{K} := \left\{ \sigma \in \mathbb{S}^3 : \frac{1}{2}\sigma^D \cdot \sigma^D \leq k^2,\ k > 0 \right\} \tag{11.23}$$

$P_\mathcal{K} : \mathbb{S}^3 \to \mathcal{K}$ represents the projection operator on $\mathcal{K}$ and σ^D is the deviatoric part of σ, i.e., $\sigma^D := \sigma - \frac{1}{3}(tr\ \sigma)I_3$.

In the antiplane framework, the constitutive law (11.22) becomes

$$\sigma := (\mu + \beta)\begin{pmatrix} 0 & 0 & u_{,1} \\ 0 & 0 & u_{,2} \\ u_{,1} & u_{,2} & 0 \end{pmatrix} - 2\beta\begin{pmatrix} 0 & 0 & (P_{\widetilde{K}}\frac{1}{2}\nabla u)_1 \\ 0 & 0 & (P_{\widetilde{K}}\frac{1}{2}\nabla u)_2 \\ (P_{\widetilde{K}}\frac{1}{2}\nabla u)_1 & (P_{\widetilde{K}}\frac{1}{2}\nabla u)_2 & 0 \end{pmatrix},$$

where $\widetilde{K} := \overline{B(0,k)}$, ($k$ given by (11.23)) and $P_{\widetilde{K}} : \mathbb{R}^2 \to \widetilde{K}$ is the projection operator on $\widetilde{K}$.

Let us define $a : \overline{\Omega} \times \mathbb{R}^2 \to \mathbb{R}^2, a(x,y) := [\mu(x) + \beta(x)]y - 2\beta(x)P_{\widetilde{K}}\frac{1}{2}y$. We assume that the following conditions are fulfilled

(H_μ) $\mu \in L^\infty(\Omega)$ and there exists $\mu^* \in \mathbb{R}$ such that $\mu(x) \geq \mu^* > 0$ a.e. $x \in \Omega$;
(H_β) $\beta \in L^\infty(\Omega)$.

Taking into account the non-expansivity of the projection map $P_{\widetilde{K}}$, under the assumptions (H_μ) and (H_β), the hypotheses (H_a) are verified with $\alpha := \|\mu\|_{L^\infty(\Omega)} + 2\|\beta\|_{L^\infty(\Omega)}$, $b \equiv 0$, and $m := \mu^*$.

Example 11.3 (Nonlinear Constitutive Law) For Hencky materials, see, e.g., Zeidler [7], the stress-strain relation is

$$\sigma := k_0(\text{tr } \varepsilon(u))I_3 + \psi(\|\varepsilon^D(u)\|^2)\varepsilon^D(u), \tag{11.24}$$

where $k_0 > 0$ is a coefficient of the material, $\psi : \mathbb{R} \to \mathbb{R}$ is a constitutive function and $\varepsilon^D(u)$ is the deviatoric part of $\varepsilon = \varepsilon(u)$. From (11.24) and (11.5) we obtain the following form for the stress field

$$\sigma = \begin{pmatrix} 0 & 0 & \frac{1}{2}\psi\left(\frac{1}{2}|\nabla u|^2\right)u_{,1} \\ 0 & 0 & \frac{1}{2}\psi\left(\frac{1}{2}|\nabla u|^2\right)u_{,2} \\ \frac{1}{2}\psi\left(\frac{1}{2}|\nabla u|^2\right)u_{,1} & \frac{1}{2}\psi\left(\frac{1}{2}|\nabla u|^2\right)u_{,2} & 0 \end{pmatrix}.$$

We define $a : \overline{\Omega} \times \mathbb{R}^2 \to \mathbb{R}^2$ by

$$a(x,y) := \frac{1}{2}\psi\left(\frac{1}{2}|y|^2\right)y \tag{11.25}$$

and we assume the following hypotheses,
(H_ψ) $\psi : \mathbb{R} \to \mathbb{R}$ is a given function satisfying:

(i) $\psi \in L^\infty(\mathbb{R}) \cap C(\mathbb{R})$;
(ii) there exists $\psi^* \in \mathbb{R}$ such that $\psi(t) \geq \psi^* > 0$ for all $t \geq 0$;
(iii) the function $t \mapsto t\psi(t^2)$ is increasing on $[0, \infty)$.

It remains to prove that conditions (H_a) are fulfilled if (H_ψ) hold. We will prove only (H_a)-(iii), the proof of the others being trivial.

Let $x \in \Omega$ and $y_1, y_2 \in \mathbb{R}^2$ be arbitrarily fixed. Keeping in mind (H_a) and (11.25), we have to prove that

$$0 \leq \frac{1}{2}\left[\psi\left(\frac{1}{2}|y_1|^2\right)y_1 - \psi\left(\frac{1}{2}|y_2|^2\right)y_2\right] \cdot (y_1 - y_2).$$

The above inequality is equivalent to

$$\psi\left(\frac{1}{2}|y_1|^2\right)|y_1|^2 + \psi\left(\frac{1}{2}|y_2|^2\right)|y_2|^2 \geq \left[\psi\left(\frac{1}{2}|y_1|^2\right) + \psi\left(\frac{1}{2}|y_2|^2\right)\right]y_1 \cdot y_2.$$

To obtain this last inequality, it suffices to prove that

$$\psi\left(\frac{1}{2}|y_1|^2\right)|y_1|^2 + \psi\left(\frac{1}{2}|y_2|^2\right)|y_2|^2 \geq \left[\psi\left(\frac{1}{2}|y_1|^2\right) + \psi\left(\frac{1}{2}|y_2|^2\right)\right]|y_1||y_2|$$

or, equivalently,

$$\left[\psi\left(\frac{1}{2}|y_1|^2\right)|y_1| - \psi\left(\frac{1}{2}|y_2|^2\right)|y_2|\right] \cdot (|y_1| - |y_2|) \geq 0. \tag{11.26}$$

Since the function $t \mapsto t\psi(t^2)$ is increasing on $[0, \infty)$, we have

$$\left[t_1\psi(t_1^2) - t_2\psi(t_2^2)\right](t_1 - t_2) \geq 0; \quad \forall t_1, t_2 \in [0, \infty).$$

Now, taking $t_1 := \frac{\sqrt{2}}{2}|y_1|$ and $t_2 := \frac{\sqrt{2}}{2}|y_2|$ we obtain (11.26).

11.4   Examples of Friction Laws

In this section we present examples of functions $h : \Gamma_3 \times \mathbb{R} \to \mathbb{R}$ and $j : \Gamma_3 \times \mathbb{R} \to \mathbb{R}$, that allow us to model the frictional contact of the cylindrical body $\mathcal{B}$ with the rigid foundation by (11.14) and, in the same time, verify the required properties in (H_h) and (H_j).

Example 11.4 (Slip Dependent Friction Law) We can consider

$$h(x, t) := k_0\left(1 + \delta e^{-|t|}\right); \quad j(x, t) := |t|, \tag{11.27}$$

with δ, $k_0 > 0$. In this case, the friction law (11.14) is equivalent with the friction law

$$|\sigma_\tau(x)| \leq h(x, u(x)), \quad |\sigma_\tau(x)| = -h(x, u(x))\frac{u(x)}{|u(x)|} \quad \text{if } u(x) \neq 0, \quad \text{on } \Gamma_3,$$

$$(11.28)$$

or, equivalently, taking into account (11.3) and (11.13), we arrive at the well known *Coulomb's law of dry friction*,

$$\|\sigma_\tau(x)\| \leq h(x, \|u_\tau(x)\|), \quad \sigma_\tau(x) = -h(x, \|u_\tau(x)\|)\frac{u_\tau(x)}{\|u_\tau(x)\|} \quad \text{if } u_\tau(x) \neq 0, \quad \text{on } \Gamma_3.$$

We note that h and j in (11.27) are non-differentiable functions. This feature leads to mathematical difficulties for optimal control or numerical reasons.

Example 11.5 (Regularized Friction Law) Let us consider the differentiable functions

$$h(x, t) := k_0 \left(1 + \delta e^{\rho - \sqrt{t^2 + \rho^2}}\right); \quad j(x, t) := \sqrt{t^2 + \rho^2} - \rho,$$

with k_0, δ, $\rho > 0$. The friction law (11.14) becomes equivalent with the friction law

$$-\sigma_\tau(x) = h(x, u(x))\frac{u(x)}{\sqrt{u(x)^2 + \rho^2}} \quad \text{on } \Gamma_3. \qquad (11.29)$$

The friction laws (11.28) and (11.29) model situation in which surfaces are dry; they are characterized by the existence of the positive function *friction bound*, h, that depends on the magnitude of the tangential displacement, see e.g., [5, 6], such that slip may occur only when the friction force reaches the critical value provided by the friction bound.

Example 11.6 (The Power Friction Law) Another choice with regularization effect is the following one

$$h(x, t) := k_0 \left(1 + \delta e^{-\frac{|t|^{\rho+1}}{\rho+1}}\right); \quad j(x, t) := \frac{|t|^{\rho+1}}{\rho + 1},$$

with k_0, $\delta > 0$ and $\rho \geq 1$. This time, the friction law (11.14) is equivalent with the power friction law

$$-\sigma_\tau(x) = \begin{cases} h(x, u(x))|u(x)|^{\rho-1}u(x) & u(x) \neq 0; \\ 0 & u(x) = 0. \end{cases} \qquad (11.30)$$

In this situations, the slip appears even for small tangential shear. Such kind of situations appear in practice when the contact surfaces are lubricated.

Example 11.7 (Non-monotone Friction Law) Let us take

$$h(x, t) \equiv 1, \quad j(x, t) := \int_0^t p(s)ds, \tag{11.31}$$

where

$$p(t) := \begin{cases} (-\alpha t_0 + k_0)e^{t_0+t} - k_0 & \text{if } t < -t_0; \\ \alpha t & \text{if } -t_0 \le t \le t_0; \\ (\alpha t_0 - k_0)e^{t_0-t} + k_0 & \text{if } t > t_0, \end{cases}$$

with α, k_0, $t_0 > 0$. In this case, the friction law (11.14) is equivalent with the friction law

$$- \sigma_\tau(x) = p(u(x)) \text{ on } \Gamma_3. \tag{11.32}$$

The friction law (11.32) is used in geomechanics or rock interface analysis; see [4] for more details.

Example 11.8 (Multivalued Friction Law) Let us consider $p : \mathbb{R} \to \mathbb{R}$ a function such that $p \in L_{loc}^\infty(\mathbb{R})$, i.e., a function essentially bounded on any bounded interval of $\mathbb{R}$. For any $\rho > 0$ and $t \in \mathbb{R}$ let us define

$$\overline{p}_\rho(t) := \text{ess inf}_{|t_1-t|\le\rho} \, p(t_1) \quad \text{and} \quad \overline{\overline{p}}_\rho(t) := \text{ess sup}_{|t_1-t|\le\rho} \, p(t_1).$$

Obviously, the monotonicity properties of $\rho \mapsto \overline{p}_\rho(t)$ and $\rho \mapsto \overline{\overline{p}}_\rho(t)$ imply that the limits as $\rho \to 0_+$ exist. Therefore, one may write that $\overline{p}(t) = \lim_{\rho\to0_+} \overline{p}_\rho(t)$ and $\overline{\overline{p}}(t) = \lim_{\rho\to0_+} \overline{\overline{p}}_\rho(t)$, and define the multivalued function $\tilde{p} : \mathbb{R} \rightsquigarrow \mathbb{R}$, $\tilde{p}(t) := [\overline{p}(t), \overline{\overline{p}}(t)]$, where $[\cdot, \cdot]$ denotes a real interval. If there exist $\lim_{s\to t_+} p(s) = p(t_+) \in \mathbb{R}$ and $\lim_{s\to t_-} p(s) = p(t_-) \in \mathbb{R}$ for each $t \in \mathbb{R}$, it can be shown (see e.g. [1]) that

$$\partial_C^2 j(x, t) = \tilde{p}(t).$$

Assume that the tangential stress satisfies the multivalued relation

$$- \sigma_\tau(x) \in \tilde{p}(u(x)) \quad \text{on } \Gamma_3. \tag{11.33}$$

We point out the fact that the multivalued friction law (11.33) is of the form (11.14) with $h(x, t) \equiv 1$ and j defined by (11.31). It is easy to check that the functions h and j verify (H_h) and (H_j), respectively, if $|p(t)| \le p_0$ for all $t \in \mathbb{R}$ with $p_0 > 0$.

A simple example of a function p which satisfies the required properties can be

$$p(t) := \begin{cases} -k_0, & \text{if } t < 0 \\ k_0, & \text{if } t \geq 0, \end{cases}$$

with $k_0 > 0$.

References

1. K.-C. Chang, Variational methods for non-differentiable functionals and their applications to partial differential equations. J. Math. Anal. Appl. **80**, 102–129 (1981)
2. N. Costea, A. Matei, Weak solutions for nonlinear antiplane problems leading to hemivariational inequalities. Nonlinear Anal. **72**, 3669–3680 (2010)
3. W. Han, M. Sofonea, *Quasistatic Contact Problems in Viscoelasticity and Viscoplasticity*. Studies in Advanced Mathematics, vol. 30 (International Press, Somerville, 2002)
4. P.D. Panagiotopoulos, Hemivariational inequalities, in *Applications in Mechanics and Engineering* (Springer, Berlin, 1993)
5. E. Rabinowicz, *Friction and Wear of Materials*, 2nd edn. (Wiley, New York, 1995)
6. C. Scholtz, *The Mechanics of Earthquakes and Faulting* (Cambridge University, Cambridge, 1990)
7. E. Zeidler, *Nonlinear Functional Analysis and Its Applications IV: Applications to Mathematical Physics* (Springer, New York, 1988)

Weak Solvability of Frictional Problems for Piezoelectric Bodies in Contact with a Conductive Foundation

12.1 The Model

The piezoelectricity is a property of a class of materials, like ceramics, characterized by the coupling between the mechanical and electrical properties. This coupling leads to the appearance of electric potential when mechanical stress is present and, conversely, mechanical stress is generated when electric potential is applied. The first effect is used in mechanical sensors and the reverse effect is used in actuators, in engineering control equipment. Models for piezoelectric materials can be found in [1, 2, 4, 5].

Before describing the problem let us first present some notations and preliminary material which will be used throughout this subsection.

Let $\Omega \subset \mathbb{R}^m$ be an open bounded subset with a Lipschitz boundary Γ and let ν denote the outward unit normal vector to Γ. We introduce the spaces

$$H := L^2(\Omega; \mathbb{R}^m), \quad \mathcal{H} := \left\{ \tau = (\tau_{ij}) : \tau_{ij} = \tau_{ji} \in L^2(\Omega) \right\} = L^2(\Omega; \mathbb{S}_m),$$

$$H_1 := \{ u \in H : \varepsilon(u) \in \mathcal{H} \} = H^1(\Omega; \mathbb{R}^m), \quad \mathcal{H}_1 := \{ \tau \in \mathcal{H} : \operatorname{Div} \tau \in H \},$$

where $\varepsilon : H_1 \to \mathcal{H}$ and $\operatorname{Div} : \mathcal{H}_1 \to H$ denote the *deformation and the divergence operators*, defined by

$$\varepsilon(u) := (\varepsilon_{ij}(u)), \quad \varepsilon_{ij}(u) = \frac{1}{2} \left(\frac{\partial u_i}{\partial x_j} + \frac{\partial u_j}{\partial x_i} \right), \quad \operatorname{Div} \tau := \left(\frac{\partial \tau_{ij}}{\partial x_j} \right),$$

N. Costea et al., *Variational and Monotonicity Methods in Nonsmooth Analysis*, Frontiers in Mathematics, https://doi.org/10.1007/978-3-030-81671-1_12

The spaces H, $\mathcal{H}$, H_1, and $\mathcal{H}_1$ are Hilbert spaces endowed with the following inner productsyd78

$$(u, v)_H := \int_\Omega u_i v_i \, dx, \quad (\sigma, \tau)_{\mathcal{H}} := \int_\Omega \sigma : \tau \, dx,$$

$$(u, v)_{H_1} := (u, v)_H + (\varepsilon(u), \varepsilon(v))_{\mathcal{H}}, \quad (\sigma, \tau)_{\mathcal{H}_1} := (\sigma, \tau)_{\mathcal{H}} + (\text{Div}\,\sigma, \text{Div}\,\tau)_H.$$

The associated norms in $H, \mathcal{H}, H_1, \mathcal{H}_1$ will be denoted by $\|\cdot\|_H$, $\|\cdot\|_{\mathcal{H}}$, $\|\cdot\|_{H_1}$ and $\|\cdot\|_{\mathcal{H}_1}$, respectively.

Given $v \in H_1$ we denote by v its trace γv on Γ, where $\gamma : H^1(\Omega; \mathbb{R}^m) \to H^{1/2}(\Gamma; \mathbb{R}^m) \subset L^2(\Gamma; \mathbb{R}^m)$ is the Sobolev trace operator. Recall that the following Green formula holds:

$$(\sigma, \varepsilon(v))_{\mathcal{H}} + (\text{Div}\,\sigma, v)_H = \int_\Gamma \sigma v \cdot v \, d\Gamma, \quad \forall v \in H_1. \tag{12.1}$$

We shall describe next the model for which we shall derive a variational formulation. Let us consider body $\mathcal{B}$ made of a piezoelectric material which initially occupies an open bounded subset $\Omega \subset \mathbb{R}^m$ ($m = 2, 3$) with smooth a boundary $\partial\Omega = \Gamma$. The body is subjected to volume forces of density f_0 and has volume electric charges of density q_0, while on the boundary we impose mechanical and electrical constraints. In order to describe these constraints we consider two partitions of Γ: the first partition is given by three mutually disjoint open parts Γ_1, Γ_2, and Γ_3 such that meas($\Gamma_1 > 0$) and the second partition consists of three disjoint open parts Γ_a, Γ_b, and Γ_c such that meas(Γ_a) > 0, $\Gamma_c = \Gamma_3$ and $\overline{\Gamma}_a \cup \overline{\Gamma}_b = \overline{\Gamma}_1 \cup \overline{\Gamma}_2$. The body is clamped on Γ_1 and a surface traction of density f_2 acts on Γ_2. Moreover, the electric potential vanishes on Γ_a and a surface electric charge of density q_b is applied on Γ_b. On $\Gamma_3 = \Gamma_c$ the body comes in frictional contact with a conductive obstacle, called foundation which has the electric potential φ_F.

Denoting by $u : \Omega \to \mathbb{R}^m$ the displacement field, by $\varepsilon(u) := (\varepsilon_{ij}(u))$ the strain tensor, by $\sigma : \Omega \to \mathbb{S}_m$ the stress tensor, by $D : \Omega \to \mathbb{R}^m$, $D = (D_i)$ the electric displacement field and by $\varphi : \Omega \to \mathbb{R}$ the electric potential we can now write the strong formulation of the problem which describes the above process:

(P) Find a displacement field $u : \Omega \to \mathbb{R}^m$ and an electric potential $\varphi : \Omega \to \mathbb{R}$ s.t.

$$\text{Div}\,\sigma + f_0 = 0 \ \text{ in } \Omega, \tag{12.2}$$

$$\text{div}\,D = q_0 \ \text{ in } \Omega, \tag{12.3}$$

$$\sigma = \mathcal{E}\varepsilon(u) + \mathcal{P}^T \nabla\varphi \ \text{ in } \Omega, \tag{12.4}$$

$$D = \mathcal{P}\varepsilon(u) - \mathcal{B}\nabla\varphi \ \text{ in } \Omega, \tag{12.5}$$

$$u = 0 \ \text{ on } \Gamma_1, \tag{12.6}$$

$$\varphi = 0 \ \text{ on } \Gamma_a, \tag{12.7}$$

$$\sigma n = f_2 \ \text{on} \ \Gamma_2, \tag{12.8}$$

$$D \cdot n = q_b \ \text{on} \ \Gamma_b, \tag{12.9}$$

$$-\sigma_n = S; \quad -\sigma_T \in \partial_C^2 j(x, u_T); \quad D \cdot n \in \partial_C^2 \phi(x, \varphi - \varphi_F) \ \text{on} \ \Gamma_3, \tag{12.10}$$

We point out the fact that once the displacement field u and the electric potential φ are determined, the stress tensor σ and the electric displacement field D can be obtained via relations (12.4) and (12.5), respectively. Similar

Let us now provide explanation of the equations and the conditions (12.2)–(12.10) in which, for simplicity, we have omitted the dependence of the functions on the spatial variable x.

First, Eqs. (12.2)–(12.3) are the *governing equations* consisting of the *equilibrium conditions*, while Eqs. (12.4)–(12.5) represent the *electro-elastic constitutive law*.

In the sequel we assume that $\mathcal{E} : \Omega \times \mathbb{S}_m \to \mathbb{S}_m$ is a *nonlinear elasticity operator*, $\mathcal{P} : \Omega \times \mathbb{S}_m \to \mathbb{R}^m$ and $\mathcal{P}^T : \Omega \times \mathbb{R}^m \to \mathbb{S}_m$ are the *piezoelectric operator* (third order tensor field) and its transpose, respectively and $\mathcal{B} : \Omega \times \mathbb{R}^m \to \mathbb{R}^m$ denotes the *electric permittivity operator* (second order tensor field) which is considered to be linear. The tensors $\mathcal{P}$ and $\mathcal{P}^T$ satisfy the equality

$$\mathcal{P}\tau \cdot y = \tau : \mathcal{P}^T y, \quad \forall \tau \in \mathbb{S}_m \ \text{and all} \ y \in \mathbb{R}^m$$

and the components of the tensor $\mathcal{P}^T$ are given by $p_{ijk}^T := p_{kij}$.

When $\tau \mapsto \mathcal{E}(x, \tau)$ is linear, $\mathcal{E}(x, \tau) := C(x)\tau$ with the elasticity coefficients $C := (c_{ijkl})$ which may be functions indicating the position in a nonhomogeneous material. The decoupled state can be obtained by taking $p_{ijk} = 0$, in this case we have purely elastic and purely electric deformations.

Conditions (12.6) and (12.7) model the fact that the displacement field and the electrical potential vanish on Γ_1 and Γ_a, respectively, while conditions (12.8) and (12.9) represent the traction and the electric boundary conditions showing that the forces and the electric charges are prescribed on Γ_2 and Γ_b, respectively.

Conditions (12.10) describe the contact, the frictional and the electrical conductivity conditions on the contact surface Γ_3, respectively. Here, S is the normal load imposed on Γ_3, the functions $j : \Gamma_3 \times \mathbb{R}^m \to \mathbb{R}^m$ and $\phi : \Gamma_3 \times \mathbb{R} \to \mathbb{R}$ are prescribed and φ_F is the electric potential of the foundation.

12.2   Variational Formulation and Existence of Weak Solutions

The strong formulation of problem (P) consists in finding $u : \Omega \to \mathbb{R}^m$ and $\varphi : \Omega \to \mathbb{R}$ such that (12.2)–(12.10) hold. However, it is well known that, in general, the strong formulation of a contact problem does not admit any solution. Therefore, we reformulate problem (P) in a weaker sense, i.e., we shall derive its variational formulation. With this end in view, we introduce the functional spaces for the displacement field and the electrical potential

$$V := \left\{ v \in H^1(\Omega; \mathbb{R}^m) : v = 0 \text{ on } \Gamma_1 \right\}, \quad W := \left\{ \varphi \in H^1(\Omega) : \varphi = 0 \text{ on } \Gamma_a \right\}$$

which are closed subspaces of H_1 and $H^1(\Omega)$. We endow V and W with the following inner products and the corresponding norms

$$(u, v)_V := (\varepsilon(u), \varepsilon(v))_{\mathcal{H}}, \quad \|v\|_V := \|\varepsilon(v)\|_{\mathcal{H}}$$

$$(\varphi, \chi)_W := (\nabla\varphi, \nabla\chi)_H, \quad \|\chi\|_W := \|\nabla\chi\|_H$$

and conclude that $(V, \|\cdot\|_V)$, $(W, \|\cdot\|_W)$ are Hilbert spaces.

Assuming sufficient regularity of the functions involved in the problem, using the Green formula (12.1), the relations (12.2)–(12.10), the definition of the Clarke subdifferntial and the equality

$$\int_{\Gamma_3} (\sigma n) \cdot v \, d\Gamma = \int_{\Gamma_3} \sigma_n v_n \, d\Gamma + \int_{\Gamma_3} \sigma_T \cdot v_T \, d\Gamma$$

we obtain the following variational formulation of problem (P) in terms of the displacement field and the electric potential:

(P_V) Find $(u, \varphi) \in V \times W$ such that for all $(v, \chi) \in V \times W$

$$\begin{cases} \left(\mathcal{E}\varepsilon(u) + \mathcal{P}^T \nabla\varphi, \varepsilon(v) - \varepsilon(u) \right)_{\mathcal{H}} + \int_{\Gamma_3} j^0_{,2}(x, u_T; v_T - u_T) d\Gamma \geq (f, v - u)_V \\ (\mathcal{B}\nabla\varphi - \mathcal{P}\varepsilon(u), \nabla\chi - \nabla\varphi)_H + \int_{\Gamma_3} \phi^0_{,2}(x, \varphi - \varphi_F; \chi - \varphi) d\Gamma \geq (q, \chi - \varphi)_W, \end{cases}$$

where $f \in V$ and $q \in W$ are the elements given by the Riesz's representation theorem as follows

$$(f, v - u)_V := \int_\Omega f_0 \cdot v \, dx + \int_{\Gamma_2} f_2 \cdot v \, d\Gamma - \int_{\Gamma_3} S v_n \, d\Gamma,$$

$$(q, \chi)_W := \int_\Omega q_0 \chi \, dx - \int_{\Gamma_b} q_2 \chi \, d\Gamma.$$

In the study of problem (P_V) we shall assume fulfilled the following hypotheses:

(H_1) The elasticity operator $\mathcal{E} : \Omega \times \mathbb{S}_m \to \mathbb{S}_m$ such that
  (i) $x \mapsto \mathcal{E}(x, \tau)$ is measurable for all $\tau \in \mathbb{S}_m$;
  (ii) $\tau \mapsto \mathcal{E}(x, \tau)$ is continuous for a.e. $x \in \Omega$;

(iii) there exist $c_1 > 0$ and $\alpha \in L^2(\Omega)$ s.t. $\|\mathcal{E}(x, \tau)\|_{\mathbb{S}_m} \leq c_1(\alpha(x) + \|\tau\|_{\mathbb{S}_m})$ for all
$\quad \tau \in \mathbb{S}_m$ and a.e. $x \in \Omega$;

$\quad(iv)$ $\tau \mapsto \mathcal{E}(x, \tau) : (\sigma - \tau)$ is weakly uu.sc for all $\sigma \in \mathbb{S}_m$ and a.e. $x \in \Omega$;

$\quad(v)$ there exists $c_2 > 0$ s.t. $\mathcal{E}(x, \tau) : \tau \geq c_2\|\tau\|_{\mathbb{S}_m}^2$ for all $\tau \in \mathbb{S}_m$.

(H_2) The piezoelectric operator $\mathcal{P} : \Omega \times \mathbb{S}_m \to \mathbb{R}^m$ is such that

$\quad(i)$ $\mathcal{P}(x, \tau) = p(x)\tau$ for all $\tau \in \mathbb{S}_m$ and a.e. $x \in \Omega$;

$\quad(ii)$ $p(x) = (p_{ijk}(x))$ with $p_{ijk} = p_{ikj} \in L^\infty(\Omega)$.

(H_3) $\mathcal{B} : \Omega \times \mathbb{R}^m \to \mathbb{R}^m$ is such that

$\quad(i)$ $\mathcal{B}(x, y) = \beta(x)y$ for all $y \in \mathbb{R}^m$ and a.e. $x \in \Omega$;

$\quad(ii)$ $\beta(x) = (\beta_{ij}(x))$ with $\beta_{ij} = \beta_{ji} \in L^\infty(\Omega)$;

$\quad(iii)$ there exists $m > 0$ s.t. $(\beta(x)y) \cdot y \geq m|y|^2$ for all $y \in \mathbb{R}^m$ and a.e. $x \in \Omega$.

(H_4) $j : \Gamma_3 \times \mathbb{R}^m \to \mathbb{R}$ is such that

$\quad(i)$ $x \mapsto j(x, y)$ is measurable for all $y \in \mathbb{R}^m$;

$\quad(ii)$ $\zeta \mapsto j(x, y)$ is locally Lipschitz for a.e. $x \in \Gamma_3$;

$\quad(iii)$ there exist $c_3 > 0$ s.t. $|\partial_C^2 j(x, y)| \leq c_3(1 + |y|)$ for all $y \in \mathbb{R}^m$;

$\quad(iv)$ there exists $c_4 > 0$ s.t. $j_{,2}^0(x, y; -y) \leq c_4|y|$ for all $y \in \mathbb{R}^m$ and a.e. $x \in \Gamma_3$;

$\quad(v)$ $y \mapsto j(x, y)$ is regular for a.e. $x \in \Gamma_3$.

(H_5) $\phi : \Gamma_3 \times \mathbb{R} \to \mathbb{R}$ is such that

$\quad(i)$ $x \mapsto \phi(x, t)$ is measurable for all $t \in \mathbb{R}$;

$\quad(ii)$ $t \mapsto \phi(x, t)$ is locally Lipschitz for a.e. $x \in \Gamma_3$;

$\quad(iii)$ there exist $c_5 > 0$ s.t. $|\partial_C^2 \phi(x, t)| \leq c_5|t|$ for all $t \in \mathbb{R}$ and a.e. $x \in \Gamma_3$;

$\quad(iv)$ $t \mapsto \phi(x, t)$ is regular for a.e. $x \in \Gamma_3$.

(H_6) $f_0 \in H$, $f_2 \in L^2(\Gamma_2; \mathbb{R}^m)$, $q_0 \in L^2(\Omega)$, $q_b \in L^2(\Gamma_2)$, $S \in L^\infty(\Gamma_3)$, $S \geq 0$, $\varphi_F \in L^2(\Gamma_3)$.

The main result of this chapter is given by the following theorem.

Theorem 12.1 ([3, Theorem 4.4]) *Assume conditions (H_1)–(H_6) hold. Then problem (P_V) possesses at least one solution.*

Proof We observe that problem (P_V) is in fact a system of two coupled hemivariational inequalities. The idea is to apply one of the existence results obtained in Sect. 8.4 with suitable choice of ψ_k, J, and F_k ($k \in \{1, 2\}$).

First, let us take $n := 2$ and define $X_1 := V$, $X_2 := W$, $Y_1 := L^2(\Gamma_3; \mathbb{R}^m)$, $Y_2 := L^2(\Gamma_3)$, $K_1 := X_1$ and $K_2 := X_2$. Next we introduce $T_1 : X_1 \to Y_1$ and $T_2 : X_2 \to Y_2$ defined by

$$T_1 := i_T \circ \gamma_m \circ i_m|_{\Gamma_3}, \quad T_2 := \gamma \circ i|_{\Gamma_3},$$

$i_m : V \to H_1 = H^1(\Omega; \mathbb{R}^m)$ is the embedding operator $\gamma_m : H_1 \to H^{1/2}(\Gamma; \mathbb{R}^m)$ is the Sobolev trace operator, $i_T : H^{1/2}(\Gamma; \mathbb{R}^m) \to L^2(\Gamma_3; \mathbb{R}^m)$ is the operator defined by

$i_T(v) := v_T$, $i : W \to H^1(\Omega)$ is the embedding operator and $\gamma : H^1(\Omega) \to H^{1/2}(\Gamma)$ is the Sobolev trace operator. Clearly T_1 and T_2 are linear and compact operators. We consider next $\psi_1 : X_1 \times X_2 \times X_1 \to \mathbb{R}$ and $\psi_2 : X_1 \times X_2 \times X_2 \to \mathbb{R}$ defined by

$$\psi_1(u, \varphi, v) := (\mathcal{E}\varepsilon(u), \varepsilon(v) - \varepsilon(u))_{\mathcal{H}} + \left(\mathcal{P}^T \nabla\varphi, \varepsilon(v) - \varepsilon(u)\right)_{\mathcal{H}},$$

$$\psi_2(u, \varphi, \chi) := (\mathcal{B}\nabla\varphi, \nabla\chi - \nabla\varphi)_H - (\mathcal{P}\varepsilon(u), \nabla\chi - \nabla\varphi)_H,$$

$J : Y_1 \times Y_2 \to \mathbb{R}$ defined by

$$J(w, \eta) := \int_{\Gamma_3} j(x, w(x)) d\Gamma + \int_{\Gamma_3} \phi(x, \eta(x) - \varphi_F(x)) d\Gamma,$$

and $F_1 : X_1 \times X_2 \to X_1^*$ and $F_2 : X_1 \times X_2 \to X_2^*$ defined by

$$F_1(u, \varphi) := f, \quad F_2(u, \varphi) := q.$$

It is easy to check from the above definitions that if (H_1)–(H_6) hold, then J is a regular locally Lipschitz functional which satisfies

$$J_{,1}^0(w, \eta; z) = \int_{\Gamma_3} j_{,2}^0(x, w(x); z(x)) d\Gamma$$

$$J_{,2}^0(w, \eta; \zeta) = \int_{\Gamma_3} \phi_{,2}^0(x, \eta(x) - \varphi_F(x); \zeta(x)) d\Gamma.$$

Moreover, all the conditions of Corollary 8.2 are fulfilled, therefore problem (P_V) possesses at least one solution. □

References

1. I. Andrei, N. Costea, A. Matei, Antiplane shear deformation of piezoelectric bodies in contact with a conductive support. J. Global Optim. **56**, 103–119 (2013)
2. R. Batra, J. Yang, Saint-Vernant's principle in linear piezoelectricity. J. Elasticity **38**, 209–218 (1995)
3. N. Costea, C. Varga, Systems of nonlinear hemivariational inequalities and applications. Topol. Methods Nonlinear Anal. **41**, 39–65 (2013)
4. T. Ikeda, *Fundamentals of Piezoelectricity* (Oxford University, Oxford, 1990)
5. J. Yang, *An Introduction to the Theory of Piezoelectricity* (Springer, Berlin, 2010)

The Bipotential Method for Contact Models with Nonmonotone Boundary Conditions

13

13.1 The Mechanical Model and Its Variational Formulation

Let us consider a body $\mathcal{B}$ which occupies the domain $\Omega \subset \mathbb{R}^m$ ($m = 2, 3$) with a sufficiently smooth boundary Γ (e.g. Lipschitz continuous) and a unit outward normal n. The body is acted upon by forces of density f_0 and it is mechanically constrained on the boundary. In order to describe these constraints we assume Γ is partitioned into three Lebesgue measurable parts $\Gamma_1, \Gamma_2, \Gamma_3$ such that Γ_1 has positive Lebesgue measure. The body is clamped on Γ_1, hence the displacement field vanishes here, while surface tractions of density f_2 act on Γ_2. On Γ_3 the body may come in contact with an obstacle which will be referred to as foundation. The process is assumed to be static and the behavior of the material is modeled by a (possibly multivalued) constitutive law expressed as a subdifferential inclusion. The contact between the body and the foundation is modeled with respect to the normal and the tangent direction respectively, to each corresponding an inclusion involving the sum between the Clarke subdifferential of a locally Lipschitz function and the normal cone of a nonempty, closed and convex set.

It is well known that the subdifferential of a convex function is a monotone set-valued operator, while the Clarke subdifferential is a set-valued operator which is not monotone in general. This is why we say that the constitutive law is monotone and the boundary conditions are nonmonotone.

The mathematical model which describes the above process is the following. For simplicity we omit the dependence of some functions of the spatial variable.

N. Costea et al., *Variational and Monotonicity Methods in Nonsmooth Analysis*, Frontiers in Mathematics, https://doi.org/10.1007/978-3-030-81671-1_13

(P) Find a displacement $u : \Omega \to \mathbb{R}^m$ and a stress tensor $\sigma : \Omega \to \mathbb{S}_m$ such that

$$\text{Div } \sigma = f_0, \text{ in } \Omega \tag{13.1}$$

$$\sigma \in \partial\phi(\varepsilon(u)), \text{ a.e. in } \Omega \tag{13.2}$$

$$u = 0, \text{ on } \Gamma_1 \tag{13.3}$$

$$\sigma\nu = f_2, \text{ on } \Gamma_2 \tag{13.4}$$

$$-\sigma_n \in \partial_C^2 j_1(x, u_n) + N_{C_1}(u_n), \text{ on } \Gamma_3 \tag{13.5}$$

$$-\sigma_T \in h(x, u_T)\partial_C^2 j_2(x, u_T) + N_{C_2}(u_T), \text{ on } \Gamma_3 \tag{13.6}$$

where $\phi : \mathbb{S}_m \to \mathbb{R}$ is convex and lower semicontinuous, $j_1 : \Gamma_3 \times \mathbb{R} \to \mathbb{R}$ and $j_2 : \Gamma_3 \times \mathbb{R}^m \to \mathbb{R}$ are locally Lipschitz with respect to the second variable and $h : \Gamma_3 \times \mathbb{R}^m \to \mathbb{R}$ is a prescribed function. Here, $C_1 \subset \mathbb{R}$ and $C_2 \subset \mathbb{R}^m$ are nonempty closed and convex subsets and N_{C_k} denotes the normal cone of C_k $(k = 1, 2)$. For a Banach space E and a nonempty, closed and convex subset $K \subset E$, recall that the normal cone of K at x is defined by

$$N_K(x) := \left\{\xi \in E^* : \langle \xi, y - x \rangle_{E^* \times E} \leq 0, \forall y \in K\right\}.$$

It is well known that

$$N_K(x) = \partial I_K(x),$$

where I_K is the indicator function of K, that is,

$$I_K(x) := \begin{cases} 0, & \text{if } x \in K, \\ +\infty, & \text{otherwise.} \end{cases}$$

Relation (13.1) represents the equilibrium equation, (13.2) is the constitutive law, (13.3)–(13.4) are the displacement and traction boundary conditions and (13.5)–(13.6) describe the contact between body and the foundation.

Relations between the stress tensor σ and the strain tensor ε of the type (13.2) describe the constitutive laws of the deformation theory of plasticity, of Hencky plasticity with convex yield function, of locking materials with convex locking functions etc. For concrete examples and their physical interpretation one can consult Sections 3.3.1 and 3.3.2 in Panagiotopoulos [7, Sections 3.3.1 & 3.3.2] (see also [8, Section 3]). A particular case of interest regarding (13.2) is when the constitutive map ϕ is Gateaux differentiable, thus the subdifferential inclusion reducing to

$$\sigma = \phi'(\varepsilon(u)), \tag{13.7}$$

which corresponds to nonlinear elastic materials.

Some classical constitutive laws which can be written in the form (13.7) are presented below:

(i) Assume that ϕ is defined by

$$\phi(\mu) := \frac{1}{2}\mathcal{E}\mu : \mu,$$

where $\mathcal{E} := (\mathcal{E}_{ijkl})$, $1 \le i, j, k, l \le m$ is a fourth order tensor which satisfies the symmetry property

$$\mathcal{E}\mu : \tau = \mu : \mathcal{E}\tau, \forall \mu, \tau \in \mathbb{S}_m,$$

and the ellipticity property

$$\mathcal{E}\mu : \mu \ge c|\mu|^2, \forall \mu \in \mathbb{S}_m.$$

In this case (13.7) reduces to *Hooke's law*, that is, $\sigma := \mathcal{E}\varepsilon(u)$, and corresponds to linearly elastic materials.

(ii) Assume that ϕ is defined by

$$\phi(\mu) := \frac{1}{2}\mathcal{E}\mu : \mu + \beta |\mu - P_{\mathcal{K}}\mu|^2,$$

where $\mathcal{E}$ is the elasticity tensor and satisfies the same properties as in the previous example, $\beta > 0$ is a constant coefficient of the material, $P : \mathbb{S}_m \to \mathcal{K}$ is the projection operator and $\mathcal{K}$ is the nonempty, closed and convex von Mises set

$$\mathcal{K} := \left\{\mu \in \mathbb{S}_m : \frac{1}{2}\mu^D : \mu^D \le a^2, a > 0\right\}.$$

Here the notation μ^D stands for the deviator of the tensor μ. In this case (13.7) becomes

$$\sigma := \mathcal{E}\varepsilon(u) + 2\beta(I - P_{\mathcal{K}})\varepsilon(u),$$

which is known in the literature as *piecewise linear constitutive law* (see, e.g., Han & Sofonea [3]).

(iii) Assume ϕ is defined by

$$\phi(\mu) := \frac{k_0}{2}Tr(\mu)I : \mu + \frac{1}{2}\varphi\left(\left|\mu^D\right|^2\right),$$

where $k_0 > 0$ is a constant and $\varphi : [0, \infty) \to [0, \infty)$ is a continuously differentiable constitutive function.

In this case (13.7) becomes

$$\sigma = k_0 Tr(\varepsilon(u))I + \varphi'\left(\left|\varepsilon^D(u)\right|^2\right)\varepsilon^D(u),$$

and this describes the behavior of the Hencky materials (see, e.g., Zeidler [9]).

Boundary conditions of the type (13.5) and (13.6) can model a large class of contact problems arising in mechanics and engineering. For the case $h \equiv 1$ many examples of nonmonotone laws of the type

$$-\sigma_n \in \partial_C j_1(u_n) \text{ and } -\sigma_T \in \partial_C j_2(u_T),$$

can be found in [8, Section 2.4], [6, Section 1.4] or [2, Section 2.8].

The case when the function h actually depends on the second variable allows the study of contact problems with *slip-dependent friction law*. This friction law reads as follows

$$-\sigma_T \le \mu(x, |u_T|), \quad -\sigma_T = \mu(x, |u_T|)\frac{u_T}{|u_T|} \text{ if } u_T \ne 0, \tag{13.8}$$

where $\mu : \Gamma_3 \times [0, +\infty) \to [0, +\infty)$ is the sliding threshold and it is assumed to satisfy

$$0 \le \mu(x, t) \le \mu_0, \text{ for a.e. } x \in \Gamma_3 \text{ and all } t \ge 0,$$

for some positive constant μ_0. It is easy to see that (13.6) can be put in the form (13.8) simply by choosing

$$h(x, u_T) := \mu(x, |u_T|) \text{ and } j_2(x, u_T) := |u_T|.$$

We point out the fact that the above example cannot be written in the form $-\sigma_T \in \partial_C j_2(u_T)$ as, in general, for two locally Lipschitz functions h, g there does not exists j such that $\partial_C j(u) = h(u)\partial_C g(u)$. We would also like to point out that many boundary conditions of classical elasticity are particular cases of (13.5) and (13.6), in most of these cases the functions j_1 and j_2 being convex, hence leading to monotone boundary conditions. We list below some examples:

(a) *The Winkler boundary condition*

$$-\sigma_n = k_0 u_n, \quad k_0 > 0.$$

This law is used in engineering as it describes the interaction between a deformable body and the soil and can be expressed in the form (13.5) by setting

$$C_1 := \mathbb{R} \text{ and } j_1(x, t) := \frac{k_0}{2} t^2,$$

More generally, if we want to describe the case when the body may lose contact with the foundation, we can consider the following law

$$\begin{cases} u_n < 0 \Rightarrow \sigma_n = 0, \\ u_n \geq 0 \Rightarrow -\sigma_n = k_0 u_n, \end{cases}$$

The first relation corresponds to the case when there is no contact, while the second models the contact case. Obviously the above law can be expressed in the form (13.5) by choosing

$$C_1 := \mathbb{R} \text{ and } j_1(x, t) := \begin{cases} 0, & \text{if } t < 0, \\ \frac{k_0}{2} t^2, & \text{if } t \geq 0, \end{cases}$$

In [5] the following nonmonotone boundary conditions were imposed to model the contact between a body and a Winkler-type foundation which may sustain limited values of efforts

$$\begin{cases} u_n < 0 \Rightarrow \sigma_n = 0, \\ u_n \in [0, a) \Rightarrow -\sigma_n = k_0 u_n, \\ u_n = a \Rightarrow -\sigma_n \in [0, k_0 a], \\ u_n > a \Rightarrow \sigma_n = 0. \end{cases}$$

This means that the rupture of the foundation is assumed to occur at those points in which the limit effort is attained. The first condition holds in the noncontact zone, the second describes the zone where the contact occurs and it is idealized by the Winkler law. The maximal value of reactions that can be maintained by the foundation is given by $k_0 a$ and it is accomplished when $u_n = a$, with k_0 being the Winkler coefficient. The fourth relation holds in the zone where the foundation has been destroyed. The above Winkler-type law can be written as an inclusion of the type (13.5) by setting

$$C_1 := \mathbb{R} \text{ and } j_1(x, t) := \begin{cases} 0, & \text{if } t < 0, \\ \frac{k_0}{2} t^2, & \text{if } 0 \leq t < a, \\ \frac{k_0}{2} a^2, & \text{if } t \geq a. \end{cases}$$

Since all of the above example only describe what happens in the normal direction, in order to complete the model we must combine these with boundary conditions concerning σ_T, u_T, or both. The simplest cases are $u_T = 0$ (which corresponds to $C_2 = \{0\}$) and $\sigma_T = S_T$, where $S_T = S_T(x)$ is given (which corresponds to $j_2(x, u_T) = -S_T \cdot u_T$).

(b) *The Signorini boundary conditions*, which hold if the foundation is rigid and are as follows

$$\begin{cases} u_n < 0 \Rightarrow \sigma_n = 0, \\ u_n = 0 \Rightarrow \sigma_n \leq 0, \end{cases}$$

or equivalently,

$$u_n \leq 0, \ \sigma_n \leq 0 \text{ and } \sigma_n u_n = 0.$$

This can be written equivalently in form (13.5) by setting

$$C_1 := (-\infty, 0] \text{ and } j_1 \equiv 0.$$

(c) In [4] the following *static version of Coulomb's law of dry friction with prescribed normal stress* was considered

$$\begin{cases} -\sigma_n(x) = F(x) \\ |\sigma_T| \leq k(x)|\sigma_n|, \\ \sigma_T = -k(x)|\sigma_n|\frac{u_T}{|u_T|}, \text{ if } u_T(x) \neq 0. \end{cases}$$

We can write the above law in the form of (13.5) and (13.6) simply by setting

$$C_1 := \mathbb{R}, \ C_2 := \mathbb{R}^m, \ j_1(x, t) := F(x)t$$

and

$$h(x, y) := k(x)|F(x)| \text{ and } j_2(x, y) := \|y\|.$$

The assumptions on the functions f_0, f_2, ϕ, h, j_1 and j_2 required to prove our main result are listed below.

(H_C) The constraint sets C_1 and C_2 are convex cones, i.e.,

$$0 \in C_k \quad \text{and} \quad \lambda C_k \subset C_k \text{ for all } \lambda > 0, \ k = 1, 2.$$

(H_f) The density of the volume forces and the traction satisfy $f_0 \in H$ and $f_2 \in L^2(\Gamma_2; \mathbb{R}^m)$.

(H_ϕ) The constitutive function $\phi : \mathbb{S}_m \to \mathbb{R}$ and its conjugate $\phi^* : \mathbb{S}_m \to (-\infty, +\infty]$
   satisfy:
       (i) ϕ is convex and lower semicontinuous;
       (ii) there exists $\alpha_1 > 0$ such that $\phi(\tau) \geq \alpha_1 |\tau|^2$, for all $\tau \in \mathbb{S}_m$;
       (iii) there exists $\alpha_2 > 0$ such that $\phi^*(\mu) \geq \alpha_2 |\mu|^2$, for all $\mu \in \mathbb{S}_m$;
       (iv) $\phi(\varepsilon(v)) \in L^1(\Omega)$, for all $v \in V$ and $\phi^*(\tau) \in L^1(\Omega)$, for all $\tau \in \mathcal{H}$.

(H_h) The function $h : \Gamma_3 \times \mathbb{R}^m \to \mathbb{R}$ is such that:
       (i) $\Gamma_3 \ni x \mapsto h(x, y)$ is measurable for each $y \in \mathbb{R}^m$;
       (ii) $\mathbb{R}^m \ni y \mapsto h(x, y)$ is continuous for a.e. $x \in \Gamma_3$;
       (iii) there exists $h_0 > 0$ such that $0 \leq h(x, y) \leq h_0$ for a.e. $x \in \Gamma_3$ and all $y \in \mathbb{R}^m$.

(H_{j_1}) The function $j_1 : \Gamma_3 \times \mathbb{R} \to \mathbb{R}$ is such that:
       (i) $\Gamma_3 \ni x \mapsto j_1(x, t)$ is measurable for each $t \in \mathbb{R}$;
       (ii) there exists $p \in L^2(\Gamma_3)$ such that for a.e. $x \in \Gamma_3$ and all $t_1, t_2 \in \mathbb{R}$

$$|j_1(x, t_1) - j_1(x, t_2)| \leq p(x)|t_1 - t_2|;$$

       (iii) $j_1(x, 0) \in L^1(\Gamma_3)$.

(H_{j_2}) The function $j_2 : \Gamma_3 \times \mathbb{R}^m \to \mathbb{R}$ is such that:
       (i) $\Gamma_3 \ni x \mapsto j_\tau(x, y)$ is measurable for each $y \in \mathbb{R}^m$;
       (ii) there exist $q \in L^2(\Gamma_3)$ such that for a.e. $x \in \Gamma_3$ and all $y_1, y_2 \in \mathbb{R}^m$

$$|j_2(x, y_1) - j_2(x, y_2)| \leq q(x)|y_1 - y_2|;$$

       (iii) $j_2(x, 0) \in L^1(\Gamma_3; \mathbb{R}^m)$.

The strong formulation of problem (P) consists in finding $u : \Omega \to \mathbb{R}^m$ and $\sigma : \Omega \to \mathbb{S}_m$, regular enough, such that (13.1)–(13.6) are satisfied. However, it is a fact that for most contact problem the strong formulation has no solution. Therefore, it is useful to reformulate problem (P) in a weaker sense, i.e., we shall derive a variational formulation. With this end in mind, we consider the following function space

$$V := \{v \in H_1 : \ v = 0 \text{ a.e. on } \Gamma_1\} \tag{13.9}$$

which is a closed subspace of H_1, hence a Hilbert space. Since the Lebesgue measure of Γ_1 is positive, it follows from Korn's inequality that the following inner product

$$(u, v)_V := (\varepsilon(u), \varepsilon(v))_{\mathcal{H}} \tag{13.10}$$

generates a norm on V which is equivalent with the norm inherited from H_1.

Let us provide a variational formulation for problem (P). To this end, we consider u a strong solution, $v \in V$ a test function and we multiply the first line of (P) by $v - u$. Using the Green formula (see (12.1)) we have

$$
\begin{aligned}
(f_0, v - u)_H &= -(\mathrm{Div}\,\sigma, v - u)_H \\
&= -\int_\Gamma (\sigma v) \cdot (v - u)\mathrm{d}\Gamma + (\sigma, \varepsilon(v) - \varepsilon(u))_H \\
&= -\int_{\Gamma_2} f_2 \cdot (v - u)\mathrm{d}\Gamma - \int_{\Gamma_3} [\sigma_n(v_n - u_n) + \sigma_T \cdot (v_T - u_T)]\mathrm{d}\Gamma \\
&\quad + (\sigma, \varepsilon(v) - \varepsilon(u))_H
\end{aligned}
$$

for all $v \in V$. Since $V \ni v \mapsto (f_0, v)_H + \int_{\Gamma_2} f_2 \cdot v\mathrm{d}\Gamma$ is linear and continuous, we can apply Riesz's representation theorem to conclude that there exists a unique element $f \in V$ such that

$$
(f, v)_V := (f_0, v)_H + \int_{\Gamma_2} f_2 \cdot v\mathrm{d}\Gamma. \tag{13.11}
$$

Consider now the following nonempty, closed and convex subset of V

$$
\Lambda := \{v \in V : v_n(x) \in C_1 \text{ and } v_T(x) \in C_2 \text{ for a.e. } x \in \Gamma_3\},
$$

which is called the *set of admissible displacement fields*.

Since C_1, C_2 are convex cones, it follows that Λ is also a convex cone. Moreover, for all $v \in \Lambda$ the following inequalities hold

$$
-\int_{\Gamma_3} \sigma_n(v_n - u_n)\mathrm{d}\Gamma \le \int_{\Gamma_3} j_1^0(x, u_n; v_n - u_n)\mathrm{d}\Gamma \tag{13.12}
$$

and

$$
-\int_{\Gamma_3} \sigma_T \cdot (v_T - u_T)\mathrm{d}\Gamma \le \int_{\Gamma_3} h(x, u_T)j_2^0(x, u_T; v_T - u_T)\mathrm{d}\Gamma. \tag{13.13}
$$

Here, and hereafter, the generalized derivatives of the functions j_1 and j_2 are taken with respect to the second variable, i.e. of the functions $\mathbb{R} \ni t \mapsto j_1(x, t)$ and $\mathbb{R}^m \ni y \mapsto j_2(x, y)$ respectively, but for simplicity we omit to mention that in fact these are partial generalized derivatives. On the other hand, according to Theorem 1.4, we can rewrite (13.2) as

$$
\varepsilon(u) \in \partial\phi^*(\sigma), \text{ a.e. in } \Omega,
$$

and which after integration over Ω leads to

$$- (\varepsilon(u), \mu - \sigma)_{\mathcal{H}} + \int_{\Omega} \phi^*(\mu) - \phi^*(\sigma) dx \geq 0, \forall \mu \in \mathcal{H}. \qquad (13.14)$$

Let us denote by $\varepsilon^* : \mathcal{H} \to V$ the adjoint of ε, i.e.,

$$(\varepsilon^*(\mu), v)_V = (\mu, \varepsilon(v))_{\mathcal{H}}, \forall v \in V \text{ and all } \mu \in \mathcal{H}.$$

Using (13.11)–(13.14) we arrive at the following system of inequalities
$(\tilde{P})$ Find $u \in \Lambda$ and $\sigma \in \mathcal{H}$ such that

$$\begin{cases} (\varepsilon^*(\sigma) - f, v - u)_V + \displaystyle\int_{\Gamma_3} \left[j_v^0(x, u_n; v_n - u_n) + h(x, u_T) j_\tau^0(x, u_T; v_T - u_T) \right] d\Gamma \geq 0, \\ -(\varepsilon(u), \mu - \sigma)_{\mathcal{H}} + \displaystyle\int_{\Omega} \left(\phi^*(\mu) - \phi^*(\sigma) \right) dx \geq 0, \end{cases}$$

for all $(v, \mu) \in \Lambda \times \mathcal{H}$.

The first inequality of $(\tilde{P})$ is related to the equilibrium relation, whereas the second inequality represents the *functional extension of the constitutive law (13.2)*. It is well-known (see, e.g., [2, Theorem 1.3.21]) that it implies $\varepsilon(u) \in \partial \phi^*(\sigma)$ a.e. in Ω.

We can connect the constitutive law, the function ϕ and its conjugate ϕ^* through the separable bipotential $a : \mathbb{S}_m \times \mathbb{S}_m \to (-\infty, +\infty]$ defined by

$$a(\tau, \mu) := \phi(\tau) + \phi^*(\mu), \forall \tau, \mu \in \mathbb{S}_m.$$

Using the bipotential a let us define $A : V \times \mathcal{H} \to \mathbb{R}$ by

$$A(v, \mu) := \int_{\Omega} a(\varepsilon(v), \mu) dx, \forall v \in V, \mu \in \mathcal{H}.$$

and note that, due to (H_ϕ), A is well defined and

$$A(v, \mu) \geq \alpha_1 \|v\|_V^2 + \alpha_2 \|\mu\|_{\mathcal{H}}^2, \forall v \in V, \mu \in \mathcal{H}.$$

Moreover,

$$A(u, \sigma) = (\varepsilon^*(\sigma), u)_V \text{ and } A(v, \mu) \geq (\varepsilon^*(\mu), v)_V, \forall v \in V, \ \mu \in \mathcal{H}. \qquad (13.15)$$

Combining the first line of $(\tilde{P})$ and (13.15) we get

$$A(v,\sigma)-A(u,\sigma)+\int_{\Gamma_3}\left[j_1^0(x,u_n;v_n-u_n)+h(x,u_T)j_2^0(x,u_T;v_T-u_T)\right]d\Gamma \geq (f,v-u)_V,$$

(13.16)

for all $v \in \Lambda$.

Let us define now the *set of admissible stress tensors with respect to the displacement* u, to be the following subset of $\mathcal{H}$

$$\Theta_u := \left\{ \mu \in \mathcal{H} \,\middle|\, \begin{matrix} (\varepsilon^*(\mu),v)_V + \int_{\Gamma_3}\left[j_1^0(x,u_n;v_n)+h(x,u_T)j_2^0(x,u_T;v_T)\right]d\Gamma \\[2mm] \geq (f,v)_V, \ \forall v \in \Lambda \end{matrix} \right\}.$$

Let $w \in \Lambda$ be fixed. Choosing $v := u + w \in \Lambda$ in the first inequality of $(\tilde{P})$ shows that $\sigma \in \Theta_u$, hence $\Theta_u \neq \varnothing$. It is easy to check that Θ_u is an unbounded, closed and convex subset of $\mathcal{H}$. Taking into account (13.15) we have

$$A(u,\mu) + \int_{\Gamma_3}\left[j_n^0(x,u_n;u_n)+h(x,u_T)j_2^0(x,u_T;u_T)\right]d\Gamma \geq (f,u)_V, \forall \mu \in \Theta_u,$$

while for $v = 0 \in \Lambda$ we have

$$-A(u,\sigma) + \int_{\Gamma_3}\left[j_n^0(x,u_n;-u_n)+h(x,u_T)j_2^0(x,u_T;-u_T)\right]d\Gamma \geq -(f,u)_V.$$

Adding the above relations, for all $\mu \in \Theta_u$ we have

$$0 \leq A(u,\mu) - A(u,\sigma) + \int_{\Gamma_3}\left[j_1^0(x,u_n;u_n)+j_1^0(x,u_n;-u_n)\right]d\Gamma \quad (13.17)$$

$$+ \int_{\Gamma_3} h(x,u_T)\left(j_2^0(x,u_T;u_T)+j_2^0(x,u_T;-u_T)\right)d\Gamma.$$

On the other hand, Proposition 2.4 and (H_h) ensure that

$$\int_{\Gamma_3}\left[j_1^0(x,u_n;u_n)+j_1^0(x,u_n;-u_n)+h(x,u_T)\left(j_2^0(x,u_T;u_T)+j_2^0(x,u_T;-u_T)\right)\right]d\Gamma \geq 0,$$

(13.18)

as

$$0 = j_1^0(x,u_n;0)+h(x,u_T)j_2^0(x,u_T;0)$$

$$= j_1^0(x,u_n;u_n-u_n)+h(x,u_T)j_2^0(x,u_T;u_T-u_T)$$

$$\leq \left(j_1^0(x,u_n;u_n)+j_1^0(x,u_n;-u_n)\right)+h(x,u_T)\left(j_2^0(x,u_T;u_T)+j_2^0(x,u_T;-u_T)\right).$$

Putting together (13.16)–(13.18) we derive the *variational formulation in terms of bipotentials* of problem (P) which reads as follows:
$\left(\mathcal{P}_{var}^{b}\right)$ Find $u \in \Lambda$ and $\sigma \in \Theta_u$ such that

$$
\begin{cases}
A(v, \sigma) - A(u, \sigma) + \displaystyle\int_{\Gamma_3} h(x, u_T) j_2^0(x, u_T; v_T - u_T) d\Gamma \\
\quad + \displaystyle\int_{\Gamma_3} j_1^0(x, u_n; v_n - u_n) d\Gamma \geq (f, v - u)_V, \ \forall v \in \Lambda, \\
A(u, \mu) - A(u, \sigma) \geq 0, \quad \forall \mu \in \Theta_u.
\end{cases}
$$

Each solution $(u, \sigma) \in \Lambda \times \Theta_u$ of problem $\left(\mathcal{P}_{var}^{b}\right)$ is called a *weak solution* for problem (P).

13.2   The Connection with Classical Variational Formulations

In this section we highlight the connection between the variational formulation in terms of bipotentials and other variational formulations such as the primal and dual variational formulations. As we have seen in the previous section, multiplying the first line of problem (P) by $v - u$, integrating over Ω and then taking the functional extension of the constitutive law, we get a coupled system of inequalities, namely problem $(\tilde{P})$. The primal variational formulation consists in rewriting $(\tilde{P})$ as an inequality which depends only on the displacement field u, while the dual variational formulation consists in rewriting $(\tilde{P})$ in terms of the stress tensor σ. The primal variational formulation can be derived by reasoning in the following way.

The second line of $(\tilde{P})$ implies that $\varepsilon(u) \in \partial \phi^*(\sigma)$ and this can be written equivalently as $\sigma \in \partial \phi(\varepsilon(u))$, hence

$$
\sigma : (\mu - \varepsilon(u)) \leq \phi(\mu) - \phi(\varepsilon(u)), \forall \mu \in \mathbb{S}_m.
$$

For each $v \in \Lambda$, taking $\mu := \varepsilon(v)$ in the previous inequality and integrating over Ω yields

$$
(\varepsilon^*(\sigma), v - u)_V \leq \int_{\Omega} \phi(\varepsilon(v)) - \phi(\varepsilon(u)) dx, \forall v \in \Lambda.
$$

Now, combining the above relation and the first line of $(\tilde{P})$ we get the following problem $\left(\mathcal{P}_{var}^{p}\right)$ Find $u \in \Lambda$ such that for all $v \in \Lambda$

$$
F(v) - F(u) + \int_{\Gamma_3} \left[j_1^0(x, u_n; v_n - u_n) + h(x, u_T) j_2^0(x, u_T; v_T - u_T) \right] d\Gamma \geq (f, v - u)_V,
$$

where $F : V \to \mathbb{R}$ is the convex and lower semicontinuous functional defined by

$$F(v) := \int_{\Omega} \phi(\varepsilon(v)) dx.$$

Problem $(\mathcal{P}^p_{var})$ is called the *primal variational formulation* of problem (P).

Conversely, in order to transform $(\tilde{P})$ into a problem formulated in terms of the stress tensor we reason in the following way. First let us define $G : \mathcal{H} \to \mathbb{R}$ by

$$G(\mu) := \int_{\Omega} \phi^*(\mu) dx,$$

and for a fixed $w \in \Lambda$ let Θ_w be the following subset of $\mathcal{H}$

$$\Theta_w := \left\{ \mu \in \mathcal{H} \; \middle| \; \begin{array}{c} (\varepsilon^*(\mu), v)_V + \int_{\Gamma_3} \left[j_1^0(x, w_n; v_n) + h(x, w_T) j_2^0(x, w_T; v_T) \right] d\Gamma \\ \geq (f, v)_V, \; \forall v \in \Lambda \end{array} \right\}.$$

Let us consider the following inclusion
$(\mathcal{P}^d_w)$ Find $\sigma \in \mathcal{H}$ such that

$$0 \in \partial G(\sigma) + \partial I_{\Theta_w}(\sigma),$$

which we call the *dual variational formulation with respect to* w.

Now, looking at the first line of $(\tilde{P})$ and keeping in mind the above notations, we deduce that $\Theta_u \neq \varnothing$ as $\sigma \in \Theta_u$. Moreover, for each $\mu \in \Theta_u$ we have

$$-(\varepsilon^*(\mu - \sigma), u)_V \leq \int_{\Gamma_3} h(x, u_T) \left(j_2^0(x, u_T; u_T) + j_2^0(x, u_T; -u_T) \right) d\Gamma$$

$$+ \int_{\Gamma_3} j_1^0(x, u_n; u_n) + j_1^0(x, u_n; -u_n) d\Gamma,$$

which combined with the second line of $(\tilde{P})$ leads to

$$G(\mu) - G(\sigma) \geq - \int_{\Gamma_3} h(x, u_T) \left(j_2^0(x, u_T; u_T) + j_2^0(x, u_T; -u_T) \right) d\Gamma \qquad (13.19)$$

$$- \int_{\Gamma_3} \left[j_1^0(x, u_n; u_n) + j_1^0(x, u_n; -u_n) \right] d\Gamma,$$

for all $\mu \in \Theta_u$.
A simple computation shows that any solution of $(\mathcal{P}^d_u)$ will also solve (13.19).

A particular case of interest regarding problem $(\mathcal{P}_w^d)$ is if the set Θ_w does not actually depend on w. In this case problem $(\mathcal{P}_w^d)$ will be simply denoted $(\mathcal{P}^d)$ and will be called *the dual variational formulation of problem* (P). For example, this case is encountered when the functions j_1 and j_2 are convex and positive homogeneous, as it is the case of examples $(a) - (c)$ presented in Sect. 13.1.

In the above particular case, problem $\left(\tilde{P}\right)$ reduces to the following system of variational inequalities
$(\tilde{P}')$ Find $u \in \Lambda$ and $\sigma \in \mathcal{H}$ such that

$$
\begin{cases}
(\varepsilon^*(\sigma), v - u)_V + H(v) - H(u) \geq (f, v - u)_V, \ \forall v \in \Lambda \\
- (\varepsilon(u), \mu - \sigma)_{\mathcal{H}} + G(\mu) - G(\sigma) \geq 0, \qquad \qquad \forall \mu \in \mathcal{H},
\end{cases}
$$

where $H := j \circ T$, $j : L^2(\Gamma_3; \mathbb{R}^m) \to \mathbb{R}$ is defined by

$$
j(y) := \int_{\Gamma_3} j_1(x, y_n) + j_2(x, y_T) d\Gamma,
$$

and $T : V \to L^2(\Gamma_3; \mathbb{R}^m)$ is given by $Tv := [(\gamma \circ i)(v)]|_{\Gamma_3}$, with $i : V \to H_1$ being the embedding operator and $\gamma : H_1 \to H^{1/2}(\Gamma; \mathbb{R}^m)$ being the trace operator. On the other hand, for each $w \in \Lambda$,

$$
\Theta_w = \Theta := \left\{ \mu \in \mathcal{H} : (\varepsilon^*(\mu), v)_V + H(v) \geq (f, v)_V, \ \forall v \in \Lambda \right\},
$$

and thus by taking $v := 2u$ and $v := 0$ in the first line of $(\tilde{P}')$ we get

$$
(\varepsilon^*(\sigma), u)_V + H(u) = (f, u)_V,
$$

hence

$$
-(\varepsilon(u), \mu - \sigma)_{\mathcal{H}} \leq 0, \forall \mu \in \Theta.
$$

Combining this and the second line of $(\tilde{P}')$ we get

$$
G(\mu) - G(\sigma) \geq 0, \forall \mu \in \Theta,
$$

which can be formulated equivalently as
$(\mathcal{P}^d)$ Find $\sigma \in \mathcal{H}$ such that

$$
0 \in \partial G(\sigma) + \partial I_\Theta(\sigma).
$$

The following proposition points out the connection between the variational formulations presented above.

Proposition 13.1 *A pair* $(u, \sigma) \in V \times \mathcal{H}$ *is a solution for* $(\mathcal{P}^b_{var})$ *if and only if u solves* $(\mathcal{P}^p_{var})$ *and* σ *solves* $(\mathcal{P}^d_u)$.

Proof " $\Rightarrow$ " Let $(u, \sigma) \in V \times \mathcal{H}$ be a solution for $(\mathcal{P}^b_{var})$. Then $u \in \Lambda$, $\sigma \in \Theta_u$ and

$$\begin{cases} A(v, \sigma) - A(u, \sigma) + \displaystyle\int_{\Gamma_3} h(x, u_T) j_2^0(x, u_T; v_T - u_T) d\Gamma \\ + \displaystyle\int_{\Gamma_3} j_1^0(x, u_n; v_n - u_n) d\Gamma \geq (f, v - u)_V, \\ A(u, \mu) - A(u, \sigma) \geq 0, \end{cases}$$

for all $(v, \mu) \in \Lambda \times \Theta_u$.

Taking into account the way A, F and G were defined we get

$$A(v, \sigma) - A(u, \sigma) = F(v) - F(u), \forall v \in V, \tag{13.20}$$

and

$$A(u, \mu) - A(u, \sigma) = G(\mu) - G(\sigma), \forall \mu \in \mathcal{H}, \tag{13.21}$$

which shows that u is a solution for $(\mathcal{P}^p_{var})$ and

$$\big[G(\mu) + I_{\Theta_u}(\mu) \big] - \big[G(\sigma) + I_{\Theta_u}(\sigma) \big] \geq 0, \forall \mu \in \mathcal{H}.$$

The last inequality can be written equivalently as

$$0 \in \partial(G + I_{\Theta_u})(\sigma).$$

On the other hand, applying Proposition 1.3.10 in [2] we deduce that

$$\partial(G + I_{\Theta_u})(\sigma) = \partial G(\sigma) + \partial I_{\Theta_u}(\sigma),$$

hence σ solves $(\mathcal{P}^d_u)$.

" $\Leftarrow$ " Assume now that $u \in V$ is a solution of $(\mathcal{P}^p_{var})$ and $\sigma \in \mathcal{H}$ solves $(\mathcal{P}^d_{var})$. The fact that σ solves $(\mathcal{P}^d_u)$ implies that $D(\partial I_{\Theta_u}) \neq \varnothing$ and

$$\sigma \in D(\partial I_{\Theta_u}).$$

On the other hand, it is well known that

$$D(\partial I_{\Theta_u}) \subseteq D(I_{\Theta_u}) = \Theta_u,$$

hence $\sigma \in \Theta_u$. Moreover,

$$\begin{cases} F(v) - F(u) + \displaystyle\int_{\Gamma_3} h(x, u_T) j_2^0(x, u_T; v_T - u_T) \mathrm{d}\Gamma \\[2mm] + \displaystyle\int_{\Gamma_3} j_1^0(x, u_n; v_n - u_n) \mathrm{d}\Gamma \geq (f, v - u)_V, \\[2mm] G(\mu) - G(\sigma) \geq 0, \end{cases}$$

for all $(v, \mu) \in \Lambda \times \Theta_u$, which combined with (13.20) and (13.21) shows that (u, σ) is a solution for problem $\left(\mathcal{P}_{var}^b\right)$.

$\square$

13.3   Weak Solvability of the Model

The main result of this chapter is given by the following theorem.

Theorem 13.1 ([1, Theorem 1]) *Assume* (H_C), (H_f), (H_h), (H_{j_1}), (H_{j_2}) *and* (H_ϕ) *hold. Then problem* $\left(\mathcal{P}_{var}^b\right)$ *has at least one solution.*

Before proving the main result we need the following Aubin-Clarke type result concerning the Clarke subdifferential of integral functions. Let us consider the function $j : L^2(\Gamma_3; \mathbb{R}^m) \times L^2(\Gamma_3; \mathbb{R}^m) \to \mathbb{R}$ defined by

$$j(y, z) := \int_{\Gamma_3} j_1(x, z_n) + h(x, y_T) j_2(x, z_T) \, \mathrm{d}\Gamma. \tag{13.22}$$

Lemma 13.1 *Assume* (H_h), $\left(H_{j_1}\right)$ *and* $\left(H_{j_2}\right)$ *are fulfilled. Then, for each* $y \in L^2(\Gamma_3; \mathbb{R}^m)$, *the function* $z \mapsto j(y, z)$ *is Lipschitz continuous and*

$$j_{,2}^0(y, z; \bar{z}) \leq \int_{\Gamma_3} j_1^0(x, z_n; \bar{z}_n) + h(x, y_T) j_2^0(x, z_T; \bar{z}_T) \, \mathrm{d}\Gamma. \tag{13.23}$$

Proof Let $y, z^1, z^2 \in L^2(\Gamma_3; \mathbb{R}^m)$ be fixed. Then

$$\left| j\left(y, z^1\right) - j\left(y, z^2\right) \right| = \left| \int_{\Gamma_3} j_1\left(x, z_n^1\right) - j_1\left(x, z_n^2\right) + h(x, y_T)\left(j_2\left(x, z_T^1\right) - j_2\left(x, z_T^2\right)\right) \mathrm{d}\Gamma \right|$$

$$\leq \int_{\Gamma_3} \left| j_1\left(x, z_n^1\right) - j_1\left(x, z_n^2\right) \right| \mathrm{d}\Gamma + h_0 \int_{\Gamma_3} \left| j_2\left(x, z_T^1\right) - j_2\left(x, z_T^2\right) \right| \mathrm{d}\Gamma.$$

The equality

$$|z|^2 = z \cdot z = z_n z_n + z_T \cdot z_T = |z_n|^2 + |z_T|^2,$$

shows that if $z \in L^2(\Gamma_3; \mathbb{R}^m)$, then $z_n \in L^2(\Gamma_3)$ and $z_T \in L^2(\Gamma_3; \mathbb{R}^m)$ and

$$\|z_n\|_{L^2(\Gamma_3)}, \|z_T\|_{L^2(\Gamma_3; \mathbb{R}^m)} \le \|z\|_{L^2(\Gamma_3; \mathbb{R}^m)}.$$

Thus, from the hypotheses and Hölder's inequality we get

$$\left| j\left(y, z^1\right) - j\left(y, z^2\right) \right| \le \|p\|_{L^2(\Gamma_3)} \left\| z_n^1 - z_n^2 \right\|_{L^2(\Gamma_3)} + h_0 \|q\|_{L^2(\Gamma_3)} \left\| z_T^1 - z_T^2 \right\|_{L^2(\Gamma_3; \mathbb{R}^m)}$$

$$\le \left(\|p\|_{L^2(\Gamma_3)} + h_0 \|q\|_{L^2(\Gamma_3)} \right) \left\| z^1 - z^2 \right\|_{L^2(\Gamma_3; \mathbb{R}^m)},$$

which shows that j is Lipschitz continuous.

In order to prove (13.23) we use Fatou's lemma and the fact that the convergence in $L^2(\Gamma_3; \mathbb{R}^m)$ implies, up to a subsequence, the pointwise convergence a.e. on Γ_3

$$j_{,2}^0(y, z; \bar{z}) = \limsup_{\substack{u \to z \\ \lambda \downarrow 0}} \frac{j(y, u + \lambda \bar{z}) - j(y, u)}{\lambda}$$

$$= \limsup_{\substack{u \to z \\ \lambda \downarrow 0}} \left(\int_{\Gamma_3} \frac{j_1(x, u_n + \lambda \bar{z}_n) - j_1(x, u_n)}{\lambda} d\Gamma \right.$$

$$\left. + \int_{\Gamma_3} h(x, y_T) \frac{j_2(x, u_T + \lambda \bar{z}_T) - j_2(x, u_T)}{\lambda} d\Gamma \right)$$

$$\le \int_{\Gamma_3} \limsup_{\substack{u \to z \\ \lambda \downarrow 0}} \frac{j_1(x, u_n + \lambda \bar{z}_n) - j_1(x, u_n)}{\lambda} d\Gamma$$

$$+ \int_{\Gamma_3} h(x, y_T) \limsup_{\substack{u \to z \\ \lambda \downarrow 0}} \frac{j_2(x, u_T + \lambda \bar{z}_T) - j_2(x, u_T)}{\lambda} d\Gamma$$

$$\le \int_{\Gamma_3} j_1^0(x, z_n; \bar{z}_n) + h(x, y_T) j_2^0(x, z_T; \bar{z}_T) d\Gamma.$$

$\square$

In order to prove Theorem 13.1 we consider the following system of nonlinear hemivariational inequalities.

$(\mathcal{S}_{K_1, K_2})$ Find $(u, \sigma) \in K_1 \times K_2$ such that

$$
\begin{cases}
\psi_1(u, \sigma, v) + J_{,1}^0(Tu, S\sigma; Tv - Tu) \geq (F_1(u, \sigma), v - u)_{X_1}, \quad \forall v \in K_1, \\
\\
\psi_2(u, \sigma, \mu) + J_{,2}^0(Tu, S\sigma; S\mu - S\sigma) \geq (F_2(u, \sigma), \mu - \sigma)_{X_2}, \quad \forall \mu \in K_2,
\end{cases}
$$

where

- $X_1 := V$, $X_2 := \mathcal{H}$, $K_i \subset X_i$ is closed and convex ($i = 1, 2$), $Y_1 := L^2(\Gamma_3; \mathbb{R}^m)$, $Y_2 := \{0\}$;
- $\psi_1 : X_1 \times X_2 \times X_1 \to \mathbb{R}$ is defined by $\psi_1(u, \sigma, v) := A(v, \sigma) - A(u, \sigma)$;
- $\psi_2 : X_1 \times X_2 \times X_2 \to \mathbb{R}$ is defined by $\psi_2(u, \sigma, \mu) := A(u, \mu) - A(u, \sigma)$;
- $T : X_1 \to Y_1$ is defined by $Tv := [(\gamma \circ i)(v)]|_{\Gamma_3}$, with $i : V \to H_1$ the embedding operator and $\gamma : H_1 \to H^{1/2}(\Gamma; \mathbb{R}^m)$ is the trace operator;
- $S : X_2 \to Y_2$ is defined by $S\tau := 0$, for all $\tau \in X_2$;
- $J : Y_1 \times Y_2 \to \mathbb{R}$ is defined by $J\left(y^1, y^2\right) = j\left(y^0, y^1\right)$, where $j : L^2(\Gamma_3; \mathbb{R}^m) \times L^2(\Gamma_3; \mathbb{R}^m) \to \mathbb{R}$ is as in (13.22) and y^0 is a fixed element of $L^2(\Gamma_3; \mathbb{R}^m)$;
- $F_1 : X_1 \times X_2 \to X_1$ is defined by $F_1(v, \mu) := f$;
- $F_2 : X_1 \times X_2 \to X_2$ is defined by $F_2(v, \mu) := 0$.

Lemma 13.2 *Assume* $(\mathbf{H_h})$, $(\mathbf{H_{j_1}})$, $(\mathbf{H_{j_2}})$ *and* $(\mathbf{H_\phi})$ *are fulfilled. Then the following statements hold:*

(i) $\psi_1(u, \sigma, u) = 0$ *and* $\psi_2(u, \sigma, \sigma) = 0$, *for all* $(u, \sigma) \in X_1 \times X_2$;

(ii) *for each* $v \in X_1$ *and each* $\mu \in X_2$ *the maps* $(u, \sigma) \mapsto \psi_1(u, \sigma, v)$ *and* $(u, \sigma) \mapsto \psi_2(u, \sigma, \mu)$ *are weakly upper semicontinuous;*

(iii) *for each* $(u, \sigma) \in X_1 \times X_2$ *the maps* $v \mapsto \psi_1(u, \sigma, v)$ *and* $\mu \mapsto \psi_2(u, \sigma, \mu)$ *are convex;*

(iv) $\liminf_{k \to +\infty}(F_1(u_k, \sigma_k), v - u_k)_{X_1} \geq (F_1(u, \sigma), v - u)_{X_1}$ *and* $\liminf_{k \to +\infty}(F_2(u_k, \sigma_k), \mu - \sigma_k)_{X_2} \geq (F_2(u, \sigma), \mu - \sigma)_{X_2}$ *whenever* $(u_k, \sigma_k) \rightharpoonup (u, \sigma)$ *as* $k \to +\infty$;

(v) *there exists* $c : \mathbb{R}_+ \to \mathbb{R}_+$ *with the property* $\lim_{t \to +\infty} c(t) = +\infty$ *such that*

$$
\psi_1(u, \sigma, 0) + \psi_2(u, \sigma, 0) \leq -c\left(\sqrt{\|u\|_{X_1}^2 + \|\sigma\|_{X_2}^2}\right)\sqrt{\|u\|_{X_1}^2 + \|\sigma\|_{X_2}^2},
$$

for all $(u, \sigma) \in X_1 \times X_2$;

(vi) *The function* $J : Y_1 \times Y_2 \to \mathbb{R}$ *is Lipschitz with respect to each variable. Moreover, for all* $\left(y^1, y^2\right), \left(z^1, z^2\right) \in Y_1 \times Y_2$ *we have*

$$
J_{,1}^0\left(y^1, y^2; z^1\right) = j_{,2}^0\left(y^0, y^1; z^1\right)
$$

and

$$J_{,2}^0 \left(y^1, y^2; z^2\right) = 0;$$

(*vii*) *There exists M > 0 such that*

$$J_{,1}^0 \left(y^1, y^2; -y^1\right) \leq M \left\|y^1\right\|_{Y_1}, \text{ for all } \left(y^1, y^2\right) \in Y_1 \times Y_2;$$

(*viii*) *there exist $m_i > 0$, $i = 1, 2$, such that $\|F_i(u, \sigma)\|_{X_i} \leq m_i$, for all $(u, \sigma) \in X_1 \times X_2$.*

Proof

(*i*) Trivial.

(*ii*) Let $v \in X_1$ be fixed and let $\{(u_k, \sigma_k)\}_k$ be a sequence such that (u_k, σ_k) converges weakly in $X_1 \times X_2$ to (u, σ) as $k \to +\infty$. Using the fact that L is linear, ϕ is convex and lower semicontinuous, hence weakly lower semicontinuous and Fatou's lemma, we have

$$\limsup_{k \to +\infty} \psi_1(u_k, \sigma_k, v) = \limsup_{k \to +\infty} [A(v, \sigma_k) - A(u_k, \sigma_k)]$$

$$= \limsup_{k \to +\infty} \int_\Omega \phi(\varepsilon(v)) - \phi(\varepsilon(u)_k)dx$$

$$\leq \int_\Omega \phi(\varepsilon(v))dx - \int_\Omega \liminf_{k \to +\infty} \phi(\varepsilon(u)_k)dx$$

$$\leq \int_\Omega \phi(\varepsilon(v)) - \phi(\varepsilon(u)) + \phi^*(\sigma) - \phi^*(\sigma)dx$$

$$= A(v, \sigma) - A(u, \sigma)$$

$$= \psi_1(u, \sigma, v),$$

which show that the map $(u, \sigma) \mapsto \psi_1(u, \sigma, v)$ is weakly upper semicontinuous.
In a similar fashion we prove that for $\mu \in X_2$ fixed, the map $(u, \sigma) \mapsto \psi_2(u, \sigma, \mu)$ is weakly upper semicontinuous.

(*iii*) Follows from the convexity of ϕ and ϕ^*;

(*iv*) Let $\{(u_k, \sigma_k)\}$ be a sequence which converges weakly to (u, σ) in $X_1 \times X_2$ as $k \to +\infty$. Then $u_k \to u$ in X_1 as $k \to +\infty$ and

$$\liminf_{k \to +\infty} (F_1(u_k, \sigma_k), v - u_k)_{X_1} = \liminf_{k \to +\infty}(f, v - u_k)_{X_1} = (f, v - u)_{X_1},$$

and

$$\liminf_{k \to +\infty} (F_2(u_k, \sigma_k), \mu - \sigma_k)_{X_2} = 0 = (F_2(u, \sigma), \mu - \sigma)_{X_2}.$$

(v) Let $(u, \sigma) \in X_1 \times X_2$. Using (H_ϕ) we get the following estimates

$$\psi_1(u, \sigma, 0) + \psi_2(u, \sigma, 0) = A(0, \sigma) - A(u, \sigma) + A(u, 0) - A(u, \sigma)$$

$$= \int_\Omega \phi(0) + \phi^*(0) - (\phi(\varepsilon(u)) + \phi^*(\sigma)) dx$$

$$\leq \tilde{c} - \min\{\alpha_1, \alpha_2\} \left(\|u\|_{X_1}^2 + \|\sigma\|_{X_2}^2 \right).$$

Choosing $c(t) := b_0 t$, with $b_0 > 0$ a suitable constant, we get the desired inequality.

(vi) It follows directly from Lemma 13.1 and the definition of J.

(vii) From (vi) and Lemma 13.1 we deduce

$$J_{,1}^0 \left(y^1, y^2; -y^1 \right) = j_{,2}^0 \left(y^0, y^1; -y^1 \right)$$

$$\leq \int_{\Gamma_3} j_\nu^0 \left(x, y_n^1; -y_n^1 \right) + h \left(x, y_T^0 \right) j_2^0 \left(x, y_T^1; -y_T^1 \right) d\Gamma$$

On the other hand, assumptions (H_{j_1}) and (H_{j_1}) imply

$$j_n^0(x, t_1; t_2) \leq p(x)|t_2|, \forall t_1, t_2 \in \mathbb{R},$$

and

$$j_2^0(x, \zeta_1; \zeta_2) \leq q(x)|\zeta_2|, \forall \zeta_1, \zeta_2 \in \mathbb{R}^m.$$

Thus, invoking Hölder's inequality we get

$$J_{,1}^0 \left(y^1, y^2; -y^1 \right) \leq \left(\|p\|_{L^2(\Gamma_3)} + h_0 \|q\|_{L^2(\Gamma_3; \mathbb{R}^m)} \right) \left\| y^1 \right\|_{L^2(\Gamma_3; \mathbb{R}^m)}.$$

(viii) Trivial. □

Proof of Theorem 13.1 The proof will be carried out in three steps as follows.

Step 1. *Let $K_1 \subset X_1$ and $K_2 \subset X_2$ be closed and convex sets. Then (S_{K_1, K_2}) admits at least one solution.*

This will be done by applying a slightly modified version of Corollary 8.2. Lemma 13.2 ensures that all the conditions of the aforementioned corollary are satisfied except the regularity of J. We point out the fact that in our case this condition needs not be imposed because the only reason it is imposed there is to ensure the following inequality

$$J^0\left(y^1, y^2; z^1, z^2\right) \le J^0_{,1}\left(y^1, y^2; z^1\right) + J^0_{,2}\left(y^1, y^2; z^2\right)$$

which in this chapter is automatically fulfilled because J does not depend on the second variable and the following equalities take place

$$J^0\left(y^1, y^2; z^1, z^2\right) = J^0_{,1}\left(y^1, y^2; z^1\right)$$

and

$$J^0_{,2}\left(y^1, y^2; z^2\right) = 0,$$

and this completes the first step.

Step 2. *Let* $K^1_1, K^2_1 \subset X_1$ *and* $K^1_2, K^2_2 \subset X_2$ *be closed and convex sets and let* $\left(u^1, \sigma^1\right)$ *and* $\left(u^2, \sigma^2\right)$ *be solutions for* $\left(S_{K^1_1, K^1_2}\right)$ *and* $\left(S_{K^2_1, K^2_2}\right)$, *respectively. Then* $\left(u^1, \sigma^2\right)$ *solves* $\left(S_{K^1_1, K^2_2}\right)$ *and* $\left(u^2, \sigma^1\right)$ *solves* $\left(S_{K^2_1, K^1_2}\right)$.

The fact that $\left(u^1, \sigma^1\right)$ solves $\left(S_{K^1_1, K^1_2}\right)$ means

$$\begin{cases} \psi_1(u^1, \sigma^1, v) + J^0_{,1}(Tu^1, S\sigma^1; Tv - Tu^1) \ge (F_1(u^1, \sigma^1), v - u^1), \\ \\ \psi_2(u^1, \sigma^1, \mu) + J^0_{,2}(Tu^1, S\sigma^1; S\mu - S\sigma^1) \ge (F_2(u^1, \sigma^1), \mu - \sigma^1), \end{cases}$$
$$(13.24)$$

for all $(v, \mu) \in K^1_1 \times K^1_2$, while the fact that $\left(u^2, \sigma^2\right)$ solves $\left(S_{K^2_1, K^2_2}\right)$ shows

$$\begin{cases} \psi_1(u^2, \sigma^2, v) + J^0_{,1}(Tu^2, S\sigma^2; Tv - Tu^2) \ge (F_1(u^2, \sigma^2), v - u^2), \\ \\ \psi_2(u^2, \sigma^2, \mu) + J^0_{,2}(Tu^2, S\sigma^2; S\mu - S\sigma^2) \ge (F_2(u^2, \sigma^2), \mu - \sigma^2), \end{cases}$$
$$(13.25)$$

for all $(v, \mu) \in K^2_1 \times K^2_2$. Putting together the first line of (13.24) and the second line of (13.25) we get

$$\begin{cases} \psi_1(u^1, \sigma^1, v) + J^0_{,1}(Tu^1, S\sigma^1; Tv - Tu^1) \ge (F_1(u^1, \sigma^1), v - u^1), \\ \\ \psi_2(u^2, \sigma^2, \mu) + J^0_{,2}(Tu^2, S\sigma^2; S\mu - S\sigma^2) \ge (F_2(u^2, \sigma^2), \mu - \sigma^2), \end{cases}$$
$$(13.26)$$

for all $(v, \mu) \in K_1^1 \times K_2^2$. On the other hand, keeping in mind the way $\psi_1, \psi_2, J, F_1, F_2$ were defined is it easy to check that for any $(v, \mu) \in K_1^1 \times K_2^2$ the following equalities hold

$$\psi_1(u^1, \sigma^1, v) = \psi_1(u^1, \sigma^2, v) \text{ and } \psi_2(u^2, \sigma^2, \mu) = \psi_2(u^1, \sigma^2, \mu),$$

$$J_{,1}^0(Tu^1, S\sigma^1; Tv - Tu^1) = J_{,1}^0(Tu^1, S\sigma^2; Tv - Tu^1)$$

$$J_{,2}^0(Tu^2, S\sigma^2; S\mu - S\sigma^2) = J_{,2}^0(Tu^1, S\sigma^2; S\mu - S\sigma^1)$$

$$F_1(u^1, \sigma^1) = F_1(u^1, \sigma^2) \text{ and } F_2(u^2, \sigma^2) = F_2(u^1, \sigma^2).$$

Using these equalities and (13.26) we obtain

$$\begin{cases} \psi_1(u^1, \sigma^2, v) + J_{,1}^0(Tu^1, S\sigma^2; Tv - Tu^1) \geq (F_1(u^1, \sigma^2), v - u^1)_{X_1}, \\ \psi_2(u^1, \sigma^2, \mu) + J_{,2}^0(Tu^1, S\sigma^2; S\mu - S\sigma^2) \geq (F_2(u^1, \sigma^2), \mu - \sigma^2)_{X_2}, \end{cases}$$

hence (u^1, σ^2) solves $\left(S_{K_1^1, K_2^2}\right)$. In a similar way we can prove that (u^2, σ^1) solves $\left(S_{K_1^2, K_2^1}\right)$.

Step 3. *There exist $u \in \Lambda$ and $\sigma \in \Theta_u$ such that (u, σ) solves $\left(\mathcal{P}_{var}^b\right)$.*

Let us choose $K_1^1 := \Lambda$ and $K_2^1 := X_2$. According to Step 1 there exists a pair (u^1, σ^1) which solves $\left(S_{K_1^1, K_2^1}\right)$. Next, we choose $K_1^2 := \Lambda$ and $K_2^2 := \Theta_{u^1}$ and use again Step 1 to deduce that there exists a pair (u^2, σ^2) which solves $\left(S_{K_1^2, K_2^2}\right)$. Then, according to Step 2, the pair (u^1, σ^2) will solve $\left(S_{K_1^1, K_2^2}\right)$. Invoking the way $\psi_1, \psi_2, J, F_1, F_2, K_1^1, K_2^2$ were defined, it is clear that the pair $(u, \sigma) := (u^1, \sigma^2) \in \Lambda \times \Theta_u$ is a solution of the system

$$\begin{cases} A(v, \sigma) - A(u, \sigma) + j_{,2}^0(y^0, Tu; Tv - Tu) \geq (f, v - u)_V, & \forall v \in \Lambda, \\ A(u, \mu) - A(u, \sigma) \geq 0, & \forall \mu \in \Theta_u, \end{cases}$$

for all $y^0 \in L^2(\Gamma_3; \mathbb{R}^m)$, since y^0 was arbitrary fixed. Choosing $y^0 := Tu$ an taking into account (13.23) we conclude that $(u, \sigma) \in \Lambda \times \Theta_u$ solves $\left(\mathcal{P}_{var}^b\right)$. □

References

1. N. Costea, M. Csirik, C. Varga, Weak solvability via bipotential method for contact models with nonmonotone boundary conditions. Z. Angew. Math. Phys. **66**, 2787–2806 (2015)
2. D. Goeleven, D. Motreanu, Y. Dumont, M. Rochdi, *Variational and Hemivariational Inequalities: Theory, Methods and Applications, Volume I: Unilateral Analysis and Unilateral Mechanics* (Kluwer Academic Publishers, Dordrecht, 2003)
3. W. Han, M. Sofonea, *Quasistatic Contact Problems in Viscoelasticity and Viscoplasticity*, vol. 30. Studies in Advanced Mathematics (International Press, Somerville, 2002)
4. A. Matei, A variational approach via bipotentials for a class of frictional contact problems. Acta Appl. Math. **134**, 45–59 (2014)
5. Z. Naniewicz, On some nonmonotone subdifferential boundary conditions in elastostatics. Ingenieur-Archiv **60**, 31–40 (1989)
6. Z. Naniewicz, P.D. Panagiotopoulos, *Mathematical Theory of Hemivariational Inequalities and Applications* (Marcel Dekker, New York, 1995)
7. P.D. Panagiotopoulos, *Inequality Problems in Mechanics and Applications. Convex and Nonconvex Energy Functions* (Birkhäuser, Basel, 1985)
8. P.D. Panagiotopoulos, *Hemivariational Inequalities. Applications in Mechanics and Engineering* (Springer, Berlin, 1993)
9. E. Zeidler, *Nonlinear Functional Analysis and its Applications IV: Applications to Mathematical Physics* (Springer, New York, 1988)

Functional Analysis

A.1 The Hahn-Banach Theorems

Theorem A.1 (Hahn-Banach, Analytic Form) *Let X be a linear space and $p : X \to \mathbb{R}$ be a Minkowski functional, i.e., a function satisfying*

$$p(\lambda u) = \lambda p(u), \quad \forall u \in X, \ \forall \lambda > 0,$$

and

$$p(u + v) \leq p(u) + p(v), \quad \forall u, v \in X.$$

Let $Y \subset X$ be a linear subspace and assume $\zeta : Y \to \mathbb{R}$ is a linear functional dominated by p, that is,

$$\zeta(u) \leq p(u), \quad \forall u \in Y.$$

Then there exists a (not necessarily unique) linear functional $\xi : X \to \mathbb{R}$ that extends ζ, i.e., $\xi(u) = \zeta(u)$, $\forall u \in Y$, and it is dominated by p, i.e.,

$$\xi(u) \leq p(u), \quad \forall u \in X. \tag{A.1}$$

Now we give some simple applications of Theorem A.1 for normed vector spaces.

© The Author(s), under exclusive license to Springer Nature Switzerland AG 2021
N. Costea et al., *Variational and Monotonicity Methods in Nonsmooth Analysis*,
Frontiers in Mathematics, https://doi.org/10.1007/978-3-030-81671-1

We denote by X^* the *dual space of* X, that is, the space of all continuous linear functionals on X. The *dual norm on* X^* is defined by

$$\|\zeta\|_* := \sup_{\substack{u \in X \\ \|u\| \leq 1}} |\zeta(u)| = \sup_{\substack{u \in X \\ \|u\| \leq 1}} \zeta(u). \tag{A.2}$$

When there is no danger of confusion we shall simply write $\|\zeta\|$ instead $\|\zeta\|_*$. Given $\zeta \in X^*$ and $u \in X$ we shall often write, $\langle \zeta, u \rangle$ instead of $\zeta(u)$; we say that $\langle \cdot, \cdot \rangle$ is the *the duality pairing* for X^* and X. It is well known that X^* is a Banach space, following by the fact that $\mathbb{R}$ is complete.

Corollary A.1 *Let $Y \subset X$ be a linear subspace. If $\zeta : Y \to \mathbb{R}$ is a continuous linear functional, then there exists $\xi \in X^*$ such that ξ extends ζ and*

$$\|\xi\|_{X^*} = \sup_{\substack{u \in Y \\ \|u\| \leq 1}} |\zeta(u)| = \|\zeta\|_{Y^*}. \tag{A.3}$$

Corollary A.2 *For every $u \in X$ we have*

$$\|u\| = \sup_{\substack{\zeta \in X^* \\ \|\zeta\| \leq 1}} |\langle \zeta, u \rangle| = \max_{\substack{\zeta \in X^* \\ \|\zeta\| \leq 1}} |\langle \zeta, u \rangle|. \tag{A.4}$$

Definition A.1 An *affine hyperplane* is a subset $\mathcal{H}$ of X of the form

$$\mathcal{H} := \{u \in X : \zeta(u) = \alpha\},$$

where ζ is a linear functional that does not vanish identically and $\alpha \in \mathbb{R}$ is a given constant. We write $\mathcal{H} : [\zeta = \alpha]$ and say that $\zeta = \alpha$ is the equation of $\mathcal{H}$.

In the previous definition we do not assume that ζ is continuous, as it is known that in every infinite-dimensional normed space there exist discontinuous linear functionals, see, e.g., Brezis [2, Exercise 1.5].

Proposition A.1 *The hyperplane $\mathcal{H} : [\zeta = \alpha]$ is closed if and only if ζ is continuous.*

Definition A.2 Let A and B be two subsets of X. We say that the hyperplane $\mathcal{H} : [\zeta = \alpha]$ *separates* A and B if

$$\zeta(u) \leq \alpha \leq \zeta(v), \quad \forall u \in A, \forall v \in B.$$

We say that $\mathcal{H}$ *strictly separates* A and B if there exists some $\varepsilon > 0$ such that

$$\zeta(u) + \varepsilon \leq \alpha \leq \zeta(v) - \varepsilon, \quad \forall u \in A, \forall v \in B.$$

Theorem A.2 (Hahn-Banach, Weak Separation Theorem) *Let X be a n.v.s. and $A, B \subset X$ be two nonempty convex subsets such that $A \cap B = \varnothing$. Assume that one of them is open. Then there exists a closed hyperplane that separates A and B.*

Theorem A.3 (Hahn-Banach, Strong Separation Theorem) *Let X be n.v.s. and $A, B \subset X$ be two nonempty convex subsets such that $A \cap B = \varnothing$. Assume that A is closed and B is compact. Then there exists a closed hyperplane that strictly separates A and B.*

A.2   Weak Topologies

In the sequel we briefly present the weak topology and weak*-topology of dual, respectively. To this end, assume X is a set and $\{Y_i\}_{i \in I}$ a collection of topological spaces. Given a collection of maps $\{\phi_i\}_{i \in I}$, with $\phi_i : X \to Y_i$, we consider the following problem:

A. Construct a topology on X that makes all the maps $\{\phi_i\}_{i \in I}$ continuous. If possible, find a topology τ_0 that is the *most economical* in the sense that it has the *fewest open sets*.

There is always a unique *smallest topology* τ_0 on X for which every map ϕ_i is continuous. It is called the *coarsest* or *weakest* topology associated to the collection $\{\phi_i\}_{i \in I}$. If $O_i \subset Y_i$ is any open set, then $\phi_i^{-1}(O_i)$ is *necessarily* an open set in τ_0. As O_i runs through the family of open sets of Y_i and i runs through I we obtain a family of subsets of X, each of which must be open in the topology τ_0. Let us denote this family by $\{U_j\}_{j \in J}$. Of course this need not to be a topology. Therefore, we are led the following problem:

B. Given a set X and a family $\{U_j\}_{j \in J}$ of subsets in X, construct the coarset topology τ_0 on X in which U_j is open for all $j \in J$.

In other words, we must find the *"smallest"* family $\mathcal{F}$ of subsets of X that is *stable* by *finite intersections* and *arbitrary unions*, and with the property that $U_j \in \mathcal{F}$, for every $j \in J$.

The construction undergoes the following steps. First, consider finite intersections of sets in $\{U_j\}_{j \in J}$, i.e., $\bigcap_{j \in J_0} U_j$ where $J_0 \subset J$ is finite. In this way we obtain a new family, called $\mathcal{G}$, of subsets of X which includes $\{U_j\}_{j \in J}$ and which is stable under finite intersections. Next, we consider the family $\mathcal{F}$ obtained by forming arbitrary unions of elements from $\mathcal{G}$. Thus, the family $\mathcal{F}$ is stable under finite intersections and arbitrary unions.

Therefore we find the open sets of the topology τ_0 in the following way. First we consider $\bigcap_{finite} \phi_i^{-1}(O_i)$ and then $\bigcup_{arbitrary}$. It follows that for every $u \in X$, we obtain a basis of neighborhoods of u for the topology τ_0 by considering sets of the form $\bigcap_{finite} \phi_i^{-1}(V_i)$, where V_i is a neighborhood of $\phi_i(u)$ in Y_i.

In the following we equip X with the topology τ_0 that is the weakest topology associated to the collection $\{\phi_i\}_{i \in I}$. We have the following simple properties of the topology τ_0.

Proposition A.2 *Let $\{u_n\}$ be a sequence in X. Then $u_n \to u$ in τ_0 if and only if $\phi_i(u_n) \to \phi_i(u)$ for every $i \in I$.*

Proposition A.3 *Let Z be a topological space and let $\Phi : Z \to X$ be a function. Then Φ is continuous if and only $\phi_i \circ \Phi$ is continuous from Z into Y_i for every $i \in I$.*

We are now in position to introduce the *weak topology* in a Banach space X and its dual X^* and present some basic properties. For this let X be a Banach space and let $\zeta \in X^*$. We denote by $\phi_\zeta : X \to \mathbb{R}$ the linear functional $\phi_\zeta(u) := \langle \zeta, u \rangle$. As ζ runs through X^* we obtain a collection $\{\phi_\zeta\}_{\zeta \in X^*}$ of maps from X into $\mathbb{R}$.

Definition A.3 The *weak topology* on X, denoted by τ_w, is the coarsest topology associated to the collection $\{\phi_\zeta\}_{\zeta \in X^*}$, with $Y_i := \mathbb{R}$ and $I := X^*$.

We shall denote the space X endowed with the τ_w topology by $w - X$.

Proposition A.4 *The space $w - X$ is Hausdorff.*

Remark A.1 The weak convergence is denoted by $\rightharpoonup$.

Theorem A.4 *Let $\{u_n\}$ be a sequence in X. Then*

(i) $u_n \rightharpoonup u \Leftrightarrow \langle \zeta, u_n \rangle \to \langle \zeta, u \rangle, \ \forall \zeta \in X^*;$

(ii) *If $u_n \to u$, then $u_n \rightharpoonup u$;*

(iii) *If $u_n \rightharpoonup u$, then $\{\|u_n\|\}$ is bounded and $\|u\| \leq \liminf_{n \to \infty} \|u_n\|$;*

(iv) *If $u_n \rightharpoonup x$ in X and $\zeta_n \to \zeta$ in X^*, then $\langle \zeta_n, u_n \rangle \to \langle \zeta, u \rangle$.*

Remark A.2 Open (resp. closed) subsets in τ_w are automatically open (resp. closed) in the strong topology.

If X is finite-dimensional, then the two topologies (weak and strong) coincide. In particular, $u_n \rightharpoonup u$ if and only if $u_n \to u$.

If X is infinite-dimensional, then the weak topology is strictly coarser than the strong topology, i.e. there exist open (resp. closed) sets in the strong topology that *are not weakly open (resp. weakly closed)*. Simple examples are as follows: the unit ball $B(0; 1)$ is not weakly open, whereas the unit sphere of $S := \{u \in X : \|u\| = 1\}$ is not weakly closed (see, e.g., Brezis [2, Examples 1 and 2]).

Theorem A.5 *Let C be a convex subset of X. Then C is weakly closed if and only if it is strongly closed.*

Corollary A.3 (Mazur) *If $u_n \rightharpoonup u$ in X, then there exists a sequence $\{v_n\}$ made up of convex combinations of the u_n' s that converges strongly to u.*

So far we have two topologies on X^*:

(*i*) the usual strong topology associated to the norm of X^*, denoted by τ_s;
(*ii*) the weak topology τ_w by performing on X^* by above construction.

A third topology on X^*, called the *weak*-topology* and denoted by τ_{w^*}, can be defined as follows: for every $u \in X$ define $\phi_u : X^* \to \mathbb{R}$ by $\phi_u(\zeta) = \langle \zeta, u \rangle$. As u runs through X we obtain a collection $(\phi_u)_{u \in X}$ mapping X^* into $\mathbb{R}$.

Definition A.4 The weak*-topology τ_{w^*}, is the coarsest topology on X^* associated to the collection $(\phi_u)_{u \in X}$.

Since $X \subset X^{**}$, it is clear that the topology τ_{w^*} is coarser than the topology τ_w, i.e., the topology τ_{w^*} has fewer open sets (resp. closed sets) than the topology τ_w, which in turn has fewer open sets (resp. closed sets) than the strong topology τ_s.

Remark A.3 Sometimes we shall denote (X^*, τ_s) by $s - X^*$, (X^*, τ_w) by $w - X^*$ and (X^*, τ_{w^*}) by $w^* - X^*$. Here and hereafter, the weak* convergence shall be denoted by $\rightharpoonup$.

Proposition A.5 *The space $w^* - X^*$ is Hausdorff.*

Regarding the weak* convergence we have the following properties.

Proposition A.6 *Let $\{\zeta_n\}$ be a sequence in X^*. Then*

(*i*) $\zeta_n \rightharpoonup \zeta \Leftrightarrow \langle \zeta_n, u \rangle \to \langle \zeta, u \rangle, \; \forall u \in X$;
(*ii*) *If $\zeta_n \to \zeta$, then $\zeta_n \rightharpoonup \zeta$, and, if $\zeta_n \rightharpoonup \zeta$, then $\zeta_n \rightharpoonup \zeta$;*
(*iii*) *If $\zeta_n \rightharpoonup \zeta$, then $\{\|\zeta_n\|\}$ is bounded and $\|\zeta\| \leq \liminf\limits_{n \to \infty} \|\zeta_n\|$;*
(*iv*) *If $\zeta_n \rightharpoonup \zeta$ and $u_n \to u$, then $\langle \zeta_n, u_n \rangle \to \langle \zeta, u \rangle$.*

Proposition A.7 *Let $\phi : X^* \to \mathbb{R}$ be a linear functional that is continuous for the weak*-topology. Then there exists some $u_0 \in X$ such that*

$$\phi(\zeta) = \langle \zeta, u_0 \rangle, \quad \forall \zeta \in X^*.$$

Corollary A.4 *Assume that* $\mathcal{H}$ *is a hyperplane in* X^* *that is closed in* $w^* - X^*$. *Then* $\mathcal{H}$ *has the form*

$$\mathcal{H} := \left\{ \zeta \in X^* : \langle \zeta, u_0 \rangle = \alpha \right\},$$

for some $u_0 \in X \setminus \{0\}$, *and some* $\alpha \in \mathbb{R}$.

The most important property of the τ_{w^*} topology is given by the following result.

Theorem A.6 (Banach-Alaoglu) *The closed unit ball of* X^*, $B_{X^*} := \{ \zeta \in X^* : \|\zeta\| \leq 1 \}$, *is weak* compact.*

A.3   Reflexive Spaces

Let X be a normed vector space and let X^* be the dual with the norm

$$\|\zeta\|_* := \sup_{\substack{u \in X \\ \|u\| \leq 1}} |\langle \zeta, u \rangle|.$$

The *bidual* X^{**} is the dual of X^* with the norm

$$\|f\|_{**} := \sup_{\substack{\zeta \in X^* \\ \|\zeta\| \leq 1}} |\langle f, \zeta \rangle|.$$

There is a *canonical injection* $I : X \to X^{**}$ defined as follows: given $u \in X$, the map $\zeta \mapsto \langle \zeta, u \rangle$ is a continuous linear functional on X^*; thus it is an element of X^{**}, which we denote by $I(u)$. We have

$$\langle I(u), \zeta \rangle_{X^{**}, X^*} := \langle \zeta, u \rangle_{X^*, X} \quad \forall u \in X, \ \forall \zeta \in X^*.$$

It is clear that I is linear and that I is an *isometry*, that is,

$$\|I(u)\|_{**} = \|u\|_X.$$

Indeed, we have

$$\|I(u)\|_{**} = \sup_{\substack{\zeta \in X^* \\ \|\zeta\|_* \leq 1}} |\langle I(u), \zeta \rangle| = \sup_{\substack{\zeta \in X^* \\ \|\zeta\|_* \leq 1}} |\langle \zeta, u \rangle| = \|u\|.$$

Definition A.5 Let X be a Banach space and let $I : X \to X^{**}$ be the canonical injection from X into X^{**}. The space X is said to be *reflexive* if I is surjective, i.e., $I(X) = X^{**}$.

Due to this bijection, for a reflexive space X, we shall identify sometimes X^{**} with X. We have the following results regarding reflexive spaces.

Theorem A.7 (Kakutani) *Let X be a Banach space. Then X is reflexive if and only if*

$$B_X := \{u \in X : \|u\| \le 1\},$$

is weakly compact.

Theorem A.8 (Eberlein-Šmulian) *A Banach space X is reflexive if and only if every bounded sequence X possesses a weakly convergent subsequence.*

Theorem A.9 *Assume that X is a reflexive Banach space and let $Y \subset X$ be a closed linear subspace of X. Then Y is reflexive.*

Corollary A.5 *A Banach space X is reflexive if and only if X^* is reflexive.*

Proposition A.8 *Let X be a reflexive Banach space and assume $K \subset X$ is a bounded, closed and convex subset of X. Then K is weakly compact.*

Set-Valued Analysis

B.1   Kuratowski Convergence

Definition B.1 (Kuratowski Convergence) Let (X, ρ) be a metric space and $\{A_n\}_{n \in \mathbb{N}}$ be a sequence of subsets of X. Then

(i) the upper limit or outer limit of the sequence $\{A_n\}_{n \in \mathbb{N}}$ is the subset of X given by

$$\limsup_{n \to \infty} A_n := \left\{ u \in X : \liminf_{n \to \infty} \text{dist}(u, A_n) = 0 \right\};$$

(ii) the lower limit or inner limit of the sequence $\{A_n\}_{n \in \mathbb{N}}$ is the subset of X given by

$$\liminf_{n \to \infty} A_n := \left\{ u \in X : \lim_{n \to \infty} \text{dist}(u, A_n) = 0 \right\}.$$

If $\limsup_{n \to \infty} A_n = \liminf_{n \to \infty} A_n$ we say that the limit of $\{A_n\}_{n \in \mathbb{N}}$ exists and

$$\lim_{n \to \infty} A_n := \limsup_{n \to \infty} A_n = \liminf_{n \to \infty} A_n.$$

Remark B.1 For a fixed set $A \subset X$, the distance function $\text{dist}(\cdot, A) : X \to \mathbb{R}$ is Lipschitz continuous. This is straightforward from the fact that

$$\text{dist}(u, A) \leq \rho(u, \bar{u}) + \text{dist}(\bar{u}, A), \ \forall u, \bar{u} \in X.$$

© The Author(s), under exclusive license to Springer Nature Switzerland AG 2021
N. Costea et al., *Variational and Monotonicity Methods in Nonsmooth Analysis*,
Frontiers in Mathematics, https://doi.org/10.1007/978-3-030-81671-1

Thus,

$$|\text{dist}(u, A) - \text{dist}(\bar{u}, A)| \le \rho(u, \bar{u}),$$

which shows that, if $u_n \to \bar{u}$ in X, then $\lim_{n\to\infty} \text{dist}(u_n, A) = \text{dist}(\bar{u}, A)$.

Proposition B.1 *Let $\{A_n\}_{n\in\mathbb{N}}$ be a sequence of subsets of a metric space X. Then*

(*i*) $\liminf_{n\to\infty} A_n \subseteq \limsup_{n\to\infty} A_n$;
(*ii*) *the sets $\limsup_{n\to\infty} A_n$ and $\liminf_{n\to\infty} A_n$ are closed in X.*

Proposition B.2 *If $\{A_n\}_{n\in\mathbb{N}}$ is a sequence of sets in a metric space X, then*

(*i*) $\limsup_{n\to\infty} A_n = \{u \in X : \exists u_{n_k} \in A_{n_k} \text{ s.t. } u_{n_k} \to u\}$;
(*ii*) $\liminf_{n\to\infty} A_n = \{u \in X : \exists u_n \in A_n \text{ s.t. } u_n \to u\}$.

That is, $\liminf_{n\to\infty} A_n$ is the collection of limits of sequences $\{u_n\}_{n\in\mathbb{N}}$, with $u_n \in A_n$; whereas $\limsup_{n\to\infty} A_n$ is the collection of cluster points of sequences $\{u_n\}_{n\in\mathbb{N}}$, with $u_n \in A_n$.

Proposition B.3 *Let $\{A_n\}_{n\in\mathbb{N}}$ be a sequence of subsets in a metric space (X, ρ). Then*

(*i*) $\limsup_{n\to\infty} A_n = \{u \in X : \forall \varepsilon > 0, \forall N \in \mathbb{N}, \exists n \ge N : B_\varepsilon(u) \cap A_n \ne \varnothing\}$;
(*ii*) $\liminf_{n\to\infty} A_n = \{u \in X : \forall \varepsilon > 0, \exists N(\varepsilon) \in \mathbb{N} : B_\varepsilon(u) \cap A_n \ne \varnothing, \forall n \ge N(\varepsilon)\}$.

Remark B.2 The statement in Propositions B.2 and B.3 can be used as alternative definitions of *inferior and superior limit* of a sequence of sets, respectively. In particular, from Proposition B.3, it follows that

(*i*) $\limsup_{n\to\infty} A_n = \bigcap_{\varepsilon > 0} \bigcap_{N \ge 1} \bigcup_{n \ge N} \mathcal{U}_\varepsilon(A_n)$;
(*ii*) $\bigcap_{n \ge 1} \text{cl}\left(\bigcup_{n \ge m} A_m\right) \subset \limsup_{m \to \infty} A_m$;
(*iii*) $\liminf_{n\to\infty} A_n = \bigcap_{\varepsilon > 0} \bigcup_{N \ge 1} \bigcap_{n \ge N} \mathcal{U}_\varepsilon(A_n)$.

Proposition B.4 *If $\{A_n\}_{n\in\mathbb{N}}$ is a sequence such that $A_{n+1} \subset A_n$, $n \in \mathbb{N}$, i.e. a decreasing sequence, then $\lim_{n\to\infty} A_n$ exists and $\lim_{n\to\infty} A_n = \bigcap_{n\in\mathbb{N}} \text{cl}(A_n)$.*

Theorem B.1 *Let $\{A_n\}_{n\in\mathbb{N}}$ and $\{B_n\}_{n\in\mathbb{N}}$ be two sequences of sets and $K \subset X$ a compact set. Assume for every neighborhood U of K, there exists $N \in \mathbb{N}$ such that $A_n \subset U$, whenever $n \ge N$. Then for every neighborhood V of $K \cap \left(\limsup_{n\to\infty} B_n\right)$, there exists $N \in \mathbb{N}$ such that $A_n \cap B_n \subset V$, whenever $n \ge N$.*

Theorem B.2 *Let* $\{A_n\}_{n\in\mathbb{N}}$ *be a sequence of sets in a metric space X and $K \subset X$. If for every neighborhood U of K, there exists $N \in \mathbb{N}$ such that $A_n \subset U$, whenever $n \geq N$, then* $\limsup_{n\to\infty} A_n \subset \mathrm{cl}(K)$.

Conversely, if X is a compact metric space, then for every neighborhood U of $\limsup_{n\to\infty} A_n$, *there exists $N \in \mathbb{N}$ such that $A_n \subset U$, whenever $n \geq N$.*

Proposition B.5 *Let* $\{A_n\}_{n\in\mathbb{N}}$ *and* $\{B_n\}_{n\in\mathbb{N}}$ *be two sequences of subsets of a metric space X. Then the following statements hold:*

(i) $\displaystyle\limsup_{n\to\infty} (A_n \cap B_n) \subset \limsup_{n\to\infty} A_n \cap \limsup_{n\to\infty} B_n;$

(ii) $\displaystyle\liminf_{n\to\infty}(A_n \cap B_n) \subset \liminf_{n\to\infty} A_n \cap \liminf_{n\to\infty} B_n;$

(iii) $\displaystyle\limsup_{n\to\infty}(A_n \cup B_n) = \limsup_{n\to\infty} A_n \cup \limsup_{n\to\infty} B_n;$

(iv) $\displaystyle\liminf_{n\to\infty}(A_n \cup B_n) \supset \liminf_{n\to\infty} A_n \cup \liminf_{n\to\infty} B_n;$

(v) $\displaystyle\limsup_{n\to\infty}(A_n \times B_n) \subset \limsup_{n\to\infty} A_n \times \limsup_{n\to\infty} B_n;$

(vi) $\displaystyle\liminf_{n\to\infty}(A_n \times B_n) = \liminf_{n\to\infty} A_n \times \liminf_{n\to\infty} B_n.$

Lemma B.1 *Let X be a real normed space, and let $\{A_n\}$, $\{B_n\}$ be two sequences of subsets of X. Then the following assertions are true:*

(i) $\displaystyle\liminf_{n\to\infty} A_n + \liminf_{n\to\infty} B_n \subset \liminf_{n\to\infty}(A_n + B_n);$

(ii) *If $A_n \subset B_n$ for all $n \in \mathbb{N}$, then* $\displaystyle\liminf_{n\to\infty} A_n \subset \liminf_{n\to\infty} B_n.$

Proposition B.6 *Let X and Y be metric spaces, $\{A_n\}_{n\in\mathbb{N}}$ and $\{B_n\}_{n\in\mathbb{N}}$ sequences of sets in X and Y, respectively. If $f : X \to Y$ is continuous function, then the following assertions hold:*

(i) $\displaystyle f\left(\limsup_{n\to\infty} A_n\right) \subset \limsup_{n\to\infty} f(A_n);$

(ii) $\displaystyle f\left(\liminf_{n\to\infty} A_n\right) \subset \liminf_{n\to\infty} f(A_n);$

(iii) $\displaystyle\limsup_{n\to\infty} f^{-1}(B_n) \subset f^{-1}\left(\limsup_{n\to\infty} B_n\right);$

(iv) $\displaystyle\liminf_{n\to\infty} f^{-1}(B_n) \subset f^{-1}\left(\liminf_{n\to\infty} B_n\right).$

B.2   Set-Valued Maps

Definition B.2 Let X and Y be topological spaces. If for each $u \in X$, there is a corresponding set $F(u) \subset Y$, then $F(\cdot)$ is called a *set-valued map* from X to Y. We denote this $F : X \rightsquigarrow Y$.

Definition B.3 Let X and Y be topological spaces and $F : X \rightsquigarrow Y$ a set-valued map. Then

- the *domain* of $F(\cdot)$, denoted by $\mathrm{Dom}(F)$, is defined as

$$\mathrm{Dom}(F) := \{u \in X : F(u) \neq \varnothing\};$$

- the *range* of $F(\cdot)$, denoted by $\mathrm{R}(F)$, is defined as

$$\mathrm{R}(F) := \bigcup_{u \in \mathrm{Dom}(F)} F(u);$$

- the *graph* of $F(\cdot)$, denoted by $\mathrm{Graph}(F)$, is defined as

$$\mathrm{Graph}(F) := \{(u, z) \in X \times Y : z \in F(u), u \in \mathrm{Dom}(F)\}.$$

A set-valued map is said to be nontrivial if it's graph is not empty, i.e. if there exists at least an element $u \in X$ such that $F(u) \neq \varnothing$.

If $K \subset X$, we denote by $F|_K$ the *restriction* of F to K, defined by

$$F|_K(u) = \begin{cases} F(u), & \text{if } u \in K \\ \varnothing, & \text{otherwise.} \end{cases} \tag{B.1}$$

Definition B.4 Let X and Y be topological spaces.

- A set-valued map $F : X \rightsquigarrow Y$ is said to be *closed valued, open valued or compact valued* if for each $u \in X$, $F(u)$ is a closed, open or compact set in Y, respectively. Furthermore, if Y is a topological linear space and $F(u)$ is a convex set in Y for each $u \in X$, the $F(\cdot)$ is called *convex valued*.
- $F : X \rightsquigarrow Y$ is said to be a *closed, open or compact set-valued map*, if $\mathrm{Graph}(F)$ is a closed, open or compact set w.r.t. the product topology of $X \times Y$. Furthermore, if X and Y are topological vector spaces, then $F(\cdot)$ called a convex set-valued map if $\mathrm{Graph}(F)$ is convex set in w.r.t. $X \times Y$.

Remark B.3 In Definition B.4 one must not confuse closed valued maps and closed set-valued maps. The former refers to the values of the map, whereas the latter refers to the graph of the map.

Definition B.5 Let X and Y be topological spaces and $F : X \rightsquigarrow Y$ a set-valued map. Then

- *the closure sv-map* associated with F is the map $\mathrm{cl}(F) : X \rightsquigarrow Y$, where $\mathrm{cl}(F)(u) := \mathrm{cl}(F(u))$, for each $u \in X$;
- the *interior* sv-map associated with F is the map $\mathrm{int}(F) : X \rightsquigarrow Y$, where $\mathrm{int}(F)(u) := \mathrm{int}(F(u))$, for each $u \in X$,
- Moreover, if Y is a topological linear space, then the *convex-hull* sv-map associated with F is the map $\mathrm{conv}(F) : X \rightsquigarrow Y$, where $\mathrm{conv}(F)(u) := \mathrm{co}(F(u))$, for each $u \in X$.

B.3   Continuity of Set-Valued Maps

Definition B.6 (Lower Inverse of a sv-Map) Let $F : X \rightsquigarrow Y$ be a set-valued map. For any $V \subset Y$ the *lower inverse image* of V under F, denoted $F^-(V)$, is defined by

$$F^-(V) := \{u \in X : \ F(u) \cap V \neq \varnothing\} = \bigcup_{v \in V} F^-(v).$$

Definition B.7 (Upper Inverse of a sv-Map) Let $F : X \rightsquigarrow Y$ be a set-valued map. For any $V \subset Y$ the *upper inverse image* of V under F, denoted $F^+(V)$, is defined by

$$F^+(V) := \{u \in X : \ F(u) \subset V\}.$$

$F^-(V)$ is called sometimes the *inverse image* of V by F, whereas $F^+(V)$ is called the *core* of V by F.

Definition B.8 Let $F : X \rightsquigarrow Y$ be a set-valued map and $\mathrm{Dom}(F) \neq \varnothing$. Then F is said to be *upper semicontinuous (u.s.c.) at* $u_0 \in X$ iff, for any open set $V \subset Y$, such that $F(u_0) \subset V$, there exists a neighborhood $U \subset X$ of u_0 such that $F(u) \subset V$, for all $u \in U$.
The map F is said to be u.s.c. on X, if it is u.s.c. at every $u \in X$.

Definition B.9 Let $F : X \rightsquigarrow Y$ be a set-valued map and $\mathrm{Dom}(F) \neq \varnothing$. Then F is said to be *lower semicontinuous (l.s.c.) at* $u_0 \in X$ iff, for any open set $V \subset Y$, such that $F(u_0) \cap V \neq \varnothing$, there exists a neighborhood $U \subset X$ of u_0 such that for every $u \in U$, we have $F(u) \cap V \neq \varnothing$.
The map F is said to be l.s.c. on X, if it is l.s.c. at every $u \in X$.

Definition B.10 A set-valued map $F : X \rightsquigarrow Y$ is called continuous if it is both lower and upper semicontinuous.

The following two propositions give a useful characterization of lower semicontinuous (upper semicontinuous) set-valued maps and are direct consequences of the above definitions.

Proposition B.7 *Let X, Y be Hausdorff topological spaces and $F \rightsquigarrow Y$ a given sv-map. Then the following statements are equivalent:*

(i) *F is l.s.c.;*
(ii) *$F^+(C)$ is closed in X whenever C is closed in Y;*
(iii) *for any pair $(u, v) \in$ Graph(F) and any sequence $\{u_n\} \subset X$ converging to u, there exists a sequence $v_n \in F(u_n)$ such that $v_n \to v$;*

Proposition B.8 *Let X, Y be Hausdorff topological spaces and $F \rightsquigarrow Y$ a given sv-map. Then the following statements are equivalent:*

(i) *F is u.s.c.;*
(ii) *$F^-(C)$ is closed in X whenever C is closed in Y;*
(iii) *For any sequence $\{u_n\} \subset X$ converging to u and any open set $V \subset Y$ such that $F(u) \subset V$, there exists a rank $n_0 \geq 1$ such that $F(u_n) \subset V$ for all $n \geq n_0$.*

Proposition B.9 *Let X, Y be two Hausdorff topological spaces and $F : X \rightsquigarrow Y$ a set-valued map. Then*

(i) *Let $F(u)$ be closed for all $u \in C \subseteq X$. If F is u.s.c. and C is closed, then Graph(F) is closed. If $\overline{F(C)}$ is compact and C is closed, then F is u.s.c. if and only if Graph(F) is closed;*
(ii) *If $K \subseteq X$ is compact, F is u.s.c. and $F(u)$ is compact for all $u \in K$, then $F(K)$ is compact.*

Remark B.4 It is clear from above that when F is single-valued, i.e. $F(u) = \{v\} \subset Y$, the notions of lower and upper semicontinuity coincide with the usual notion of continuity of a map between two Hausdorff topological spaces.

In general the notion of lower and upper semicontinuity are distinct. In order to see this let us consider $F : \mathbb{R} \rightsquigarrow \mathbb{R}$ be defined by

$$F(u) := [\phi_1(u), \phi_2(u)],$$

where $\phi_i : \mathbb{R} \to \mathbb{R}$ are prescribed functionals such that $\phi_1(u) \leq \phi_2(u)$ for all $u \in X$. Then

- ϕ_1 u.s.c. and ϕ_2 l.s.c. $\Rightarrow F$ is l.s.c.;
- ϕ_1 l.s.c. and ϕ_2 u.s.c. $\Rightarrow F$ is u.s.c.;
- ϕ_1, ϕ_2 continuous $\Rightarrow F$ continuous.

Here, the l.s.c. (u.s.c.) of the functionals ϕ_i is understood in the sense of Section 1.2.

Definition B.11 Let X and Y be normed spaces and $F : X \rightsquigarrow Y$ a set-valued map. We say that F is *Lipschitz around* $u \in X$, if there exists a positive constant L and a neighborhood $U \subset \mathrm{Dom}(F)$ of u such that

$$\forall u_1, u_2 \in U : \quad F(u_1) \subset F(u_2) + L\|u_1 - u_2\| B_Y(0, 1).$$

In this case F also called *Lipschitz or L-Lipschitz on* U.

F is said to be *pseudo-Lipschitz around* $(u, v) \in \mathrm{Graph}(F)$ if there exists a positive constant L, a neighborhood $U \subset \mathrm{Dom}(F)$ of u and a neighborhood V of v such that

$$\forall u_1, u_2 \in U : \quad F(u_1) \cap V \subset F(u_2) + L\|u_1 - u_2\| B_Y(0, 1).$$

In particular, if $F : X \rightsquigarrow \mathbb{R}$ is a set-valued mapping, we say that F is Lipschitz around $u \in X$ if there exists a positive constant L and a neighborhood U of u such that for every $u_1, u_2 \in U$ we have

$$F(u_1) \subset F(u_2) + L\|u_1 - u_2\|[-1, 1].$$

For a nonempty subset K of X, we say that F is K-*locally Lipschitz* if it is Lipschitz around all $u \in K$.

Proposition B.10 *If $F : X \rightsquigarrow \mathbb{R}$ is a K-locally Lipschitz sv-map, then the restriction $F|_K : K \rightsquigarrow \mathbb{R}$ is continuous on K.*

B.4   Monotonicity of Set-Valued Operators

Unless otherwise stated, throughout this subsection X denotes a real Banach space with dual X^*. A set-valued map $A : X \rightsquigarrow X^*$ shall often be called *set-valued operator*. In order to increase the clarity of the exposition, we shall denote the elements of $A(u)$ by u^* instead of using Greek letters (as we have done so far when referring to elements of X^*).

If $A(u)$ is a singleton, then we shall often identify $A(u)$ with its unique element. The *inverse* $A^{-1} : X^* \rightsquigarrow X$ of A is defined as

$$A^{-1}(u^*) := \left\{ u \in X : u^* \in A(u) \right\}.$$

Obviously $\text{Dom}(A^{-1}) = R(A)$, $R(A^{-1}) = \text{Dom}(A)$ and $\text{Graph}(A^{-1}) = \{(u^*, u) \in X^* \times X : (u, u^*) \in \text{Graph}(A)\}$.

Definition B.12 A set-valued operator $A : X \rightsquigarrow X^*$ is said to be *monotone* if

$$\langle v^* - u^*, v - u \rangle \geq 0, \quad \forall (u, u^*), (v, v^*) \in \text{Graph}(A). \tag{B.2}$$

A monotone operator $A : X \rightsquigarrow X^*$ is called *maximal motonone* if $\text{Graph}(A)$ is not properly contained in the graph of any other monotone operator $A' : X \rightsquigarrow X^*$.

We point out the fact that A is said to be *strictly monotone* if (B.2) holds with strict inequality whenever $u \neq v$. Moreover, if there exists $m > 0$ such that the stronger inequality holds

$$\langle v^* - u^*, v - u \rangle \geq m \|v - u\|^2, \quad \forall (u, u^*), (v, v^*) \in \text{Graph}(A), \tag{B.3}$$

then A is called *strongly monotone*. Actually, (B.3) means that $A - mJ$ is monotone, with J being the normalized duality mapping.

We present next some generalizations of the monotonicity concept.

Definition B.13 Let $\eta : K \times K \to X$ and $\alpha : X \to \mathbb{R}$ be two single-valued maps. A set-valued map $A : K \rightsquigarrow X^*$ is said to be

- *relaxed $\eta - \alpha$ monotone*, if

$$\langle v^* - u^*, \eta(v, u) \rangle \geq \alpha(v - u), \quad \forall (u, u^*), (v, v^*) \in \text{Graph}(A); \tag{B.4}$$

- *relaxed $\eta - \alpha$ pseudomonotone*, if

$$[\exists u^* \in A(u) : \langle u^*, \eta(v, u) \rangle \geq 0] \Rightarrow [\langle v^*, \eta(v, u) \rangle \geq \alpha(v - u), \forall v^* \in A(v)];$$
$$\tag{B.5}$$

- *relaxed $\eta - \alpha$ quasimonotone*, if

$$[\exists u^* \in A(u) : \langle u^*, \eta(v, u) \rangle > 0] \Rightarrow [\langle v^*, \eta(v, u) \rangle \geq \alpha(v - u), \forall v^* \in A(v)];$$
$$\tag{B.6}$$

If $\eta(v - u) := v - u$, then

(B.4) reduces to

$$\langle v^* - u^*, v - u \rangle \geq \alpha(v - u), \quad \forall (u, u^*), (v, v^*) \in \mathrm{Graph}(A);$$

and A is said to be *relaxed α monotone*;
(B.5) reduces to

$$[\exists u^* \in A(u) : \langle u^*, v - u \rangle \geq 0] \Rightarrow [\langle v^*, v - u \rangle \geq \alpha(v - u), \ \forall v^* \in A(v)];$$

and A is said to be *relaxed α pseudomonotone*;
(B.6) reduces to

$$[\exists u^* \in A(u) : \langle u^*, v - u \rangle > 0] \Rightarrow [\langle v^*, v - u \rangle \geq \alpha(v - u), \ \forall v^* \in A(v)];$$

and A is said to be *relaxed α quasimonotone*.

Geometry of Banach Spaces

C.1   Smooth Banach Spaces

Definition C.1 A Banach space X is called *smooth* if for every $u \neq 0$ there exists a unique $\zeta \in X^*$ such that $\|\zeta\| = 1$ and $\langle \zeta, u \rangle = \|u\|$.

Proposition C.1 *For every $u \neq 0$ one has*

$$\partial \|u\| = \left\{ \zeta \in X^* : \langle \zeta, u \rangle = \|u\|, \|\zeta\| = 1 \right\}.$$

From this proposition and the fact that any proper convex continuous functional φ is Gateaux differentiable at $u \in \mathrm{int}(D(\varphi))$ if and only if $\partial \varphi(u)$ is a singleton (see, e.g., Ciorănescu [4, Corollary 2.7]) we have the following characterization of smooth spaces.

Theorem C.1 *A Banach space X is smooth if and only if $\| \cdot \|$ is Gateaux differentiable on $X \setminus \{0\}$.*

Definition C.2 For a Banach space X the function $\rho : (0, \infty) \to (0, \infty)$ defined by

$$\rho(t) := \frac{1}{2} \sup_{\|u\|=\|v\|=1} (\|u + tv\| + \|u - tv\| - 2)$$

is called the *modulus of smoothness* of X.
The *modulus of smoothness at $u \in X$* is defined as

$$\rho(t, u) := \frac{1}{2} \sup_{\|v\|=1} (\|u + tv\| + \|u - tv\| - 2\|u\|).$$

© The Author(s), under exclusive license to Springer Nature Switzerland AG 2021
N. Costea et al., *Variational and Monotonicity Methods in Nonsmooth Analysis*,
Frontiers in Mathematics, https://doi.org/10.1007/978-3-030-81671-1

Definition C.3 A Banach space X is called *uniformly smooth* if

$$\lim_{t\searrow 0} \frac{\rho(t)}{t} = 0.$$

The space X is said to be *locally uniformly smooth* if

$$\lim_{t\searrow 0} \frac{\rho(t, u)}{t} = 0, \quad \forall u \in X \setminus \{0\}.$$

Proposition C.2 *For a Banach spaces X the following implications hold*

$$X \text{ is uniformly smooth} \Rightarrow X \text{ is locally uniformly smooth} \Rightarrow X \text{ is smooth.}$$

Theorem C.2

(i) *A Banach space X is locally uniformly smooth if and only if $\| \cdot \|$ is Fréchet differentiable on $X \setminus \{0\}$;*

(ii) *A Banach space X is uniformly smooth if and only if $\| \cdot \|$ is uniformly Fréchet differentiable on the unit sphere, i.e.,*

$$\lim_{t\searrow 0} \sup_{\|u\|=\|v\|=1} \left| \frac{\|u + tv\| - \|u\|}{t} - \langle \| \cdot \|'(u), v\rangle \right| = 0.$$

C.2   Uniform Convexity, Strict Convexity and Reflexivity

Definition C.4 A Banach space is called *uniformly convex* if for any $\varepsilon \in (0, 2]$ there exists $\delta = \delta(\varepsilon) > 0$ such that for $u, v \in X$ satisfying $\|u\| = \|v\| = 1$ and $\|u - v\| \geq \varepsilon$ one has $\left\|\frac{u+v}{2}\right\| \leq 1 - \delta$.

In other words, X is uniformly convex if for any two distinct points on the sphere u, v the midpoint of the line segment joining u and v is never on the sphere, but inside the unit ball.

Example C.1 The Lebesgue spaces L^p, $1 < p < \infty$, are uniformly convex.
   This is a simple consequence of Clarkson's inequalities (see, e.g., Diestel [5]):

$$\left\|\frac{u+v}{2}\right\|^p + \left\|\frac{u-v}{2}\right\|^p \leq \frac{\|u\|^p + \|v\|^p}{2}, \quad \forall u.v \in L^p, \, p \in [2, \infty), \tag{C.1}$$

and

$$\left\|\frac{u+v}{2}\right\|^p + \left\|\frac{u-v}{2}\right\|^p \le \left(\frac{\|u\|^p + \|v\|^p}{2}\right)^{1/(p-1)}, \quad \forall u.v \in L^p, \, p \in (1,2]. \quad (C.2)$$

Let $\varepsilon \in (0, 2]$ and $u, v \in X$ be such that $\|u\| = \|v\| = 1$ and $\|u-v\| \ge \varepsilon$. Then Clarkson's inequalities ensure that

$$\left\|\frac{u+v}{2}\right\| \le \left[1 - \left(\frac{\varepsilon}{2}\right)^p\right]^{1/p},$$

hence we can pick $\delta := 1 - \left[1 - (\varepsilon/2)^p\right]^{1/p}$.

Example C.2 Any Hilbert space is uniformly convex.

In order to see this, fix $\varepsilon \in (0, 2]$ and $u, v \in H$ such that $\|u\| = \|v\| = 1$ and $\|u - v\| \ge \varepsilon$. Then, according to the parallelogram law

$$\left\|\frac{u+v}{2}\right\|^2 + \left\|\frac{u-v}{2}\right\|^2 = \frac{\|u\|^2 + \|v\|^2}{2},$$

and thus

$$\left\|\frac{u+v}{2}\right\| \le 1 - \delta, \text{ with } \delta := 1 - \sqrt{1 - \frac{\varepsilon^2}{4}}.$$

Definition C.5 A Banach space X is called *strictly convex* if for $u, v \in X$ satisfying $u \ne v$, $\|u\| = \|v\| = 1$ one has

$$\|\lambda u + (1 - \lambda)v\| < 1, \quad \forall \lambda \in (0, 1).$$

Proposition C.3 *The following statements are equivalent:*

  (i) X is strictly convex;
 (ii) The unit sphere contains no line segments;
(iii) If $u \ne v$ and $\|u\| = \|v\| = 1$, then $\|u + v\| < 2$;
 (iv) If $u, v, w \in X$ are such that $\|u - v\| = \|u - w\| + \|w - v\|$, then there exists $\lambda \in [0, 1]$ such that $w = \lambda u + (1 - \lambda)v$;
  (v) Any $\zeta \in X^*$ assumes its supremum in at most one point of the unit ball.

It is clear from this proposition that any uniformly convex space is strictly convex. However, not all Banach spaces are strictly convex and there exist strictly convex spaces that are not uniformly convex as it can be seen from the following examples.

Example C.3 The space l^1 is not strictly convex.

   To see this take $u := (1, 0, 0, 0, \ldots) \in l^1$, $v := (0, -1, 0, 0, \ldots) \in l^1$. Clearly, $\|u\|_1 = \|v\|_1 = 1$ and $\|u + v\|_1 = 2$, which shows that l^1 is not strictly convex.

Example C.4 The space l^∞ is not strictly convex.

   Choose $u := (1, 1, 0, 0, \ldots)$ and $v := (-1, 1, 0, 0, \ldots)$. Again $u, v \in l^\infty$, $\|u\|_\infty = \|v\|_\infty = 1$ and $\|u + v\|_\infty = 2$, showing that l^∞ is not strictly convex.

Example C.5 (Goebel and Kirk [8])

(*i*) The space $(C[0, 1], \| \cdot \|_\infty)$ is not strictly convex, where $\| \cdot \|_\infty$ is the standard "sup norm";
(*ii*) The space $(C[0, 1], \| \cdot \|_\mu)$ is strictly convex, but not uniformly convex, where for $\mu > 0$

$$\|u\|_\mu := \|u\|_\infty + \mu \left(\int_0^1 u^2(x) \mathrm{d}x \right)^{1/2}.$$

Definition C.6 Let X be a Banach space with $\dim X \geq 2$. The *modulus of convexity* of X is the function $\Delta : (0, 2] \to [0, 1]$ defined by

$$\Delta(\varepsilon) := \inf \left\{ 1 - \left\| \frac{u + v}{2} \right\| : \|u\| = \|v\| = 1, \|u - v\| \geq \varepsilon \right\}.$$

Theorem C.3 *A Banach space is uniformly convex if and only if $\Delta(\varepsilon) > 0$ for all $\varepsilon \in (0, 2]$.*

Theorem C.4 (Milman-Pettis) *If X is a uniformly convex Banach space, then X is reflexive.*

Remark C.1 Uniform convexity is a *geometric* property of the norm: endowed with an equivalent norm the space might not be uniformly convex. On the other hand, reflexivity is a *topological* property: a reflexive space remains reflexive for an equivalent norm.

Proposition C.4 *Let X be a uniformly convex Banach space and $\{u_n\} \subset X$ be such that $u_n \rightharpoonup u$ and $\limsup_{n \to \infty} \|u_n\| \leq \|u\|$. Then $u_n \to u$.*

   In order to establish the connection between the strict/uniform convexity and the differentiability of the norm on a Banach space, we have the following duality results.

Theorem C.5 *For a Banach space X the following implications hold:*

(*i*) X^* *is smooth* $\Rightarrow$ X *is strictly convex;*
(*ii*) X^* *is strictly convex* $\Rightarrow$ X *is smooth.*

Corollary C.1 (Weak Duality) *If X is reflexive, then X is strictly convex (respectively smooth) if and only if X^* is smooth (respectively strictly convex).*

Theorem C.6 (Strong Duality) *Let X be a Banach space. Then*

(*i*) *X is uniformly smooth if and only if X^* is uniformly convex;*
(*ii*) *X is uniformly convex if and only if X^* is uniformly smooth.*

Corollary C.2 *If X is a uniformly smooth Banach space, then X is reflexive.*

C.3   Duality Mappings

Definition C.7 A continuous and strictly increasing function $\phi : [0, \infty) \to [0, \infty)$ such that $\phi(0) = 0$ and $\lim_{t \to \infty} \phi(t) = \infty$ is called a *normalization function*.

Lemma C.1 *Let ϕ be a normalization function and*

$$\Phi(t) := \int_0^t \phi(s)\mathrm{d}s.$$

Then Φ is convex function on $[0, \infty)$.

Definition C.8 Given a normalization function ϕ, the set-valued map $J_\phi : X \rightsquigarrow X^*$ defined by

$$J_\phi u := \left\{ \zeta \in X^* : \langle \zeta, u \rangle = \|\zeta\|\|u\|, \|\zeta\| = \phi(\|u\|) \right\}$$

is called the *duality mapping* corresponding to ϕ.
   The duality mapping corresponding to $\phi(t) = t$ is called the *normalized duality mapping*.

Remark C.2 For any normalization function ϕ, $J_\phi u \neq \varnothing$ for every $u \in X$, hence $D(J_\phi) = X$.

Proposition C.5 *If H is a Hilbert space, then the normalized duality mapping is the identity operator.*

Theorem C.7 (Asplund) *If J_ϕ is the duality mapping corresponding to the normalization function ϕ, then $J_\phi u = \partial \Phi(\|u\|)$, for all $u \in X$.*

Corollary C.3 *Let X be a Banach space and ϕ a normalization function. Then X is smooth if and only if J_ϕ is single-valued. In this case*

$$\langle J_\phi u, v \rangle = \frac{\mathrm{d}}{\mathrm{d}t} \Phi(\|u + tv\|)\Big|_{t=0}, \qquad \forall u, v \in X. \tag{C.3}$$

Theorem C.8 *If J_ϕ is the duality mapping corresponding to the normalization function ϕ, then*

(i) *For every $u \in X$ the set $J_\phi u$ is nonempty, convex and weak* closed in X^*;*
(ii) *J_ϕ is monotone;*
(iii) *$J_\phi(-u) = -J_\phi(u)$, for all $u \in X$;*
(iv) *For every $u \in X \setminus \{0\}$ and every $\lambda > 0$ we have*

$$J_\phi(\lambda u) = \frac{\phi(\lambda\|u\|)}{\phi(\|u\|)} J_\phi(u);$$

(v) *If ϕ^{-1} is the inverse of ϕ, then ϕ^{-1} is a normalization function and $\zeta \in J_\phi u$ whenever $u \in J^*_{\phi^{-1}}\zeta$, with $J^*_{\phi^{-1}}$ being the duality mapping corresponding to ϕ^{-1} on X^*;*
(vi) *If J_ψ is the duality mapping corresponding to the normalization function ψ, then*

$$J_\psi u = \frac{\psi(\|u\|)}{\phi(\|u\|)} J_\phi u, \qquad \forall u \in X \setminus \{0\}.$$

Proposition C.6 *If X is uniformly convex and smooth, then J_ϕ satisfies the (S_+) property, i.e., if $u_n \rightharpoonup u$ and $\limsup_{n\to\infty}\langle J_\phi u_n, u_n - u \rangle \leq 0$, then $u_n \to u$.*

Proposition C.7 *If X is reflexive and smooth, then J_ϕ is demicontinuous, i.e., if $u_n \to u$ in X, then $J_\phi u_n \rightharpoonup J_\phi u$ in X^*.*

KKM-Type Theorems, Fixed Point Results and Minimax Principles

D

D.1   Variants of the KKM Lemma and Fixed Point Results

In this subsection we present some variants of the KKM lemma. We begin with the well known result of Knaster, Kuratowski and Mazurkiewicz.

Lemma D.1 (KKM Lemma [9]) *Let* $P_0 P_1 \ldots P_n \subset \mathbb{R}^n$ *be a closed simplex and let* $K_0, K_1, \ldots, K_n$ *be compact subsets of* $\mathbb{R}^n$ *such that*

$$P_{i_0} P_{i_1} \ldots P_{i_k} \subset \bigcup_{s=0}^{k} K_{i_s},$$

for every face of $P_0 P_1 \ldots P_n$. *Then*

$$\bigcap_{i=0}^{n} K_i \neq \varnothing.$$

Theorem D.1 (Fan [7, Theorem 4]) *In a Hausdorff topological vector space* E, *let* C *be a convex set and* $\varnothing \neq K \subset C$. *Let* $F : K \rightsquigarrow C$ *be a set-valued mapping such that*

(*i*) *for each* $u \in K$, $F(u)$ *a relatively closed subset of* C;
(*ii*) F *is a KKM mapping, i.e., for any finite subset* $\{u_1, u_2, \ldots, u_n\} \subset K$ *one has*

$$\mathrm{co}\{u_1, \ldots, u_n\} \subset \bigcup_{i=1}^{n} F(u_i);$$

© The Author(s), under exclusive license to Springer Nature Switzerland AG 2021
N. Costea et al., *Variational and Monotonicity Methods in Nonsmooth Analysis*,
Frontiers in Mathematics, https://doi.org/10.1007/978-3-030-81671-1

(*iii*) *there exists a nonempty subset K_0 of K such that the intersection $\bigcap_{u \in K_0} F(u)$ is compact and K_0 is contained in a compact convex subset K_1 of C.*

Then $\bigcap_{u \in K} F(u) \neq \varnothing$.

The following result represents a generalization of the KKM-lemma and it was originally proved by Ky Fan in [6]. Here it is stated here as a particular case of the previous theorem.

Corollary D.1 (Fan-KKM) *Let K be an arbitrary set in a Hausdorff topological vector space E and $F : K \rightsquigarrow E$ such that:*

(*i*) *for each $u \in K$, $F(u)$ is closed in E;*
(*ii*) *F is a KKM mapping;*
(*iii*) *there exists $u_0 \in K$ such that $F(u_0)$ is compact.*

Then $\bigcap_{u \in K} F(u) \neq \varnothing$.

Theorem D.2 (Lin [10]) *Let K be a nonempty convex subset of a Hausdorff topological vector space E. Let $A \subseteq K \times K$ be a subset such that*

(*i*) *for each $u \in K$, the set $\{v \in K : (u, v) \in A\}$ is closed in K;*
(*ii*) *for each $v \in K$, the set $\{u \in K : (u, v) \notin A\}$ is convex or empty;*
(*iii*) *$(u, u) \in A$ for each $u \in K$;*
(*iv*) *K has a nonempty compact subset K_0 such that the set*

$$B := \{v \in K : (u, v) \in A, \ \forall u \in K_0\}$$

is compact.

Then there exists a point $v_0 \in B$ such that $K \times \{v_0\} \subset A$.

Theorem D.3 (Tarafdar [17]) *Let K be a nonempty convex subset of a topological vector space. Let $f : K \rightsquigarrow K$ be a set valued mapping such that:*

(*i*) *for each $u \in K$, $f(u)$ is a nonempty convex subset of K;*
(*ii*) *for each $v \in K$, $f^{-1}(v) := \{u \in K : v \in f(u)\}$ contains a relatively open subset O_v of K (O_v may be empty for some v);*
(*iii*) *$\bigcup_{u \in K} O_u = K$*

(iv) *there exists a nonempty subset $K_0 \subset K$ such that K_0 is contained in a compact convex subset K_1 of K and the set $D := \bigcap_{u \in K_0} O_u^c$ is compact, (D could be empty and O_u^c denotes the complement of O_u in K).*

Then there exists a point $u_0 \in K$ such that $u_0 \in f(u_0)$.

Theorem D.4 (Ansari and Yao [1]) *Let K be a nonempty closed and convex subset of a Hausdorff topological vector space X and let $S, T : K \rightsquigarrow K$ be two set-valued maps. Assume that:*

(i) *for each $u \in K$, $S(u)$ is nonempty and $\mathrm{co}\{S(u)\} \subseteq T(u)$;*
(ii) $K = \bigcup_{v \in K} \mathrm{int}_K S^{-1}(v)$;
(iii) *if K is not compact, assume that there exists a nonempty compact convex subset C_0 of K and a nonempty compact subset C_1 of K such that for each $u \in K \setminus C_1$ there exists $\bar{v} \in C_0$ with the property that $u \in \mathrm{int}_K S^{-1}(\bar{v})$.*

Then there exists $u_0 \in K$ such that $u_0 \in T(u_0)$.

D.2   Minimax Results

In this subsection we present some minimax inequalities due to Ky Fan [6, 7], a variant proved by Brezis, Nirenberg & Stampacchia [3] and an alternative due to Mosco [12] on vector spaces, as well as minimax results on topological spaces due to McClendon [11] and Ricceri [13–15].

Definition D.1 Let K be a convex set of a topological vector space E and $f : K \rightarrow \mathbb{R}$ be a function. The function f is said to be *quasi-convex* if for every real number t the set $\{x \in K : f(u) < t\}$ is convex. The function f is said to be *quasi-concave* if $-f$ is quasi-convex.

Theorem D.5 (Fan Minimax Principle [7]) *Let K be a nonempty convex set in a Hausdorff topological vector space and let be $f : K \times K \rightarrow \mathbb{R}$ be a function such that:*

(i) *For each fixed $u \in K$, $v \mapsto f(u, v)$ is lower semicontinuous;*
(ii) *For each fixed $v \in K$, $u \mapsto f(u, v)$ is quasi-concave;*
(iii) $f(u, u) \leq 0$, *for all $u \in K$;*

(*iv*) *K has a nonempty compact convex subset* K_0 *such that the set*

$$\bigcap_{u \in K_0} \{v \in K : f(u, v) \leq 0\}$$

  is compact.

Then there exists a point $\hat{v} \in K$, *such that* $f(u, \hat{v}) \leq 0$, *for all* $u \in K$.

Corollary D.2 *Let K be a nonempty convex compact set in a Hausdorff topological vector space and let be* $f : K \times K \to \mathbb{R}$ *be a function such that:*

  (*i*) *For each fixed* $u \in K$, $v \mapsto f(u, v)$ *is lower semicontinuous;*
  (*ii*) *For each fixed* $v \in K$, $u \mapsto f(u, v)$ *is quasi-concave;*
  (*iii*) $f(u, u) \leq 0$, *for all* $u \in K$;

Then there exists a point $\hat{v} \in K$, *such that* $f(u, \hat{v}) \leq 0$, *for all* $u \in K$.

Theorem D.6 (Brezis et al. [3]) *Let X be a Hausdorff linear topological vector space and K a convex subset in E. Let* $f : K \times K \to \mathbb{R}$ *be a real function satisfying:*

(1) $f(u, u) \leq 0$ *for all* $u \in K$;
(2) *For every fixed* $u \in K$, *the set* $\{v \in K : f(u, v) > 0\}$ *is convex;*
(3) *For every fixed* $v \in K$, $f(\cdot, v)$ *is a lower semicontinuous function on the intersection of K with any finite dimensional subspace of E.*
(4) *Whenever* $u, v \in K$ *and* u_α *is a filter on K converging to u, then* $f(u_\alpha, (1-t)u+tv) \leq 0$ *for every* $t \in [0, 1]$ *implies* $f(u, v) \leq 0$.
(5) *There exists a compact subset L of E and* $v_0 \in L \cap K$ *such that* $f(u, v_0) > 0$ *for every* $u \in K, u \notin L$.

Then there exists $u_0 \in L \cap K$ *such that*

$$f(u_0, v) \leq 0, \quad \forall v \in K.$$

In particular,

$$\inf_{u \in K} \sup_{v \in K} f(u, v) \leq 0.$$

  Now, let F be a Hausdorff topological vector space and let G be a vector space and let $A \subset F, B \subset G$ be convex sets.

Theorem D.7 *Let* $f : A \times B \to \mathbb{R}$ *be a function satisfying*

(i) *For each fixed* $v \in B$, *the function* $f(\cdot, v)$ *is quasi-convex and lower semicontinuous on* A.

(ii) *For each fixed* $u \in A$, *the function* $f(u, \cdot)$ *is quasi-concave on* B *and lower semicontinuous on the intersection of* B *with any finite dimensional subspace.*

(iii) *For some* $\tilde{v} \in B$ *and some* $\lambda > \sup_{v \in B} \inf_{u \in A} f(u, v)$, *the set* $\{u \in A : f(u, \tilde{v}) \leq \lambda\}$ *is compact.*

Then

$$\sup_{v \in B} \inf_{u \in A} f(u, v) = \inf_{u \in A} \sup_{v \in B} f(u, v).$$

Theorem D.8 (Mosco's Alternative [12]) *Let* K *be a nonempty, compact and convex subset of a Hausdorff topological vector space* E *and let* $\varphi : E \to (-\infty, \infty]$ *be a proper, convex and lower semicontinuous functional such that* $D(\varphi) \cap K \neq \emptyset$. *Assume* $f, g : E \times E \to \mathbb{R}$ *are two functions that satisfy:*

(i) $f(u, v) \leq g(u, v)$, *for all* $u, v \in E$;

(ii) *for each* $u \in E$, $v \mapsto f(u, v)$ *is lower semicontinous;*

(iii) *for each* $v \in E$, $u \mapsto g(u, v)$ *is concave.*

Then for every $\lambda \in \mathbb{R}$ *holds the alternative:*

(A_1) *there exists* $v_\lambda \in D(\phi) \cap K$ *such that*

$$f(u, v_\lambda) + \varphi(v_\lambda) - \varphi(u) \leq \lambda, \quad \forall u \in E;$$

(A_2) *there exists* $u_0 \in E$ *such that* $g(u_0, u_0) > \lambda$.

We conclude this part by a KyFan-type minimax result on topological spaces given by McClendon [11]. To complete this, we need two notions.

Definition D.2

(a) An *ANR* (absolute neighborhood retract) is a separable metric space X such that whenever X is embedded as a closed subset into another separable metric space Y, it is a retract of some neighborhood in Y.

(b) A nonempty set X is *acyclic* if it is connected and its Čech homology (coefficients in a fixed field) is zero in dimensions greater than zero.

The following result can be viewed as a topological version of the Fan minimax principle (see Corollary D.2):

Theorem D.9 (McClendon [11, Theorem 3.1]) *Suppose that X is a compact acyclic finite-dimensional ANR. Suppose $h : X \times X \to \mathbb{R}$ is a function such that $\{(x, y) : h(y, y) > h(x, y)\}$ is open and $\{x : h(y, y) > h(x, y)\}$ is contractible or empty for all $y \in X$. Then there is a $y_0 \in X$ with $h(y_0, y_0) \leq h(x, y_0)$ for all $x \in X$.*

Theorem D.10 (Ricceri [14, Theorem 4]) *Let X be a real, reflexive Banach space, let $\Lambda \subseteq \mathbb{R}$ be an interval, and let $\varphi : X \times \Lambda \to \mathbb{R}$ be a function satisfying the following conditions:*

1. *$\lambda \mapsto \varphi(u, \lambda)$ is concave for all $u \in X$;*
2. *$u \mapsto \varphi(u, \lambda)$ is continuous, coercive and sequentially weakly lower semicontinuous in for all $\lambda \in \Lambda$;*
3. *$\beta_1 := \sup_{\lambda \in \Lambda} \inf_{u \in X} \varphi(u, \lambda) < \inf_{u \in X} \sup_{\lambda \in \Lambda} \varphi(u, \lambda) =: \beta_2$.*

Then, for each $\sigma > \beta_1$ there exists a nonempty open set $\Lambda_0 \subset \Lambda$ with the following property: for every $\lambda \in \Lambda_0$ and every sequentially weakly lower semicontinuous function $\phi : X \to \mathbb{R}$, there exists $\mu_0 > 0$ such that, for each $\mu \in (0, \mu_0)$, the function $\varphi(\cdot, \lambda) + \mu\phi(\cdot)$ has at least two local minima lying in the set $\{u \in X : \varphi(u, \lambda) < \sigma\}$.

Theorem D.11 (Ricceri [15, Theorem 1]) *Let X be a topological space, $I \subseteq \mathbb{R}$ an open interval and $\psi : X \times I \to \mathbb{R}$ a function satisfying the following conditions:*

(*i*) *for each $u \in X$, the function $\lambda \mapsto \psi(u, \lambda)$ is quasi-concave and continuous;*
(*ii*) *for each $\lambda \in I$, the function $u \mapsto \psi(u, \lambda)$ has compact and closed sub-level sets;*
(*iii*) *one has*

$$\sup_{\lambda \in I} \inf_{u \in X} \psi(u, \lambda) < \inf_{u \in X} \sup_{\lambda \in I} \psi(u, \lambda).$$

Then there exists $\lambda^* \in I$, such that the function $u \mapsto \psi(u, \lambda^*)$ has at least two global minimum points.

Theorem D.12 (Ricceri [13, Theorem 1 and Remark 1]) *Let (X, τ) be a topological space, I a real interval and $\psi : X \times I \to \mathbb{R}$ a functional satisfying:*

(c_1) *for every $u \in X$, the function $\lambda \mapsto \psi(u, \lambda)$ is quasi-concave and continuous;*
(c_2) *for each $\lambda \in I$, the function $u \mapsto \psi(u, \lambda)$ is l.s.c. and each of its local minima is a global minimum;*

(c_3) *there exist $\rho > \sup_{\lambda \in I} \inf_{u \in X} \psi(u, \lambda)$ and $\lambda_0 \in I$ such that the set*

$$\{u \in X : \ \psi(u, \lambda_0) \leq \rho\}$$

   is compact.

Then the following equality holds

$$\sup_{\lambda \in I} \inf_{u \in X} \psi(u, \lambda) = \inf_{u \in X} \sup_{\lambda \in I} \psi(u, \lambda).$$

Linking Sets

In this section we introduce various concepts of *linking sets* and highlight the connection between different types of "linking" used in the literature to find, classify and locate the critical points of a given smooth or nonsmooth functional.

Definition E.1 Let X be a topological space and let $D \subseteq X$ be a nonempty subset. We say that D is *contractible*, if there exists a continuous function $h : [0, 1] \times D \to X$ (the so-called *homotopy*) and a point $u_0 \in X$, such that $h(0, u) = u$ and $h(1, u) = u_0$ for all $u \in D$.

Definition E.2 Let X be a Banach space and $A, C \subseteq Z$ two nonempty subset. We say that A *links* C if and only if $A \cap C = \emptyset$ and A is not contractible in $X \setminus C$.

In many books appears the following definition of the notion of linking.

Definition E.3 Let X be a Banach space and A and C be two nonempty subsets of X. We say that A *and* C link if and only if there exists a closed set $B \subseteq Z$ such that $A \subseteq B$, $A \cap C = \emptyset$ and for any map $\theta \in C(B, Z)$ with $\theta|_A = id_A$, we have $\theta(B) \cap C \neq \emptyset$.

In some conditions the Definitions E.2 and E.3 are equivalent. We have the following result.

Theorem E.1 *Let X be a Banach space, A is relative boundary of a nonempty bounded convex set $B \subseteq X$. Then the definitions E.2 and E.3 are equivalent.*

Lemma E.1 *If Y is a finite dimensional Banach space, $U \subseteq Y$ is a nonempty, bounded, open set and $y_0 \in U$, then ∂U is not contractible in $Y \setminus \{y_0\}$.*

N. Costea et al., *Variational and Monotonicity Methods in Nonsmooth Analysis*, Frontiers in Mathematics, https://doi.org/10.1007/978-3-030-81671-1

Example E.1 Let X be a Banach space, $A := \{u_1, u_2\}$, $C := S_R(u_1) = \{u \in X : \|u - u_1\| = R\}$ with $R > 0$ and $\|u_1 - u_2\| > R$. It is clear that A is not contractible in $X \setminus C$. Note that if we use Theorem E.1 follows that the sets A and C also link in the sense of Definition E.3.

Example E.2 Let X be a Banach space, $X := Y \oplus Z$ with $\dim Y < +\infty$, $A := \{u \in Y : \|u\|_X = R\}$ with $R > 0$ and $C = Z$. Then the set A links C.

By contradiction, suppose the statement is false. Then we can find $h : [0, 1] \times A \to X \setminus C$, a contraction of A in $X \setminus C$. Let $P_Y : X \to Y$ be the projection operator to the finite dimensional subspace Y and let

$$\psi(t, x) := (P_Y \circ h)(t, u), \quad (t, u) \in [0, 1] \times A.$$

Then ψ is a contraction of A in $Y \setminus \{0\}$, which contradicts Lemma E.1 (take $U := B_R = \{u \in Y : \|u\|_X < R\}$). Using Theorem E.1 follows that the sets A and C link in the sense of Definition E.3.

Example E.3 Let X be a Banach space, $X := Y \oplus Z$, with $\dim Y < +\infty$, $v_0 \in Z$ $\|v_0\|_X = 1$ and $0 < r < R$. Let

$$B := \{v + tv_0 : 0 \le t \le R, \ \|v\|_X \le R\}$$

and let A be the boundary of B, hence

$$A = \{v + tv_0 : t \in \{0, R\}, \ \|v\|_X \le R \text{ or } t \in [0, R], \ \|v\|_Z = R\}$$

and let

$$C := \{u \in Z : \|u\|_X = r\}.$$

Then the set A links C.

We proceed by contradiction. Suppose that $h : [0, 1] \times A \to X \setminus C$ is a contraction of A in $X \setminus C$. Consider the projections $P_Y : X \to Y$ and $P_Z : X \to Z$ and set

$$\psi(t, u) := (P_Y \circ h)(t, u) + \|(P_Z \circ h)(t, u)\|_X v_0.$$

Then ψ is a contraction of A in $(Y \oplus \mathbb{R}) \setminus \{rv_0\}$, which contradicts Lemma E.1. As before the two sets A and C link in the sense of Definition E.3.

Next we present the linking notion introduced by Schechter, see [16]. Let X be a Banach space. We introduce the set of *admissible deformations* $\Phi \subset C([0, 1] \times X, X)$ whose elements Γ satisfy the following properties:

(a) for each $t \in [0, 1]$, $\Gamma(t, \cdot)$ is a homeomorphism of X into itself and $\Gamma^{-1}(t, \cdot)$ is continuous on $[0, 1) \times X$;

(b) $\Gamma(0, \cdot) = \mathrm{id}$;

(c) for each $\Gamma \in \Phi$ there is a $u_\Gamma \in X$, such that $\Gamma(1, u) = u_\Gamma$ for all $u \in X$ and $\Gamma(t, u) \to u_\Gamma$ as $t \to 1$ uniformly on bounded subsets of X.

Definition E.4 Let X be a Banach space and $A, B \subset X$. We say A *links* B *w.r.t.* Φ if $A \cap B = \emptyset$ and, for each $\Gamma \in \Phi$, there is a $t \in (0, 1]$ such that $\Gamma(t, A) \cap B \neq \emptyset$.

If there is no danger of confusion we shall simply say that A links B. In the following we present some properties of this notion of linking and some examples.

Proposition E.1 *Let A, B be two closed, bounded subsets of X such that $X \setminus A$ is path connected. If A links B, then B links A.*

Proposition E.2 *Let A, B be subsets of X such that A links B. Let $S(t)$ be a family of homeomorphisms of X onto itself such that $S(0) = I$, $S(t)$, $S(t)^{-1}$ are in $C([0, 1] \times X, X)$ and*

$$S(t)A \cap B = \emptyset, \quad 0 \le t \le T. \tag{E.1}$$

Then $A_1 := S(T)A$ links B.

Proposition E.3 *Under the same hypotheses as in Proposition E.2, A links $B_1 := S^{-1}(T)B$.*

Proposition E.4 *If $H : X \to X$ is a homeomorphism and A links B, then HA links HB.*

The next result gives a very useful method of checking the linking of two sets.

Proposition E.5 *Let $F : E \to \mathbb{R}^n$ be a continuous map, and let $Q \subset E$ be such that $F_0 = F|_Q$ is a homeomorphisms of Q onto the closure of a bounded open subset Ω of $\mathbb{R}^n$. If $p \in \Omega$, then $F_0^{-1}(\partial\Omega)$ links $F^{-1}(p)$.*

Remark E.1 The examples of linking sets given above, i.e., E.1, E.2 and E.3 are also valid in the sense of Definition E.4.

Few more examples due to Schechter [16] are provided below.

Example E.4 Let X be a Hilbert space and Y, Z two closed subspaces such that $\dim Y < \infty$ and $X = Y \oplus Z$ and let $B_R := \{u \in X : \|u\| < R\}$. Let $v_0 \neq 0$ be an element of Y. We write $Y := \{v_0\} \oplus Y'$. We take

$$A := \{v' \in Y' : \|v'\| \leq R\} \cup \{sv_0 + v' : v' \in Y', s \geq 0, \|sv_0 + v'\| = R\}$$

$$B := \{w \in Z : \|w\| \geq r\} \cup \{sv_0 + w : w \in Z, s \geq 0, \|sv_0 + w\| = r\},$$

where $0 < r < R$. Then A links B.

To see this let

$$Q := \{sv_0 + v' : v' \in Y', s \geq 0, \|sv_0 + v'\| \leq R\}.$$

For simplicity, we assume that $\|v_0\| = 1$. Because X is a Hilbert space follows that the splitting $X := Y' \oplus \{v_0\} \oplus Z$ is orthogonal. If

$$u := v' + w + sv_0, \; v' \in Y', w \in Z, s \in \mathbb{R}, \tag{E.2}$$

we define

$$F(u) := \begin{cases} v' + \left(s + r + \sqrt{r^2 - \|w\|^2}\right) v_0, & \text{if } \|w\| \leq r \\ v' + (s + r)v_0, & \text{if } \|w\| > r. \end{cases} \tag{E.3}$$

Note that $F|_Q = I$ while $F^{-1}(rv_0)$ is precisely the set B. Then we can conclude via Proposition E.5 that A links B.

Example E.5 This is the same as Example E.4 with A replaced by $A := \partial B_R \cap Y$. The proof is the same with Q replaced by $Q := \overline{B}_R \cap Y$.

Example E.6 Let Y, Z be as in Example E.4. Take $A := \partial B_r \cap Y$, and let v_0 be any element in $\partial B_1 \cap Y$. Take B to be the set of all u of the form

$$u := w + sv_0, w \in Z$$

satisfying any of the following

  (i) $\|w\| \leq R, s = 0$;
 (ii) $\|w\| \leq R, s = 2R_0$;
(iii) $\|w\| = R, 0 \leq s \leq 2R_0$,

where $0 < r < \min\{R, R_0\}$. Then A and B link each other.

To see this take $Y := \{v_0\} \oplus Y'$. Then any $u \in X$ can be written in the form (E.2). Define

$$F(u) := \left(R_0 - \max \left\{ \frac{R_0}{R} \|w\|, |s - R_0| \right\} \right) v_0 + v'$$

and $Q := \overline{B}_r \cap Y$.

We may identify Y with some $\mathbb{R}^n$. Then $F \in C(X, Y)$ with $F|_Q = I$. Moreover, $A = F^{-1}(0)$. Hence A links B by Proposition E.5. Since $X \setminus A$ is path connected, B links A by Proposition E.1.

We end this appendix with the notion of Schechter's definition of linking for the ball $\overline{B}_R$. To this end we introduce the family of admissible deformations to be the set $\Phi_R \subset C([0, 1] \times \overline{B}_R, \overline{B}_R)$ whose elements $\Gamma \in \Phi_R$ satisfy:

(a) For each $t \in [0, 1)$, $\Gamma(t, \cdot) : \overline{B}_R \to \overline{B}_R$ is a homeomorphism;
(b) $\Gamma(0, \cdot) = I$;
(c) For each $\Gamma \in \Phi_R$, there exists $u_\Gamma \in \overline{B}_R$ such that $\Gamma(1, u) = u_\Gamma$ for all $u \in \overline{B}_R$ and $\Gamma(t, u) \to u_\Gamma$ uniformly as $t \to 1$.

Definition E.5 We say that $A \subset \overline{B}_R$ *links* $B \subset \overline{B}_R$ w.r.t. Φ_R if

(L_1) $A \cap B = \varnothing$;
(L_2) For every $\Gamma \in \Phi_R$ there exists $t \in (0, 1]$ such that $\Gamma(t, A) \cap B \neq \varnothing$.

Using the above examples one can easily construct linking sets in a ball.

References

1. Q.H. Ansari, J.-C. Yao, A fixed point theorem and its application to a system of variational inequalities. Bull. Aust. Math. Soc. **59**, 433–442 (1999)
2. H. Brezis, *Functional Analysis, Sobolev Spaces and Partial Differential Equations* (Springer, Berlin, 2011)
3. H. Brezis, L. Nirenberg, G. Stampacchia, A remark on Ky Fan's minimax principle. Boll. Unione Mat. Ital. **6**, 293–300 (1972)
4. I. Ciorănescu, *Geometry of Banach Spaces, Duality Mappings and Nonlinear Problems*, vol. 62. Mathematics and Its Applications (Kluwer Academic Publishers, Berlin, 1990)
5. J. Diestel, *Geometry of Banach spaces – Selected Topics*. Lecture Notes in Mathematics (Springer, Berlin, 1975)
6. K. Fan, A generalization of Tychonoff's fixed point theorem. Math. Ann. **142**, 305–310 (1961)
7. K. Fan, Some properties of convex sets related to fixed point theorems. Math. Ann. **266**, 519–537 (1984)

8. K. Goebel, W. Kirk, *Topics in Metric Fixed Point Theory* (Cambridge University Press, Cambridge, 1990)
9. B. Knaster, C. Kuratowski, S. Mazurkiwicz, Ein beweis des fixpunktsatzes für *n*-dimensional simplexe. Fund. Math. **14**, 132–137 (1929)
10. T.-C. Lin, Convex sets, fixed points, variational and minimax inequalities. Bull. Aust. Math. Soc. **34**, 107–117 (1986)
11. J.F. McClendon, Minimax and variational inequalities for compact spaces. Proc. Amer. Math. Soc. **89**, 717–721 (1983)
12. U. Mosco, Implicit variational problems and quasivariational inequalities, in *Nonlinear Operators and the Calculus of Variations*, ed. by J.P. Gossez, E.J. Lami-Dozo, J. Mahwin, L. Waelbroeck (Springer, Berlin, 1976)
13. B. Ricceri, A further improvement of a minimax theorem of Borenshtein and Shul'man. J. Nonlin. Convex Anal. **2**, 279–283 (2001)
14. B. Ricceri, Minimax theorems for limits of parametrized functions having at most one local minimum lying in a certain set. Topology Appl. **153**, 3308–3312 (2006)
15. B. Ricceri, Multiplicity of global minima for parametrized functions. Atti Accad. Naz. Lincei Cl. Sci. Fis. Mat. Natur. Rend. Lincei (9) Mat. Appl. **21**, 47–57 (2010)
16. M. Schechter, *Linking Methods in Critical Point Theory* (Birkhäuser, Basel, 1999)
17. E. Tarafdar, A fixed point theorem equivalent to Fan-Knaster-Kuratowski-Mazurkiewic's theorem. J. Math. Anal. Appl. **128**, 475–479 (1987)

Index

© The Author(s), under exclusive license to Springer Nature Switzerland AG 2021
N. Costea et al., *Variational and Monotonicity Methods in Nonsmooth Analysis*,
Frontiers in Mathematics, https://doi.org/10.1007/978-3-030-81671-1

Printed in the United States
by Baker & Taylor Publisher Services